A Guide to the Complete Interpretation of Infrared Spectra of Organic Structures

A Guide to the Complete Interpretation of Infrared Spectra of Organic Structures

NOËL P. G. ROEGES

Katholieke Industriele Hogeschool O-VI
Campus Rabot, Gent, Belgium

JOHN WILEY & SONS
Chichester · New York · Brisbane · Toronto · Singapore

Telephone: *National* Chichester (0243) 779777
International +44 243 779777

Other Wiley Editorial Offices

John Wiley & Sons, Inc., 605 Third Avenue,
New York, NY 10158-0012, USA

Jacaranda Wiley Ltd, 33 Park Road, Milton,
Queensland 4064, Australia

John Wiley & Sons (Canada) Ltd, 22 Worcester Road,
Rexdale, Ontario M9W 1L1, Canada

John Wiley & Sons (SEA) Pte Ltd, 37 Jalan Pemimpin #05-04,
Block B, Union Industrial Building, Singapore 2057

Library of Congress Cataloging-in-Publication Data

Roeges, Noel P. G.
A guide to the complete interpretation of infrared spectra of organic structures / Noel P.G. Roeges.
p. cm.
Includes bibliographical references and index.
[illegible]BN 0-471-93998-6
[illegible] Orga[illegible] compounds—Spectra—Handbooks, manuals, etc.
2. Infrared spectroscopy—[illegible]tc. 3. Chemistry,
Org[illegible]andbooks, manuals, etc. I. Title.
QC462.85.R64 1994 [illegible]
547.046—dc20 [illegible] 94-2445
CIP

British Library Cataloguing in Publication Data

A catalogue record for this book is available from the British Library

ISBN 0 471 93998 6

Typeset in 10/12pt Times from author's disks by Production Technology Department, John Wiley & Sons Ltd, Chichester
Printed and bound in Great Britain by Biddles Ltd, Guildford, Surrey

Dedicated to the memory of W.A. Seth Paul

Contents

Preface xi

Acknowledgements xiii

Abbreviations xv

1 Introduction **1**

1.1 Fundamental vibrations of a molecular fragment 1
1.2 Fundamental vibrations of a molecule 2
1.3 Interpretation of spectra by means of tables with absorption regions of molecular fragments 3

2 Normal Vibrations and Absorption Regions of CX_3 . . . . . **10**

2.1 Methyl $—CH_3$ 10
2.1.1 α-Saturated 11
2.1.2 α-Unsaturated and aromatic 13
2.1.3 Nitrogen-bonded methyl 15
2.1.4 Sulfur-bonded methyl 17
2.1.5 Mercury- and silicon-bonded methyl 19
2.1.6 Acetyl 20
2.1.7 Methyl esters 23
2.1.8 Methyl ethers 24
2.2 Trihalogenomethyl 28
2.2.1 Trifluoromethyl $—CF_3$ 29
2.2.2 Trichloromethyl $—CCl_3$ 34
2.2.3 Tribromomethyl $—CBr_3$ 36
2.3 Tertiary butyl $—C(CH_3)_3$ 40

3 Normal Vibrations and Absorption Regions of CH_2X 48

3.1 Halogenomethyl 48
3.1.1 Fluoromethyl —CH_2F 48
3.1.2 Chloromethyl —CH_2Cl 51
3.1.3 Bromomethyl —CH_2Br 57
3.1.4 Iodomethyl —CH_2I 62
3.2 Oxymethyl 64
3.2.1 Hydroxymethyl —CH_2OH 64
3.2.2 Methoxymethyl —CH_2OMe 68
3.3 Sulfur-bonded methylene 72
3.3.1 Mercaptomethyl —CH_2SH 72
3.3.2 Methylthiomethyl —CH_2SMe 75
3.3.3 Thiocyanatomethyl —CH_2SCN 78
3.4 Nitrogen-bonded methylene 80
3.4.1 Aminomethyl —CH_2NH_2 80
3.4.2 Ammoniomethyl —$CH_2NH_3^+$ 83
3.4.3 Isocyanatomethyl —CH_2NCO 87
3.4.4 Isothiocyanatomethyl —CH_2NCS 88
3.5 Carbon-bonded methylene 89
3.5.1 Ethyl —CH_2CH_3 89
3.5.2 Trichloroethyl —CH_2CCl_3 94
3.5.3 Chloroethyl —CH_2CH_2Cl 95
3.5.4 Hydroxyethyl —CH_2CH_2OH 98
3.5.5 n-Propyl —$CH_2CH_2CH_3$ 100
3.5.6 Propynyl —$CH_2C\equiv CH$ 104
3.5.7 Cyanomethyl —$CH_2C\equiv N$ 108

4 Normal Vibrations and Absorption Regions of CHX_2 113

4.1 Dihalogenomethyl 113
4.1.1 Difluoromethyl —CHF_2 114
4.1.2 Dichloromethyl —$CHCl_2$ 115
4.1.3 Dibromomethyl —$CHBr_2$ 117
4.2 Isopropyl —$CH(CH_3)_2$ 119

5 Normal Vibrations and Absorption Regions of CHX 125

5.1 Halogenomethylene 125
5.1.1 Fluoromethylene —CHF— 125
5.1.2 Chloromethylene —CHCl— 126
5.1.3 Bromomethylene —CHBr— 126
5.2 Cyanomethylene —CH(CN)— 127
5.3 Hydroxymethylene —CH(OH)— 128

6 Normal Vibrations and Absorption Regions of CX_2 132

6.1 Difluoromethylene $—CF_2—$ 132
6.2 Dichloromethylene $—CCl_2—$ 133
6.3 Dibromomethylene $—CBr_2—$ 134

7 Normal Vibrations and Absorption Region of C(=X)Y 137

7.1 Carbonyl compounds 137
7.1.1 Formyl —C(=O)H 137
7.1.2 Fluoroformyl —C(=O)F 145
7.1.3 Chloroformyl —C(=O)Cl 148
7.1.4 Bromoformyl —C(=O)Br 152
7.1.5 Acetyl —C(=O)Me 155
7.1.6 Propionyl —C(=O)Et 160
7.1.7 Carboxyl —C(=O)OH 163
7.1.8 Methoxycarbonyl —C(=O)OMe 169
7.1.9 Ethoxycarbonyl —C(=O)OEt 174
7.2 Amino(thio)carbonyl compounds 178
7.2.1 Carbamoyl $—C(=O)NH_2$ 179
7.2.2 Thiocarbamoyl $—C(=S)NH_2$ 184
7.3 Methylamino(thio)carbonyl compounds 189
7.3.1 Methylcarbamoyl —C(=O)NHMe 190
7.3.2 Methylthiocarbamoyl —C(=S)NHMe 195
7.4 Carboxylate $—CO_2^-$ 198

8 Normal Vibrations and Absorption Regions of Alkenes and Alkynes 204

8.1 Alkenes 204
8.1.1 Vinyl $—CH=CH_2$ 204
8.1.2 Vinylidene $>C=CH_2$ 212
8.1.3 Vinylene —CH=CH— 218
8.2 Alkynes 225
8.2.1 Ethynyl —C≡CH 225
8.2.2 Chloroethynyl —C≡CCl 227
8.2.3 Bromoethynyl —C≡CBr 228
8.2.4 Iodoethynyl —C≡CI 229

9 Normal Vibrations and Absorption Regions of Nitrogen Compounds 233

9.1 Amino $—NH_2$ 233
9.2 Methylamino —NHMe 239
9.3 Acetylamino —NHC(=O)Me 242

9.4 Dimethylamino —NMe_2 ... 246
9.5 Nitro —NO_2 ... 250

10 Normal Vibrations and Absorption Regions of Oxy Compounds . 258

10.1 R′-oxy compounds ... 258
10.1.1 Hydroxy —OH ... 258
10.1.2 Methoxy —OMe ... 263
10.1.3 Ethoxy —OEt ... 269
10.2 R′-yloxy compounds ... 272
10.2.1 Formyloxy —OC(=O)H ... 272
10.2.2 Chloroformyloxy —OC(=O)Cl ... 273
10.2.3 Acetyloxy —OC(=O)Me ... 274

11 Normal Vibrations and Absorption Regions of Sulfur Compounds 277

11.1 Thio compounds ... 277
11.1.1 Methylthio —SMe ... 277
11.1.2 Ethylthio —SEt ... 281
11.2 Methylsulfinyl —S(=O)Me ... 284
11.3 Sulfonyl compounds ... 286
11.3.1 Methylsulfonyl —$S(=O)_2Me$ ... 286
11.3.2 Fluorosulfonyl —$S(=O)_2F$ ... 289
11.3.3 Chlorosulfonyl —$S(=O)_2Cl$ ... 291
11.3.4 R′-oxysulfonyl —$S(=O)_2O$— ... 292
11.3.5 Aminosulfonyl —$S(=O)_2NH_2$ ... 294
11.3.6 R′-aminosulfonyl —$S(=O)_2NH$— ... 295
11.3.7 Sulfonyl —$S(=O)_2$— ... 296

12 Normal Vibrations and Absorption Regions of Ring Structures . 301

12.1 Cyclopropyl —cPr ... 301
12.2 Oxiranyl —Ox ... 303
12.3 Aziridinyl —Az ... 303
12.4 Phenyl ... 306
12.4.1 Monosubstituted benzene derivatives —Ph ... 313
12.4.2 Disubstituted benzene derivatives —Ph— ... 314
12.4.3 Trisubstituted benzene derivatives —Ph< ... 318
12.5 Pyridyl —Py ... 329
12.6 Pyrimidinyl —Pym ... 331
12.7 Thienyl —Th ... 332
12.8 Furyl —Fu ... 334

Index ... 337

Preface

When interpreting infrared spectra of organic compounds, the chemist will restrict himself to the treatment of the typical group vibrations that produce vibrational bands in a characteristic spectral region : the 'group frequencies' or 'characteristic frequencies'. This procedure is widely known and can be considered as a 'qualitative interpretation' of an infrared spectrum. This work has a tentative assignment of all normal vibrations of a molecule in view: a 'quantitative interpretation'. To attain this objective, the $3N - 6$ absorption frequencies of the most important functional groups in distinct surroundings have been compared and classified in tables. In this way complete interpretations of the absorption frequencies of a number of molecular fragments can be made, based on data of vibrational analysis. When these molecular fragments join to form a molecule, it is possible to interpret the infrared spectrum of this molecule for the most part.

In the first place this work tries to be a guide for the organic chemist to interpret infrared spectra as extensively as possible. A preliminary knowledge of the interpretation of spectra is desirable but not necessary. Much information is provided to the spectroscopist on starting a vibrational analysis. In this case a description of the types of vibration is useful information.

Acknowledgements

I wish to express my sincere appreciation to W.A. Seth Paul for his large contribution in the initial stage of this work. His untimely decease brought an unexpected end to our fertile collaboration.

Special mention should be made of the contribution of Mrs Hilde Lauwereys, who has critically read over the entire text and carried out the necessary language corrections.

I gratefully acknowledge the permission given by the Aldrich Chemical Co. to use the data of the spectra published in 'The Aldrich Library of FT-IR Spectra' by Charles J.Pouchert.

N.P.G. Roeges
Autumn 1993

Abbreviations and Symbols

Az	aziridinyl	m	moderate
br	broad	Me	methyl
iBu	isobutyl	Naph	naphthyl
nBu	normal butyl	Ox	oxiranyl
sBu	secondary butyl	Pent	pentyl
tBu	tertiary butyl	Ph	phenyl
def	deformation	Pr	propyl
Et	ethyl	Py	pyridyl
ext	external	Pym	pyrimidinyl
Fu	furyl	Pyr	pyrrolyl
g	*gauche*	s	strong
cHex	cyclohexyl	sh	shoulder
HW	high wavenumber	sk	skeletal
Im	imidazolyl	Th	thienyl
LW	low wavenumber	*tr*	*trans*
ν_a	antisymmetric stretch	w	weak
ν_s	symmetric stretch	ω	wagging vibration
δ	in-plane deformation	ρ	rocking vibration
γ	out-of-plane deformation	τ	twisting vibration

(*italic*,upright): in case of extreme values, only the value in upright type is extreme.

1

Introduction

1.1 FUNDAMENTAL VIBRATIONS OF A MOLECULAR FRAGMENT

In the infrared a chemical compound displays $3N - 6$ fundamental vibrations, in which N represents the number of atoms from which the compound is built up. For each molecular fragment also $3N - 6$ vibrations are assigned if one external bond is calculated as an atom. The fundamental vibrations are easily derived from the following models:

—CX: three vibrations νCX, δCX, γ or ωCX (or torsion in a terminal multi-atomic fragment).

—CX_2: six vibrations $\nu_a CX_2$, $\nu_s CX_2$, δCX_2, ωCX_2, ρCX_2, τCX_2 (or torsion).

—CX_3: nine vibrations $\nu_a CX_3$, $\nu'_a CX_3$, $\nu_s CX_3$, $\delta_a CX_3$, $\delta'_a CX_3$, δ_s CX_3, ρ CX_3, $\rho' CX_3$, torsion.
or five vibrations in the case of C_{3v} symmetry: $\nu_a CX_3$, $\nu_s CX_3$, $d_a CX_3$, $\delta_s CX_3$, ρCX_3.

For the X-substituted n ring $3n - 6$ ring vibrations are expected.

X—Oxirane (cC_2H_3OX): 15 vibrations $\nu_a CH_2$, $\nu_s CH_2$, δCH_2, τCH_2, ωCH_2, ρCH_2, νCH, δCH, $\gamma(\omega)$CH, δCX, γCX, torsion (or νCX), νring, $\delta_a$ring, $\delta_s$ring.

X—Aziridine (cC_2H_4NX): 18 vibrations two $\nu_a CH_2$, two $\nu_s CH_2$, two δCH_2, two τCH_2, two ωCH_2, two ρCH_2, δCX, γCX, torsion (or νCX), νring, $\delta_a$ring, $\delta_s$ring.

X—Cyclopropane (cC_3H_5X): 21 vibrations	two $\nu_a CH_2$, two $\nu_s CH_2$, two δCH_2, two τCH_2, two ωCH_2, two ρCH_2, νCH, δCH, γCH, δCX, γCX, torsion (or νCX), νring, $\delta_a$ring, $\delta_s$ring.
X—Phenyl (Ph—X): 30 vibrations	five νCH, five δCH, five γCH, νCX (or torsion), δCX, γCX, six νring, three δring, three γring.

The following examples illustrate how the fundamental vibrations of a molecular fragment are found.

The nine vibrations of CH_3 ($\nu_a CH_3$, $\nu'_a CH_3$, $\nu_s CH_3$, $\delta_a CH_3$, $\delta'_a CH_3$, $\delta_s CH_3$, ρCH_3, $\rho' CH_3$, torsion) together with the three vibrations of OC (νO—C, δ—O—C, torsion) make up the 12 vibrations of $—OCH_3$.

For the $—C(=O)OCH_3$ fragment another six vibrations (νC=O, δC=O, γC=O, and νC—O, δ—C—O, torsion) are added.

$CH_3CH_2—$ possesses 18 vibrations: nine of CH_3, six of CH_2 and νC—C, δ—C—C, torsion.

The 12 vibrations of the $H_2C{=}CH—$ fragment are: six of CH_2, three of CH, νC=C, δ—C=C and torsion.

1.2 FUNDAMENTAL VIBRATIONS OF A MOLECULE

Since the external C–X stretching vibration in a molecular fragment, except for aromatic structures, depends largely upon X, the wavenumbers of the torsions rather than the stretching vibrations, are collected in tables. If, for the construction of a molecule, the coupling of two fragments leads to two torsions, one torsion has to be replaced by a C–C stretching vibration. In the case of fragments with two free bonds, one of them is coupled to an atom or another group by which three vibrations are added: a stretching vibration, an in-plane deformation and an out-of-plane deformation (or torsion).

The following examples illustrate the determination of the fundamental vibrations of a molecule.

When $ClCH_2CH_3$ splits into Cl— and $—CH_2CH_3$ one finds 17 of the 18 vibrations in the table for ethyl (Section 3.5.1). Evidently there is no ethyl torsion and the missing Cl—CH_2 stretching vibration is located in the table for $—CH_2Cl$ (Section 3.1.2). By splitting into $ClCH_2—$ and $—CH_3$, eight vibrations in the table for $—CH_2Cl$ and nine vibrations in the table for methyl (Section 2.1) need to be considered for the interpretation. The missing νC—C takes the place of a torsion and is found in the table for ethyl.

$H_2C{=}CH—C(=O)H$ possesses 18 normal vibrations: the 12 vibrations of $H_2C{=}CH—$ together with the six vibrations of $—C(=O)H$ but one of the two torsions has to be substituted by one C—C stretching vibration.

The 21 normal vibrations of Me—CH=CH—Cl are deduced in a similar way from the nine vibrations of Me and the 12 vibrations of —CH=CH—Cl. Those of —CH=CH—Cl are found by adding to the nine vibrations of —CH=CH— a C—Cl stretching vibration, an C=C—Cl in-plane deformation and a C=C—Cl torsion.

The 30 vibrations of the monosubstituted benzene ring (Section 12.4) with the 12 vibrations of $—C(=O)NH_2$ (Section 7.2.1) give the 42 vibrations of $Ph—C(=O)NH_2$.

1.3 INTERPRETATION OF SPECTRA BY MEANS OF TABLES WITH ABSORPTION REGIONS OF MOLECULAR FRAGMENTS

Example 1: $Me—C(=O)—CH_2OH$

The 27 normal vibrations of $Me—C(=O)—CH_2OH$ are divided into 15 vibrations of Me—C(=O)— and 12 vibrations of $—CH_2OH$. The C—C stretching vibration between the two groups takes the place of one of the two torsions. The remaining torsion is this one between the two groups.

No.	Wavenumber	Me—C(=O)— (Section 7.1.5)	$C(=O)—CH_2$	$—CH_2OH$ (Section 3.2.1)
1	3390 br			νOH. . .O
2	2990 w	ν_aMe, ν_a'Me		ν_aCH_2
3	2897 m	ν_sMe		ν_sCH_2
4	1721 s	νC=O		
5	1416 m	δ_aMe, δ_a'Me		δCH_2, δOH. . .O
6	1359 s	δ_sMe		
7	1286 m			ωCH_2
8	1228 m			τCH_2
9	1187 m		νC-C	
10	1083 s	ρMe		νC-O
11	971 m	ρ'Me		
12	871 m			ρCH_2
13	813 m	νC-C		
14	613 s	δC=O		
15	575 m,br			γOH. . .O
15′	503 m	γC=O		
16	409 m			δ-C-O
17	277 m	δC-C-		
	—	torsion Me		
	—		torsion	

Spectrum 1.1

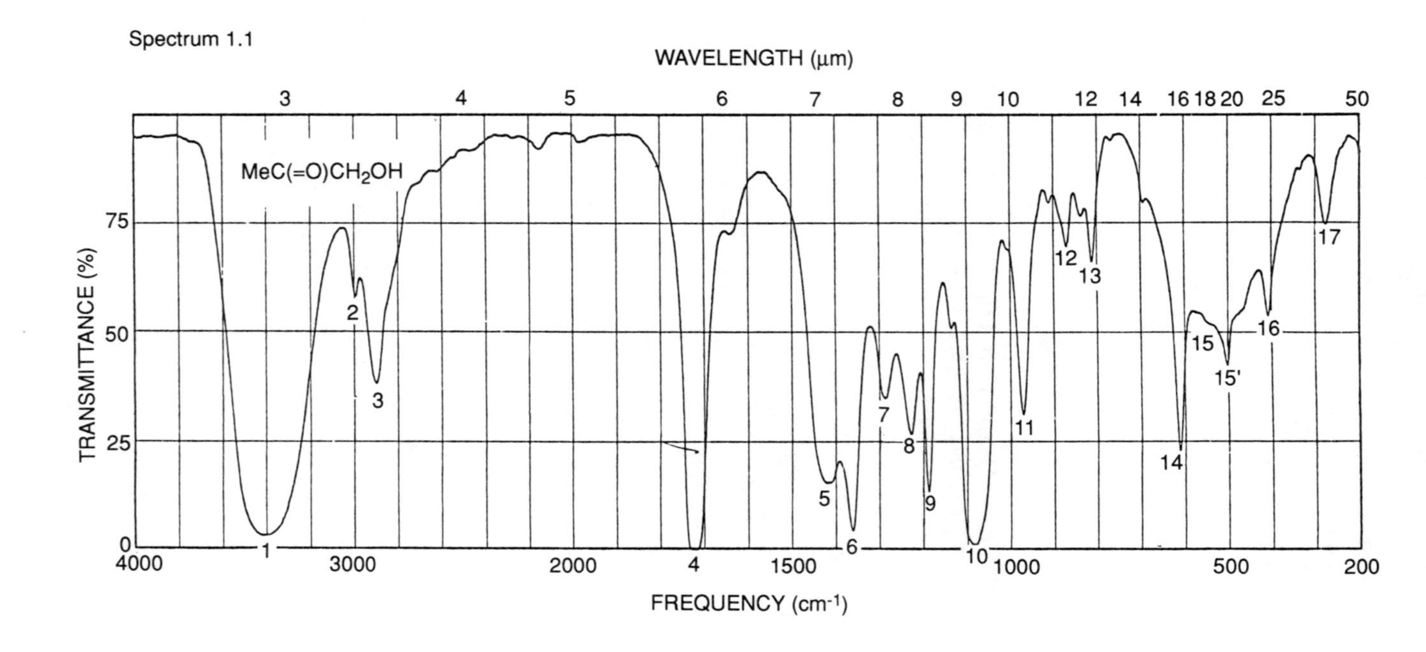
WAVELENGTH (μm)
3
4
5
6
7
8
9
10
12
14
16
18
20
25
50
MeC(=O)CH2OH
1
2
3
4
5
6
7
8
9
10
11
12
13
14
15
15'
16
17
TRANSMITTANCE (%)
0
25
50
75
4000
3000
2000
1500
1000
500
200
FREQUENCY (cm-1)

Example 2: $HOCH_2CH_2OMe$

The molecule $HOCH_2CH_2OMe$ (33 vibrations) can split into $HOCH_2$— (12 vibrations) and —CH_2OMe (21 vibrations), or into $HOCH_2CH_2$— (21 vibrations) and —OMe (12 vibrations).The former splitting gives the C—O—C stretchings but the latter describes the CH_2 vibrations and some skeletal vibrations more precisely.

No.	Wavenumber	$HOCH_2$— (Section 3.2.1)	CH_2—CH_2	—CH_2OMe (Section 3.2.2)
1	3400 br	νOH. . .O		
2	2980 sh	$\nu_a CH_2$		ν_aMe, ν'_aMe
3	2928 s			$\nu_a CH_2$
4.	2879 s	$\nu_s CH_2$		$\nu_s CH_2$
5	2827 m			ν_sMe
6	1470 sh	δCH_2		δ_aMe
7	1457 m			δCH_2, δ'_aMe, δ_sMe
8	1408 br	δOH. . .O		
9	1368 m	ωCH_2		
10	1326 w			ωCH_2
11	1286 w	τCH_2		
12	1233 m			τCH_2
13	1193 m			ρMe
14	1155 sh			ρ'Me
15	1123 s			ν_aC—O—C
16	1066 s	ν—O—C		
17	1017 m		νC—C	
18	963			ν_sC—O—C
19	891 m	ρCH_2		
20	834 m			ρCH_2
21	615 br	γOH. . .O		
22	539 m	δO—C		
23	463 w			δC—O—C
24	376 w			δC—C—O
25	225 w			torsion
	—			torsion
	—		torsion	

No.	Wavenumber	$HOCH_2CH_2$— (Section 3.5.4)	CH_2—O	—OMe (Section 10.1.2)
1	3400 br	νOH. . .O		
2	2980 sh	$\nu_a CH_2$		ν_aMe, ν'_aMe
3	2928 s	$\nu_a CH_2OH$		
4	2879 s	$\nu_s CH_2$, $\nu_s CH_2OH$		
5	2827 m			ν_sMe
6	1470 sh	δCH_2OH		δ_aMe
7	1457 m	δCH_2		δ'_aMe, δ_sMe
8	1408 br	δOH. . .O		

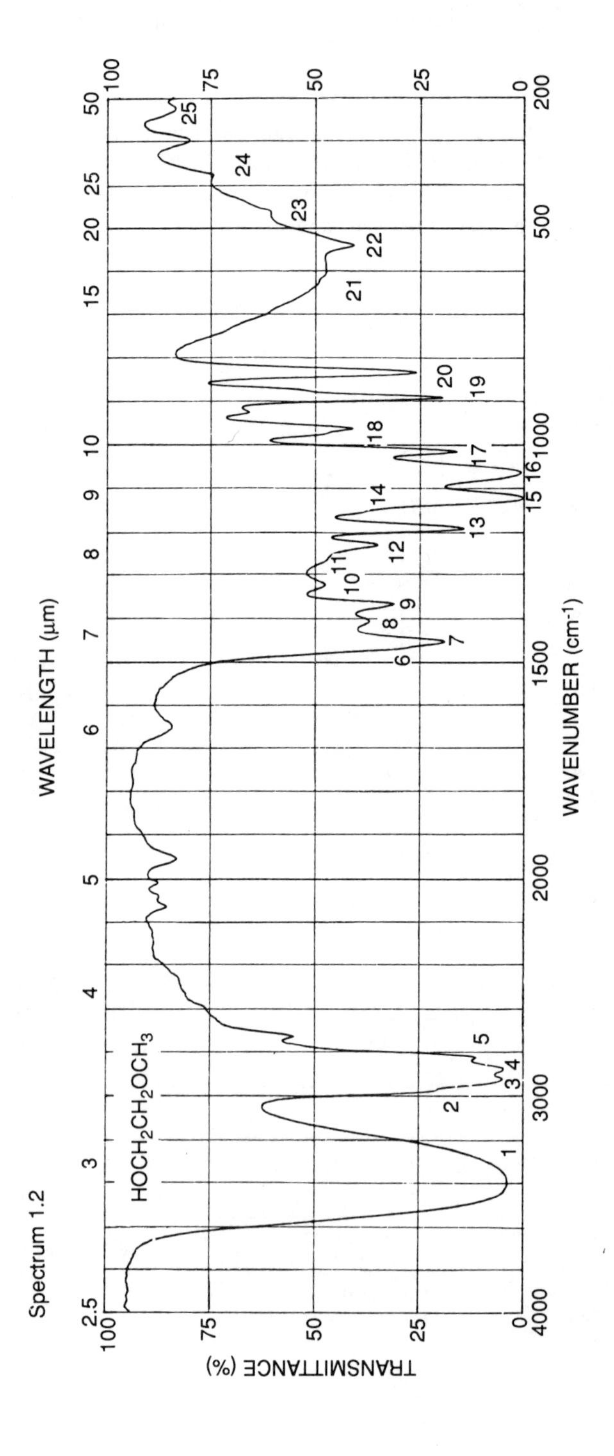
Spectrum 1.2
HOCH2CH2OCH3
WAVELENGTH (µm)
2.5
3
4
5
6
7
8
9
10
15
20
25
50
WAVENUMBER (cm-1)
4000
3000
2000
1500
1000
500
200
TRANSMITTANCE (%)
100
75
50
25
0
1
2
3
4
5
6
7
8
9
10
11
12
13
14
15
16
17
18
19
20
21
22
23
24
25

No.	Wavenumber (Section 3.5.4)	$HOCH_2CH_2$—	CH_2—O (Section 10.1.2)	—OMe
9	1368 m	ωCH_2OH		
10	1326 w	ωCH_2		
11	1286 w	τCH_2OH		
12	1233 m	τCH_2		
13	1193 m			ρMe
14	1155 sh			ρ'Me
15	1123 s		ν—O	
16	1066 s	νO—C		
17	1017 m	νC—C		
18	963 m			νO—C
19	891 m	ρCH_2OH		
20	834 m	ρCH_2		
21	615 br	γOH...O		
22	539 m	δO—C—C		
23	463 w			δ—O—C
24	376 w	δC—C—		
25	225 w			torsion
	—		torsion	
	—	torsion		

Example 3: Ph—C(=O)Et

The 54 normal vibrations of Ph—C(=O)Et are divided into 30 vibrations for the benzene ring and 24 vibrations for the —C(=O)Et group.

No.	Wavenumber	Ph— (Section 12.4.1)	Wilson	—C(=O)CH_2CH_3 (Section 7.1.6)
1	3085 w	νCH	20a	
2	3062 m	νCH	20b, 2	
3	3038 sh	νCH	13	
4	3028 w	νCH	7b	
5	2980 m			ν_aMe, ν'_aMe
5′	2950 m			$\nu_a CH_2$
6	2910 w			ν_sMe
6′	2880 w			$\nu_s CH_2$
7	1686 s			νC=O
8	1600 m	νPh	8a	
9	1582 m	νPh	8b	
10	1490 w	νPh	19a	
11	1460 m			δ_aMe, δ'_aMe
12	1448 s	νPh	19b	
13	1413 m			δCH_2
14	1377 m			δ_sMe
15	1352 m	νPh	14	

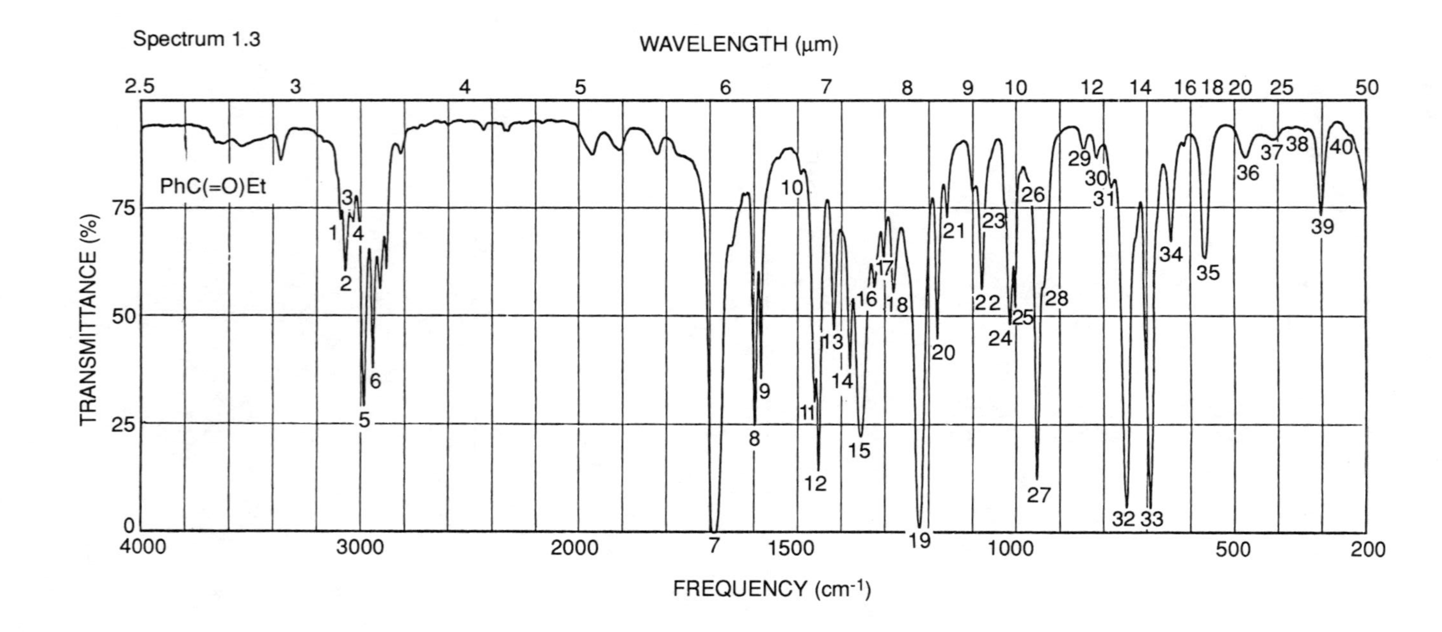

Spectrum 1.3
PhC(=O)Et
WAVELENGTH (μm)
FREQUENCY (cm-1)
TRANSMITTANCE (%)

No.	Wavenumber	Ph— (Section 12.4.1)	Wilson	—C(=O)CH_2CH_3 (Section 7.1.6)
16	1320 w			ωCH_2
17	1301 w	δCH	3	
18	1278 m			τCH_2
19	1219 s	νPh—C	7a (X)	
20	1180 m	δCH	9a	ρMe
21	1159 w	δCH	9b	
22	1077 m	δCH	15	ρ'Me
23	1025 sh	δCH	18a	
24	1013 m			νC(=O)—C
25	1002 m	νPh	1	
26	974 sh	γCH	5	
27	951 s	γCH	17a	νC—Me
28	935 w	γCH	17b	
29	847 w	γCH	10a	
30	817 w			ρCH_2
31	782 w	δPh	12 (X)	
32	745 s	γCH	11	
33	690 s	γPh	4	
34	640 m	δPh	6b	
35	565 m	γPh	16b (X)	δC=O
36	475 m	δPh	6a (X)	γC=O
37	414 w	γPh	16a	
38	339 w			δC(=O)C—C
39	302 m			δ—C(=O)—C
40	238 w	δPh—C	18b (X)	torsion
	—	γPh—C	10b (X)	
	—			torsion
	—			torsion

2

Normal Vibrations and Absorption Regions of CX_3

2.1 METHYL

The simplest molecules containing a methyl group are the CH_3X molecules (X = F, Cl, Br, CN, CCH, ...) which belong to point group C_{3v}. The methyl group possesses **five** normal vibrations, i.e. two stretchings, two deformations and a rocking:

$$\nu_a\text{Me},\ \nu_s\text{Me},\ \delta_a\text{Me},\ \delta_s\text{Me and } \rho\text{Me}$$

The methyl group accounts for **nine** normal vibrations when the symmetry of the molecule is lowered from C_{3v} to C_s or C_1. Stated differently, a methyl group possesses an antisymmetric stretching, a deformation and a rocking vibration with C_{3v} symmetry and two of each when the symmetry is lowered from C_{3v} to C_s, where the degeneracy is removed. Moreover, the methyl torsion now becomes a normal vibration. With C_{3v}, $\{C_s$ or $C_1\}$ molecules the torsion is a rotation {external deformation} which is not active {active} in the IR. The normal vibrations of methyl may be described as follows:

	ν_a	ν'_a	ν_s	δ_a	δ'_a	δ_s	ρ	ρ'	τ
C_{3v}	e	—	a_1	e	—	a_1	e	—	—
C_s	a″	a′	a′	a″	a′	a′	a″	a′	a″

With the a′ {a″}-vibrations the dipole change proceeds in {is perpendicular to} the plane of symmetry.

Instead of describing three stretchings (deformations) in terms of two antisymmetric and one symmetric mode of vibration, these modes can also be described in terms of three independent CH stretchings (deformations). These approximations are derived from the fact that in a methyl group the CH bonds are not always equal (about 0.11 nm) but depend on the surroundings. For instance, a halogen atom in the *gauche* position of the methyl group often causes **dissimilar CH bonds**. Despite the fact that a methyl group provides two antisymmetric modes of vibrations these fundamental vibrations will not always be observed separately. In practice, and in particular with larger molecules, these vibrations often coincide with or are superposed on the much stronger absorptions of the CH_2 vibrations. The two rockings are often separately observed but are generally strongly coupled to skeletal vibrations and vice versa, which is reflected in the wide absorption regions and the description CC stretch/Me rocking or Me rocking/CC stretch. One also refers to the parallel {perpendicular} methyl rocking when the dipole changes parallel {perpendicular} to the direction of the local C_3 axis. Generally, the parallel rocking is strongly coupled to the CC stretching vibrations.

2.1.1 α-Saturated

The group consists of Me—R molecules in which R is a substituent and the α-carbon atom is fully or partly substituted or not substituted (see Sections 3.5.1 and 3.5.5). Differentiation has been made between the molecules MeCHXY, $MeCX_2Y$ and $MeCX_3$ in which X equals F, Cl, Br, OH, OD and CN.

Stretching vibrations

For molecules of the type MeCHXY, the methyl antisymmetric stretchings are found in the neighbourhood of 2970 and 2960 cm^{-1}. The absorptions are generally situated at higher wavenumbers, i.e. in the vicinity of 3005 and 2980 cm^{-1} for molecules of the type $MeCX_2Y$, and this tendency continues for $MeCX_3$ (X = F, Cl and Br), whereas the methyl antisymmetric stretching absorbs somewhere near 3010 cm^{-1}.

Behaviour similar to that of the antisymmetric stretchings has been observed for the symmetric mode. For MeCHXY molecules this vibration is active in the neighbourhood of 2910 cm^{-1}, whereas molecules of the type $MeCX_2Y$ absorb in the vicinity of 2940 cm^{-1}. This region becomes 2955±20 cm^{-1} for $MeCX_3$ (X = F, Cl and Br). Table 2.1 illustrates the influence of the fluorine on the wavenumber

Table 2.1

Molecule	ν_aMe	ν'_aMe	ν_sMe
Me—CHF—Me	2989	2935	2893
Me—CF_2—Me	3018	3018	2959
Me—CF_3	3035	–	2975

of the methyl stretchings. The intensity varies from strong for MeCHXY, medium for $MeCX_2Y$ to weak for $MeCX_3$ molecules.

Deformations

The wavenumber regions in which these vibrations are active clearly indicate that in many cases the wavenumbers of both antisymmetric deformations often or always coincide.

For molecules of the type MeCHXY the antisymmetric deformations are observed around 1455 and 1450 cm^{-1} and for $MeCX_2Y$ molecules both deformations absorb in the neighbourhood of 1450 cm^{-1} with a medium to strong intensity. For $MeCX_3$ (X = F, Cl and Br) the antisymmetric deformations absorb weakly around the value 1445 cm^{-1}.

With the symmetric deformation, again the wavenumber tends to increase with degree of substitution and the nature of the substituents. If the region is 1370 ± 25 cm^{-1} for MeCHXY molecules, for $MeCX_2Y$ it is 1380 ± 20 cm^{-1} and it increases up to 1390 ± 20 cm^{-1} for $MeCX_3$. The intensity varies from medium to strong. Ethane displays the symmetric deformation at 1379 cm^{-1}. In the spectra of CH_3X molecules (X = F, Cl, Br and I) this vibration is observed at 1475, 1355, 1305 and 1252 cm^{-1} respectively.

Rockings

The wavenumbers of these modes occur over such broad regions that one can hardly speak of group wavenumbers. The Me rockings are, like the symmetric deformation, mass sensitive. Because of coupling with skeletal modes, these rockings occur in larger regions of wavenumber compared with the stretchings and deformations. The intensity, however, is only weak to medium. Ethane displays both rockings at 1155 (Raman) and 821 cm^{-1}. Hydrocarbons display one of the rockings very weakly in the vicinity of 1138 cm^{-1}.

For many MeCHXY molecules one of the rockings, namely ρMe, is observed in the region 1150 ± 35 cm^{-1}, which is more or less acceptable as a group-wavenumber. However, molecules of the type $MeCX_2Y$ display this methyl rocking in the region 1080 ± 115 cm^{-1}. Low values have been observed in the spectra of 1-chloro-2,2-difluoropropane which absorbs at 968 cm^{-1}, and 2,2-difluoropropane, which displays rockings at 991 and 988 cm^{-1}. When these values are not considered, the region becomes more user friendly: 1135 ± 60 cm^{-1}.

MeCHXY molecules display the ρ'Me in the region 1010 ± 120 cm^{-1}. 2-Fluoropropane absorbs at 1129 cm^{-1}. If this value is neglected the absorption region is reduced to 975 ± 85 cm^{-1}. $MeCX_2Y$ molecules absorb at 1000 ± 95 cm^{-1} and $MeCX_3$ molecules in the region 1020 ± 60 cm^{-1}. Undoubtedly, the coupling with CC and CX stretchings seriously affects and broadens the regions in which these rockings have been observed.

Torsion

Torsion has been observed at relatively high values, that is 250 ± 65 cm^{-1}, although values lower than 185 cm^{-1} may not be excluded. Many MeCHXY {$MeCX_2Y$} molecules display the torsion in the region 245 ± 30 {280 ± 35} cm^{-1}. The highest {lowest} value: 314 {187} cm^{-1} has been observed for 2,2-difluoropropane {2,2-dibromopropane}.

Table 2.2 Absorption regions (cm^{-1}) of Methyl in alkane fragments

	$—CH_2Me$	$—CH_2CH_2Me$	—CHXMe	$—CX_2Me$	CX_3Me
ν_aMe	2985 ± 25	2975 ± 15	2970 ± 30	3005 ± 25	3010 ± 25
ν'_aMe	2970 ± 30	2965 ± 15	2960 ± 35	2980 ± 40	–
ν_sMe	2905 ± 65	2890 ± 50	2910 ± 50	2940 ± 50	2955 ± 20
δ_aMe	1465 ± 20	1465 ± 10	1455 ± 20	1450 ± 20	1445 ± 15
δ'_aMe	1445 ± 25	1450 ± 15	1450 ± 15	1450 ± 20	–
δ_sMe	1380 ± 20	1375 ± 15	1370 ± 25	1380 ± 20	1390 ± 20
ρMe	1100 ± 95	1105 ± 45	1150 ± 35	1080 ± 115	–
ρ'Me	1080 ± 80	1055 ± 45	1010 ± 120	1000 ± 95	1020 ± 60
τMe	230 ± 105	210 ± 70	245 ± 30	250 ± 65	–

R—Me molecules

R = $R'CH_2$— (see Section 3.5.1), $R'CH_2CH_2$— (see Section 3.5.5)

R = Me—CHF— [1–4], Me—CHCl— [5–7], $ClCH_2$—CHCl— [8–10], Cl_2CH—CHCl— [10], Et—CHCl— [11–13], Me—CHBr— [5, 6, 18], $ClCH_2$—CHBr— [9], $BrCH_2$—CHBr— [9], F_3C—CH(OH)—, F_3C—CH(OD)—, F_3C—CD(OH)— and F_3C—CD(OD)— [19], F_3C—CH(CN)—, FCHCl— [14], Cl_2CH— [15–17], Me—CF_2— [20], $ClCH_2$—CF_2— [21], Me—CH_2—CCl_2— [22], Me—CCl_2— [23–25], Cl_2CH—CCl_2— [10], F—CCl_2— [14], Me—$(CClBr)_2$— [26], Me—CBr_2— [27], Me—$(CBr_2)_2$— [28], F_3C— [29–32], Cl_3C— [15, 31, 33–36], Br_3C— [31, 37].

2.1.2 α-Unsaturated and aromatic

Stretching vibrations

The band intensity of the stretchings varies from medium to strong. A small part of the region covers that of the =CH stretching modes. One should therefore be cautious in assigning bands within this region. In contrast to saturated hydrocarbons, both antisymmetric stretching vibrations are regularly displayed in separate regions. In the region of the symmetric stretch a second smaller band is due to an overtone of the methyl bending vibration, intensified by Fermi resonance.

The ν_aMe vibration is observed in the region 2980 ± 30 cm^{-1}, generally in the neighbourhood of 3000 cm^{-1} or lower. Exceptions occur in the spectra of HC≡CMe, DC≡CMe and H_2C=CFMe which absorbs ≈3010 cm^{-1}. The lowest value (2950 cm^{-1}) has been assigned in the spectra of MeDC=NOD and 3-NaO_2C—Ph—Me.

The ν'_aMe vibration displays a band in the region 2950 ± 45 cm^{-1}. The highest {lowest} value, i.e. 2995 {2905} cm^{-1}, has been assigned in the spectrum of 4-NaO_2C—Ph—Me {MeO—CH=CH—Me}.

The symmetric stretching vibration appears in the region 2895 ± 50 cm^{-1}. A molecule such as 3-methylphenol absorbs at the HW side at 2944 cm^{-1}, followed by H_2C=CFMe (2942 cm^{-1}) and Me—CH=CH—C(=O)H with 2938 cm^{-1}. Me—CH=NOH with 2847 cm^{-1} absorbs at the LW side followed by MeO—CH=CH—Me, for which the methyl stretching vibration has been assigned at 2855 cm^{-1}.

Deformations

The antisymmetric deformations are observed with band intensities that are weak, mostly medium or sometimes strong. The overlap between the two regions is quite considerable so that for many molecules the deformations often coincide. In the HW region (δ_aMe: 1457 ± 27 cm^{-1}) the highest {lowest} value is assigned to the δ_aMe in the spectrum of 1,3-dimethylbenzene {Me—C≡C—C(=O)Cl}. In the LW region (δ'_aMe: 1435 ± 35 cm^{-1}) these values are 1470 and 1400 cm^{-1}, respectively occurring in the spectra of H_2=C(OMe)Me and cC_3H_3Me.

In many molecules the symmetric deformation (δ_sMe: 1380 ± 25 cm^{-1}) appears with an intensity varying from medium to strong. The wavenumber 1401 cm^{-1} is the highest value assigned in the spectrum of 2-fluoropropene. The lowest value for this vibration is 1357 cm^{-1}, observed in the spectrum of *cis*-Me—CH=CH—Me.

Rockings

Aromatic molecules clearly display a methyl rock in the neighbourhood of 1045 cm^{-1}. The second rock in the region 970 ± 70 cm^{-1} is more difficult to find among the =CH out-of-plane deformations.

The ρMe vibration provides a band with a weak to medium, rarely strong intensity, in the region 1065 ± 65 cm^{-1}. DON=CH—Me occupies the top place with 1130 cm^{-1}, followed by DON=CD—Me (1122 cm^{-1}) and HON=CH—Me (1110 cm^{-1}). The values 1000, 1005 and 1007 are the lowest which have been found in the spectra of 2,6-Me_2—Ph—OH, 2,6-Me_2—Ph—F and 2,3-Me_2—Ph—F respectively. In the series Me—C≡C—R in which R equals C(=O)Cl, C(=O)F, I, Br and Cl, this rocking absorbs respectively at 1016, 1023, 1021, 1027 and 1033 cm^{-1}.

The ρ'Me is expected to give a band with a weak to medium intensity in the region 980 ± 80 cm^{-1}. The highest wavenumber, namely 1060 cm^{-1}, is assigned in the spectrum of $H_2C{=}C(CH_2CN)Me$ followed by 1050 in that of $H_2C{=}CMe_2$. In the series Me—CH=CH—R where R equals C(=O)H, C(=O)OMe, C(=O)OEt and C(=O)Me this rocking has respectively been observed at 1042, 1030, 1030 and 1020 cm^{-1}. The lowest values have been assigned in the spectra of 2,3-Me_2—Ph—OH (*990*, 901 cm^{-1}), 2,6-Me_2—Ph—OH (*987*, 912 cm^{-1}) and Ph—CH=CH—Me (910 cm^{-1}).

Torsion

Very little information is available as to the torsion although it is expected that many of these torsions absorb at 185 ± 65 cm^{-1}.

Table 2.3 Absorption regions (cm^{-1}) of methyl in alkenes, alkynes and phenyl

	=C—Me	≡C—Me	—Ph—Me
ν_aMe	2980 ± 30	2985 ± 25	2975 ± 25
ν'_aMe	2940 ± 35	–	2960 ± 35
ν_sMe	2895 ± 50	2910 ± 30	2925 ± 20
δ_aMe	1455 ± 25	1445 ± 15	1460 ± 25
δ'_aMe	1435 ± 35	–	1435 ± 35
δ_sMe	1380 ± 25	1375 ± 15	1375 ± 15
ρMe	1070 ± 60	1035 ± 20	1045 ± 45
ρ'Me	985 ± 75	–	970 ± 70
τMe	185 ± 65	–	–

R—Me molecules

R = $H_2C{=}CR'$— (see Section 8.1.2), R′—CH=CH— (see Section 8.1.3), X—C≡C— (X = H, Cl, Br, I) (see Section 8.2);

R = $H_2C{=}CH$—, $H_2C{=}CD$— and $D_2C{=}CH$— [38, 39], Me—CH=CD— and Me—CD=CD— [40], HDC=CBr— and $D_2C{=}CBr$— [41], cC_3H_3— [42], HON=CH—, DON=CH— and DON=CD— [43, 44], D—C≡C— [45, 46], Me—C≡C— [92], Et—C≡C— [47], F—C(=O)—C≡C— [48], Cl—C(=O)—C≡C— [49], N≡C— [50–55];

R = (substituted) phenyl (see Section 12.4).

2.1.3 Nitrogen-bonded methyl

Molecules of type —NHMe and of type X—Me in which X = H_2N—, D_2N—, $(SiH_3)_2N$—, O=C=N—, S=C=N—, O_2N—, N_3— and CN— will be considered in this section.

Stretching vibrations

Me—NO_2 (3060 cm^{-1}) and Me—N_3 (3023 cm^{-1}) provide the highest values for the methyl antisymmetric stretching. Methyl stretchings of N—methyl substituted amines or amides absorb below 3000 cm^{-1}. The highest value (2995 cm^{-1}) has been observed in the spectrum of MeSC(=S)NHMe, followed by 2990 cm^{-1} in the spectra of MeOC(=S)NHMe and KOC(=O)C(=O)NHMe.

With the exception of the X—Me molecules, the symmetric stretching absorbs sharply in the neighbourhood of 2850 cm^{-1}, with a band intensity varying from medium to strong. The highest values are shown by $MeNHNO_2$ (2923 cm^{-1}), MeC(=O)NHMe (2915 cm^{-1}) where the methyl group is *trans* with respect to the C=O group, and MeSC(=S)NHMe (2910 cm^{-1}). Dimethylamine, which absorbs at 2791 cm^{-1}, provides the lowest wavenumber, followed by 2810 cm^{-1} in the spectrum of, for instance, MeNHC(=O)—C(=S)NHMe and 2824 cm^{-1} in that of NC—C(=O)NHMe. If these extreme values are neglected, the region of this ν_sMe is reduced to 2855 $\pm$ 40 cm^{-1}. This band is often quite useful in identifying NMe or OMe groups in unknown molecules.

Deformations

With Me—N molecules, a separate observation of both Me antisymmetric deformations (1455 $\pm$ 35 and 1445 $\pm$ 35 cm^{-1}) is an exception rather than a rule.

Dimethylamine displays the antisymmetric deformations at 1483, 1467, 1463 and 1445 cm^{-1}. $D_3CS(=O)_2NHMe$ absorbs at 1472 and 1466 cm^{-1}. HC(=O)NHMe displays the antisymmetric deformations at 1467 and 1458 cm^{-1}. A wavenumber of 1480 cm^{-1} is found for both antisymmetric deformations of $H_2NC(=O)NHMe$ and 1420 cm^{-1} for those of R′C(=O)NHMe (R′= Cl and Br).

The symmetric deformations of dimethylamine have been assigned at 1441 and 1412 cm^{-1}, but δ_sMe of nitromethane absorbs at a lower value: 1376 cm^{-1}. The region of this symmetric deformation (1410 $\pm$ 35 cm^{-1}) narrows to 1410 $\pm$ 20 cm^{-1} for most nitrogen-substituted methyl compounds.

Rockings

With Me—N molecules both rockings (1150 $\pm$ 50 and 1090 $\pm$ 75 cm^{-1}) are more or less strongly coupled to the C—N stretching vibration, often assigned in the region 1010 $\pm$ 90 cm^{-1}. Therefore, one of the methyl rockings also absorbing in this region often receives the notation ρMe/νCN or νCN/ρMe, which is reflected in the large overlap of both ranges. The assignment of νCN/ρMe is more difficult when the α-atom of the R-substituent comprises a saturated C atom because the —CNC— skeleton now provides two C—N stretchings. For some R′-NHMe molecules Table 2.4 summarizes the C—N stretching and methyl rockings and

Table 2.4

R	ρMe	ρ'Me	νC—N
$H_2NC(=O)$—	1169	1106	1106
ClC(=O)—	1163	1163	1003
MeC(=O)—	1161	1114	1045
MeNHC(=O)—	1170	1132	1038
	1148	1035	1020
HC(=O)—	1148	1040	951
Me—	1158	1079	930
	1145	1022	925

Table 2.5 Absorption regions (cm^{-1}) of nitrogen-bonded methyl

	—NHMe	$—SO_2NHMe$	—C(=O)NHMe	—C(=S)NHMe	X—Me
ν_aMe	2965 ± 25	2965 ± 25	2970 ± 30	2970 ± 30	3005 ± 55
ν'_aMe	2950 ± 25	2950 ± 25	2945 ± 45	2945 ± 25	(2990 ± 55)
ν_sMe	2855 ± 70	2855 ± 70	2870 ± 45	2875 ± 45	2895 ± 75
δ_aMe	1470 ± 15	1470 ± 15	1450 ± 30	1450 ± 25	1445 ± 20
δ'_aMe	1460 ± 15	1460 ± 15	1445 ± 35	1435 ± 25	(1435 ± 15)
δ_sMe	1410 ± 35	1410 ± 35	1400 ± 25	1400 ± 25	1405 ± 25
ρMe	1150 ± 30	1135 ± 15	1155 ± 30	1145 ± 45	1160 ± 40
ρ'Me	1085 ± 65	1070 ± 15	1100 ± 65	1075 ± 40	(1075 ± 50)
τMe	230 ± 30	230 ± 30	230 ± 30	195 ± 50	–

also includes high and low wavenumbers of each series. The highest methyl rocks, in the neighbourhood of 1200 cm^{-1}, are assigned in the spectra of $MeN(SiH_3)_2$, $MeNH_2$ and Me_3N.

R—Me molecules
R = R′NH— (see Section 9.2), $R'SO_2NH$— (see Section 9.2), R′C(=O)NH— (see Section 7.3.1), R′C(=S)NH— (see Section 7.3.2).

X—Me molecules
X = H_2N— [56, 57], D_2N— [56, 57], $(SiH_3)_2N$— [58, 59], O=C=N— [60–62], S=C=N— [63–68], O_2N— [69–73], N_3— [74, 75], CN— [52, 76].

2.1.4 Sulfur-bonded methyl

Stretching vibrations

The region 3015 ± 35 cm^{-1} is assigned to ν_aMe and the region 2995 ± 55 cm^{-1} to ν'_aMe of the sulfur substituted methyl group. The intensity of the band is weak or medium. Some sulfonyl compounds of type $MeS(=O)_2R'$ (R′ = Me, Ph, PhNH,

NH_2, ND_2, F, Cl, Br) show both methyl antisymmetric stretchings in the region 3030 ± 20 and 3025 ± 25 cm^{-1} with the highest values in the vicinity of 3045 cm^{-1} for R′ = F, Cl and Br. Both vibrations are not always observed separately. Most R—SMe molecules display ν_aMe at 3005 ± 25 cm^{-1} with the highest value (3030 cm^{-1}) for NCSMe and EtSSMe and the lowest for MeSMe (2982 cm^{-1}). They display the ν'_aMe at 2980 ± 45 cm^{-1} with 3022 cm^{-1} for NCSMe and 2940 cm^{-1} for $H_2NC(=S)SMe$.

The symmetric stretching mode absorbs around 2930 ± 30 cm^{-1} with an intensity varying from weak to medium. The highest value observed for ν_sMe is 2958 cm^{-1} in the spectrum of $FS(=O)_2Me$. R—SMe molecules absorb around 2925 ± 20 cm^{-1}, that is, about 100 cm^{-1} higher than the corresponding mode in OMe. For instance, methoxymethane {methylthiomethane} displays the ν_sMe at 2820 {2917} cm^{-1} and Ph—OMe {Ph—SMe} displays the same vibration at 2834 {2922} cm^{-1}.

Deformations

The absorption regions of both antisymmetric deformations (1435 ± 35 and 1430 ± 30 cm^{-1}) overlap each other appreciably. The highest values of δ_aMe, namely 1468 and 1457 cm^{-1}, are assigned in the spectra of methylthio-2-pyrimidine and $ClCH_2CH_2OS(=O)_2Me$ and the lowest, namely 1404 and 1409 cm^{-1}, in those of $iPr_2NS(=O)_2Me$ and 4-HOC(=O)PhS$(=O)_2$Me. Many δ_aMe are centered around 1435 cm^{-1}.

The highest {lowest} wavenumber for δ'_aMe, namely 1460 {1400} cm^{-1} has also been assigned to methylthio-2-pyrimidine {4-HOC(=O)PhS$(=O)_2$Me}. ClSMe absorbs around 1408 cm^{-1}, although most of the δ'_aMe absorptions occur in the vicinity of 1425 cm^{-1}.

The methyl symmetric deformation is observed in the low wavenumber region 1320 ± 30 cm^{-1} with a band intensity between medium and strong. The highest value, 1350 cm^{-1}, has been observed in the spectrum of $EtOS(=O)_2Me$, the lowest, 1295 cm^{-1}, in that of MeS(=O)Me. Whereas ν_sMe of SMe absorbs nearly 100 cm^{-1} higher compared with OMe, the reverse is true for δ_sMe, where the difference amounts to about 125 cm^{-1}. Methoxymethane {methylthiomethane} shows δ_sMe at 1449 and 1432 {1328 and 1303} cm^{-1} and methoxybenzene {methylthiobenzene} absorbs at 1442 {1317} cm^{-1}. The methyl symmetric deformation for sulfonyl compounds absorbs also at lower wavenumbers (1325 ± 25 cm^{-1}) than those usually observed for paraffins. This is attributed to the influence (including mass) of the S atom.

Rockings

The methyl rockings of S-attached methyl occur in the regions 990 ± 45 and 940 ± 45 cm^{-1}. The overlap is considerable, both rockings therefore often coincide.

Again, the regions are lower (by about 200 cm^{-1}) than those of the corresponding O-attached methyl groups.

Leaving aside the high value for ρMe (1065 cm^{-1}) in the spectrum of MeSH, most of the ρMe have been observed in the range 990 $\pm$ 45 cm^{-1}. High values are furnished by MeSMe (1032 cm^{-1}) and MeS(=O)Me (1022 cm^{-1}) and low values by ClS(=O)Me (948 cm^{-1}) and the disulfides MeSSMe, EtSSMe and tBuSSMe ($\cong$955 cm^{-1}).

In the region 940 $\pm$ 45 cm^{-1} the highest {lowest} wavenumber for ρ'Me, 981 {896} cm^{-1}, is observed in the spectrum of $FS(=O)_2Me$ {MeS(=O)Me} followed by 976 {900} cm^{-1} in that of ClC(=O)SMe {4-ClPhNHS$(=O)_2$Me}.

R—Me molecules

R = R′S— (see Section 11.1.1), R′S(=O)— (see Section 11.2), R′S$(=O)_2$— (see Section 11.3.1).

2.1.5 Mercury- and silicon-bonded methyl

Stretching vibrations

The three stretchings are found in the region 2945 $\pm$ 55 cm^{-1} in which the band intensities vary from weak to medium. The two molecules MeHgMe and MeSHgMe do not show separate antisymmetric Me stretchings. The number of investigated molecules, however, is too small to attach much value to the observed regions.

Deformations

The antisymmetric deformations are weakly active in the region 1420 $\pm$ 30 cm^{-1}. The symmetric deformation absorbs in the lowest of all regions in which this vibrational mode has been observed: 1265 $\pm$ 25 {1180 $\pm$ 10} cm^{-1} for Si{Hg}-bound methyl. The band intensity is mostly medium. In the spectrum of F_3SiMe {I_3SiMe} δ_sMe has been assigned at 1286 {1247} cm^{-1}. If methyl is attached to an element other than carbon, the wavenumber region of the methyl symmetric deformation is largely determined by the electronegativity of the element and its position in the Periodic Table. It has already been mentioned that for X—Me molecules (X = F, Cl, Br and I) this mode absorbs at 1475, 1355, 1305 and 1252 cm^{-1}. O{N}-attached methyl displays this mode around 1450 {1410} cm^{-1}, S{P}-attached methyl in the neighbourhood of 1320 {1280} cm^{-1} and Si{Hg}-attached methyl around 1265 {1180} cm^{-1}.

Rockings

Both methyl rockings absorb in the lowest of all regions in which these vibrations are found: 840 $\pm$ 50 and 805 $\pm$ 65 cm^{-1} for Si-attached methyl. For R—HgMe

Table 2.6 Absorption regions (cm^{-1}) of sulfur-, silicon- and mercury-bonded methyl

	—SMe	—S(=O)Me	—S(=O)$_2$Me	—SiMe	—HgMe
ν_aMe	3005 ± 25	3005 ± 20	3030 ± 20	2965 ± 35	2975 ± 10
ν'_aMe	2980 ± 45	2995 ± 15	3025 ± 25	2950 ± 25	2975 ± 10
ν_sMe	2925 ± 20	2925 ± 25	2940 ± 20	2910 ± 20	2915 ± 10
δ_aMe	1445 ± 25	1425 ± 15	1430 ± 30	1425 ± 15	1425 ± 25
δ'_aMe	1430 ± 30	1415 ± 15	1415 ± 15	1415 ± 25	1420 ± 25
δ_sMe	1320 ± 20	1305 ± 15	1325 ± 25	1265 ± 25	1180 ± 10
ρMe	995 ± 40	985 ± 40	985 ± 35	840 ± 50	745 ± 45
ρ'Me	940 ± 40	930 ± 35	940 ± 40	805 ± 65	745 ± 45
τMe	175 ± 55	–	–	–	–

molecules the two rockings are not separated in the region 745 ± 45 cm^{-1}. Usually the intensity varies from medium to strong.

R—Me molecules
R = H_3Si—, HD_2Si— and D_3Si— [77], H_2C=CH—SiH_2— [78], $MeOSiH_2$— and $MeOSiD_2$— [79], $MeSSiH_2$— [80], MeSiHX— and MeSiDX— (X = F, Cl, Br, I) [81], $MeSi(OH)_2$— [82], $MeSiCl_2$— [83, 88], $MeSiBr_2$— [89], iPr—$SiCl_2$— [84], F_3Si— [85, 87], Cl_3Si— [88], Br_3Si— [89], I_3Si— [86], MeHg— [90], MeSHg— [91].

2.1.6 Acetyl

The absorption regions of the normal vibrations of the methyl group in acetyl molecules of type R′C(=O)Me have been grouped with R′ representing saturated, unsaturated and aromatic fractions. Acetamides, acetates and thioacetates are also considered in this section.

Stretching vibrations

Typical for acetyl molecules is the weak intensity of the Me stretching vibrations. Despite the strong overlap of the wavenumber regions of the antisymmetric stretchings, both modes are observed separately.

Acetyl substituents often display ν_aMe in the region 3005 ± 40 cm^{-1}. The HW side of this region is limited by 3043 cm^{-1} from the spectrum of acetyl fluoride followed by 3029 cm^{-1} assigned in the spectra of acetyl chloride and 1,1,1-trifluoro-2-propanone. At the LW side one finds 2966 and 2974 cm^{-1} in the spectra of cPrC(=O)Me and $Me_2C(OH)CH_2C$(=O)Me respectively. Aromatic acetyl substituents absorb in a narrow region (3010 ± 10 cm^{-1}), an absorption which sometimes coincides with a CH stretching mode of the ring.

Acetyl substituents display the ν'_aMe in the region 2975 ± 45 cm^{-1}. Acetyl chloride, bromide and iodide absorb near 3015 cm^{-1}, cPrC(=O)Me at 2954 cm^{-1} and 4-X-PhNHC(=O)Me (X = H, Et, Br, HO) in the neighbourhood of 2930 cm^{-1}.

The methyl symmetric stretching has been traced in the region 2905 ± 65 cm^{-1}. This large region is caused by the absorption of ν_sMe in Fermi resonance with the overtone of the Me bending. The highest value (2970 cm^{-1}) has been observed in the spectrum of acetyl fluoride and the lowest (2846 cm^{-1}) in that of DC(=O)Me, followed by 2895 cm^{-1} for $Me_2C(OH)CH_2C(=O)Me$. α-Unsaturated acetyl molecules absorb in the range 2900 ± 50 cm^{-1} with 2947, 2937 or 2850 cm^{-1} for $H_2C=CH-C(=O)Me$ and 2858 cm^{-1} for Me—HC=CH—Me. For acetamides this range is 2900 ± 45 cm^{-1} except for dimethylacetamide in which sometimes the value 2816 or 2810 cm^{-1} is assigned to this mode.

Deformations

One of the two deformations absorbs at 1445 ± 35 cm^{-1}, the other spans the region 1430 ± 40 cm^{-1} with band intensities being weak or medium. Consequently the two bands often coincide.

Maximum values for δ_aMe are observed at 1480 cm^{-1} for 4-EtOPhNHC(=O)Me and at 1470 cm^{-1} for PhNHC(=O)Me and 3-PyC(=O)Me. With 1411, 1416 and 1420 cm^{-1} respectively, the molecules BrC(=O)C(=O)Me, $HOCH_2C(=O)Me$ and cPrC(=O)Me occupy the lower edges of the region.

For δ'_aMe the highest values (1470, 1460 and 1447 cm^{-1}) have been observed in the spectra of 3-PyC(=O)Me, $Me_2NC(=O)Me$ and EtOC(=O)Me respectively, whereas molecules such as NaOC(=O)Me (1390 cm^{-1}), BrC(=O)C(=O)Me (1406 cm^{-1}), MeC(=O)SC(=O)Me (1412 cm^{-1}) and HC(=O)Me with 1414 cm^{-1} absorb in the lowest part of the region.

The δ_sMe of acetyl absorbs at 1365 ± 25 cm^{-1} with an intensity varying from medium to strong, although the weakest bands have been observed in the spectra of acetyl halides. The highest value (1388 cm^{-1}) has been assigned in the spectrum of cPrC(=O)Me, which coincides with a CH bend of the ring. $F_3CC(=O)Me$ follows with 1381 cm^{-1}. The lowest value (1342 cm^{-1}) comes from the spectrum of DC(=O)Me followed by HOC(=O)C(=O)Me and $2-O_2NPhC(=O)Me$, both absorbing at 1349 cm^{-1}. If these extreme values are neglected, the region is reduced to 1365 ± 15 cm^{-1}, which makes this mode in acetyl an appropriate group wavenumber. 2-Propanone absorbs at 1359 and 1355 cm^{-1}.

Rockings

The methyl rockings of the acetyl group generally appear in the regions 1085 ± 70 and 985 ± 85 cm^{-1} as a weak, moderate or sometimes strong band, the wavenumber of which is coupled to the CC stretching vibration, which occurs in the neighbourhood of 900 cm^{-1}. The modes are not considered to provide reliable group wavenumbers. With acetates the rockings are clearly separated and show weak to medium activity in the regions 1050 ± 30 and 980 ± 45 cm^{-1}. However, many esters also show the C—O and C—C stretching bands in the 1050 cm^{-1} region so that the first mentioned rocking cannot always be assigned unambiguously; the band might be hidden behind these (sometimes strong) vibrational bands.

High values for ρMe, namely 1155, 1151, 1140 and 1135 cm^{-1}, have been assigned in the spectra of $ClCH_2C(=O)Me$, $Cl_2CHC(=O)Me$, MeSC(=O)Me and MeC(=O)SC(=O)Me. The lower region is occupied by $CD_3OC(=O)Me$ and $Me_2C(OH)CH_2C(=O)Me$ (1020 cm^{-1}) and by $MeC(=O)CH_2C(=O)Me$ with 1020 and 1040 cm^{-1}.

For ρ'Me, molecules such as IC(=O)Me (1070 cm^{-1}), HSC(=O)Me (1065 cm^{-1}) and DSC(=O)Me with 1060 cm^{-1} absorb at the HW side, whereas MeC(=O)Me (902 cm^{-1}), $Me_2C(OH)CH_2C(=O)Me$ (913 cm^{-1}) and HC(=O)Me with 918 cm^{-1} are found to absorb at the LW side of the range. Both regions of the methyl rockings are too broad to label these modes as suitable group wavenumbers.

Torsion

The torsion is often found in the region 190 ± 80 cm^{-1}.

Table 2.7 Absorption regions (cm^{-1}) of methyl in acetyl

	—C(=O)Me saturated	—C(=O)Me unsaturated	—C(=O)Me aromatic	>NC(=O)Me acetamides	—OC(=O)Me acetates	—SC(=O)Me thioacetates
ν_aMe	3005 ± 40	3000 ± 30	3010 ± 10	2990 ± 20	3010 ± 30	3000 ± 10
ν'_aMe	2990 ± 30	2960 ± 30	2975 ± 25	2965 ± 35	2965 ± 35	2990 ± 10
ν_sMe	2905 ± 65	2900 ± 50	2925 ± 15	2900 ± 45	2910 ± 50	2920 ± 10
δ_aMe	1440 ± 25	1425 ± 15	1445 ± 25	1450 ± 30	1440 ± 25	1435 ± 15
δ'_aMe	1425 ± 15	1415 ± 25	1445 ± 25	1440 ± 20	1430 ± 20	1420 ± 10
δ_sMe	1365 ± 25	1355 ± 10	1360 ± 15	1365 ± 10	1370 ± 20	1355 ± 10
ρMe	1085 ± 70	1060 ± 40	1070 ± 25	1080 ± 50	1050 ± 30	1120 ± 20
ρ'Me	985 ± 85	1000 ± 25	1020 ± 20	995 ± 55	980 ± 45	1000 ± 65
τMe	200 ± 70	–	205 ± 20	–	160 ± 50	–

R—Me molecules

R = R′C(=O)— (see Section 7.1.5), R′HNC(=O)— (see Section 9.3), R′R″NC(=O)— (see Section 7.1.5), R′OC(=O)— (see Section 10.2.3), R′SC(=O)— (see Section 7.1.5).

2.1.7 Methyl esters

Discussed will be mainly methyl esters of the type R′C(=O)OMe, but methyl esters of the type R′C(=S)OMe, R′S(=O)OMe, R′R″P(=O)OMe and R′R″P(=S)OMe are also mentioned.

Stretching vibrations

In the spectra of methyl esters the overlap of the regions in which both antisymmetric stretchings absorb with a weak to medium intensity (3020 ± 30 and 2990 ± 40 cm^{-1}) is not large. Both normal modes are usually observed separately. Methyl formate displays the stretching vibrations in the neighbourhood of 3037 and 3010 cm^{-1} and methyl acetate at 3030 and 3001 cm^{-1}. With methyl esters of unsaturated carboxylic acids the ν_aMe cannot always be attributed unambiguously because the =CH stretching modes also occur in the vicinity of 3000 cm^{-1}. A molecule such as Cl_2P(=S)OMe displays both antisymmetric stretches at 3030 and 3004 cm^{-1}.

Although methyl esters absorb weakly or with medium intensity in the region 2920 ± 80 cm^{-1}, the literature is not always unanimous concerning the assignment. The discrepancies result because, for OMe substituents, overtones and combination bands of the symmetric and antisymmetric bends can adopt considerable intensities by Fermi resonance so that these absorptions may wrongly be interpreted as normal vibrations. This concept is gradually being abandoned. The extra bands are probably the result of individual CH stretching modes of the methyl group capable of taking different positions relative to the free pair of electrons of the O atom. The highest wavenumbers for this ν_sMe are found in the spectra of KOC(=O)OMe (2996 cm^{-1}), F_2P(=O)OMe (2978 cm^{-1}) and FC(=O)OMe (2974 cm^{-1}). If these high values are neglected the region is reduced to 2905 ± 65 cm^{-1}.

Deformations

With methyl esters the overlap of the regions in which the methyl antisymmetric deformations are active (1460 ± 25 and 1455 ± 20 cm^{-1}) is quite strong, which leads to many coinciding wavenumbers. This is obvious, not only for the antisymmetric deformations, but also for the symmetric deformation mostly displayed at the LW side (1435 ± 35 cm^{-1}). The symmetric deformation of oxygen-attached methyl absorbs at higher wavenumbers than are normally shown by saturated hydrocarbons. The intensity of these absorptions is only weak to moderate.

Rockings

Both rockings are active in the region 1170 ± 50 cm^{-1} with a band intensity which is sometimes weak although the band can also be of medium intensity or sometimes even be strong. The vibrational modes are generally observed separately, for instance 1202 and 1159 for ClC(=O)OMe, 1194 and 1163 for MeC(=O)OMe and 1172 and 1145 cm^{-1} for $ClCH_2C$(=O)OMe. This is not the case for Me_2P(=O)OMe, Cl_2P(=S)OMe and MeS(=O)OMe, where both rockings have been placed respectively at 1187, 1175 and 1173 cm^{-1}.

Table 2.8 Absorption regions (cm^{-1}) of methyl in methyl esters

	—C(=O)OMe	—C(=S)OMe	—S(=O)OMe	>P(=O)OMe and >P(=S)OMe
ν_aMe	3020 ± 30	3015 ± 25	3015 ± 25	3020 ± 30
ν'_aMe	2990 ± 40	2995 ± 15	3000 ± 25	2985 ± 35
ν_sMe	2920 ± 80	2940 ± 20	2940 ± 25	2930 ± 30
δ_aMe	1460 ± 25	1460 ± 15	1465 ± 20	1460 ± 15
δ'_aMe	1450 ± 15	1450 ± 15	1455 ± 10	1455 ± 15
δ_sMe	1435 ± 15	1415 ± 15	1445 ± 15	1445 ± 25
ρMe	1185 ± 35	1175 ± 25	1195 ± 25	1185 ± 15
ρ'Me	1155 ± 35	1145 ± 25	1165 ± 25	1165 ± 25
τMe	225 ± 65	250 ± 40	–	220 ± 50

R—Me molecules

R = R′C(=O)O— (see Section 7.1.8), R′C(=S)O— (see Section 10.1.2), R′S(=O)O— (see Section 10.1.2), R′R″P(=O)O— and R′R″P(=S)O— (see Section 10.1.2).

2.1.8 Methyl ethers

Molecules of the type R′—OMe, in which R′ is a saturated or an unsaturated substituent or a (substituted) aromatic ring, have served to establish the absorption regions of the OMe group.

Stretching vibrations

The antisymmetric stretchings are observed in the regions 2990 ± 40 and 2955 ± 35 cm^{-1}. The band intensity is generally weak or medium. The highest values for the methyl antisymmetric stretchings have been attributed in the spectra of Cl_2HCCF_2OMe (3027 and 2974 cm^{-1}), FCH_2OMe (3024 and 2976 cm^{-1}) and $BrCH_2OMe$ (3021 and 2965 cm^{-1}). Low values for these modes are found in the spectra of MeOCH=CHOMe (both stretchings at 2955 cm^{-1}), H_2C=CHOMe (2959 and 2927 cm^{-1}) and MeOMe (2988 and 2922 cm^{-1}). Most of the aromatic ethers give these stretchings in the neighbourhood of 2985 and 2950 cm^{-1}.

The methyl symmetric stretching in ethers absorbs in the neighbourhood of 2850 cm^{-1} as a sharp discrete peak with a medium intensity, that is about 50 cm^{-1} lower than in methyl esters and typical for methyl ethers. It seems that the methyl stretching vibrations absorb at the HW side of the above-mentioned wavenumber regions if OMe is adjacent to a partly or completely halogen-substituted carbon atom. If more bands are present than expected, this is almost certainly due to the various positions of the CH bonds with respect to the free pair of electrons of the O atom.

Deformations

Both methyl antisymmetric (1465 $\pm$ 20 and 1455 $\pm$ 20 cm^{-1}) and symmetric (1450 $\pm$ 20 cm^{-1}) deformations absorb in narrow wavenumber regions with varying band intensities. In addition, the overlap is considerable, so that coinciding wavenumbers frequently occur. It is obvious that the antisymmetric deformations of methyl esters and ethers absorb in almost similar regions although the symmetric deformation in esters generally absorbs about 15 cm^{-1} lower than in ethers.

Rockings

The methyl rockings of saturated, unsaturated and aromatic methyl ethers absorb in the regions 1190 $\pm$ 45 and 1155 $\pm$ 35 cm^{-1}. With saturated ethers, the antisymmetric COC stretching vibration (1110 $\pm$ 70 cm^{-1}) is also active in the above-mentioned region. This band is strong and mostly occurs at the LW side of both rockings. With aromatic methyl ethers some planar CH bends absorb in the above-mentioned regions and the possibility that one of the rockings coincides with these absorptions is therefore substantial. Methoxybenzene displays one Me rocking at 1180 cm^{-1}. The other is not assigned unless this mode accidentally coincides with the first or with a CH planar bend.

Methyl ethers display the highest ρMe at 1190 $\pm$ 45 cm^{-1}, mostly centred around 1200 cm^{-1}. The high value of 1250 cm^{-1} observed in the spectrum of methoxymethane is outside the above-mentioned range. High values are also scored by $MeOCH_2OMe$ (1232 cm^{-1}), 2-MeOPhOMe (1231 cm^{-1}), cBuOMe (1228 cm^{-1}), $BrCH_2OMe$ (1227 cm^{-1}) and 3-MeOPhOMe (1210 cm^{-1}). Low rockings have been assigned in the spectra of 2-ThOMe (1151 cm^{-1}), MeOOMe (1156 cm^{-1}) and Cl_2CHCF_2OMe (1159 cm^{-1}).

The highest value for ρ'Me is furnished by $BrCH_2OMe$ (1190 cm^{-1}) followed by MeOMe and ICH_2OMe, both absorbing at 1180 cm^{-1}. The lowest values are 1121 cm^{-1} from 2-BrPhOMe and 1129 cm^{-1} from 2-ClPhOMe. The remaining observed rocking modes gather around 1155 cm^{-1}. The rocking modes of methyl esters and ethers absorb in almost similar wavenumber regions.

Table 2.9 Absorption regions (cm^{-1}) of methyl in methyl ethers

	—OMe saturated	—OMe unsaturated	—OMe aromatic
ν_aMe	3000 ± 25	2985 ± 35	2985 ± 20
ν'_aMe	2955 ± 35	2945 ± 20	2955 ± 20
ν_sMe	2845 ± 25	2850 ± 30	2845 ± 15
δ_aMe	1465 ± 20	1470 ± 15	1465 ± 10
δ'_aMe	1455 ± 20	1460 ± 15	1460 ± 15
δ_sMe	1445 ± 15	1450 ± 10	1450 ± 10
ρMe	1195 ± 40	1190 ± 30	1190 ± 45
ρ'Me	1160 ± 30	1145 ± 25	1150 ± 30
τMe	225 ± 40	235 ± 15	–

R—Me molecules
R = R′O— (see Section 10.1.2).

References

1. J.H. Griffiths, N.L. Owen and J. Sheridan, *J. Chem. Soc., Faraday Trans. 2*, **9**, 1359 (1973).
2. G.A. Crowder and T. Koger, *J. Mol. Struct.*, **23**, 311 (1974).
3. J. Gustavsen and P. Klaeboe, *Spectrochim. Acta, Part A*, **32A**, 755 (1976).
4. G.A. Crowder and T. Koger, *J. Mol. Struct.*, **29**, 233 (1975).
5. P. Klaeboe, *Spectrochim. Acta, Part A*, **26A**, 87 (1970).
6. J.R. Durig, C.M. Player, Y.S. Li, J. Bragin and C.W. Hawley, *J. Chem. Phys.*, **57**, 4544 (1972).
7. C.G. Opascar and S. Krimm, *Spectrochim. Acta, Part A*, **23A**, 2261 (1967).
8. G.A. Crowder, *J. Mol. Struct.*, **100**, 415 (1983).
9. J. Thorbjørnsrud, H.O. Ellestad, P. Klaboe and T. Torgrimsen, *J. Mol. Struct.*, **15**, 45 (1973).
10. A.B. Dempster, K. Price and N. Sheppard, *Spectrochim. Acta, Part A*, **27A**, 1563 (1971).
11. A.J. Barnes, M.L. Evans and H.E. Hallam, *J. Mol. Struct.*, **99**, 235 (1983).
12. W.H. Moore and S. Krimm, *Spectrochim. Acta, Part A*, **29A**, 2025 (1973).
13. E. Benedetti and P. Cecchi, *Spectrochim. Acta, Part A*, **28A**, 1007 (1972).
14. J.R. Durig, C.J. Murrey, W.E. Bucy and A.E. Sloan, *Spectrochim. Acta, Part A*, **32A**, 175 (1976).
15. S. Suzuki and A. Dempster, *J. Mol. Struct.*, **32**, 339 (1976).
16. D.C. McKean, J.C. Lavalley, O. Saur, H.G.M. Edwards and V. Fawcett, *Spectrochim. Acta, Part A*, **33A**, 914 (1977).
17. L.W. Daasch, C.Y. Liang and J.R. Nielsen, *J. Chem. Phys.*, **22**, 1293 (1954).
18. P. Klaboe, A. Linde and B.N. Cyvin, *Spectrochim. Acta, Part A*, **30A**, 1513 (1974).
19. J. Murto, A. Kivinen, K. Edelman and E. Hassinen, *Spectrochim. Acta, Part A*, **31A**, 479 (1975).
20. G.A. Crowder and D. Jackson, *Spectrochim. Acta, Part A*, **27A**, 2505 (1971).
21. G.A. Crowder, *J. Mol. Struct.*, **15**, 351 (1973).
22. G.A. Crowder and W.Y. Lin, *J. Mol. Struct.*, **62**, 7 (1980).

23. G.A. Crowder, *Spectrochim. Acta, Part A*, **42A**, 1079 (1986).
24. M.S. Wu, P.C. Painter and M.M. Coleman, *Spectrochim. Acta, Part A*, **35A**, 823 (1979).
25. J.H.S. Green and D.J. Harrison, *Spectrochim. Acta, Part A*, **27A**, 1217 (1971).
26. A.O. Diallo, *Spectrochim. Acta, Part A*, **34A**, 235 (1978).
27. P. Klaboe, *Spectrochim. Acta, Part A*, **26A**, 977 (1970).
28. A.O. Diallo, *Spectrochim. Acta, Part A*, **32A**, 295 (1976).
29. J.R. Nielsen, H.H. Claassen and D.C. Smith, *J. Chem. Phys.*, **18**, 1471 (1950).
30. H.W. Thompson and R.B. Temple, *J. Chem. Soc.*, 1428 (1948).
31. T.R. Stengle and R.C. Taylor, *J. Mol. Spectrosc.*, **34**, 33 (1970).
32. H. Bürger, H. Nispel and G. Pawelke, *Spectrochim. Acta, Part A*, **36A**, 7 (1980).
33. P. Venkateswarlu, *J. Chem. Phys.*, **20**, 1810 (1952).
34. M.Z. El-Saban, A.G. Meister and F.F. Cleveland, *J. Chem. Phys.*, **19**, 855 (1951).
35. D.C. Smith, G.M. Brown, J.R. Nielsen, R.M. Smith and C.Y. Liang, *J. Chem. Phys.*, **20**, 473 (1952).
36. S.G. Frankiss and D.J. Harrison, *Spectrochim. Acta, Part A*, **31A**, 29 (1975).
37. J.R. Durig, S.M. Craven, C.W. Hawley and J. Bragin, *J. Chem. Phys.*, **57**, 131 (1972).
38. B. Silvi, P. Labarbe and J.P. Perchard, *Spectrochim. Acta, Part A*, **29A**, 263 (1973).
39. I. Tokue, T. Fukuama and K. Kuchitsu, *J. Mol. Struct.*, **17**, 207 (1973).
40. D.C. McKean, M.W. Mackenzie, A.R. Morrison, J.C. Lavalley, A. Janin, V. Fawcett and H.G.M. Edwards, *Spectrochim. Acta, Part A*, **41A**, 435 (1985).
41. R. Meyer and H.S. Günthard, *Spectrochim. Acta, Part A*, **23A**, 2341 (1967).
42. R.W. Mitchell and J.A. Merritt, *Spectrochim. Acta, Part A*, **25A**, 1881 (1969).
43. G. Geiseler, H. Böhlig and J. Fruwert, *J. Mol. Struct.*, **18**, 43 (1973).
44. J. Kidric, D. Hadzi and B. Barlic, *J. Mol. Struct.*, **22**, 45 (1974).
45. A. Natarajan and J.S.P. Ebenezer, *Can. J. Spectrosc.*, **31**, 158 (1986).
46. J.C. Whitmer, *J. Mol. Struct.*, **21**, 173 (1974).
47. J.C. Lavalley, J. Saussey and J. Lamotte, *Spectrochim. Acta, Part A*, **35A**, 696 (1979).
48. W.J. Balfour, K. Beveridge and J.C.M. Zwinkels, *Spectrochim. Acta, Part A*, **35A**, 163 (1979).
49. E. Augdahl, E. Kloster-Jensen and A. Rogstad, *Spectrochim. Acta, Part A*, **30A**, 399 (1974).
50. K. Hamada and H. Morishita, *Spectrosc. Lett.*, **13**, 15 (1980).
51. D. Milligan and J. Jacox, *J. Mol. Spectrosc.*, **8**, 129 (1962).
52. T. Freedman and E.R. Nixon, *Spectrochim. Acta, Part A*, **28A**, 1388 (1972).
53. A. Givan and A. Loewenschuss, *J. Mol. Struct.*, **98**, 234 (1983).
54. D.C. McKean, *Spectrochim. Acta, Part A*, **30A**, 1169 (1974).
55. J.L. Duncan, *Spectrochim. Acta*, **20**, 1197 (1964).
56. P. Pulay and F. Török, *J. Mol. Struct.*, **29**, 239 (1975).
57. A.P. Gray and R.C. Lord, *J. Chem. Phys.*, **26**, 690 (1957).
58. J.R. Durig and P. Cooper, *J. Mol. Struct.*, **41**, 188 (1977).
59. M.J. Buttler, D.C. McKean, R. Taylor and L.A. Woodward, *Spectrochim. Acta*, **21**, 1384 (1965).
60. E.H. Eyster and R.H. Gilette, *J. Chem. Phys.*, **8**, 369 (1940).
61. W.G. Fately and F.A. Miller, *Spectrochim. Acta*, **17**, 857 (1961).
62. R.P. Hirschmann, R.N. Kniseley and V.A. Fassel, *Spectrochim. Acta*, **21**, 2125 (1965).
63. A.G. Moritz, *Spectrochim. Acta*, **22**, 1021 (1966).
64. A. Costoulas and R.L. Werner, *Aust. J. Chem.*, **12**, 601 (1959).
65. N.S. Ham and J.B. Willis, *Spectrochim. Acta*, **16**, 279 (1960).
66. J.R. Durig, J.F. Sullivan, H.L. Heusel and S. Cradock, *J. Mol. Struct.*, **100**, 248 (1983).
67. R.N. Kniseley, R.P. Hirschmann and V.A. Fassel, *Spectrochim. Acta, Part A*, **23A**, 109 (1967).

68. F.A. Miller and W.B. White, *Z. Elektrochem.*, **64**, 701 (1960).
69. E. Tannenbaum, R.J. Meyers and W.D. Gwinn, *J. Chem. Phys.*, **25**, 42 (1956).
70. G. Malewski, M. Pfeiffer and P. Reich, *J. Mol. Struct.*, **3**, 420 (1969).
71. C. Trinquecoste, M. Rey-Lafon and M-T. Forel, *Spectrochim. Acta, Part A*, **30A**, 813 (1974).
72. D.C. McKean and R.A. Watt, *J. Mol. Spectrosc.*, **61**, 184 (1976).
73. J.R. Hill, D.S. Moore, S.C. Schmidt and C.B. Storm, *J. Phys. Chem.*, **95**, 3037 (1991).
74. F.A. Miller and D. Bassi, *Spectrochim. Acta*, **19**, 565 (1963).
75. W.T. Thompson and W.H. Fletcher, *Spectrochim. Acta*, **22**, 1907 (1966).
76. R.L. Williams, *J. Chem. Phys.*, **25**, 656 (1956).
77. D.F. Ball, T. Carter, D.C. McKean and L.A. Woodward, *Spectrochim. Acta*, **20**, 1721 (1964).
78. J.R. Durig, J.F. Sullivan and M.A. Qtaitat, *J. Mol. Struct.*, **243**, 239 (1991).
79. K. Ohno, K. Taga and H. Murata, *J. Mol. Struct.*, **55**, 7 (1979).
80. K. Taga, K. Ohno and H. Murata, *J. Mol. Struct.*, **67**, 199 (1980).
81. A.J.F. Clark, J.E. Drake, R.T. Hemmings and Q. Shen, *Spectrochim. Acta, Part A*, **39A**, 127 (1983).
82. T.D. Ho, *Appl. Spectrosc.*, **40**, 29 (1986).
83. A.L. Smith and D.R. Andersen, *Appl. Spectrosc.*, **38**, 822 (1984).
84. K. Ohno, K. Taga, I. Yoshiva and H. Murata, *Spectrochim. Acta, Part A*, **35A**, 883 (1979).
85. A.J.F. Clark and J.E. Drake, *Spectrochim. Acta, Part A*, **37A**, 391 (1981).
86. A.J.F. Clark and J.E. Drake, *Spectrochim. Acta, Part A*, **32A**, 1419 (1976).
87. R.L. Collins and J.R. Nielsen, *J. Chem. Phys.*, **23**, 351 (1955).
88. A.L. Smith, *J. Chem. Phys.*, **21**, 1997 (1953).
89. H. Murata and S. Hayashi, *J. Chem. Phys.*, **19**, 1217 (1951).
90. I.S. Butler and M.L. Newbury, *Spectrochim. Acta, Part A*, **33A**, 671 (1977).
91. R.A. Nyquist and J.R. Mann, *Spectrochim. Acta, Part A*, **28A**, 511 (1972).
92. P.L. Stanghelli and R. Rossetti, *Inorg. Chem.*, **29**, 2047 (1990).

2.2 TRIHALOGENOMETHYL

The CX_3 group (X = F, Cl, Br and I) possesses as many normal vibrations as methyl and, as halogen atoms are much heavier than hydrogen, these modes are expected at considerably lower values. The band intensities are strongest with the stretching vibrations in the sequence F>Cl>Br>I. Stated differently, the intensity correlates more or less with the difference in electronegativity between the X and the C atom.

For most of the R—Me molecules the antisymmetric stretching vibrations and deformations absorb at the HW side, and the symmetric mode at the LW side of the region in which these vibrations are found. Thus, the antisymmetric modes generally possess a higher wavenumber than the symmetric ones. This is not always true for R—CX_3 molecules with C_{3v} symmetry where for R—CF_3 the $\nu_s\{\delta_a\}$ absorbs at higher wavenumbers than $\nu_a\{\delta_s\}$. For molecules with C_s symmetry one of the antisymmetric stretchings does not always absorb at the highest wavenumber. This would, as with methyl, indicate a dissimilarity in the CX bonds. Yet it is not always straightforward to assign unambiguously the symmetric and antisymmetric vibrations if the molecule cannot be studied in the gaseous phase or, if it can, does not provide decisive band structures. On the other hand, because of the non-equivalence of the CX bonds, one cannot be certain whether a band arises from either a symmetric or an antisymmetric CX_3 vibration or from an isolated CX' vibration. Moreover, as with methyl, overtones and/or combination bands of deformations whether or not intensified may occur in the stretching mode area, which obstructs a proper assignment. Therefore in this text the nine CX_3 vibrations will be designated as follows:

$$\nu CX \geq \nu' CX \geq \nu'' CX > \delta CX \geq \delta' CX \geq \delta'' CX > \rho CX \geq \rho' CX > \tau CX$$

Using this notation it is quite possible that for instance $\nu''CX$ represents an antisymmetric mode with one molecule although it more approaches the symmetric mode in another. For C_{3v} molecules, for example $R'—C{\equiv}C—CX_3$, the notations ν_s, ν_a, δ_s, δ_a and ρ are still appropriate.

2.2.1 Trifluoromethyl

The three CF stretching modes absorb in broad regions and are considered as valuable group vibrations merely because of their sometimes moderate but mostly strong intensity. The intensity of the deformations and rockings varies from weak

Table 2.10 Absorption regions (cm^{-1}) of α-di- and α-tri-halogen substituted $—CF_3$

	Total region	Molecules absorbing at high wavenumbers	Molecules absorbing at low wavenumbers	Region of remaining molecules
νCF	1315 ± 105	$F_3C—CF_3$ (1417) $F_3C—CF_2—CF_3$ (1370) $Cl—CF_2—CF_3$ (1354)	$X_3C—CF_3$ (X=I, Br, Cl, F) (1210, 1234, 1249, 1250)	1295 ± 45
$\nu' CF$	1270 ± 80	$F_3C—CF_2—CF_3$ (1350) $Cl(Br)CH—CF_3$ (1265)	$I_3C—CF_3$ (1190)	1240 ± 25
$\nu'' CF$	1190 ± 80	$F_3C—CF_2—CF_3$ (1268) $F_3C—CF_3$ (1250)	$F_3C—CF_3$ (1117) $Cl(Br)CH—CF_3$ (1126)	1195 ± 30
δCF	715 ± 95	$Cl(Br)CH—CF_3$ (808) $F_3C—CF_3$ (807) $Cl_3C—CF_3$ (793)	$F_3C—CF_3$ (619) $I_3C—CF_3$ (668)	730 ± 50
$\delta' CF$	620 ± 100	$F_3C—CF_3$ (714)	$F_3C—CF_3$ (520)	590 ± 60
$\delta'' CF$	565 ± 30	$ClCF_2—CF_3$ (593)	$F_3C—CI_2—CF_3$ (535)	565 ± 25
ρCF	340 ± 120	$F_3C—CF_2—CF_3$ (460) $BrCF_2—CF_3$ (438) $ICF_2—CF_3$ (433)	$F_3C—CF_2—CF_3$ (220) $I_3C—CF_3$ (303)	355 ± 35
$\rho' CF$	275 ± 55	$F_3C—CCl_2—CF_3$ (325) $F_3C—CBr_2—CF_3$ (321)	$BrCF_2—CF_3$ (221) $ICF_2—CF_3$ (223)	285 ± 25

to medium. The CF_3 torsion is expected to absorb in the neighbourhood of 100 cm^{-1}.

R—CF_3 molecules
R = F_2CH— [1], F_3C—CF_2— [2, 3], $ClCF_2$— [4, 5], $BrCF_2$— [4], ICF_2— [4], Cl_2CH— [6], Cl(Br)CH— [7], F_3C—CCl_2—, F_3C—CBr_2— and F_3C—CI_2— [3], F_3C—, Cl_3C—, Br_3C— and I_3C— [8].

Table 2.11 Absorption regions (cm^{-1}) of —CF_3 in trifluoroacetyl

	Total region	Molecules absorbing at high wavenumbers	Molecules absorbing at low wavenumbers	Region of remaining molecules
νCF	1290 ± 85	$F_3CC(=O)N(Me)C(=O)CF_3$ (1374) $MeOC(=O)—CF_3$ (1368)	$AgOC(=O)—CF_3$ (1206) $NaOC(=O)—CF_3$ (1213) $BrCH_2C(=O)—CF_3$ (1217)	1290 ± 60
ν'CF	1210 ± 50	$F_3C—OC(=O)—CF_3$ (1258) $FC(=O)—CF_3$ (1254)	$MeC(=O)—CF_3$ (1163) $BrCH_2C(=O)—CF_3$ (1168) $PhC(=O)—CF_3$ (1180)	1220 ± 30
ν''CF	1165 ± 55	$F_3CC(=O)—CF_3$ (1219) $FC(=O)—CF_3$ (1214)	$NaOC(=O)—CF_3$ (1140) $MeOC(=O)—CF_3$ (1144)	1175 ± 30
δCF	700 ± 85	$H_3SiOC(=O)—CF_3$ (785)	$MeC(=O)—CF_3$ (615) (region of δC=O)	735 ± 45
δ'CF	590 ± 80	$PhC(=O)—CF_3$ (670) $F_3CC(=O)—CF_3$ (633) $AgOC(=O)—CF_3$ (606)	$H_3SiOC(=O)—CF_3$ (517) $F_3COC(=O)—CF_3$ (520)	570 ± 20
δ''CF	515 ± 20	$EtSC(=O)—CF_3$ (536)	$HSC(=O)—CF_3$ (496)	515 ± 15
ρCF	355 ± 130	$PhC(=O)—CF_3$ (484) $H_2NC(=O)—CF_3$ (435) $F_3COC(=O)—CF_3$ (433)	$HSC(=O)—CF_3$ (228) $MeC(=O)—CF_3$ (246)	335 ± 80
ρ'CF	230 ± 40	$H_2NC(=O)—CF_3$ (265) $F_3CC(=O)—CF_3$ (257)	$F_3CC(=O)—CF_3$ (193)	230 ± 25

R—CR_3 molecules
R = R′C(=O);
R′ = H— [9–14], D— [9–11], Me— [15], MeC(=O)CH_2—, MeC(=O)CD_2—, F_3C—C(=O)CH_2— and F_3C—C(=O)CD_2— [16], 2-Th—CH_2— [17], $BrCH_2$— [18, 19], F_3C— [2, 20–22], $F_3CC(=O)$—, Ph—, H_2N— [23–26], $F_3CC(=O)NMe$—, HO— [11, 27–34], DO— [11, 33, 34], MeO— [35–37], EtO—, F_3CO— [38], NaO— [39–41], AgO— [40], H_2C=CH—O— [36], 4-O_2N—PhO—, $F_3CC(=O)O$— [11, 33], H_3SiO— and D_3SiO— [37], HS— [11, 42], EtS—, F— [11, 13, 14, 40, 43, 44], Cl— [11, 13, 14, 42, 45, 46], Br— [45, 46], I— [47].
R = FC(=S)— [48].

Table 2.12 Absorption regions (cm^{-1}) of α-saturated —CF_3

	Total region	Molecules absorbing at high wavenumbers	Molecules absorbing at low wavenumbers	Region of remaining molecules
νCF	1295 ± 65	MeCD(OH)—CF_3 (1360) F_3CCH(OH)—CF_3 (1350)	Cl_3C(OH)C(CF_3)$_2$ (1238) F_3CCD(OH)—CF_3 (1255) F_3CCH(OH)—CF_3 (1258)	1300 ± 40
ν'CF	1215 ± 65	DHN—CH_2—CF_3 (1275) D_2N—CH_2—CF_3 (1273) BrCH_2—CF_3 (1274) ICH_2—CF_3 (1259)	HCF_3 (1152)	1210 ± 40
ν''CF	1120 ± 70	F_3C—CH(OH)—CF_3 (1188) F_3C—CH_2—CF_3 (1178) DHN—CH_2—CF_3 (1175)	H_2NCH_2—CF_3.HCl (1054) Br—CF_3 (1085) F_3C—CH_2—CF_3 (1088)	1130 ± 40
δCF	695 ± 95	(H_2N)$_2$C(CF_3)$_2$ (785) Cl—CF_3 (783) Br—CF_3 (762)	HO—C(Me)$_2$—CF_3 (602) F_3C—CD(OD)—CF_3 (605)	660 ± 50
δ'CF	585 ± 65	MeCH(OH)—CF_3 (650) (H_2N)$_2$C(CF_3)$_2$ (638)	BrCH_2—CF_3 (526) F_3C—CH(OH)—CF_3 (520)	575 ± 40
δ''CF	530 ± 45	HOC(Me)$_2$—CF_3 (575) Cl—CF_3 (562)	Cl_3C(OD)C(CF_3)$_2$ (487)	530 ± 30
ρCF	360 ± 125	4-MeS(=O)$_2$OCH_2—CF_3 (484) H_2NCH_2—CF_3 (422) DHNCH_2—CF_3 (407) D_2NCH_2—CF_3 (398)	F_3C—CH(OH)—CF_3 (237) F_3C—CD(OH)—CF_3 (240) F_3C—CD(OD)—CF_3 (240)	325 ± 65
ρ'CF	275 ± 115	H_2NCH_2—CF_3 (383) DHNCH_2—CF_3 (367) D_2NCH_2—CF_3 (364)	F_3C—CH(OH)—CF_3 (167) Cl_3C(OH)C(CF_3)$_2$ (169) F_3C—CD(OH)—CF_3 (170) MeC(OH)(CF_3)$_2$ (170)	260 ± 40
τCF	100 ± 30			

R—CF_3 molecules

R = H— [49, 50], D— [49], Me— [8, 51–54], F_3CCH_2— [55–57], H_2NC(=O)CH_2—, HOCH_2— [58–61], MeOCH_2— [62], H_2C=CHOCH_2— [63], 4-MePhS(=O)$_2$OCH_2—, H_2NCH_2—, DHNCH_2— and D_2NCH_2— [64], ClH$_3$NCH_2—, FCH_2— [65, 66], ClCH_2— [6, 66], BrCH_2— [65–67], ICH_2— [65, 66], MeCH(OH)— [68, 69], MeCH(OD)—, MeCD(OH)— and MeCD(OD)— [68], MeCH(OMe)— [72],

$F_3CCH(OH)—$ [70, 71, 129], $F_3CCH(OD)—$, $F_3CCD(OH)—$ and $F_3CCD(OD)—$ [70, 71], $F_3CCH(OMe)—$ [73], $Cl_3CCH(OH)—$ [74], $EtOCH(OH)—$, $HOC(Me)_2—$ [75], $MeC(OH)(CF_3)—$ [76], $Cl_3C(OH)C(CF_3)—$ and $Cl_3C(OD)C(CF_3)—$ [77], $(H_2N)_2C(CF_3)—$ and $(D_2N)_2C(CF_3)—$ [78], Cl— [79–81], Br— [80, 81].

Table 2.13 Absorption regions (cm^{-1}) of α-unsaturated $—CF_3$

	Total region	Molecules absorbing at high wavenumbers	Molecules absorbing low wavenumbers	Region of remaining molecules
νCF	1285 ± 105^a	$(CF_3)_2C{=}NH$ (1389) $F—C{\equiv}C—CF_3$ (1376) $N{=}N{=}C(CF_3)_2$ (1361)	$F_3C—C{\equiv}C—C{\equiv}C—CF_3$ (1185 and 1232) $N{\equiv}C—CF_3$ (1227)	1295 ± 50
$\nu' CF$	1195 ± 20			1195 ± 20
$\nu'' CF$	1135 ± 80^a	$NC—CF_3$ (1214) $F_3C—C{\equiv}C—CF_3$ (1198)	$F_3C—C{\equiv}C—C{\equiv}C—CF_3$ (1056)	1160 ± 25
δCF	685 ± 75	*cis* $F_3C—CH{=}CH—CF_3$ (759)	*cis* $F_3C—CH{=}CH—CF_3$ (610)	675 ± 50
$\delta' CF$	575 ± 65^a		$F_2C{=}C(OH)—CF_3$ (515)	605 ± 35
$\delta'' CF$	505 ± 65^a	$F_3C—C{\equiv}C—C{\equiv}C—CF_3$ (568)	$O{=}C{=}C(CF_3)_2$ (448) $N{=}N{=}C(CF_3)_2$ (450) $F_2C{=}C(OH))—CF_3$ (456)	515 ± 35
ρCF	405 ± 95^a	*cis* $F_3C—CH{=}CH—CF_3$ (480, 496) $F_3C—C{\equiv}C—C{\equiv}C—CF_3$ (494)	$N{=}N{=}C(CF_3)_2$ (314) $O{=}C{=}C(CF_3)_2$ (314)	425 ± 50
$\rho' CF$	320 ± 40			320 ± 40
τCF	65 ± 35			

[a] Absorption regions of C_{3v} molecules: $\nu_s CF_3$, $\nu_a CF_3$, $\delta_a CF_3$, $\delta_s CF_3$ and ρCF_3.

$R—CF_3$ molecules

R = $H_2C{=}CH—$ [82, 83], *cis* and *trans* $F_3C—CH{=}CH—$ [84], $EtOC({=}O)CH{=}CH—$, $HN{=}C(CF_3)—$ [69, 85], $N{=}N{=}C(CF_3)—$ and $O{=}C{=}C(CF_3)—$ [85], $F_2{=}C(OH)—$ [86], $H—C{\equiv}C—$ [87–90], $D—C{\equiv}C—$ [88–90], $F_3C—C{\equiv}C—$ [87, 88], $F_3C—C{\equiv}C—C{\equiv}C—$ [91], $N{\equiv}C—$ [92, 93], $F—C{\equiv}C—$ [90], $Cl—C{\equiv}C—$, $Br—C{\equiv}C—$ and $I—C{\equiv}C—$ [90, 94].

Table 2.14 Absorption regions (cm^{-1}) of $—CF_3$ in substituted benzenes

	Total region	Molecules absorbing at high wavenumbers	Molecules absorbing at low wavenumbers	Region of remaining molecules
νCF	1305 ± 40	2-F, 5-$H_2NPh—CF_3$ (1343) 2-MeO, 5-$O_2NPh—CF_3$ (1342) 2-HO(O=)C, 3-$F_3CPh—CF_3$ (1342)	Ph-$d_5—CF_3$ (1270) 3-F_3C, 5-$H_2NPh—CF_3$ (1279) 3-F_3C, 5-$HOPh—CF_3$ (1279)	1315 ± 25
$\nu' CF$	1160 ± 30		Ph-$d_5—CF_3$ (1130) 2-MeO, 5-$O_2NPh—CF_3$ (1140)	1170 ± 20
$\nu'' CF$	1135 ± 30	4-$H_2NPh—CF_3$ (1162) 2, 4-Cl_2, 5-$O_2NPh—CF_3$ (1161)	2-$H_2NPh—CF_3$ (1107) 2-$ClPh—CF_3$ (1107) 4-$O_2NPh—CF_3$ (1110)	1135 ± 20
δCF	650 ± 70	4-$F_3CPh—CF_3$ (718) 3-O_2N, 4-$HOPh—CF_3$ (699) 3-O_2N, 4-$H_2NPh—CF_3$ (696)	4-$O_2NPh—CF_3$ (586) 4-HO(O=)$CPh—CF_3$ (588) 4-$BrPh—CF_3$ (590) 4-$H_2NPh—CF_3$ (601)	660 ± 30
$\delta' CF$	590 ± 55	3-$HOPh—CF_3$ (643)	4-HO(O=)$CPh—CF_3$ (541) 2-Br, 5-$H_2NPh—CF_3$ (557)	610 ± 30
$\delta'' CF$	525 ± 85	3-$O_2NPh—CF_3$ (606) 3-Cl(O=)$CPh—CF_3$ (592) 4-$FPh—CF_3$ (592)	4-$BrPh—CF_3$ (441) 4-$O_2NPh—CF_3$ (485) 4-HO(O=)$CPh—CF_3$ (487)	540 ± 50
ρCF	405 ± 65	3-Me(O=)$CCH_2Ph—CF_3$ (466) 2-HO(O=)CCH=$CHPh—CF_3$ (465) 3-HO(O=)CCH=$CHPh—CF_3$ (464)	2-$ClPh—CF_3$ (340) 3-$ClPh—CF_3$ (342) 3-$IPh—CF_3$ (345)	400 ± 50
$\rho' CF$	310 ± 50	4-$O_2NPh—CF_3$ (358)	3-$BrPh—CF_3$ (253)	305 ± 45

$R—CF_3$ molecules
R = substituted phenyl (see Section 12.4)

Table 2.15 Absorption regions (cm^{-1}) of $—CF_3$ adjacent to N, O, S, P and Si

	N	O	S	P and Si
νCF	1245 ± 45	1305 ± 25	1230 ± 50	1210 ± 25
$\nu' CF$	1210 ± 40	1250 ± 10	1195 ± 50	1205 ± 30
$\nu'' CF$	1160 ± 60	1185 ± 25	1130 ± 55	1135 ± 55
δCF	750 ± 50	705 ± 45	760 ± 20	730 ± 30
$\delta' CF$	600 ± 50	615 ± 45	570 ± 40	565 ± 55
$\delta'' CF$	560 ± 60	570 ± 40	530 ± 30	555 ± 35
ρCF	410 ± 90	430 ± 60	310 ± 50	240 ± 60
$\rho' CF$	310 ± 70	350 ± 60	250 ± 50	210 ± 80
τCF	60 ± 40		125 ± 50	

$R—CF_3$ molecules

R = O=N— [95, 96], O_2N— [97–99], F_2N— [100], F_2S=N— [101], *cis* and *trans* F_3C—N=N— [84, 102], CN— [103], HO— [104], F_3CO— [105], F_3COO— [106–108], FC(=O)O— [109], F_3COOO— [107, 108], F_3COC(=O)O— and F_3CC(=O)O— [38], HS— [110–113], DS— [113], F_3CS— [114, 115], F_3CSS— [114], ClS— [116], ClSS— [117], F_3CSF_2— [118], F_3S— [118], F_5S— [119], F_3CS(=O)— [114], HOS$(=O)_2$— [120], ClS$(=O)_2$—, $^-$OS$(=O)_2$— [121], 2,6-Et_2PhNHS-$(=O)_2$—, NCS— [169], H_2P— [122], Me_2P— [123], Cl_2P— [124], F_2P(=O)— [125], Cl_2P(=O)— [126], H_3Si— [127], Me_3Si— [128], $(CD_3)_3Si$— [128].

2.2.2 Trichloromethyl

Because chlorine is heavier than fluorine the stretching vibrations obviously absorb at lower wavenumbers compared with those of the trifluoromethyl group. The intensity of the bands varies from moderate to strong. One easily appreciates that, in contrast to similar $R—CF_3$ molecules with C_{3v} symmetry, for which the symmetric CF_3 stretching vibrations absorb at a higher value than the antisymmetric mode, the CCl_3 antisymmetric stretching vibration absorbs at a higher value than the symmetric mode. The intensity of the deformation, however, is normally very weak.

Table 2.16 Absorption regions (cm^{-1}) of $—CCl_3$

	Total region	Molecules absorbing at high wavenumber	Molecules absorbing at low wavenumber	Region of remaining molecules
νCCl	805 ± 95	$Cl_2CHC(=O)—CCl_3$ (895) $Cl_2CHCH(Cl)—CCl_3$ (870) $Cl_2CHCCl_2—CCl_3$ (870)	$MeCCl_3$ (713) $DCCl_3$ (744) $Cl_3Si—CCl_3$ (751)	815 ± 55
$\nu' CCl$	730 ± 85	$OCN—C(=O)—CCl_3$ (814) $ClS(=O)_2—CCl_3$ (813) $ClCH_2CH_2—CCl_3$ (797)	$Cl_3C—C(=O)—CCl_3$ (661, 645) $Cl_2CHC(=O)—CCl_3$ (648) $H_2C(=O)—CCl_3$	730 ± 60
$\nu'' CCl$	555 ± 125	$Cl_3C—CCl_3$ (678) $H—CCl_3$ (668) $D—CCl_3$ (655)	$Cl_3C—CCl_3$ (430) $F_3C—CCl_3$ (430) $O_2N—CCl_3$ (439)	540 ± 90
δCCl	365 ± 70	$nPrOC(=O)—CCl_3$ (433) $nBuOC(=O)—CCl_3$ (432) $ClC(=O)—CCl_3$ (428)	$F_3Si—CCl_3$ (299) $Cl_3Si—CCl_3$ (300) $HOC(=O)—CCl_3$ (302)	365 ± 50
$\delta' CCl$	325 ± 60	$ClMe_2CCH_2—CCl_3$ (382)	$DC(=O)—CCl_3$ (268) $HC(=O)—CCl_3$ (270)	325 ± 45
$\delta'' CCl$	290 ± 65	$MeCH(Cl)CH_2—CCl_3$ (353) $F—CCl_3$ (349) $ClMe_2CCH_2—CCl_3$ (345)	$Cl_3C—CCl_3$ (225) $DC(=O)—CCl_3$ (228) $Cl_2CH—CCl_3$ (238)	290 ± 50
ρCCl	225 ± 35	$ClCH_2CH_2—CCl_3$ (252)	$MeCH(Cl)CH_2—CCl_3$ (191) $(CF_3)_3COH$ (192) $(CF_3)_3COD$ (193)	225 ± 25
$\rho' CCl$	150 ± 80	$nPrOC(=O)—CCl_3$ (230) $MeOC(=O)—CCl_3$ (217) $nBuOC(=O)—CCl_3$ (209)	$DC(=O)—CCl_3$ (70) $HC(=O)—CCl_3$ (84)	155 ± 45
τCCl	100 ± 50			100 ± 50

$R—CCl_3$ molecules

R = H— [50, 130–133, 168], D— [130, 131, 168], Me— [54, 134–138], Et— [139, 140], $ClCH_2CH_2$— [139, 141, 142], $BrCH_2CH_2$— [139, 142], $ClCH_2$— [134, 143], $MeCH(Cl)CH_2$— $MeCH(Br)CH_2$ and $MeCD(Br)CH_2$— [142], $ClMe_2CCH_2$— and $BrMe_2CCH_2$— [142], $HOCH_2$— [58, 59], $ClC(=O)OCH_2$—, Cl_2POCH_2—, $Cl_2P(=O)OCH_2$—, MeCH(OH)—, Cl_2CH— [134], $ClCH_2—CHCl$—, $Cl_2CH—CHCl$— and $Cl_2CH—CCl_2$— [141], $ClC(=O)OC(Me)_2$—, $Cl_2POC(Me)_2$—, $HO(CF_3)_2$— and $DO(CF_3)_2$— [77], $Me_2C(OH)$—, $HO(O=)C—C(OH)_2$—, F_3C— [5, 8], Cl_3C— [134, 144], HC(=O)— and DC(=O)— [145, 146,

170], $Cl_2CHC(=O)$—, $Cl_3CC(=O)$—, $H_2NC(=O)$— [23], HOC(=O)— [147–149], $^{-}O_2C$— [39, 171], ClC(=O)— [147], O=C=N—C(=O)—, MeOC(=O)— and $CD_3OC(=O)$— [147, 150], EtOC(=O)— [151], nPrOC(=O)— and nBuOC(=O)— [152], $Cl_3CC(=O)OC(=O)$—, HSC(=O)— [153], H_2C=CH— [154], Cl_2C=C(Cl)—, NC— [155], Ph— [156–158], 2-FPh—, 2-ClPh—, 4-ClPh—, 4-Cl_3CPh—, 2-Cl, 6-FPh—, O_2N— [171], $Cl_3CS(=O)_2$— [172], $HOS(=O)_2$— [159], ClS—, $ClS(=O)_2$—, F— [160], Br— [132], H_3Si—, D_3Si—, F_3Si— and Cl_3Si— [161].

2.2.3 Tribromomethyl

The number of investigated molecules is too small to attach a lot of value to the observed regions. Moreover, many assignments are still uncertain.

Table 2.17 Absorption regions (cm^{-1}) of —CBr_3

νCBr	705 ± 80	δ'CBr	235 ± 25
ν'CBr	645 ± 25	δ''CBr	190 ± 50
ν''CBr	410 ± 160	ρCBr	165 ± 55
δCBr	240 ± 40	ρ'CBr	145 ± 35

The broad region of ν''CBr is due to the low values assigned in the spectra of Br_3C—CBr_3 (*558* and 256 cm^{-1}), F_3C—CBr_3 (282 cm^{-1}) and HC(=O)—CBr_3 (324 cm^{-1}). If these values are neglected, the region of ν''CBr is reduced to 490 ± 80 cm^{-1}. The CBr_3 torsion probably absorbs in the neighbourhood of 100 cm^{-1}.

R—CBr_3 molecules
R = H— and D— [131, 162], Me— [54, 163], $HOCH_2$— [59], $BrCH_2$—, $Cl_2P(=O)OCH_2$—, F_3C— [8], Br_3C— [164], HC(=O)— [145, 146, 166], HOC(=O)—, $^{-}O_2C$— [39], ClC(=O)— [165], NC— [167], $Br_3CS(=O)_2$— [172], F—.

References

1. J.R. Nielsen, H.H. Claassen and N.B. Moran, *J. Chem. Phys.*, **23**, 329 (1955).
2. E.L. Pace, A.C. Plaush and H.V. Samuelson, *Spectrochim. Acta*, **22**, 993 (1966).
3. H. Bürger and G. Pawelke, *Spectrochim. Acta, Part A*, **35A**, 525 (1979).
4. O. Risgin and R.C. Taylor, *Spectrochim. Acta*, **15**, 1036 (1959).
5. J.R. Nielsen, C.Y. Liang, R.M. Smith and D.C. Smith, *J. Chem. Phys.*, **21**, 383 (1953).
6. J.R. Nielsen, C.Y. Liang and D.C. Smith, *J. Chem. Phys.*, **21**, 1060 (1953).
7. R. Theimer and J.R. Nielsen, *J. Chem. Phys.*, **27**, 887 (1957).

8. H. Bürger, H. Niepel and G. Pawelke, *Spectrochim. Acta, Part A*, **36A**, 7 (1980).
9. R.E. Dodd, H.L. Roberts and L.A. Woodward, *J. Chem. Soc.*, 2783 (1957).
10. C.V. Berney, *Spectrochim. Acta, Part A*, **25A**, 793 (1969).
11. R.L. Redington, *Spectrochim. Acta, Part A*, **31A**, 1699 (1975).
12. J.R. Durig, G.A. Guirgis and B.J. Van der Veken, *J. Raman Spectrosc.*, **18**, 549 (1987).
13. E. Ottavianell, E.A. Castro and A.H. Jubert, *J. Mol. Struct.*, **254**, 279 (1992).
14. J.S. Francisco and I.H. Williams, *Spectrochim. Acta, Part A*, **48A**, 1115 (1992).
15. J.R. Durig and J.S. Church, *Spectrochim. Acta, Part A*, **36A**, 957 (1980).
16. S.F. Tayyari, T.Zeegers-Huyskens and J.L. Wood, *Spectrochim. Acta, Part A*, **35A**, 1265 (1979).
17. P.C. Metha, S.S.L. Surana and S.P. Tandon, *Can. J. Spectrosc.*, **18**, 56 (1973).
18. G.A. Crowder and P. Pruettiangkura, *J. Mol. Struct.*, **15**, 161 (1973).
19. J.R. Durig, T.G. Sheehan and J.A. Hardin, *J. Mol. Struct.*, **243**, 275 (1991).
20. C.V. Berney, *Spectrochim. Acta*, **21**, 1809 (1965).
21. M. Perttilä, *Acta Chem. Scand., Ser. A*, **28A**, 934 (1974).
22. D.A.C. Compton, J.D. Goddard, S.C. Hsi, W.F. Murphy and D.M. Rayner, *J. Phys. Chem.*, **88**, 356 (1984).
23. D. Troitino, E. Sanchez de la Blanca and M.V. Garcia, *Spectrochim. Acta, Part A*, **46A**, 1281 (1990).
24. E.K. Murthy and G.R. Rao, *J. Raman Spectrosc.*, **19**, 359 (1988).
25. E.K. Murthy and G.R. Rao, *J. Raman Spectrosc.*, **19**, 439 (1988).
26. E.K. Murthy and G.R. Rao, *J. Raman Spectrosc.*, **20**, 409 (1989).
27. N. Fuson, M.L. Josien, E.A. Jones and J.R. Lawson, *J. Chem. Phys.*, **20**, 1627 (1952).
28. M.L. Josien, N. Fuson, J.R. Lawson and E.A. Jones, *Compt. Rend.*, **234**, 1163 (1952).
29. N. Fuson and M.L. Josien, *J. Opt. Soc. Am.*, **43**, 1102 (1953).
30. R.E. Kagarise, *J. Chem. Phys.*, **27**, 519 (1957).
31. J.R. Barcelo and C. Otero, *Spectrochim. Acta*, **18**, 1231 (1962).
32. T.S.S.R. Murty and K.S. Pitzer, *J. Phys. Chem.*, **73**, 1426 (1969).
33. R.L. Redington and K.C. Lin, *Spectrochim. Acta, Part A*, **27A**, 2445 (1971).
34. C.V. Berney, *J. Am. Chem. Soc.*, **95**, 708 (1973).
35. G.A. Crowder and D.A. Jackson, *Spectrochim. Acta, Part A*, **27A**, 1873 (1971).
36. G.A. Crowder, *Spectrochim. Acta, Part A*, **28A,** 1625 (1972).
37. A.G. Robiette and J.C. Thompson, *Spectrochim. Acta*, **21**, 2023 (1965).
38. E.L. Varetti and P.J. Aymonino, *J. Mol. Struct.*, **7**, 155 (1971).
39. E. Spinner, *J. Chem. Soc.*, 4217 (1964).
40. K.O. Christe and D. Naumann, *Spectrochim. Acta, Part A*, **29A**, 2017 (1973).
41. W. Klemperer and G.C.P imentel, *J. Chem. Phys.*, **22**, 1399 (1954).
42. G.A. Crowder, *Appl. Spectrosc.*, **27**, 440 (1973).
43. K.R. Loos and R.C. Lord, *Spectrochim. Acta*, **21**, 119 (1965).
44. C.V. Berney, *Spectrochim. Acta, Part A*, **27A**, 663 (1971).
45. C.V. Berney, *Spectrochim. Acta*, **20**, 1437 (1964).
46. C.V. Berney and A.D. Cormier, *Spectrochim. Acta, Part A*, **33A**, 929 (1977).
47. C.O. Della Védova and P.J. Aymonino, *J. Raman Spectrosc.*, **20**, 135 (1989).
48. L.M. Osborne and D.J. Clouthier, *Spectrochim. Acta, Part A*, **43A**, 1075 (1987).
49. A. Ruoff, H. Bürger and S. Bierdermann, *Spectrochim. Acta, Part A*, **27A**, 1359 (1971).
50. S.L. Paulson and A.J. Barnes, *J. Mol. Struct.*, **80**, 151 (1982).
51. H.W. Thompson and R.B. Temple, *J. Chem. Soc.*, 1428 (1948).
52. J.R. Nielsen, H.H. Claassen and D.C. Smith, *J. Chem. Phys.*, **18**, 1471 (1950).
53. R.D. Cowan, G. Herzberg and S.P. Sinha, *J. Chem. Phys.*, **18**, 1538 (1950).
54. Th.R. Stengle and R.C. Taylor, *J. Mol. Spectrosc.*, **34**, 33 (1970).
55. J.R. Nielsen, C.H. Richards and H.L. McMurry, *J. Chem. Phys.*, **16**, 67 (1948).

56. J.R. Nielsen and C.W. Gullikson, *J. Chem. Phys.*, **21**, 1416 (1953).
57. H. Bürger and G. Pawelke, *Spectrochim. Acta, Part A*, **35A**, 517 (1979).
58. M. Perttilä, *Spectrochim. Acta, Part A*, **35A**, 585 (1979).
59. J. Travert and J.C. Lavalley, *Spectrochim. Acta, Part A*, **32A**, 637 (1976).
60. O. Schrems and W.A.P. Luck, *J. Mol. Struct.*, **80**, 477 (1982).
61. J.R. Durig and R.A. Larsen, *J. Mol. Struct.*, **238**, 195 (1990).
62. Y.-S. Li, F.O. Cox and J.R. Durig, *J. Phys. Chem.*, **91**, 1334 (1987).
63. S.W. Charles, F.C. Cullen and N.L. Owen, *J. Chem. Soc., Faraday Trans.*, **70**, 483 (1974).
64. H. Wolff, D. Horn and H.G. Rollar, *Spectrochim. Acta, Part A*, **29A**, 1835 (1973).
65. W.F. Edgell, T.R. Riethof and C. Ward, *J. Mol. Spectrosc.*, **11**, 92 (1963).
66. G.A. Crowder, *J. Fluorine Chem.*, **3**, 125 (1973/74).
67. J.R. Nielsen and R. Theimer, *J. Chem. Phys.*, **27**, 891 (1957).
68. J. Murto, A. Kivinen, K. Edelmann and E. Hassinen, *Spectrochim. Acta, Part A*, **31A**, 479 (1975).
69. J.R. Durig, F.O. Cox, P. Groner and B.J. Van der Veken, *J. Phys. Chem.*, **91**, 3211 (1987).
70. J. Murto, A. Kivinen, R. Viitala and J. Hyömäki, *Spectrochim. Acta, Part A*, **29A**, 1121 (1973).
71. S.J. Cyvin, J. Brunvoll and M. Perttilä, *J. Mol. Struct.*, **17**, 17 (1973).
72. J.R. Durig, R.A. Larsen, R. Kelley and F.-Y. Sun, *J. Raman Spectrosc.*, **21**, 109 (1990).
73. Y.S. Li, R.A. Larsen, F.O. Cox and J.R. Durig, *J. Raman Spectrosc.*, **20**, 1 (1989).
74. J. Murto, A. Kivinen and P. Saarinen, *Acta Chem. Scand., Ser. A*, **30A**, 448 (1976).
75. J. Korppi-Tomola, *Acta Chem. Scand., Ser. A*, **A31**, 563 (1977).
76. J. Korppi-Tomola, *Acta Chem. Scand., Ser. A*, **A31**, 568 (1977).
77. J. Murto, A. Kivinen, K. Kajander, J. Hyömäki and J. Korppi-Tomola, *Acta Chem. Scand., Ser. A*, **27**, 96 (1973).
78. K.E. Blick, F.C. Nahm and K. Niedenzu, *Spectrochim. Acta, Part A*, **27A**, 777 (1971).
79. A. Ruoff and H. Bürger, *Spectrochim. Acta, Part A*, **33A**, 775 (1977).
80. P.R. McGee, F.F. Cleveland, A.G. Meister and C.E. Decker, *J. Chem. Phys.*, **21**, 242 (1953).
81. R.C. Taylor, *J. Chem. Phys.*, **22**, 714 (1954).
82. G.A. Crowder and N. Smyrl, *J. Chem. Phys.*, **53**, 4102 (1970).
83. I. Tokue, T. Fukuyama and K. Kuchitsu, *J. Mol. Struct.*, **17**, 207 (1973).
84. H. Bürger, G. Pawelke and H. Oberhammer, *J. Mol. Struct.*, **84**, 49 (1982).
85. F.A. Miller and F.E. Kiviat, *Spectrochim. Acta, Part A*, **25A**, 1577 (1969).
86. J. Murto, A. Kivinen, R. Henriksson, A. Aspiala and J. Partanen, *Spectrochim. Acta, Part A*, **36A**, 607 (1980).
87. V. Galasso and A. Bigotto, *Spectrochim. Acta*, **21**, 2085 (1965).
88. C.V. Berney, L.R. Cousins and F.A. Miller, *Spectrochim. Acta*, **19**, 2019 (1963).
89. R.H. Sanborn, *Spectrochim. Acta, Part A*, **23A**, 1999 (1967).
90. H.B. Friedrich, D.J. Burton and P.A. Schemmer, *Spectrochim. Acta, Part A*, **45A**, 181 (1989).
91. D.H. Lemmon, *J. Mol. Struct.*, **49**, 71 (1978).
92. J.A. Faniran and H.F. Shurvell, *Spectrochim. Acta, Part A*, **27A**, 1945 (1971).
93. W.F. Edgell and R.M. Potter, *J. Chem. Phys.*, **24**, 80 (1956).
94. E. Augdahl, E.Kloster-Jensen, V. Devarajan and S.J. Cyvin, *Spectrochim. Acta, Part A*, **29A**, 1329 (1973).
95. R. Demuth, H. Bürger, G. Pawelke and H. Willner, *Spectrochim. Acta, Part A*, **34A**, 113 (1978).
96. H.F. Shurvell, S.C. Dass and R.D. Gordon, *Can. J. Chem.*, **52**, 3149 (1974).

97. J. Mason and J. Dunderdale, *J. Chem. Soc.*, 759 (1956).
98. A. Castelli, A. Palm and C. Alexander, *J. Chem. Phys.*, **44**, 1577 (1966).
99. B. Vizi, B.N. Cyvin and S.J. Cyvin, *Acta Chim. Acad. Sc. Hung.*, **83**, 303 (1974).
100. H. Oberhammer, H. Gunther, H. Bürger, F. Heyder and G. Pawelke, *J. Phys. Chem.*, **86**, 664 (1982).
101. J.E. Griffiths and D.F. Sturman, *Spectrochim. Acta, Part A*, **25A**, 1355 (1969).
102. R.A. Hayden, E.C. Tuazon and W.G. Fateley, *J. Mol. Struct.*, **16**, 35 (1973).
103. J. Lee and B.G. Willoughby, *Spectrochim. Acta, Part A*, **33A**, 395 (1977).
104. J.S. Francisco, *Spectrochim. Acta, Part A*, **40A**, 923 (1984).
105. H. Bürger and G. Pawelke, *Spectrochim. Acta, Part A*, **31A**, 1965 (1975).
106. A.J. Arvia and P.J. Aymonino, *Spectrochim. Acta*, **18**, 1299 (1962).
107. D.W. Wertz and J.R. Durig, *J. Mol. Spectrosc.*, **25**, 467 (1968).
108. R.P. Hirschmann, W.B. Fox and L.R. Anderson, *Spectrochim. Acta, Part A*, **25A**, 811 (1969).
109. E.L. Varetti and P.J. Aymonino, *J. Mol. Struct.*, **1**, 39 (1967).
110. J. Borrajo, E.L. Varetti and P.J. Aymonino, *J. Mol. Struct.*, **29**, 163 (1975).
111. M. Perttilä, *Spectrochim. Acta, Part A*, **32A**, 1011 (1976).
112. R.E. Dininny and E.L. Pace, *J. Chem. Phys.*, **31**, 1630 (1959).
113. R.L. Redington, *J. Mol. Spectrosc.*, **9**, 469 (1962).
114. H.A. Carter, C.S.C. Wang and J.M. Shreeve, *Spectrochim. Acta, Part A*, **29A**, 1479 (1973).
115. H. Oberhammer, W. Gombler and H. Willner, *J. Mol. Struct.*, **70**, 273 (1981).
116. D. Bielefeldt and H. Willner, *Spectrochim. Acta, Part A*, **36A**, 989 (1980).
117. C.O. Della Védova, *J. Raman Spectrosc.*, **20**, 279 (1989).
118. D.A. Coe and J.M. Shreeve, *Spectrochim. Acta, Part A*, **33A**, 965 (1977).
119. D.F. Eggers Jr., H.E. Wright and D.W. Robinson, *J. Chem. Phys.*, **35**, 1045 (1961).
120. E.L. Varetti, *Spectrochim. Acta, Part A*, **44A**, 733 (1988).
121. E.L. Varetti, E.L. Fernandez and A.B. Altabef, *Spectrochim. Acta, Part* , **47A**, 1767 (1991).
122. H. Bürger, J. Cichon, R. Demuth and J. Grobe, *Spectrochim. Acta, Part A*, **29A**, 943 (1973).
123. R. Demuth, J. Apel and J. Grobe, *Spectrochim. Acta, Part A*, **34A**, 357 (1978).
124. J.E. Griffiths, *Spectrochim. Acta, Part A*, **21A**, 1135 (1965).
125. J.E. Griffiths, *Spectrochim. Acta, Part A*, **24A**, 115 (1968).
126. J.E. Griffiths, *Spectrochim. Acta, Part A*, **24A**, 303 (1968).
127. H. Beckers, H. Bürger, R. Eujen, B. Rempfer and H. Oberhammer, *J. Mol. Struct.*, **140**, 281 (1986).
128. R. Eujen, *Spectrochim. Acta, Part A*, **43A**, 1165 (1987).
129. J.R. Durig, R.A. Larsen, F.O. Fox and B.J. Van der Veken, *J. Mol. Struct.*, **172**, 183 (1988).
130. J. Morcillo, J.F. Biarge, J.M. Heredia and A. Medina, *J. Mol. Struct.*, **3**, 77 (1969).
131. H. Ratajczak, T.A. Ford and W.J. Orville-Thomas, *J. Mol. Struct.*, **14**, 281 (1972).
132. J.R. Madigan and F.F. Cleveland, *J. Chem. Phys.*, **19**, 119 (1951).
133. A. Ruoff and H. Bürger, *Spectrochim. Acta, Part A*, **26A**, 989 (1970).
134. S. Suzuki and A.B. Dempster, *J. Mol. Struct.*, **32**, 339 (1976).
135. S.G. Frankiss and D.J. Harrison, *Spectrochim. Acta, Part A*, **31A**, 29 (1975).
136. D.C. Smith, G.M. Brown, J.R. Nielsen, R.M. Smith and C.Y. Liang, *J. Chem. Phys.*, **20**, 473 (1952).
137. M.-Z. El-Saban, A.G.Meister and F.F.Cleveland, *J. Chem. Phys.*, **19**, 855 (1951).
138. P. Venkateswarlu, *J. Chem. Phys.*, **20**, 1810 (1952).
139. A. Goursot-Leray, M. Carles-Lorjou, G. Pouzard and H. Bodot, *Spectrochim. Acta,*

Part A, **29A**, 1497 (1973).
140. K. Ohno, K. Taga, I. Yoshida and H. Murata, *Spectrochim. Acta, Part A*, **36A**, 721 (1980).
141. A.B. Dempster, K. Price and N. Sheppard, *Spectrochim. Acta, Part A*, **27A**, 1563 (1971).
142. M. Carles-Lorjou, A. Goursot-Leray, H. Bodot and R. Gaufrès, *Spectrochim. Acta, Part A*, **29A**, 329 (1973).
143. G. Allen and H.J. Bernstein, *Can. J. Chem.*, **32**, 1124 (1954).
144. F. Watari and K. Aïda, *J. Mol. Spectrosc.*, **24**, 503 (1967).
145. G. Lucazeau and A. Novak, *Spectrochim. Acta, Part A*, **25A**, 1615 (1969).
146. G. Hagen, *Acta Chem. Scand.*, **25**, 813 (1971).
147. R. Fausto and J.J.C. Teixeira-Dias, *J. Mol. Struct.*, **144**, 141 (1986).
148. J. Adams and H. Kim, *Spectrochim. Acta, Part A*, **29A**, 675 (1973).
149. M.D.P. Jorge and J.R. Barcelo, *An. Soc. Esp. Fis. Quim.*, **B53**, 339 (1957).
150. Y. Mido, K. Suzuki, N. Komatsu and M. Hashimoto, *J. Mol. Struct.*, **144**, 329 (1986).
151. Y. Mido, T. Kawashita, K. Suzuki, J. Morcillo and M.V. Garcia, *J. Mol. Struct.*, **162**, 169 (1987).
152. Y. Mido, N. Komatsu, J. Morcillo and M.V. Garcia, *J. Mol. Struct.*, **172**, 49 (1988).
153. H.S. Randhawa, C.O. Meese and W. Walter, *J. Mol. Struct.*, **36**, 25 (1977).
154. E.R. Shull, *J. Chem. Phys.*, **27**, 399 (1957).
155. H.F. Shurvell, S.E. Gransden, J.A. Faniran and D.W. James, *Spectrochim. Acta, Part A*, **32A**, 559 (1976).
156. C.V. Stephenson and W.C. Coburn, *J. Chem. Phys.*, **42**, 35 (1965).
157. A.L. Smith, *Spectrochim. Acta, Part A*, **24A**, 695 (1968).
158. R.J.A. Ribeiro-Claro, A.M. d'A Rocha Gonsalves and J.J.C. Teixeira —Dias, *Spectrochim. Acta, Part A*, **41A**, 1055 (1985).
159. H.G.M. Edwards and D.N. Smith, *J. Mol. Struct.*, **263**, 11 (1991).
160. R.B. Bernstein, J.P. Zietlow and F.F. Cleveland, *J. Chem. Phys.*, **21**, 1778 (1953).
161. I.V. Kochikov, G. Kuramshina, S.V. Syn'ko and Yu.A. Pentin, *J. Mol. Struct.*, **172**, 299 (1988).
162. H. Bürger and J. Cichon, *Spectrochim. Acta, Part A*, **27A**, 2191 (1971).
163. J.R. Durig, S.M. Craven, C.W. Hawley and J. Bragin, *J. Chem. Phys.*, **57**, 131 (1972).
164. R.D. McLachlan and V.B. Carter, *Spectrochim. Acta, Part A*, **26A**, 2247 (1970).
165. H.S. Randhawa and W. Walter, *J. Mol. Struct.*, **35**, 303 (1976).
166. M.I. Suero, F. Marquez and M.J. Martin-Delgado, *Spectrosc. Lett.*, **23**, 771 (1990).
167. J.F. Arenas, J.I. Marcos and M.I. Suero, *J. Raman Spectrosc.*, **15**, 132 (1984).
168. N.B. Colthup, L.H. Daly and S.E. Wiberley, *Introduction to Infrared and Raman Spectroscopy*, Academic Press, New York (1964).
169. A.B. Altabef, E.H. Cutin and C.O. Della Védova, *J. Raman Spectrosc.*, **22**, 297 (1991).
170. J.R. Durig and W.J. Natter, *J. Raman Spectrosc.*, **11**, 32 (1981).
171. M.S. Soliman, *Spectrochim. Acta, Part A*, **49A**, 189 (1993).
172. E. Vajda, C.J. Nielsen, P. Klaboe, R. Seip and J. Brunvoll, *Acta Chem. Scand., Ser. A*, **37A**, 341 (1983).

2.3 TERTIARY BUTYL

The simplest tBu molecules are those of type Me_3CX (X = halogen, C≡N, C≡CH, etc.). In the sterically favoured C_{3v} structure, a hydrogen atom of each methyl group is in a plane of symmetry while the two other hydrogen atoms of each Me group are symmetric with respect to this plane. For instance, X—tBu (X = halogen) is of C_{3v} symmetry so that the 24 normal vibrations differentiate between $8a_1 + 4a_2 + 12e$ vibrational modes. From these, 23 ($7a_1 + 4a_2 + 12e$) can be attributed to the tBu substituent, 19 ($7a_1 + 12e$) of which are IR active. The remaining vibration is the νCX (a_1).

The correlation between C_{3v} and C_s is as follows: $8a_1 = 8a'$; $4a_2 = 4a''$ and $12e = 12a' + 12a''$, so that with C_s symmetry of X—tBu, the 36 normal vibrations are distributed in $20a' + 16a''$ vibrational species. Substitution of the νCX (a′) by a torsion results in $19a' + 17a''$ normal vibrations for tBu. Molecules with two tBu substituents in gauche position, those responding to D_{3d} symmetry, often occur in the liquid state. The correlation D_{3d} to C_{3v} is as follows: a_{1g}(R) and a_{2u}(IR) = a_1; e_g(R) and e_u(ir) = e and a_{2g} and a_{1u} = a_2. Similarly to the a_2 vibrations with C_{3v}, the a_{2g} and a_{1u} modes are inactive. With molecules belonging to a point group other than C_{3v} (C_s, C_3, ...) the tBu substituent is often described in terms of C_{3v}. Especially for those molecules where the C_{3v} symmetry is more or less perturbed (HO—tBu, HS—tBu), the e modes do not split significantly. The a_2 modes, on the other hand, are weakly active. The absorptions intensify with descending symmetry or in the solid state of the molecule. In practice this often results in 23 vibrations plus a torsion. Even if the e vibrations occur as a doublet, the wavenumber values are normally so closely packed that, in considering the tBu substituent, only 24 regions instead of 36 will be discussed.

Methyl stretching vibrations

ν_aMe(a″, e) ν_aMe(a′)	ν_aMe(a″, e) ν_aMe(a′)	ν_aMe(a′, a_1) ν_aMe(a″, a_2)	ν_sMe(a′, e) ν_sMe(a″)	ν_sMe(a′, a_1)
2980 ± 30	2960 ± 30	2945 ± 45	2905 ± 45	2885 ± 35 cm^{-1}

Although the tBu substituent can provide six methyl antisymmetric stretching vibrations, generally only three are observed as moderate or strong bands. The a_2 vibration is generally very weak or inactive. Molecules such as F—tBu (3010 cm^{-1}) and O_2N—tBu (3006 cm^{-1}) occupy the HW region while aromatic molecules usually absorb at the LW part of the region. For instance, 4-HOPh—tBu displays ν_aMe at 2910 cm^{-1} and 4-tBuPh—tBu at 2905 cm^{-1}. Most tBu molecules give these methyl antisymmetric stretching vibrations between 2990 and 2930 cm^{-1}:

ν_aMe 2970 ± 20 2970 ± 15 2950 ± 20 cm^{-1}

The methyl symmetric stretchings are active between 2950 and 2850 cm^{-1}. Aromatic molecules usually display these symmetric stretching vibrations between 2915 and 2860 cm^{-1}.

Methyl deformations

δ_aMe(a′, a_1)	δ_aMe(a″, e) δ_aMe(a′)	δ_aMe(a″, a_2)	δ_aMe(a″, e) δ_aMe(a′)
1470 ± 25	1465 ± 25	1460 ± 25	1460 ± 25

The methyl antisymmetric deformations absorb between 1495 and 1435 cm^{-1}. Usually, the a_1 and e vibrations appear spectrally as moderate to strong bands. Molecules such as O_2N—tBu (1493 cm^{-1}), NC—tBu (1486 or 1480 cm^{-1}), P-attached tBu (1485–1475 cm^{-1}) and tBu—C≡C—C≡C—tBu (1483, 1480 cm^{-1}) absorb at the HW side of the region. The latter molecule (1425 cm^{-1}) and Cl—tBu (1438 cm^{-1}, calculated for the a_1 vibration) together provide the lowest values. The remaining molecules display δ_aMe between 1480 and 1440 cm^{-1} in regions given below :

δ_aMe	1465 ± 15	1460 ± 15	(1460 ± 15)	1455 ± 15 cm^{-1}

Although three (2a′ + a″) methyl symmetric bending vibrations are expected {two (a_1 + e) in the case of C_{3v} symmetry} often only two emerge.

δ_sMe(a′, a_1)	δ_sMe(a′, e) δ_sMe(a″)
1395 ± 25	1375 ± 25 cm^{-1}

As far as the a_1(a′) mode is concerned, which absorbes in the range 1395 ± 25 cm^{-1}, the highest values have been observed for HOC(=O)—tBu (1414 cm^{-1}), P-attached tBu (1400 ± 10 cm^{-1}), O_2N—tBu (1406 cm^{-1}) and HC(=O)—tBu with 1405 cm^{-1}. The remaining molecules mostly absorb in the narrow region 1385 ± 15 cm^{-1}. The e mode, which is supposed to split up into two (a″ + a′) counterparts, mostly appears as a single rather than a twin band. The strongest band δ_sMe(a″) at the LW side, sometimes accompanied by the a′ mode (1375 ± 20 cm^{-1}), appears in the region 1365 ± 15 cm^{-1}. The highest values are displayed by F_2P(=O)—tBu and F—tBu, both absorbing at 1374 cm^{-1}, O_2N—tBu (1373 cm^{-1}) and CN—tBu (1372 cm^{-1}). The lowest wavenumber for this mode has been observed for DC(=O)—tBu (1350 cm^{-1}) and the remaining investigated molecules absorb in the extremely narrow region of 1365 ± 5 cm^{-1}, a range which is much narrower than observed for the same mode for acetyl (Section 2.1.6).

Skeletal stretchings and methyl rocks

Despite the fact that all three CC_3 stretching vibrations severely couple with the (six) methyl rocking modes, six absorption regions more or less overlapping one another can be assigned. Normally, the three e modes, each of which is expected to split into a′ + a″ counterparts, do not always appear as doublets. The a_2 mode, forbidden for C_{3v} molecules, sometimes emerges as an ill defined absorption band.

$\nu_a CC_3$(a″, e) $\nu_a CC_3$(a′)	$\nu_s CC_3$(a′, a_1)
1235 ± 60	800 ± 90 cm^{-1}

The antisymmetric CC_3 stretch (a″, e) normally appears in the above-mentioned region with high values (1293 cm^{-1}) for both $H_2C{=}C(Me)CH_2$—tBu and $H_2C{=}C(Me)$—tBu and also for $HOCH_2$—tBu (1288 cm^{-1}). Low values, coincident with the first methyl rocking mode, have been observed for $Cl_2P({=}S)$—tBu (1183 cm^{-1}), H—tBu (1189 cm^{-1}), Ph—tBu (1189 cm^{-1}) and 4-BrPh—tBu (1203 cm^{-1}). The remaining investigated molecules display $\nu_a CC_3$(a″) in the region 1240 ± 35 cm^{-1}. The $\nu_a CC_3$(a′) absorbs in the region 1220 ± 40 cm^{-1}. Disregarding some high {low} values around 1255 {1185} cm^{-1}, this mode mostly appears in the region 1215 ± 25 cm^{-1}, often but not always coincident with $\nu_a CC_3$(a″) and sometimes with ρMe(a′, a_1).

Sometimes the $\nu_s CC_3$(a′, a_1) is of reasonable intensity, but normally this vibration mode absorbs weakly in the region 800 ± 90 cm^{-1}. Even considering the fact that after disregarding some high and low values of this range most molecules display $\nu_s CC_3$ at 795 ± 60 cm^{-1}, it is still a too large region for purposes of identification of the tBu group.

ρMe(a′, a_1)	ρMe(a″, e) ρMe(a′)	ρMe(a″, a_2)	ρMe(a″, e) ρMe(a′)
1160 ± 55	1035 ± 55	990 ± 50	925 ± 35 cm^{-1}

Most of the investigated molecules display the first methyl rocking (a′, a_1) in the region 1150 ± 35 cm^{-1}. The ρMe(a′ and a″) both often emerge in the region 1035 ± 55 cm^{-1} as separate bands or as one band of weak or moderate intensity. Disregarding some high values in the neighbourhood of 1085 cm^{-1} and the value 984 cm^{-1} for H—tBu, these rockings mainly absorb in the region 1045 ± 30 cm^{-1}. The last three methyl rocking modes appear in the region 965 ± 75 cm^{-1}, which tentatively may be subdivided into 990 ± 50 cm^{-1} for the a″, a_2-mode and 925 ± 25 cm^{-1} for the a′, a″(e)-modes.

Skeletal deformations

The tBu group gives rise to five (3a′ + 2a ″) skeletal deformations absorbing in three regions:

$\delta_a CC_3$(a″, e) $\delta_a CC_3$(a′)	$\delta_s CC_3$(a′, a_1)	ρCC_3(a″, e) ρCC_3(a′)
435 ± 85	335 ± 80	300 ± 80 cm^{-1}

These modes normally produce bands of weak or medium intensity. The highest {lowest} values for $\delta_a CC_3$ are observed around 510 {355} cm^{-1}. Most of the $\delta_a CC_3$ modes have been assigned in the region 435 ± 65 cm^{-1}. The $\delta_s CC_3$ is weakly observed in the region 335 ± 80 cm^{-1} where molecules such as tBuI and tBuBr respectively are responsible for the low values of 259 and 303 cm^{-1}. Roughly estimated, because many assignments are not available, this mode appears in the range 370 ± 50 cm^{-1} for the remaining molecules, which also applies to the CC_3 rocking modes absorbing weakly in the region 300 ± 80 cm^{-1}. However, these skeletal deformations are not significant for identification purposes.

Torsions

If the symmetry of the molecule lowers from C_{3v} to C_s, the degenerate e mode splits up into a′ and a″ modes absorbing in regions around 250 cm^{-1}. The third methyl torsion, which is of species a_2 for C_{3v} or a″ for C_s molecules, is weakly active around 220 cm^{-1} or much lower.

The following R—tBu compounds have been taken into account:

R = H— [1–7], F— [1], Cl— [1–5], Br— [1, 2, 5, 6], I— [1, 2, 5], Et— [8], $HOCH_2CH_2$—, nBu— [9], $tBuCH_2$— [10], $ClCH_2$— [11, 12], $BrCH_2$— [13], iPr—, iBu— and Et(CH)Me— [14], $H_2C{=}C(Me)CH_2$—, $MeC({=}O)CH_2$—, $HOC({=}O)CH_2$—, H_2NCH_2—, $HOCH_2$—, HC(=O)—[15, 16], $H_2NC({=}O)$—, HOC(=O)— [17], NaOC(=O)— [48], MeOC(=O)— and $CD_3OC({=}O)$— [18], EtOC(=O)—, ClC(=O)—, $H_2C{=}CH$—, $H_2C{=}C(Me)$—, tBu—CH=CH—, $Me_2C{=}CH$—, tBu—C≡C— [19], tBu—C≡C—C≡C— [4], N≡C— [20, 21], Ph— [22], 3- and 4-MePh— [22], 4-$ClCH_2Ph$—, 4-tBuPh— [22, 23], 4-HC(=O)Ph—, 4-ClC(=O)Ph—, 4-H_2NPh—, 4-SCNPh— [22], 2-, 3- and 4-HOPh— [24], 4-MeOPh— [22], 4-EtOPh— [22], 4-BrPh—, 4-Py—, EtNH— [47], MeNHC(=O)NH—, EtNHC(=O)NH—, nPrNHC(=O)NH—, nBuNHC(=O)NH— and tBuNHC(=O)NH— [25], cHexNHC(=O)NH— [26], $R'S({=}O)_2NH$— [27], O_2N— [28], OCN— [16, 29], SCN— [16, 30], H_2N— [16], CN— [31], HO— [1, 16, 32, 33], DO— [32], MeO— [16, 31], CD_3O— [16], EtO—, HC(=O)— and DC(=O)O— [34], MeC(=O)O—, EtC(=O)O—, $ClCH_2C({=}O)O$—, $BrCH_2C({=}O)O$—, $NCCH_2C({=}O)O$—, $H_2NC({=}O)O$—, tBuOO— [35], tBuO—N=N—O— [36], HS— [1, 16, 37, 38], MeS— [37], EtS— [37, 39], MeSS— [40, 41], tBuSS— [40, 42, 43], H_2P— [44], F_2P— [45],

Table 2.18 Absorption regions (cm^{-1}) of the normal vibrations of —tBu

adjacent to			Hal. and H	Sat. C	Unsat. C	C=O	Ph	N	O	S	P	Total
ν_aMe	a″	e	2990 ± 25	2965 ± 15	2975 ± 20	2975 ± 15	2965 ± 5	2985 ± 25	2985 ± 10		2980 ± 10	2980 ± 30
ν_aMe	a′			2965 ± 15	2970 ± 10	2975 ± 10	2965 ± 5	2975 ± 20	2985 ± 10		2980 ± 10	2970 ± 25
ν_aMe	a″	e	2975 ± 10	2960 ± 10	2970 ± 15	2970 ± 15	2965 ± 10	2960 ± 30	2980 ± 10		2955 ± 10	2960 ± 30
ν_aMe	a′			2955 ± 15	2965 ± 15	2950 ± 15	2950 ± 20	2955 ± 25	2960 ± 30		2955 ± 10	2960 ± 30
ν_aMe	a′	a_1	2970 ± 15	2950 ± 15	2950 ± 20	2950 ± 15	2945 ± 25	2935 ± 25	2945 ± 15		2945 ± 15	2945 ± 40
ν_aMe	a″	a_2		2940 ± 25	2945 ± 25	2935 ± 15	2935 ± 35	2935 ± 10	2935 ± 10		2945 ± 15	2935 ± 35
ν_sMe	a′	e	2920 ± 10	2925 ± 25	2920 ± 20	2915 ± 10	2905 ± 10	2925 ± 25	2920 ± 20		2910 ± 10	2920 ± 30
ν_sMe	a″		2895 ± 15	2895 ± 30	2890 ± 25	2900 ± 25	2890 ± 25	2890 ± 15	2905 ± 35		2910 ± 10	2905 ± 40
ν_sMe	a′	a_1	2870 ± 10	2865 ± 10	2875 ± 15	2865 ± 15	2880 ± 20	2875 ± 10	2895 ± 25		2910 ± 10	2885 ± 35
δ_aMe	a′	a_1	1475 ± 5	1475 ± 10	1480 ± 5	1480 ± 10	1480 ± 10	1470 ± 25	1480 ± 20	1475 ± 5	1480 ± 5	1470 ± 25
δ_aMe	a″	e		1475 ± 10	1475 ± 10	1475 ± 10	1475 ± 15	1465 ± 15	1470 ± 10	1465 ± 10	1465 ± 10	1470 ± 20
δ_aMe	a′		1460 ± 10	1475 ± 10	1470 ± 15	1470 ± 15	1470 ± 15	1455 ± 15	1470 ± 10	1460 ± 10	1465 ± 10	1465 ± 25
δ_aMe	a″	a_2		1465 ± 20	1465 ± 10	1460 ± 5	1460 ± 10	1460 ± 10	1450 ± 15	1450 ± 10	1465 ± 10	1460 ± 25
δ_aMe	a″	e	1450 ± 10	1465 ± 20	1460 ± 20	1460 ± 10	1455 ± 15	1455 ± 15	1450 ± 15	1450 ± 10	1450 ± 10	1460 ± 25
δ_aMe	a′			1465 ± 20	1450 ± 25	1435 ± 30	1455 ± 15	1455 ± 15	1450 ± 15	1445 ± 10	1450 ± 10	1460 ± 25
δ_sMe	a′	a_1	1385 ± 15	1395 ± 5	1395 ± 10	1395 ± 20	1390 ± 10	1395 ± 15	1385 ± 15	1390 ± 5	1400 ± 10	1395 ± 25
δ_sMe	a′	e	1370 ± 5	1375 ± 15	1375 ± 20	1365 ± 10	1365 ± 10	1375 ± 15	1375 ± 20	1365 ± 5	1370 ± 5	1375 ± 20
δ_sMe	a″		1370 ± 5	1365 ± 5	1365 ± 5	1365 ± 10	1365 ± 5	1365 ± 10	1360 ± 10	1365 ± 5	1370 ± 5	1365 ± 15
$\nu_a CC_3$	a″	e	1210 ± 30	1270 ± 25	1250 ± 45	1245 ± 25	1235 ± 40	1245 ± 15	1240 ± 30	1210 ± 10	1205 ± 25	1235 ± 60
$\nu_a CC_3$	a′			1215 ± 35	1220 ± 20	1215 ± 20	1200 ± 20	1220 ± 20	1225 ± 30	1210 ± 10	1205 ± 25	1220 ± 40
ρMe	a′	a_1	1155 ± 20	1165 ± 50	1165 ± 45	1140 ± 25	1140 ± 30	1160 ± 50	1165 ± 25	1165 ± 5	1155 ± 35	1160 ± 55
ρMe	a″	e		1035 ± 30	1055 ± 25	1035 ± 10	1050 ± 40	1050 ± 20	1035 ± 15	1025 ± 15	1020 ± 10	1045 ± 45
ρMe	a′		1005 ± 25	1020 ± 20	1030 ± 10	1030 ± 10	1025 ± 15	1045 ± 15	1035 ± 15	1025 ± 10	1020 ± 10	1020 ± 40
ρMe	a″	a_2		995 ± 25	1005 ± 35	990 ± 50	995 ± 45	1005 ± 35	990 ± 40	980 ± 30	1020 ± 10	990 ± 50
ρMe	a″	e	905 ± 15	930 ± 10	925 ± 20	935 ± 10	940 ± 20	940 ± 15	935 ± 25	930 ± 10	940 ± 10	925 ± 35
ρMe	a′			920 ± 20	915 ± 25	925 ± 25	915 ± 25	930 ± 20	910 ± 20	930 ± 10	940 ± 10	920 ± 30
$\nu_s CC_3$	a′	a_1	780 ± 30	770 ± 60	845 ± 45	765 ± 30	790 ± 55	800 ± 60	765 ± 45	815 ± 15	810 ± 15	800 ± 90
$\delta_a CC_3$	a″	e	425 ± 40	465 ± 55	435 ± 65	450 ± 70	440 ± 30	460 ± 55	460 ± 25	410 ± 10	390 ± 15	445 ± 75
$\delta_a CC_3$	a′			390 ± 25	390 ± 30	380 ± 20	380 ± 30	415 ± 35	425 ± 15	400 ± 10	390 ± 15	400 ± 50
$\delta_s CC_3$	a′	a_1	335 ± 80	370 ± 30		360 ± 40		365 ± 25	375 ± 40	365 ± 10	350 ± 25	335 ± 80
ρCC_3	a″	e	280 ± 55	310 ± 30		330 ± 30		315 ± 25	360 ± 20	330 ± 20	310 ± 20	300 ± 80
ρCC_3	a′			285 ± 40		300 ± 25		305 ± 30	295 ± 30	310 ± 20	280 ± 10	290 ± 45
τMe	a′	e	285 ± 15	250 ± 40		275 ± 15		290 ± 10	270 ± 30		255 ± 45	255 ± 45
τMe	a″		260 ± 40	245 ± 35		255 ± 10		270 ± 30	245 ± 35			255 ± 45
τMe	a″	a_2		225 ± 15		225 ± 10		240 ± 15	220 ± 50			220 ± 50
τCC_3	a″			155 ± 55		140 ± 80			160 ± 65			145 ± 85

Cl_2P— [46], $F_2P(=O)$— [45], $Cl_2P(=O)$— [46], $F_2P(=S)$— [45], $Cl_2P(=S)$— [46].

References

1. D.E. Mann, N. Acquista and D.R. Lide, *J. Mol. Spectrosc.*, **2**, 575 (1958).
2. M.C. Tobin, *J. Am. Chem. Soc.*, **75**, 1788 (1953).
3. J.C. Evans and G.I.-S. Lo, *J. Am. Chem. Soc.*, **88**, 2118 (1966).
4. J. Nakovich Jr., S.D. Shook, F.A. Miller, D.R. Parnell and R.E. Sacher, *Spectrochim. Acta, Part A*, **35A**, 495 (1979).
5. N.T. McDevitt, A.L. Rozek, F.F. Bentley and A.D. Davidson, *J. Chem. Phys.*, **42**, 1173 (1965).
6. J.E. Bertie and S. Sunder, *Spectrochim. Acta, Part A*, **30A**, 1373 (1974).
7. B. Schrader, J. Pacansky and U. Pfeiffer, *J. Phys. Chem.*, **88**, 4069 (1984).
8. K. Ohno, K. Taga, I. Yoshida and H. Murata, *Spectrochim. Acta, Part A*, **36A**, 721 (1980).
9. G.A. Crowder and R.M.P. Jaiswal, *J. Mol. Struct.*, **102**, 145 (1983).
10. H. Matsuura, K. Fukuhara, K. Takashima and M. Sakakibara, *J. Mol. Struct.*, **239**, 43 (1990).
11. P. Klaboe, C.J. Nielsen and D.L. Powell, *Spectrochim. Acta, Part A*, **41A**, 1315 (1985).
12. G.A. Crowder and W.-L. Lin, *J. Mol. Struct.*, **64**, 193 (1980).
13. G.A. Crowder, C. Harper and M.R. Jalilian, *J. Mol. Struct.*, **49**, 403 (1978).
14. G.A. Crowder and L. Gross, *J. Mol. Struct.*, **102**, 257 (1983).
15. G.A. Crowder, *J. Chem. Soc., Perkin Trans.*, 2, 1241 (1973).
16. J.R. Durig, S.M. Craven, J.H. Mulligan and C.W. Hawley, *J. Chem. Phys.*, **58**, 1281 (1973).
17. W. Longueville, H. Fontaine and G. Vergoten, *J. Raman Spectrosc.*, **13**, 213 (1982).
18. R.M. Moravie and J. Corset, *J. Mol. Struct.*, **30**, 113 (1976).
19. P. Klaeboe, D. Bougeard, B. Schrader, P. Paetzold and C. von Ploto, *Spectrochim. Acta, Part A*, **41A**, 53 (1985).
20. K. Kumar, *Spectrochim. Acta, Part A*, **28A**, 459 (1972).
21. G.A. Crowder, *J. Phys. Chem.*, **75**, 2806 (1971).
22. G. Varsanyi, *Assignments for Vibrational Spectra of Seven Hundred Benzene Derivatives*, J. Wiley and Sons, New York (1974).
23. S. Dobos, A. Szabo and B. Zelei, *Spectrochim. Acta, Part A*, **32A**, 1401 (1976).
24. R. Soda, *Bull. Chem. Soc. Jpn.*, **34**, 1482 (1961).
25. Y. Mido, *Spectrochim. Acta, Part A*, **28A**, 1503 (1972).
26. Y. Mido, *Spectrochim. Acta, Part A*, **32A**, 1105 (1976).
27. M. Goldstein, M.A. Russell and H.A. Willis, *Spectrochim. Acta, Part A*, **25A**, 1275 (1969).
28. J.R. Durig, F. Sun and Y.S. Li, *J. Mol. Struct.*, **101**, 79 (1983).
29. D.F. Koster, *Spectrochim. Acta, Part A*, **24A**, 395 (1968).
30. R.N. Kniseley, R.P. Hirschmann and V.A. Fassel, *Spectrochim. Acta, Part A*, **23A**, 109 (1967).
31. D.A. Edwards, S.M. Tetrick and R.A. Walton, *J. Organomet. Chem.*, **349**, 383 (1988).
32. J. Korppi-Tommola, *Spectrochim. Acta, Part A*, **34A**, 1077 (1978).
33. J.G. Pritchard and H.M. Nelson, *J. Phys. Chem.*, **64**, 795 (1960).
34. Y. Omura, J. Corset and R.M. Moravie, *J. Mol. Struct.*, **52**, 175 (1979).
35. D.C. McKean, *Spectrochim. Acta, Part A*, **23A**, 605 (1967).
36. C.A. Ogle, K.A. Vanderkooi, G.D. Mendenhall, V. Lorprayoon and B.C. Cornilsen, *J.*

Am. Chem. Soc., **104**, 5119 (1982).
37. D.W. Scottt and J.P. McCullough, *J. Am. Chem. Soc.*, **80**, 3554 (1958).
38. J.P. McCullough, D.W. Scott, H.L. Finke, W.N. Hubbard, M.E. Gross, C. Katz, R.E. Pennington, J.F. Messerly and G. Waddington, *J. Am. Chem. Soc.*, **75**, 1818 (1953).
39. M. Sakakibara, I. Harada, H. Matsuura and T. Shimanouchi, *J. Mol. Struct.*, **49**, 29 (1978).
40. H. Sugeta, *Spectrochim. Acta, Part A*, **31A**, 1729 (1975).
41. H. Sugeta, A. Go and T. Miyazawa, *Bull. Chem. Soc. Jpn.*, **46**, 3407 (1973).
42. K.G. Allum, J.A. Creighton, J.H.S. Green, G.J. Minkoff and L.J.S. Prince, *Spectrochim. Acta, Part A*, **24A**, 927 (1968).
43. J.H.S. Green, D.J. Harrison, W. Kynaston and D.W. Scott, *Spectrochim. Acta, Part A*, **25A**, 1313 (1969).
44. J.R. Durig and A.W. Cox, Jr., *J. Mol. Struct.*, **38**, 77 (1977).
45. R.R. Holmes and M. Fild, *Spectrochim. Acta, Part A*, **27A**, 1525 (1971).
46. R.R. Holmes and M. Fild, *Spectrochim. Acta, Part A*, **27A**, 1537 (1971).
47. S. Konaka, J. Hirose, A. Suwa, H. Takeuchi, T. Egawa, K. Siam, J.D. Ewbank and L. Schäfer, *J. Mol. Struct.*, **244**, 1 (1991).
48. E. Spinner, *J. Chem. Soc.*, 4217 (1964).

3

Normal Vibrations and Absorption Regions of $-CH_2X$

3.1 HALOGENOMETHYL

Similarly to methyl the $—CH_2X$ substituent (X = F, Cl, Br and I) yields nine normal modes which may be described as follows: ν_aCH_2, ν_sCH_2, δCH_2, ωCH_2, τCH_2, ρCH_2, νCX, δCX and torsion, where ν, δ, ω, τ and ρ respectively indicate for stretch, deformation (bend or scissor), wag, twist and rock. The presence of CH_2X in a chemical structure usually leads to the existence of several conformers, each possessing a different energy content where the X atom takes up different positions with respect to the rest of the molecule. In practice, multiple bands are observed for one and the same vibration.

3.1.1 Fluoromethyl

In the spectrum of fluoroethane the 5a′ + 4a″ modes of CH_2F in the C_s conformation are observed at the following wavenumbers:

a′	cm^{-1}	a″	cm^{-1}
ν_sCH_2	2960	ν_aCH_2	3003
δCH_2	1479	τCH_2	1277
ωCH_2	1365	ρCH_2	811
νCF	1103	torsion	243
δCF	415		

In CH_2F compounds both methylene stretching vibrations are always observed separately, with intensities varying from weak to medium. The methylene deformation produces a band with moderate intensity but a differentiation of δCH_2F and other methylene and/or methyl deformations of the molecule is not always feasible. The methylene wag and twist are also well separated but their intensities are merely weak to medium.

The CF stretch yields the most intense absorption of all normal modes of CH_2F. On account of its intensity and small absorption region, the band is considered to represent a group frequency, but esters, ethers and alcohols also absorb in this region.

Table 3.1 Absorption regions (cm^{-1}) of —CH_2F

	Total region	Molecules absorbing at high wavenumbers	Molecules absorbing at low wavenumbers	Region of remaining molecules
$\nu_a CH_2$	3000 ± 50	$ClCH_2F$ (3048) FCH_2F (3015) $NaOC(=O)CH_2F$	$FCH_2C(=O)CH_2F$ (2964)	2995 ± 20
$\nu_s CH_2$	2965 ± 30	$ClCH_2F$ (2993) FCH_2CH_2F (2984, *2959*)	$BrC(=O)CH_2F$ (2935)	2955 ± 20
δCH_2	1455 ± 55	FCH_2F (1510) $MeOCH_2F$ (1496)	$NaOC(=O)CH_2F$ (1412) $BrC(=O)CH_2F$ (1413) *g*-FCH_2CH_2F (*1458*, 1409) *tr*-FCH_2CH_2F (*1455*, 1415)	1455 ± 25
ωCH_2	1350 ± 85	FCH_2F (1435) $EtOC(=O)CH_2F$ (1410) $MeOCH_2F$ (1405)	*tr*-FCH_2CH_2F (*1371*, 1278) *g*-FCH_2CH_2F (*1376*, 1285)	1365 ± 30
τCH_2	1205 ± 90	$ClC(=O)CH_2F$ (1295)	*tr*-FCH_2CH_2F (*1240*, 1116) *g*-FCH_2CH_2F (*1244*, 1119)	1235 ± 45
νCF	1050 ± 60	$MeOCH_2F$ (1104) $MeCH_2F$ (1103)	$PhCH_2F$ (990) $ClCH_2F$ (1004) $FCH_2C\equiv CCH_2F$ (1008)	1050 ± 30
ρCH_2	895 ± 95	$MeOCH_2F$ (990) $^+H_3NCH_2F$ (980) $FCH_2C\equiv CCH_2F$ (986)	$PhCH_2F$ (808) $MeCH_2F$ (811)	920 ± 50
δCF	420 ± 150	$N\equiv CCH_2F$ (567) $HC\equiv CCH_2F$ (544)	$FCH_2C(=O)CH_2F$ (275) $EtCH_2F$ (315) $H_2C=C(Br)CH_2F$ (318)	425 ± 90
torsion	180 ± 70	$MeCH_2F$ (243)		150 ± 40

R-CH_2F molecules

R = Me— [1, 2], Et— [2], nPr—, nBu— and nPe— [3], $Me(CH_2)_n$— (*n* = 5, 6, 7, 8, 12) [56], FCH_2CH_2— [4], $^+H_3NCH_2$—, H_2NCH_2— [5],

$HOCH_2$— [6–10], $DOCH_2$— [6–8], *tr* and *g* FCH_2— [11–14], $ClCH_2$— [15], $BrCH_2$— [16], $FCH_2CH(OH)$— [17], Ox— [18], CF_3— [19, 20], HC(=O)— [21], FC(=O)— [22–24], Cl(=O)— [24–27], BrC(=O)— [24, 25], MeC(=O)— [28–30], $FCH_2C(=O)$— [31], EtOC(=O)— [32], NaOC(=O)— [33], $H_2NC(=O)$— [34–39], $H_2C=CH$— [40–43], $H_2C=C(Cl)$— and $H_2C=C(Br)$— [44], HC≡C— [45–47], $FCH_2C≡C$— [48], $ClCH_2C≡C$— [57], N≡C— [49–51], Ph— [55], MeO— [52, 58], F— [53, 54], Cl— and Br— [53].

References

1. D.C. Smith, R.A. Saunders, J.R. Nielsen and E.E. Ferguson, *J. Chem. Phys.*, **20**, 847 (1952).
2. G.A. Crowder and H.K. Mao, *J. Mol. Struct.*, **18**, 33 (1973).
3. G.A. Crowder and H.K. Mao, *J. Mol. Struct.*, **23**, 161 (1974).
4. P. Klaboe, D.L. Powell, R. Stoelevik and V. Oeyvind, *Acta Chem. Scand., Ser. A*, **36A**, 471 (1982).
5. J.A.S. Smith and V.F. Kalasinsky, *Spectrochim. Acta, Part A*, **42A**, 157 (1986).
6. E. Wyn-Jones and W.J. Orville-Thomas, *J. Mol. Struct.*, **1**, 79 (1967).
7. M. Pertillä, J. Murto, A. Kivinen and K. Turunen, *Spectrochim. Acta, Part A*, **34A**, 9 (1978).
8. G. Davidovics, J. Pourcin, M. Carles, L. Pizzala and H. Bodot, *J. Mol. Struct.*, **99**, 165 (1983).
9. G. Davidovics, J. Pourcin, M. Monnier, P. Verlaque and H. Bodot, *J. Mol. Struct.*, **116**, 39 (1984).
10. G.A. Crowder and D. Tennant, *J. Fluorine Chem.*, **6**, 279 (1975).
11. P. Klaboe and J.R. Nielsen, *J. Chem. Phys.*, **33**, 1764 (1960).
12. W.C. Harris, J.R. Holtzclaw and V.F. Kalasinsky, *J. Chem. Phys.*, **67**, 3330 (1977).
13. B. Beagle and D.E. Brown, *J. Mol. Struct.*, **54**, 175 (1979).
14. L.M. Sverdlov, M.A. Kovner and E.P. Krainov, *Vibrational Spectra of Polyatomic Molecules*, J.Wiley and Sons, New York (1974).
15. J.R. Durig, J. Liu and T.S. Little, *J. Phys. Chem.*, **95**, 4664 (1991).
16. J.R. Durig, J. Liu and T.S. Little, *J. Mol. Struct.*, **248**, 25 (1991).
17. G.A. Crowder and T. Douglas, *J. Fluorine Chem.*, **7**, 537 (1976).
18. R.A. Nyquist, C.L. Putzig and N.E. Skelly, *Applied Spectrosc.*, **40**, 821 (1986).
19. W.F. Edgell, T.R. Riethof and C. Ward, *J. Mol. Spectrosc.*, **11**, 92 (1963).
20. G.A. Crowder, *J. Fluorine Chem.*, **3**, 125 (1973/74).
21. H.V. Phan and J.R. Durig, *J. Mol. Struct.*, **209**, 333 (1990).
22. R. Fausto, J.J.C. Teixeira-Dias and M.N. Ramos, *Spectrochim. Acta, Part A*, **44A**, 47 (1988).
23. J.R. Durig, H.V. Phan, J.A. Hardin, R.J. Berry and T.S. Little, *J. Mol. Struct.*, **198**, 365 (1989).
24. A.Y. Khan and N. Jonathan, *J. Chem. Phys.*, **52**, 147 (1970).
25. J.E.F. Jenkins and J.A. Ladd, *J. Chem. Soc. B*, 1237 (1968).
26. J.R. Durig, H.V. Phan, J.A. Hardin and T.S. Little, *J. Chem. Phys.*, **90**, 6840 (1989).
27. M. Monnier, G. Davidovics and A. Allouche, *J. Mol. Struct.*, **243,** 13 (1991).
28. G.A. Crowder and P. Pruettiangkura, *J. Mol. Struct.*, **15**, 197 (1973).
29. J.R. Durig, J.A. Hardin, H.V. Phan and T.S. Little, *Spectrochim. Acta, Part A*, **45A**, 1239 (1989).

30. G.A. Crowder and B.R. Cook, *J. Chem. Phys.*, **47**, 367 (1967).
31. G.A. Crowder and P. Pruettiangkura, *J. Mol. Struct.*, **18**, 177 (1973).
32. M.A. Raso, M.V. Garcia and J. Morcillo, *J. Mol. Struct.*, **115**, 449 (1984).
33. E. Spinner, *J. Chem. Soc.*, 4217 (1964).
34. S. Samdal and R. Seip, *J. Mol. Struct.*, **52**, 195 (1979).
35. F.M. Abid, J.M. Al-Katti and M.F. El-Bermani, *J. Mol. Struct.*, **67**, 169 (1980).
36. D. Troitino, E. Sanchez De La Blanca and M.V. Garcia, *Spectrochim. Acta, Part A*, **46A**, 1281 (1990).
37. E.K. Murthy and G.R. Rao, *J. Raman Spectrosc.*, **19**, 359 (1988).
38. E.K. Murthy and G.R. Rao, *J. Raman Spectrosc.*, **19**, 419 (1988).
39. E.K. Murthy and G.R. Rao, *J. Raman Spectrosc.*, **20**, 409 (1989).
40. R.D. McLachlan and R.A. Nyquist, *Spectrochim. Acta, Part A*, **24A**, 103 (1968).
41. J.R. Durig, T.J. Geyer, T.S. Little and D.T. Durig, *J. Mol. Struct.*, **172**, 165 (1988).
42. J. Nieminen, J. Murto and M. Räsänen, *Spectrochim. Acta, Part A*, **47A**, 1495 (1991).
43. J.R. Durig, M. Zhen, H.L. Heusel, P.J. Joseph, P. Grower and T.S. Little, *J. Phys. Chem.*, **89**, 2877 (1985).
44. P. Klaboe, T. Torgrimsen and D.H. Christensen, *J. Mol. Struct.*, **23**, 15 (1974).
45. J.C. Evans and R.A. Nyquist, *Spectrochim. Acta*, **19**, 1153 (1963).
46. R.A. Nyquist and W.W. Muelder, *J. Mol. Struct.*, **2**, 465 (1968).
47. R.A. Nyquist, *Spectrochim. Acta, Part A*, **27A**, 2513 (1971).
48. A. Karlsson, P. Klaeboe, K.-M. Marstokk, H. Møllendal and C.J. Nielsen, *Acta Chem. Scand., Ser. A*, **40A**, 374 (1986).
49. R.G. Jones and W.J. Orville-Thomas, *J. Chem. Soc.*, 4632 (1965).
50. G.A. Crowder, *Mol. Phys.*, **23**, 707 (1972).
51. J.R. Durig and D.W. Wertz, *Spectrochim. Acta, Part A*, **24A**, 21 (1968).
52. D.C. McKean, I. Torto and A.R. Morrison, *J. Mol. Struct.*, **99**, 101 (1983).
53. R.G. Jones and W.J. Orville-Thomas, *Spectrochim. Acta*, **20**, 291 (1964).
54. I. Suzuki and T. Shimanouchi, *J. Mol. Spectrosc.*, **46**, 130 (1973).
55. G.A. Crowder and M.J. Townsend, *J. Fluorine Chem.*, **10**, 181 (1977).
56. G.A. Crowder and J.M. Lightfoot, *J. Mol. Struct.*, **99**, 77 (1983).
57. A. Karlsson, P. Klaboe and C.J. Nielsen, *J. Raman Spectrosc.*, **23**, 167 (1992).
58. J.R. Durig, J. Liu, G.A. Guirgis and B.J. Van der Veken, *Struct. Chem.*, **4**, 103 (1993).

3.1.2 Chloromethyl

The methylene symmetric stretching is always well separated from its antisymmetric counterpart. Nevertheless the weak to medium absorptions of these methylene stretching vibrations often disappear within the stronger absorptions of saturated and unsaturated CH stretching modes. The methylene scissor absorbs, with mostly medium band intensity, coincident with or at the LW side of aliphatic methylene deformations. The methylene wag, often appearing as a moderate to strong band, is helpful to the identification of —CH_2Cl compounds. In many cases the wavenumber of this band depends upon the conformation. The methylene twist, normally appearing as a medium band at the LW side of the wag, is not always sensitive to conformational changes of the molecule. The intensity of the methylene rocking vibration is mostly moderate but the band sometimes matches that of the C—Cl stretch, which is awkward in assigning both modes. Normally, the rock is at the HW side of the C—Cl stretch. The ν_sCOC in chloromethyl ethers or esters also can interfere with this rock, so that the latter is assigned at higher wavenumbers. The C—Cl stretch is the most characteristic band of all normal vibrations of the CH_2Cl substituent. The band is often, but not always, sensitive to conformation. The band intensity varies from medium to strong, depending upon which conformation predominates. The position of the δCCl often depends upon conformation. This deformation easily couples with the skeletal modes and is therefore of minor importance as a diagnostic tool. The band intensity is normally weak or sometimes medium.

Table 3.2 Absorption regions (cm^{-1}) of α-C saturated —CH_2Cl

	Total region	Molecules absorbing at high wavenumbers	Molecules absorbing at low wavenumbers	Region of remaining molecules
$\nu_a CH_2$	3015 ± 25	$MeCF_2CH_2Cl$ (3039)	$tBuCH_2Cl$ (2990) $MeOCH_2CH_2Cl$ (2990)	3015 ± 15
$\nu_s CH_2$	2960 ± 20		$iPrCH_2Cl$ (2943) $nPrCH_2Cl$ (2945)	2965 ± 15
δCH_2	1430 ± 20		$4\text{-}FPhC(=O)(CH_2)_2Cl$ (1411) $EtC(=O)(CH_2)_2Cl$ (1412) $NCCH_2CH_2Cl$ (1416)	1435 ± 15
ωCH_2	1265 ± 50		$ClC(=O)CHClCH_2Cl$ (1215) $H_2C=CHCHClCH_2Cl$ (1221) $MeCHBrCH_2Cl$ (1223) $MeCHClCH_2Cl$ (1234)	1280 ± 35
τCH_2	1195 ± 55	$MeCH_2Cl$ (1250) $HC\equiv CCH_2CH_2Cl$ (1247)	*g*-$MeCF_2CH_2Cl$ (1141) $ClCH_2CH_2Cl$ (1143 and *1233*)	1195 ± 50
ρCH_2	885 ± 105	$HC\equiv CCH_2CH_2Cl$ (989) *tr*-$ClCH_2CH_2Cl$ (982) $ClCH_2C(Me)_2CH_2Cl$ (978 and *857*)	$MeCH_2Cl$ (786) $ClCH_2(CHCl)_2CH_2Cl$ (792 and *838*) $(ClCH_2)_4C$ (794, 818, *860*) $iPrCH_2Cl$ (816)	890 ± 70
νCCl	700 ± 70	$ClCH_2(CHCl)_2CH_2Cl$ (767 and *720*) $ClCH_2CCl_2CH_2Cl$ (766 and *700*) $MeCF_2CH_2Cl$ (760)	D_3CCH_2Cl (631) DCH_2CH_2Cl (642) $Cl(CH_2)_3Cl$ (641 and *679*)	705 ± 50
δCCl	285 ± 80	Ox—CH_2Cl (363) $ClCH_2C(Me)_2CH_2Cl$ (363 and *316*) $Cl(CH_2)_3Cl$ (362 and *348*)	$MeCF_2CH_2Cl$ (205) $cPrCH_2Cl$ (205) $tBuCH_2Cl$ ((210)	290 ± 65
torsion	140 ± 65	$(ClCH_2)_4$ (204, *123, 88*) $EtC(=O)CH_2CH_2Cl$ (201)	$ClCH_2CHClCH_2Cl$ (75, *154*) $ClCH_2C(Me)_2CH_2Cl$ (79 and *193*)	140 ± 55

R—CH_2Cl molecules
R = $R'CH_2$— (see Section 3.5.3);
R = Me— [1–7], CD_3— [1], iPr— [7–9], cPr— [10–12], Ox— [12, 13], MeCHOH—, $ClCH_2CHOH$—, MeCHCl— [15, 16, 18], $ClCH_2CHCl$— [14, 17], $Cl_2CHCHCl$— [17], Cl_3CCHCl— [18], $H_2C{=}CHCHCl$—, ClC(=O)CHCl—, Cl_2CH— [6, 20–22], $ClCH_2(CHCl)_2$— [19], MeCHBr— [15], tBu— [8, 23], $PhC(Me)_2$—, $ClCH_2C(Me)_2$— [24], $(ClCH_2)_2MeC$— [118], $(ClCH_2)_3C$— [25], $MeCF_2$— [26], $ClCH_2CCl_2$— [17, 24], Cl_2CHCCl_2— [17], F_3C— [27, 28], Cl_3C— [6, 29].

Table 3.3 Absorption regions (cm^{-1}) of —CH_2Cl

attached to	Ph	C=C	C≡C	C=O	N, O	S
$\nu_a CH_2$	3005 ± 15	3000 ± 15	3000 ± 15	3000 ± 15	3035 ± 35	3020 ± 15
$\nu_s CH_2$	2970 ± 15	2955 ± 15	2960 ± 10	2960 ± 20	2975 ± 30	2950 ± 20
δCH_2	1450 ± 10	1435 ± 20	1430 ± 15	1410 ± 30	1440 ± 25	1400 ± 10
ωCH_2	1265 ± 10	1265 ± 15	1270 ± 10	1275 ± 45	1315 ± 35	1245 ± 25
τCH_2	1205 ± 20	1185 ± 30	1180 ± 10	1185 ± 45	1240 ± 35	1140 ± 20
ρCH_2	745 ± 20	905 ± 50	905 ± 10	880 ± 55	960 ± 60	845 ± 40
νCCl	690 ± 30	705 ± 55	715 ± 25	755 ± 45	690 ± 60	700 ± 55
δCCl	270 ± 40	310 ± 80	395 ± 55	300 ± 70	310 ± 60	–
torsion	–	150 ± 65	120 ± 80	120 ± 85	150 ± 50	–

R—CH_2Cl molecules
R = (substituted)phenyl (see Section 12.4);
R = $H_2C{=}CH$— [30–35], $H_2C{=}CMe$— [30], $H_2C{=}C(CH_2Cl)$— [36, 37], $H_2C{=}CCl$— [38–40, 117], $H_2C{=}CBr$— [40, 41], *cis*-$ClCH_2CH{=}CH$— [42], *tr*-ClCH=CH— [43], HC≡C— [44–46], DC≡C— [46, 47], ClC≡C— [45, 48], $FCH_2C{\equiv}C$— [119], $ClCH_2C{\equiv}C$— [49–51], N≡C— [52–54], HC(=O)— [55, 56], MeC(=O)— [57–59], $ClCH_2C(=O)$— [60, 61], 2-ThC(=O)—, 4-FPhC(=O)— [62], 2, 4-$Cl_2PhC(=O)$— [63], $H_2NC(=O)$— [64–68], MeHNC(=O)— [68], HOC(=O)— [68–72], DOC(=O)— [70, 71], MeOC(=O)— [72–74], $CD_3OC(=O)$— [72, 73], EtOC(=O)— [75–77], tBuOC(=O)—, $Me(CH_2)_nOC(=O)$— [68], HSC(=O)— [78], FC(=O)— [68, 79, 83], ClC(=O)— [72, 79–82, 120], BrC(=O)— [64, 79, 120], OCN—, O_2N— [84], MeO— [64, 85–88], D_3CO— [85, 87, 88], $ClCH_2O$— [89], HC(=O)O— [90–92], MeC(=O)O— [93, 94], $CD_3C(=O)O$— [94], tBuC(=O)O—, $PhCH_2O$—, 4-ClPhO—, FC(=O)O— [95], ClC(=O)O— [96], MeS— [64], PhS—, NCS— [97, 98], $MeS(=O)_2$— [99], $FS(=O)_2$— and $ClS(=O)_2$— [100].

Table 3.4 Absorption regions (cm^{-1}) of —CH_2Cl

attached to	P	Si	Hg	Cl, Br and I
$\nu_a CH_2$	3000 ± 10	2990 ± 15	3010 ± 10	3050 ± 10
$\nu_s CH_2$	2935 ± 15	2940 ± 15	2945 ± 10	2985 ± 05
δCH_2	1400 ± 10	1400 ± 10	1400 ± 10	1410 ± 15
ωCH_2	1210 ± 25	1185 ± 15	1195 ± 45	1225 ± 40
τCH_2	1110 ± 40	1110 ± 10	1080 ± 35	1130 ± 20
ρCH_2	790 ± 35	795 ± 50	720 ± 10	845 ± 55
νCCl	705 ± 25	715 ± 55	685 ± 15	720 ± 20
δCCl	–	–	–	≈285
torsion	–	–	–	–

R—CH_2Cl molecules

R = F_2P— [101, 102, 121], Cl_2P— [103–105], F_2P(=O)— [102, 106], Cl_2P(=O)— [105, 107], F_2P(=S)— [102, 108], Cl_2P(=S)— [105, 107], $(HO)_2P$(=O)—, (HO)(KO)P(=O)— and $(EtO)_2P$(=O)— [109], H_3Si—, D_3Si—, F_3Si— and Cl_3Si— [110], Me_2HSi— and Me_2DSi— [111], $ClCH_2Hg$— [112], ClHg— [113], F— [64], Cl— [1, 64, 114, 115], Br— [64, 116], I— [64].

References

1. S. Mizushima, T. Shimanouchi, I. Nakagawa and A. Miyake, *J. Chem. Phys.*, **21**, 215 (1953).
2. L.W. Daasch, C.Y. Liang and J.R. Nielsen, *J. Chem. Phys.*, **22**, 1293 (1954).
3. N.T. McDevitt, A.L. Rozek, F.F. Bentley and A.D. Davidson, *J. Chem. Phys.*, **42**, 1173 (1965).
4. R.G. Snyder and J.H. Schachtschneider, *J. Mol. Spectrosc.*, **30**, 290 (1969).
5. F.A. Miller and F.E. Kiviat, *Spectrochim. Acta, Part A*, **25A**, 1363 (1969).
6. S. Suzuki and A.B. Dempster, *J. Mol. Struct.*, **32**, 339 (1976).
7. A.J. Barnes, M.L. Evans and H.E. Hallam, *J. Mol. Struct.*, **99**, 235 (1983).
8. G.A. Crowder and W.L. Lin, *J. Mol. Struct.*, **64**, 193 (1980).
9. J.R. Durig, J.F. Sullivan and S.E. Godbey, *J. Mol. Struct.*, **146**, 213 (1986).
10. T. Hirokawa and H. Murata, *J. Sci. Hiroshima Univ. Ser A*, **38**, 271 (1974).
11. F.G. Fujiwara, J.C. Chang and H. Kim, *J. Mol. Struct.*, **41**, 177 (1977).
12. V.F. Kalasinsky and C.J. Wurrey, *J. Raman Spectrosc.*, **9**, 315 (1980).
13. R.A. Nyquist, C.L. Putzig and N.E. Skelly, *Appl. Spectrosc.*, **40**, 821 (1986).
14. J. Thorbjørnsrud, O.H. Ellestad, P. Klaboe, T. Torgrimsen and D.H. Christensen, *J. Mol. Struct.*, **17**, 5 (1973).
15. J. Thorbjørnsrud, O.H. Ellestad, P. Klaboe and T. Torgrimsen, *J. Mol. Struct.*, **15**, 45 (1973).
16. G.A. Crowder, *J. Mol. Struct.*, **100**, 415 (1983).
17. A.B. Dempster, K. Price and N. Sheppard, *Spectrochim. Acta, Part A*, **27A**, 1579 (1971).
18. A.B. Dempster, K. Price and N. Sheppard, *Spectrochim. Acta, Part A*, **27A**, 1563 (1971).

19. A.B. Dempster, K. Price and N. Sheppard, *Spectrochim. Acta, Part A*, **31A**, 331 (1975).
20. K. Kuratani and S.-I. Mizushima, *J. Chem. Phys.*, **22**, 1403 (1954).
21. R.H. Harrison and K.A. Kobe, *J. Chem. Phys.*, **26**, 1411 (1957).
22. S.D. Christian, J. Grundnes, P. Klaboe, C.J. Nielsen and T. Woldbaek, *J. Mol. Struct.*, **34**, 33 (1976).
23. P. Klaboe, C.J. Nielsen and D.L. Powell, *Spectrochim. Acta, Part A*, **41A**, 1315 (1985).
24. D.L. Powell, P. Klaboe, K. Saebø and G.A. Crowder, *J. Mol. Struct.*, **98**, 55 (1983).
25. P. Klaboe, B. Klewe, K. Martinsen, C.J. Nielsen, D.L. Powell and D.J. Stubbles, *J. Mol. Struct.*, **140**, 1 (1986).
26. G.A. Crowder, *J. Mol. Struct.*, **15**, 351 (1973).
27. J.R. Nielsen, C.Y. Liang and D.C. Smith, *J. Chem. Phys.*, **21**, 1060 (1953).
28. G.A. Crowder, *J. Fluorine Chem.*, **3**, 125 (1973/74).
29. G. Allen and H.J. Bernstein, *Can. J. Chem.*, **32**, 1124 (1954).
30. D.A.C. Compton, S.C. Hsi, H.H. Mantsch and W.F. Murphy, *J. Raman Spectrosc.*, **13**, 30 (1982).
31. R.D. McLachlan and R.A. Nyquist, *Spectrochim. Acta, Part A*, **24A**, 103 (1968).
32. C. Sourisseau and B. Pasquier, *J. Mol. Struct.*, **12**, 1 (1972).
33. B. Silvi and C. Sourisseau, *Spectrochim. Acta, Part A*, **31A**, 565 (1975).
34. A.J. Barnes, S. Holroyd, W.O. George, J.E. Goodfield and W.F. Maddams, *Spectrochim. Acta, Part A*, **38A**, 1245 (1982).
35. J.R. Durig, D.T. Durig, M.R. Jalilian, M. Zhen and T.S. Little, *J. Mol. Struct.*, **194**, 259 (1989).
36. R. Gaufrès and C. Roulph, *J. Mol. Struct.*, **9**, 107 (1971).
37. G.A. Crowder, *J. Mol. Struct.*, **10**, 294 (1971).
38. G.A. Crowder, *J. Mol. Spectrosc.*, **20**, 430 (1966).
39. T. Torgrimsen and P. Klaboe, *J. Mol. Struct.*, **20**, 229 (1974).
40. S.H. Schei, *Spectrochim. Acta, Part A*, **39A**, 1043 (1983).
41. S.H. Schei and P. Klaboe, *J. Mol. Struct.*, **96**, 9 (1982).
42. P. Piaggio, G. Viviano and G. Dellepiane, *J. Mol. Struct.*, **20**, 243 (1974).
43. J.R. Durig, T.G. Costner, T.S. Little and D.T. Durig, *J. Phys. Chem.*, **96**, 7194 (1992).
44. J.C. Evans and R.A. Nyquist, *Spectrochim. Acta*, **19**, 1153 (1963).
45. R.A. Nyquist, A.L. Johnson and Y.S. Lo, *Spectrochim. Acta*, **21**, 77 (1965).
46. R.A. Nyquist, T.L. Reder, G.R. Ward and G.J. Kallos, *Spectrochim. Acta, Part A*, **27A**, 541 (1971).
47. R.A. Nyquist, *Spectrochim. Acta, Part A*, **27A**, 2513 (1971).
48. D. Christen, F. Gleisberg, G. Kremer and W. Zeil, *J. Mol. Spectrosc.*, **70**, 179 (1978).
49. G.A. Crowder, *J. Mol. Struct.*, **12**, 302 (1972).
50. S. Suzuki, *J. Mol. Struct.*, **46**, 155 (1978).
51. A. Karlsson, P. Klaeboe and C.J. Nielsen, *J. Raman Spectrosc.*, **18**, 461 (1987).
52. R.G. Jones and W.J.Orville-Thomas, *J. Chem. Soc.*, 4632 (1965).
53. G.A. Crowder, *Mol. Phys.*, **23**, 707 (1972).
54. J.R. Durig and D.W. Wertz, *Spectrochim. Acta*, **24**, 21 (1968).
55. G. Lucazeau and A. Novak, *J. Chim. Phys. Phys.-Chim. Biol.*, **67**, 1614 (1970).
56. S. Dyngeseth, H. Schei and K. Hagen, *J. Mol. Struct.*, **102**, 45 (1983).
57. K. Tanabe and S. Saëki, *J. Mol. Struct.*, **25**, 243 (1975).
58. J.R. Durig, J. Lin, C.L. Tolley and T.S. Little, *Spectrochim. Acta, Part A*, **47A**, 105 (1991).
59. S. Mizushima, T. Shimanouchi, T. Miyazawa, I. Ichishima, K. Kuratani, I. Nakagawa and N. Shido, *J. Chem. Phys.*, **21**, 815 (1953).
60. L.J. Bellamy and R.L. Williams, *J. Chem. Soc.*, 4294 (1957).
61. L.W. Daasch and R.E. Kagarise, *J. Am. Chem. Soc.*, **77**, 6156 (1955).

62. W.A. Seth Paul, *Bull. Soc. Chim. Belg.*, **85**, 187 (1976).
63. T.V.K. Sarma, *Acta Chim. Hung.*, **115**, 89 (1984).
64. R.G. Jones and W.J. Orville-Thomas, *Spectrochim. Acta*, **20**, 291 (1964).
65. S. Samdal and R. Seip, *J. Mol. Struct.*, **52**, 195 (1979).
66. F.M. Abid, J.M.L. Katti and M.F. El-Bermani, *J. Mol. Struct.*, **67**, 169 (1980).
67. D. Troitino, E. Sanchez de la Blanca and M.V. Garcia, *Spectrochim. Acta, Part A*, **46A**, 1281 (1990).
68. Y. Mido, T. Shono and H. Matsuura, *J. Mol. Struct.*, **265**, 75 (1992).
69. J.D. Barceló, M.P. Jorge and C. Otero, *J. Chem. Phys.*, **28**, 1230 (1958).
70. D. Sinha, J.E. Katon and R.J. Jakobsen, *J. Mol. Struct.*, **20**, 381 (1974).
71. D. Sinha, J.E. Katon and R.J. Jakobsen, *J. Mol. Struct.*, **24**, 279 (1975).
72. R. Fausto and J.J.C. Teixeira-Dias, *J. Mol. Struct.*, **144**, 225 (1986).
73. J.E. Katon and D. Sinha, *Spectrochim. Acta, Part A*, **33A**, 45 (1977).
74. Y. Mido and M. Hashimoto, *J. Mol. Struct.*, **129**, 253 (1985).
75. M.A. Raso, M.V. Garcia and J. Morcillo, *J. Mol. Struct.*, **115**, 449 (1984).
76. N. Dube, P. Porwal and R. Prasad, *J. Raman Spectrosc.*, **19**, 189 (1988).
77. Y. Mido, N. Kakizawa, H. Matsuura, M.A. Raso, M.V. Garcia and J. Morcillo, *J. Mol. Struct.*, **220**, 169 (1990).
78. H.S. Randhawa and W. Walter, *J. Mol. Struct.*, **38**, 89 (1977).
79. A.Y. Khan and N. Jonathan, *J. Chem. Phys.*, **50**, 1801 (1969).
80. I. Nakagawa, I. Ichishima, K. Kuratani, T. Miyazawa, T. Shimanouchi and S. Mizushima, *J. Chem. Phys.*, **20**, 1720 (1952).
81. K. Tanabe and S. Saëki, *Spectrochim. Acta, Part A*, **28A**, 1083 (1972).
82. G. Davidovics, A. Allouche and M. Monnier, *J. Mol. Struct.*, **243**, 1 (1991).
83. J.R. Durig, W. Zhao, D. Lewis and T.S. Little, *J. Chem. Phys.*, **89**, 1285 (1988).
84. P. Gluzinsky and Z. Eckstein, *Spectrochim. Acta, Part A*, **24A**, 1777 (1968).
85. D.C. McKean, I. Torto and A.R. Morrisson, *J. Mol. Struct.*, **99**, 101 (1983).
86. H.F. Hameka, *J. Mol. Struct.*, **226**, 241 (1991).
87. R.G. Jones and W.J. Orville-Thomas, *J. Chem. Soc.*, 692 (1964).
88. H.R. Linton and E.R. Nixon, *Spectrochim. Acta*, **15**, 146 (1959).
89. S.W. Charles, F.C. Cullen and N.L. Owen, *Spectrochim. Acta, Part A*, **32A**, 1171 (1976).
90. M.G. Dahlqvist, *Spectrochim. Acta, Part A*, **36A**, 37 (1980).
91. M. Räsänen, H. Kunttu, J. Murto and M. Dahlqvist, *J. Mol. Struct.*, **159**, 65 (1987).
92. F. Daeyaert and B.J. Van der Veken, *J. Mol. Struct.*, **213**, 97 (1989).
93. S.W. Charles, G.I.L. Jones, N.L. Owen and L.A. West, *J. Mol. Struct.*, **32**, 111 (1976).
94. F. Daeyaert, H.O. Desseyn and B.J. Van der Veken, *Spectrochim. Acta, Part A*, **44A**, 1165 (1988).
95. F. Daeyaert and B.J. Van der Veken, *Spectrochim. Acta, Part A*, **45A**, 993 (1989).
96. F. Daeyaert and B.J. Van der Veken, *J. Mol. Struct.*, **198**, 239 (1989).
97. G.A. Crowder, *J. Chem. Phys.*, **47**, 3080 (1967).
98. G.A. Crowder, *Texas J. Sci.*, **26**, 565 (1975).
99. A.B. Remizov, F.S. Bilalov and I.S. Pominov, *Spectrochim. Acta, Part A*, **43A**, 309 (1987).
100. R. Aroca, J. Ali and E.A. Robinson, *J. Mol. Struct.*, **116**, 9 (1984).
101. P. Coppens, B.J. Van der Veken and J.R. Durig, *J. Mol. Struct.*, **142**, 367 (1986).
102. P. Coppens, Thesis, UIA, Antwerp, 1987.
103. A.I. Fishman, A.B. Remisov, I.Ya. Kuramshin and I.S. Pominov, *Spectrochim. Acta, Part A*, **32A**, 651 (1976).
104. B.J. Van der Veken, R.S. Sanders and J.R. Durig, *J. Mol. Struct.*, **216**, 113 (1990).
105. R.A. Nyquist, *Appl. Spectrosc.*, **22**, 452 (1968).

106. B.J. Van der Veken, P. Coppens, R.D. Johnson and J.R. Durig, *J. Chem. Phys.*, **83**, 1517 (1985).
107. E. Steger, J. Rehak and H.F. Faltus, *Z. Phys. Chem.*, **229**, 110 (1965).
108. B.J. Van der Veken, P. Coppens, R.D. Johnson and J.R. Durig, *J. Phys. Chem.*, **90**, 4537 (1986).
109. B.J. Van der Veken, *J. Mol. Struct.*, **25**, 75 (1975).
110. I.V. Kochikov, G.M. Kuramshina, S.V. Syn'ko and Yu.A. Pentin, *J. Mol. Struct.*, **172**, 299 (1988).
111. K. Ohno, K. Suehiro and H. Murata, *J. Mol. Struct.*, **98**, 251 (1983).
112. Y. Imai and K. Aida, *Spectrochim. Acta, Part A*, **28A**, 517 (1972).
113. H.G.M. Edwards, *Spectrochim. Acta, Part A*, **42A**, 427 (1986).
114. J. Morcillo, L.J. Zamorano and J.M.V. Heredia, *Spectrochim. Acta*, **22**, 1969 (1966).
115. T.A. Ford, *J. Mol. Spectrosc.*, **58**, 185 (1975).
116. S. Giorgianni, R. Visinoni, A. Baldacci, A. Gambi and S. Ghersetti, *Spectrochim. Acta, Part A*, **44A**, 463 (1988).
117. D.T. Durig, G.A. Guirgis and J.R. Durig, *J. Raman Spectrosc.*, **23**, 37 (1992).
118. K. Martinsen, D.L. Powell, C.J. Nielsen and P. Klaboe, *J. Raman Spectrosc.*, **17**, 437 (1986).
119. A. Karlsson, P. Klaboe and C.J. Nielsen, *J. Raman Spectrosc.*, **23**, 167 (1992).
120. A.A. El-Bindary, P. Klaboe and C.J. Nielsen, *Acta Chem. Scand.*, **45**, 877 (1991).
121. B.J. Van der Veken, P. Coppens, R.S. Sanders, F.F. Daeyaert and J.R. Durig, *J. Mol. Struct.*, **272**, 305 (1992).

3.1.3 Bromomethyl

The methylene stretching vibrations are observed with band intensities from very weak to medium.The antisymmetric methylene stretch appears partly in the same region as for unsaturated and aromatic CH stretching modes and the symmetric stretch sometimes disappears into other aliphatic CH stretching vibrations. The methylene deformation is active, with a medium intensity, at the LW side of other aliphatic methylene and/or methyl deformations. The methylene wag is often recognized by its intensity, which varies from moderate to strong. The band is often sensitive to conformation. The wavenumber differences between *trans* and *gauche* conformers may even rise up to 50 cm^{-1}. The methylene twist is located with medium intensity at the LW side of the wag. The intensity of the methylene rock varies from weak to moderate. If the molecule is in different conformational states, the wavenumber differences between the rocking modes may be as great as 100 cm^{-1}. The CBr stretch provides the most characteristic band of all CH_2Br modes of vibration. Normally the band is moderate or strong and therefore easy to recognize in this part of the spectrum. Weak CBr bands usually originate from various conformational states of the molecule, whereas band separations up to 100 wavenumbers are sometimes observed between CBr stretches of different conformers. The —CBr deformation is often assigned as a skeletal deformation. This weak to medium band is usually not important as a diagnostic tool.

Table 3.5 Absorption regions (cm^{-1}) of α-C saturated $—CH_2Br$

	Total region	Molecules absorbing at high wavenumbers	Molecules absorbing at low wavenumbers	Region of remaining molecules
ν_aCH_2	3020 ± 30	F_3CCH_2Br (3049) $EtOC(=O)CH(Br)CH_2Br$ (3043)	$(MeO)_2CHCH_2Br$ (2994) $(MeO)_2C(Me)CH_2Br$ (2995)	3020 ± 20
ν_sCH_2	2945 ± 45	F_3CCH_2Br (2990) $EtOC(=O)CH(Br)CH_2Br$ (2983)	$BrCH_2CHBrCH_2Br$ (2902 and *2948*) $MeCHBrCH_2Br$ (2929)	2960 ± 20
δCH_2	1430 ± 20	CD_3CH_2Br (1449)	Br_3CCH_2Br (1410) $(BrCH_2)_4C$ (1411, *1415*, *1422*)	1430 ± 15
ωCH_2	1250 ± 50	$NCCH(Br)CH_2Br$ (1298) $MeOC(=O)CH_2CH_2Br$ (1294) $Br(CH_2)_3Br$ (1293)	*tr*-$ClCH_2CH_2Cl$ (1203) $HOCH_2CH_2Br$ (1215)	1255 ± 35
τCH_2	1175 ± 70	$PhCH_2CH_2Br$ (1242) $MeOC(=O)CH_2CH_2Br$ (1240) $Br(CH_2)_3Br$ (1239 and *1194*)	*g*-$BrCH_2CH_2Br$ (1105) $H_2C=CH(CH_2)_2Br$ (1108) $BrCH_2CHBrCH_2Br$ (1110 and *1159*)	1170 ± 55
ρCH_2	830 ± 115	$Br(CH_2)_3Br$ (943 and *853*) $HO(CH_2)_3Br$ (912)	$(BrCH_2)_4C$ (716, *780*, *835*) $(MeO)_2C(Me)CH_2Br$ (745)	835 ± 75
νCBr	635 ± 85	F_3CCH_2Br (720) $(BrCH_2)_4C$ (700 and *611*)	*tr*-$BrCH_2CH_2Br$ (552)	625 ± 65
δCBr	265 ± 90	*g*-$BrCH_2CH_2Br$ (355)	$tBuCH_2Br$ (179) $(BrCH_2)_4C$ (180 and *235*)	260 ± 75
torsion	130 ± 60	$Br(CH_2)_3Br$ (186)	$(BrCH_2)_4C$ (73, *148*) *tr*-Br_2HCCH_2Br (80)	120 ± 30

$R—CH_2Br$ molecules

R = Me— [1–4], CD_3— [2], Et— [4–11], nPr— [10, 11], $MeCH(Br)CH_2CH_2$—, $ClC(=O)CH_2CH_2$—, $H_2C=CH—CH_2CH_2$— [12], $NCCH_2CH_2$—, $HOCH_2CH_2$—, $PhOCH_2CH_2$—, $ClCH_2CH_2$— and $BrCH_2CH_2$— [35], $MeCH(Br)CH_2$—, Cl_3CCH_2— [13, 14], $MeOC(=O)CH_2$—, $ClC(=O)CH_2$—, $H_2C=CHCH_2$— [12, 15], $PhCH_2$— [16], $NCCH_2$— [17, 18], $^+H_3NCH_2$—, $HOCH_2$— [19–21], $MeOCH_2$— [22], $EtOCH_2$—, $PhOCH_2$—, 4-$BrPhOCH_2$—, $HSCH_2$— [23], $MeSCH_2$— [24], FCH_2— [25], $ClCH_2$— [26], $BrCH_2$— [1, 3, 26–28], iPr— [4, 29, 30], cPr— [33], Ox— [32, 33], Et(Me)CH— [29], PhCH(Me)—,

$BrCH_2CH(OH)$—, $(MeO)_2CH$— and $(CD_3O)_2CH$— [43], $(EtO)_2CH$—, MeCH(Br)— [34, 37, 38], EtCH(Br)— [37], NCCH(Br)—, EtOC(=O)CH(Br)—, $BrCH_2C(Me)_2$— [39], $(BrCH_2)_3$— [40], $(MeO)_2C(Me)$—, $BrCH_2CHBr$— [36], Br_2HC— [41, 42], tBu— [31], F_3C— [44–46], Br_3C—.

Table 3.6 Absorption regions (cm^{-1}) of $—CH_2Br$

attached to	Ph	C=C	C≡C	C=O	N, O, S	Si, Hg	Cl, Br
ν_aCH_2	3015 ± 20	3000 ± 20	3010 ± 15	3010 ± 20	3035 ± 30	3000 ± 15	≈3060
ν_sCH_2	2970 ± 20	2955 ± 15	2965 ± 15	2960 ± 20	2975 ± 20	2945 ± 10	≈2990
δCH_2	1435 ± 15	1430 ± 15	1425 ± 15	1410 ± 25	1415 ± 25	1380 ± 10	≈1400
ωCH_2	1225 ± 15	1220 ± 20	1215 ± 15	1250 ± 45	1240 ± 40	1120 ± 15	≈1210
τCH_2	1190 ± 25	1160 ± 30	1150 ± 10	1170 ± 40	1145 ± 45	1060 ± 10	≈1130
ρCH_2	880 ± 20	870 ± 35	865 ± 15	860 ± 40	870 ± 80	730 ± 40	≈830
νCBr	620 ± 20	620 ± 70	630 ± 20	635 ± 60	620 ± 60	590 ± 40	≈620
δCBr	320 ± 70	260 ± 70	340 ± 60	210 ± 60	220 ± 80	–	≈175
torsion	–	≈100	–	–	–	–	–

$R—CH_2Br$ molecules

R = (substituted)phenyl (see Section 12.4)

R = $H_2C{=}CH$— [49–52], $H_2C{=}CMe$— [53, 54], $H_2C{=}CBr$— [55–57], $H_2C{=}C(COOEt)$—, $H_2C{=}C(CH_2Br)$— [58], $BrCH_2CH{=}CH$—, HC≡C— [59–62], DC≡C— [62, 63], BrC≡C— [61], $BrCH_2C{\equiv}C$— [64, 90], N≡C— [65, 68], MeC(=O)— [69, 91], $BrCH_2C(=O)$— [70], $F_3CC(=O)$— [71, 72], $H_2NC(=O)$— [48], 2-HOPhC(=O)— [74], HOC(=O)— and DOC(=O)— [75], NaOC(=O)— [73], MeOC(=O)—, EtOC(=O)— [76, 77], tBuOC(=O)—, HSC(=O)— [78], FC(=O)— [80], ClC(=O)— [48, 79, 80], BrC(=O)— [48, 79–81], O_2N— [82], MeO— [83, 89], $MeS(=O)_2$— [84], $FS(=O)_2$— and $ClS(=O)_2$— [85], Me_2HSi— and Me_2DSi— [86], $BrCH_2Hg$— [87], Cl— [47, 48], Br— [1, 48, 88].

References

1. S. Mizushima, T. Shimanouchi, I. Nakagawa and A. Miyake, *J. Chem. Phys.*, **21**, 215 (1953).
2. R. Gaufrès and M. Bejaud-Bianchi, *Spectrochim. Acta, Part A*, **27A**, 2249 (1971).
3. F.F. Bentley, N.T. McDevitt and A.L. Rozek, *Spectrochim. Acta*, **20**, 105 (1964).
4. N.T. McDevitt, A.L. Rozek, F.F. Bentley and A.D. Davidson, *J. Chem. Phys.*, **42**, 1173 (1965).
5. J.K. Brown and N. Sheppard, *Trans. Faraday Soc.*, **50**, 1164 (1954).
6. C. Komaki, I. Ichishima, K. Kutanani, T. Miyazawa, T. Shimanouchi and S. Mizushima, *Bull. Chem. Soc. Jpn.*, **25**, 330 (1955).
7. K. Radcliffe and J.L. Wood, *Trans. Faraday Soc.*, **62**, 1678 (1966).

8. M. Hayashi, K. Ohno and H. Murata, *Bull. Chem. Soc. Jpn.*, **46**, 2332 (1973).
9. K. Tanabe and S. Saëki, *J. Mol. Struct.*, **27**, 79 (1975).
10. G.A. Crowder and M.-H. Jalilian, *Can. J. Spectrosc.*, **22**, 1 (1977).
11. Y. Ogawa, S. Imazeki, H. Yamaguchi, H. Matsuura, I. Harada and T. Shimanouchi, *Bull. Chem. Soc. Jpn.*, **51**, 748 (1978).
12. G.A. Crowder, *J. Mol. Struct.*, **10**, 290 (1971).
13. A. Goursot-Leray, M. Carles-Lorjou, G. Pouzard and H. Bodot, *Spectrochim. Acta, Part A*, **29A**, 1497 (1973).
14. M. Carles-Lorjou, A. Goursot-Leray, H. Bodot and R. Gaufrès, *Spectrochim. Acta, Part A*, **29A**, 329 (1973).
15. G.A. Crowder and N. Smyrl, *J. Mol. Struct.*, **8**, 255 (1971).
16. J.E. Saunders, J.J. Lucier and J.N. Willis Jr. *Spectrochim. Acta, Part A*, **24A**, 2027 (1968).
17. P. Klaboe and J. Grundnes, *Spectrochim. Acta, Part A*, **24A**, 1905 (1968).
18. T. Fujiyama, *Bull. Chem. Soc. Jpn.*, **44**, 3317 (1971).
19. E. Wyn-Jones and W.J. Orville-Thomas, *J. Mol. Struct.*, **1**, 79 (1967).
20. L. Homanen, *Spectrochim. Acta, Part A*, **39A**, 77 (1983).
21. P. Buckley, P.A. Giquère and M. Schneider, *Can. J. Chem.*, **47**, 901 (1969).
22. H. Matsuura, M. Kono, H. Iizuka, Y. Ogawa, I. Harada and T. Shimanouchi, *Bull. Chem. Soc. Jpn.*, **50**, 2272 (1977).
23. M. Hayashi, Y. Shiro, M. Murakami and H. Murata, *Bull. Chem. Soc. Jpn.*, **38**, 1740 (1965).
24. H. Matsuura, N. Miyauchi, H. Murata and M. Sakakibara, *Bull. Chem. Soc. Jpn.*, **52**, 344 (1979).
25. J.R. Durig, J. Liu and T.S. Little, *J. Mol. Struct.*, **248**, 25 (1991).
26. K. Tanabe, *Spectrochim. Acta, Part A*, **28A**, 407 (1972).
27. K. Tanabe, *Spectrochim. Acta, Part A*, **30A**, 1901 (1974).
28. K. Tanabe, J. Hiraishi and T. Tamura, *J. Mol. Struct.*, **33**, 19 (1976).
29. G.A. Crowder and M.-R. Jalilian, *Spectrochim. Acta, Part A*, **34A**, 707 (1978).
30. J.R. Durig, J.F. Sullivan and S.E. Godbey, *J. Mol. Struct.*, **146**, 213 (1986).
31. G.A. Crowder, C. Harper and M.R. Jalilian, *J. Mol. Struct.*, **49**, 403 (1978).
32. R.A. Nyquist, C.L. Putzig and N.E. Skelly, *Appl. Spectrosc.*, **40**, 821 (1986).
33. C.J. Wurrey, R. Krishnamoorti, S. Pechsiri and V.F. Kalasinski, *J. Raman Spectrosc.*, **12**, 95 (1982).
34. J. Thorbjørnsrud, H.O. Ellestad, P. Klaboe and T. Torgrimsen, *J. Mol. Struct.*, **15**, 45 (1973).
35. J. Thorbjørnsrud, H.O. Ellestad, P. Klaboe and T. Torgrimsen, *J. Mol. Struct.*, **15**, 61 (1973).
36. J. Thorbjørnsrud, H.O. Ellestad, P. Klaboe and T. Torgrimsen, *J. Mol. Struct.*, **17**, 5 (1973).
37. G.A. Crowder, *J. Mol. Struct.*, **100**, 415 (1983).
38. J. Som and G.S. Kastha, *Indian J. Phys.*, **47**, 494 (1973).
39. A. Gatial, P. Klaeboe, C.J. Nielsen and D.L. Powell, *Croat. Chem. Acta*, **61**, 375 (1988).
40. P. Klaeboe, B. Klewe, K. Martinsen, C.J. Nielsen, D.L. Powell and D.J. Stubbles, *J. Mol. Struct.*, **140**, 1 (1986).
41. T. Torgrimsen and P. Klaeboe, *Acta Chem. Scand.*, **24**, 1145 (1970).
42. S. Suzuki and G. Vergoten, *Spectrochim. Acta, Part A*, **37A**, 37 (1981).
43. J.E. Katon and Ph.D. Miller, *Appl. Spectrosc.*, **29**, 501 (1975).
44. J.R. Nielsen and R. Theimer, *J. Chem. Phys.*, **27**, 891 (1957).
45. W.F. Edgell, T.R. Riethof and C. Ward, *J. Mol. Spectrosc.*, **11**, 92 (1963).
46. G.A. Crowder, *J. Fluorine Chem.*, **3**, 125 (1973/74).

47. S. Giorgianni, R. Visinoni, A. Baldacci, A. Gambi and S. Ghersetti, *Spectrochim. Acta, Part A*, **44A**, 463 (1988).
48. R.G. Jones and W.J. Orville-Thomas, *Spectrochim. Acta*, **20**, 291 (1964).
49. R.D. McLachlan and R.A. Nyquist, *Spectrochim. Acta, Part A*, **24A**, 103 (1968).
50. C. Sourisseau and P. Pasquier, *J. Mol. Struct.*, **12**, 1 (1972).
51. J.R. Durig and M.R. Jalilian, *J. Phys. Chem.*, **84**, 3543 (1980).
52. J.R. Durig, Q. Tang and T.S. Little, *J. Mol. Struct.*, **269**, 257 (1992).
53. A.O. Diallo, *Spectrochim. Acta, Part A*, **36A**, 799 (1980).
54. A.O. Diallo, *Spectrochim. Acta, Part A*, **39A**, 327 (1983).
55. G.A. Crowder, *J. Mol. Spectrosc.*, **23**, 1 (1967).
56. T. Torgrimsen and P. Klaeboe, *J. Mol. Struct.*, **20**, 229 (1974).
57. S.H. Schei, *Spectrochim. Acta, Part A*, **39A**, 1043 (1983).
58. R. Gaufrès and C. Roulph, *J. Mol. Struct.*, **9**, 107 (1971).
59. E. Hirota and Y. Morino, *Bull. Chem. Soc. Jpn.*, **34**, 341 (1961).
60. J.C. Evans and R.A. Nyquist, *Spectrochim. Acta*, **19**, 1153 (1963).
61. R.A. Nyquist, A.L. Johnson and Y.S. Lo, *Spectrochim. Acta*, **21**, 77 (1965).
62. R.A. Nyquist, T.L. Reder, F.F. Stec and G.J. Kallos, *Spectrochim. Acta, Part A*, **27A**, 897 (1971).
63. R.A. Nyquist, *Spectrochim. Acta, Part A*, **27A**, 2513 (1971).
64. O.H. Ellestad and K. Kveseth, *J. Mol. Struct.*, **25**, 175 (1975).
65. R.G. Jones and W.J. Orville-Thomas, *J. Chem. Soc.*, 4623 (1965).
66. F. Watari and K. Aida, *Spectrochim. Acta, Part A*, **23A**, 2951 (1967).
67. J.R. Durig and D.W. Wertz, *Spectrochim. Acta, Part A*, **24A**, 21 (1968).
68. G.A. Crowder, *Mol. Phys.*, **23**, 707 (1972).
69. G.A. Crowder and B.R. Cook, *J. Chem. Phys.*, **47**, 367 (1967).
70. G.A. Crowder and N. Smyrl, *J. Mol. Struct.*, **7**, 478 (1971).
71. G.A. Crowder and P. Pruettiangkura, *J. Mol. Struct.*, **15**, 161 (1973).
72. J.R. Durig, T.G. Sheehan and J.A. Hardin, *J. Mol. Struct.*, **243**, 275 (1991).
73. J.E. Katon and R.L. Kleinlein, *J. Mol. Struct.*, **17**, 239 (1973).
74. W.A.L.K. Al-Rashid and M.F. Bermani, *Spectrochim. Acta, Part A*, **47A**, 35 (1991).
75. J.E. Katon and R.L. Kleinlein, *Spectrochim. Acta, Part A*, **29A**, 791 (1973).
76. M.A. Raso, M.V. Garcia and J. Morcillo, *J. Mol. Struct.*, **115**, 449 (1984).
77. N. Dube, P. Porwal and R. Prasad, *J. Raman Spectrosc.*, **19**, 189 (1988).
78. H.S. Randhawa and W. Walter, *Bull. Chem. Soc. Jpn.*, **51**, 1579 (1978).
79. I. Nakagawa, I. Ichisma, K. Kuratini, T. Miyazawa, T. Shimanouchi and S. Mizushima, *J. Chem. Phys.*, **20**, 1720 (1952).
80. J.R. Durig, H.V. Phan and T.S. Little, *J. Mol. Struct.*, **212**, 187 (1989).
81. K. Tanabe and S. Saëki, *Spectrochim. Acta, Part A*, **28A**, 1083 (1972).
82. P. Gluzińsky and Z. Eckstein, *Spectrochim. Acta, Part A*, **24A**, 1777 (1968).
83. D.C. McKean, I. Torto and A.R. Morrisson, *J. Mol. Struct.*, **99**, 101 (1983).
84. A.B. Remizov, F.S. Bilalov and I.S. Pominov, *Spectrochim. Acta, Part A*, **43A**, 309 (1987).
85. R. Aroca, J. Ali and E.A. Robinson, *J. Mol. Struct.*, **116**, 9 (1984).
86. K. Ohno, K. Suehiro and H. Murata, *J. Mol. Struct.*, **98**, 251 (1983).
87. Y. Imai and K. Aida, *Spectrochim. Acta, Part A*, **28A**, 517 (1972).
88. T.A. Ford, *J. Mol. Struct.*, **58**, 185 (1975).
89. B.J. Van der Veken, G.A. Guirgis, J. Liu and J.R. Durig, *J. Raman Spectrosc.*, **23**, 205 (1992).
90. A. Karlsson, G.D. Bauza, P. Klaboe, C.J. Nielsen and D. Sülzle, *J. Raman Spectrosc.*, **23**, 391 (1992).
91. J.R. Durig, J. Lin and H.V. Phan, *J. Raman Spectrosc.*, **23**, 253 (1992).

3.1.4 Iodomethyl

The weak to medium absorptions of the methylene stretchings, which often coincide with other CH stretching vibrations, are difficult to assign unambiguously. The methylene scissoring vibration absorbs at the LW side of aliphatic methylene deformations with a weak to moderate intensity. With an intensity varying from medium to strong, the methylene wag is often sensitive to rotational isomerism, yielding band separations up to 80 wavenumbers, although normally the differences are much smaller. The twist, which occur with weak intensity at the LW side of the wag, is likewise sensitive to conformational changes of the molecule. The methylene rock gives rise to weak or moderate absorptions and is sometimes difficult to detect among other rocks or aromatic out-of-plane deformations. The CI stretch often yields a conformation-sensitive band with a medium to strong intensity. A weak intensity normally indicates that the molecule is not inclined to stick to that conformation. The CI stretch and the methylene wag are the two most characteristic

Table 3.7 Absorption regions (cm^{-1}) of —CH_2I

	Total region	Molecules absorbing at high wavenumbers	Molecules absorbing at low wavenumbers	Region of remaining molecules
$\nu_a CH_2$	3015 ± 35	$ClCH_2I$ (3050) CH_2I_2 (3046) F_3CCH_2I (3044)	$H_2NC(=O)CH_2I$ (2980) $H_2C=CHCH_2I$ (2980)	3015 ± 25
$\nu_s CH_2$	2960 ± 25	F_3CCH_2I (2985) $HOC(=O)CH_2I$ (2985)	ICH_2HgCH_2I (2935) Me_2HSiCH_2I (2936)	2960 ± 20
δCH_2	1395 ± 45	*tr*-$H_2C(=O)CH_2I$ (1440) $MeCH_2I$ (1439) $H_2C=CHCH_2I$ (1437)	ICH_2HgCH_2I (1350) CH_2I_2 (1352) Me_2HSiCH_2I (1374)	1410 ± 25
ωCH_2	1165 ± 115	$H_2NC(=O)CH_2I$ (1280) $ICH_2C(=O)CH_2I$ (1268 and *1248*)	ICH_2HgCH_2I (1059 and 1050) Me_2HSiCH_2I (1080) CH_2I_2 (1106)	1195 ± 65
τCH_2	1100 ± 100	$cPrCH_2I$ (1197)	ICH_2HgCH_2I (1000) Me_2HSiCH_2I (1012) CH_2I_2 (1030)	1120 ± 70
ρCH_2	760 ± 100	$MeOCH_2I$ (860) $HOC(=O)CH_2I$ (850) $Cl(CH_2)_3I$ (849)	ICH_2HgCH_2I (660 and 672) CH_2I_2 (717) $H_2NC(=O)CH_2I$ (720)	785 ± 60
νCI	560 ± 100	CF_3CH_2I (660) $NaOC(=O)CH_2I$ (653) $HOC(=O)CH_2I$ (635)	$H_2NC(=O)CH_2I$ (465) CH_2I_2 (485) *g*-ICH_2CH_2I (485 and *507*)	555 ± 65
δCI	220 ± 100	*g*-ICH_2CH_2I (313)	CH_2I_2 (125) Me_2HSiCH_2I (131) *g*-ICH_2CH_2I (134)	220 ± 70
torsion	≈120			≈120

bands of all CH_2I normal modes. Not only the CI stretch but also the CH_2 wag, twist and rock absorb in a range which is about 50 cm^{-1} lower than for the comparable R—CH_2Br molecules. The —CI deformation, sometimes referred to as a skeletal deformation, is not significant as a diagnostic tool.

R—CH_2I molecules
R = Me— [1–4], CD_3— [3, 4], Et— [2, 5–8], nPr— [1, 6, 8], nBu— [40], $ClCH_2CH_2$—, ICH_2CH_2— [1, 6, 9], $I(CH_2)_3$— [41], HOC(=O)CH_2—, $PhCH_2$— [10], $HOCH_2$— [11, 12], $MeOCH_2$— [13], ICH_2— [1, 14, 15], iPr— [2], cPr— [16], Ox— [16, 17], F_3C— [18, 19], MeC(=O)— [20], ICH_2C(=O)— [21], H_2NC(=O)— [22–25], HOC(=O)— and NaOC(=O)— [26], EtOC(=O)—, H_2C=CH— [27, 38], H_2C=C(Me)— [28], HC≡C— [29], ICH_2C≡C— [39], N≡C— [30–32], MeO— [33], Me_2HSi— and Me_2DSi— [34], ICH_2Hg— [35], Cl— [36], I— [36, 37].

References

1. F.F. Bentley, N.T. McDevitt and A.L. Rozek, *Spectrochim. Acta*, **20**, 105 (1964).
2. N.T. McDevitt, A.L. Rozek, F.F. Bentley and A.D. Davidson, *J. Chem. Phys.*, **42**, 1173 (1965).
3. G.A. Crowder, *J. Mol. Spectrosc.*, **48**, 467 (1973).
4. J.R. Durig, J.W. Thompson, V.W. Thyagesan and J.D. Witt, *J. Mol. Struct.*, **24**, 41 (1975).
5. J.K. Brown and N. Sheppard, *Trans. Faraday Soc.*, **50**, 1164 (1954).
6. G.A. Crowder and S. Ali, *J. Mol. Struct.*, **25**, 377 (1975).
7. K. Tanabe and S. Saëki, *J. Mol. Struct.*, **27**, 79 (1975).
8. Y. Ogawa, S. Imazeki, H. Yamaguchi, H. Matsuura, I. Harada and T. Shimanouchi, *Bull. Chem. Soc. Jpn.*, **51**, 748 (1978).
9. J. Thorbjørnsrud, O.H. Ellestad, P. Klaeboe and T. Torgrimsen, *J. Mol. Struct.*, **15**, 61 (1973).
10. J.E. Saunders, J.J. Lucier and J.N. Willis Jr., *Spectrochim. Acta, Part A*, **24A**, 2027 (1968).
11. E. Wyn-Jones and W.J. Orville-Thomas, *J. Mol. Struct.*, **1**, 79 (1967).
12. L. Homanen, *Spectrochim. Acta, Part A*, **39A**, 77 (1983).
13. H. Matsuura, M. Kono, H. Iizuka, Y. Ogawa, I. Harada and T. Shimanouchi, *Bull. Chem. Soc. Jpn.*, **50**, 2272 (1977).
14. K. Tanabe, *Spectrochim. Acta, Part A*, **30A**, 1901 (1974).
15. K. Tanabe, J. Hiraishi and T. Tamura, *J. Mol. Struct.*, **33**, 19 (1976).
16. C.J. Wurrey, Y.Y. Yeh, R. Krishnamoorthi, R.J. Berry, J.E. Dewitt and V.F. Kalasinsky, *J. Phys. Chem.*, **88**, 4059 (1984).
17. R.A. Nyquist, C.L. Putzig and N.E. Skelly, *Appl. Spectrosc.*, **40**, 821 (1986).
18. W.F. Edgell, T.R. Riethof and C. Ward, *J. Mol. Spectrosc.*, **11**, 92 (1963).
19. G.A. Crowder, *J. Fluorine Chem.*, **3**, 125 (1973/74).
20. B.R. Cook and G.A. Crowder, *J. Chem. Phys.*, **47**, 1700 (1967).
21. G.A. Crowder and N. Smyrl, *J. Mol. Struct.*, **7**, 478 (1971).
22. F.M. Abid, J.M. Al-Katti and M.F. El-Bermani, *J. Mol. Struct.*, **67**, 169 (1980).
23. E.K. Murthy and G.R. Rao, *J. Raman Spectrosc.*, **19**, 359 (1988).

24. E.K. Murthy and G.R. Rao, *J. Raman Spectrosc.*, **19**, 419 (1988).
25. E.K. Murthy and G.R. Rao, *J. Raman Spectrosc.*, **20**, 409 (1989).
26. J.E. Katon and T.P. Carll, *J. Mol. Struct.*, **7**, 391 (1971).
27. R.D. McLachlan and R.A. Nyquist, *Spectrochim. Acta, Part A*, **24A**, 103 (1968).
28. F. Northam, J. Oliver and G.A. Crowder, *J. Mol. Spectrosc.*, **25**, 436 (1968).
29. J.C. Evans and R.A. Nyquist, *Spectrochim. Acta, Part A*, **19**, 1153 (1963).
30. R.G. Jones and J. Orville-Thomas, *J. Chem. Soc.*, 4632 (1965).
31. J.R. Durig and D.W. Wertz, *Spectrochim. Acta*, **24**, 21 (1968).
32. G.A. Crowder, *Mol. Phys.*, **23**, 707 (1972).
33. D.C. McKean, I. Torto and A.R. Morrisson, *J. Mol. Struct.*, **99**, 101 (1983).
34. K. Ohno, K. Suehiro and H. Murata, *J. Mol. Struct.*, **98**, 251 (1983).
35. Y. Imai and K. Aida, *Spectrochim. Acta, Part A*, **28A**, 517 (1972).
36. R.G. Jones and W.J. Orville-Thomas, *Spectrochim. Acta*, **20**, 291 (1964).
37. T.A. Ford, *J. Mol. Spectrosc.*, **58**, 185 (1975).
38. J.R. Durig, Q. Tang and T.S. Little, *J. Raman Spectrosc.*, **23**, 653 (1992).
39. A. Karlsson, G.D. Bauza, P. Klaboe, C.J. Nielsen and D. Sülzle, *J. Raman Spectrosc.*, **23**, 391 (1992).
40. G.A. Crowder and M. Jalilian, *J. Mol. Struct.*, **33**, 127 (1976).
41. G.A. Crowder and S. Ali, *J. Mol. Struct.*, **27**, 43 (1975).

3.2 OXYMETHYL

3.2.1 Hydroxymethyl

Ethanol is an example of a simple compound containing a $—CH_2OH$ fragment in the structure. The 21 normal vibrations differentiate between 13a′ plus 8 a″ vibrational modes in the C_s conformation. After subtraction of 8 methyl vibrations (5a′ + 3a″) and a CC stretching vibration (a′), 7a′ + 5a″ normal vibrations remain, so that they can be assigned to the modes of vibration of CH_2OH:

a′: νOH, $\nu_s CH_2$, δCH_2, ωCH_2, δOH, νCO and δ-CO;
a″: $\nu_a CH_2$, τCH_2, ρCH_2, γOH and torsion.

The OH stretching vibration

In the associated state the OH stretching vibration absorbs as a strong, broad band in the region 3300 ± 120 cm^{-1}. It is the most characteristic absorption of all CH_2OH normal vibrations. This broad band, caused by intermolecular hydrogen bridges, disappears in dilute solutions or in the gaseous state, and returns as the undisturbed OH stretching vibration round about 3620 cm^{-1} in the form of a sharp peak. Compounds such as $EtOCH_2CH_2OH$ (3420 cm^{-1}), $NCCH_2CH_2OH$ (3415 cm^{-1}), $MeOCH_2CH_2OH$ (3410 cm^{-1}) and $HC(=O)CH_2OH$ (3410 cm^{-1}) absorb at the HW side of the above-mentioned region. At the LW side one finds the OH absorptions of polyols and amino-alcohols insofar as they are not sterically hindered. Nevertheless most of the primary alcohols absorb in the region 3370 ± 50 cm^{-1}.

Methylene stretching vibrations

The $\nu_a CH_2$ vibration is found in the region 2945 ± 45 cm^{-1} for such compounds as, for instance, Cl_2CHCH_2OH and F_2CHCH_2OH (2988 cm^{-1}), and for FCH_2CH_2OH (2967 cm^{-1}) at the HW side (the $\nu_a CH_2F$ absorbs in the neighbourhood of 2995 cm^{-1}) and $MeC(=O)CH_2OH$ (2900 cm^{-1}, coincident with ν_sMe), $H_2C=C(Me)CH_2OH$ and 2-$ClPhCH_2OH$ (2918 cm^{-1}) at the LW side of the above-mentioned region. The remaining studied compounds absorb at 2940 cm^{-1}, a range situated distinctly lower than that of $\nu_a CH_2X$ (X = F, Cl, Br, I and CN), with a band intensity that fluctuates between weak and medium.

The $\nu_s CH_2$ is active in the range 2885 ± 45 cm^{-1} with, for instance, $MeCH_2OH$ (2928 cm^{-1}) and FCH_2CH_2OH (2924 cm^{-1}) at the HW side and such compounds as $MeNHCH_2CH_2OH$ and H_2NNHCH_2OH at 2840 cm^{-1} at the LW side. Generally one finds the $\nu_s CH_2$ in primary alcohols in the region 2880 ± 20 cm^{-1} with a band intensity which varies from weak to medium. That is about 80 cm^{-1} lower than the $\nu_s CH_2X$ (X = F, Cl, Br or I), which appears in the neighbourhood of 2960 cm^{-1}.

Methylene deformation

The δCH_2 vibration absorbs in the region 1445 ± 35 cm^{-1} with a band intensity which varies between weak and medium. Compounds such as $HC\equiv CCH_2CH_2OH$ (1474 and the second deformation at 1426 cm^{-1}), $iPrCH(NH_2)CH_2OH$ and $iPrCH_2CH(NH_2)CH_2OH$ (1468 cm^{-1} together with the Me deformation) absorb in the HW region. The deformations of 2-$ClPhCH_2OH$ (1410 cm^{-1}) and $MeC(=O)CH_2OH$ (1416 cm^{-1}, together with δ_a and δ'_aMe) represent the lower values of this region. Most of the primary alcohols show this deformation at 1445 ± 20 cm^{-1}.

OH in-plane deformation

The associated OH in—plane deformation is assigned in the region 1400 ± 40 cm^{-1}. This absorption possesses a weak to moderate intensity and arises mostly as a broad band on which other deformations, such as methylene wagging and twisting vibrations, are superimposed. In addition, this vibration couples easily with the wag and twist and therefore it is difficult to decide which vibration plays the biggest part in this absorption, so that not infrequently the methylene wag is assigned in the above-mentioned range. The δOH is sensitive to conformation and shifts to the lower wavenumbers (1200 cm^{-1}) in dilute solutions or for the gaseous form. Compounds such as $EtCH=CHCH_2CH_2OH$ and $H_2C=C(Me)CH_2CH_2OH$ produce this δOH at 1440 cm^{-1} but the value for ICH_2CH_2OH is 1370 cm^{-1}. A measurement in solution or a deuteration of OH gives a supplementary security for this assignment.

Methylene wagging vibration

This wag is not always easily detected in the region 1335 ± 55 cm^{-1}, which partly coincides with the absorption regions of OH deformation and methylene twist, the normal vibrations with which the wag readily is coupled. The band intensity varies between weak and medium. In addition, this vibration can be sensitive to the conformational state so that every factor is present to prevent an unambiguous assignment. Neglecting the assignments at low wavenumber in the spectra of $HC(=O)CH_2OH$ and $MeC(=O)CH_2OH$ (about 1280 cm^{-1}), these waggings are located in the more favourable region of 1365 ± 25 cm^{-1}.

Methylene twisting vibration

Problems similar to those with the CH_2 wag and OH in-plane deformation arise with the assignment of the CH_2 twist, which is weakly to moderately active in the region 1240 ± 60 cm^{-1}. Molecules such as $D_2C{=}CHCH_2OH$ (1190 cm^{-1}), $HC{\equiv}CCH_2CH_2OH$ (1189 cm^{-1}) and $EtC{\equiv}CCH_2CH_2OH$ (1186 cm^{-1}) absorb at low wavenumbers. The remaining compounds give this twist in the region 1240 ± 40 cm^{-1}, and this band is mostly sensitive to conformational changes. In the unbonded state this τCH_2 absorbs at 30 cm^{-1} lower wavenumber.

The CO stretching vibration

The CO stretching vibration absorbs moderately to strongly in the range 1045 ± 45 cm^{-1} and is, as for the OH stretch, a highly characteristic absorption for the primary alcohols. The CO stretch of the molecules X_3CCH_2OH (X = F, Cl and Br) (1090 cm^{-1}) and $HOCH_2CH_2OH$ (1087 and 1043 cm^{-1} for the other νCO) absorbs on the HW side of this region. 2-Unsaturated primary alcohols absorb on the LW side, that is 1025 ± 25 cm^{-1}, with a few representatives: $H_2C{=}CHCH_2OH$ (1028 cm^{-1}), $PhCH_2OH$ (1016 cm^{-1}), 2-$FuCH_2OH$ and 2-$ThCH_2OH$ (1010 cm^{-1}). The remaining molecules give this absorption in the region 1055 ± 25 cm^{-1}, so that primary alcohols are easily characterized by this mode of vibration.

Methylene rocking vibration

The methylene rocking vibration gives rise to absorptions in the region 890 ± 90 cm^{-1} with band intensities varying from weak to moderate. On the HW side $MeCH(NO_2)CH_2OH$ (980 cm^{-1}) and O_2NCH_2OH, $O_2NCH_2CH_2OH$ and Cl_2CHCH_2OH (976 cm^{-1}) absorb. On the LW side one finds the rock of $MeCH_2OH$ (804 cm^{-1}), $HC(=O)CH_2OH$ (813 cm^{-1}), $MeC(=O)CH_2OH$ (813 cm^{-1}) and $PhCH_2OH$ (815 cm^{-1}). Most of the compounds produce this rock at 895 ± 65 cm^{-1}, so that this absorption does not score well as a group vibration.

OH out-of-plane deformation

The γOH absorption is active in the region 640 $\pm$ 70 cm^{-1}. The band intensity varies from weak to medium, rarely strong. Usually this band is very broad and easy to recognize. Sometimes the exact wavenumber is difficult to locate because other bands are superimposed on it. Just as νOH and δOH, the γOH is sensitive to conformation and dilution. $NCCH_2CH_2OH$ displays this vibration at 579 cm^{-1}, MeC(=O)CH_2OH and F_2CHCH_2OH in the neighbourhood of 575 cm^{-1} and $HOCH_2CBr(NO_2)CH_2OH$ at 708 cm^{-1}, but most of the investigated molecules absorb in the region 635 $\pm$ 35 cm^{-1}. In the unbonded state this γOH disappears to make way for the free torsion in the vicinity of 300 cm^{-1} [24].

CO in-plane deformation

The δCO with a range of 475 $\pm$ 80 cm^{-1} is not a model of a good group vibration. The intensity of this band is weak or moderate. This broad region contains bands caused by the following compounds: HC≡CCH_2OH (551 cm^{-1}), $H_2NNHCH_2CH_2OH$ (546 cm^{-1}), $MeOCH_2CH_2OH$ (539 cm^{-1}), $MeNHCH_2CH_2OH$ (535 cm^{-1}) and H_2C=C(Me)CH_2OH (534 cm^{-1}). The outsiders on the LW side are HC(=O)CH_2OH (397 cm^{-1}) and MeC(=O)CH_2OH (409 cm^{-1}). For the remaining molecules the CO in-plane deformation is concentrated in the region 465 $\pm$ 55 cm^{-1}.

Table 3.8 Absorption regions (cm^{-1}) of the normal vibrations of associated —CH_2OH

Vibration	C_s	Region	Vibration	C_s	Region
νOH	a′	3300 $\pm$ 120	τCH_2	a″	1240 $\pm$ 60
$\nu_a CH_2$	a″	2945 $\pm$ 45	νCO	a′	1045 $\pm$ 45
$\nu_s CH_2$	a′	2885 $\pm$ 45	ρCH_2	a″	890 $\pm$ 90
δCH_2	a′	1445 $\pm$ 35	γOH	a″	640 $\pm$ 70
δOH	a′	1400 $\pm$ 40	δ—C—O	a′	475 $\pm$ 80
ωCH_2	a′	1335 $\pm$ 55	torsion	a″	–

The following R—CH_2OH compounds have been taken into account:

R = R′CH_2— (see Section 3.5.4)
R = Me— [1–6], iPr—, cPr— [7, 8], Ox—, EtCH(Me)—, MeCH(NH_2)—, EtCH(NH_2)—, iPrCH(NH_2)—, iPrCH_2CH(NH_2)—, MeCH(NO_2)—, MeCH(OH)—, EtCH(OH)—, tBu—, $HOCH_2CBr(NO_2)$—, $HOCH_2CBr(NO_2)$—, $Me_2C(NH_2)$—, Ph_2CH— [25], F_2CH— [9], Cl_2CH— [9], EtC($CH_2OH)_2$—, F_3C— [3, 11–13], Cl_3C— [11, 12],

Br_3C— [11], HC(=O)— [14, 15], MeC(=O)—, HOC(=O)— [10], MeOC(=O)— and CD_3OC(=O)— [10], H_2C=CH— and D_2C=CH— [16], H_2C=C(Me)—, *cis* and *trans* nPrCH=CH—, HC≡C— [17, 18], O_2N—, Ph— [23], 4-EtPh—, 3-$HOCH_2Ph$—, 2-, 3- and 4-F_3CPh—, 4-MeOPh— [19], 4-EtOPh—, 4-NCPh— [23], 2- and 4-FPh— [20, 21], 2-ClPh—, 3-IPh—, 2-Fu— [22], 2-Th—, 2-HC(=O)Fu—5-, 2-$HOCH_2$Fu-5—.

References

1. Y. Mikawa, J.W. Brasch and R.J. Jakobsen, *Spectrochim. Acta, Part A*, **27A**, 529 (1971).
2. J.R. Durig, W.E. Bucy, C.J. Wurrey and L.A. Carreira, *J. Phys. Chem.*, **79**, 988 (1975).
3. O. Schrems and W.A.P. Luck, *J. Mol. Struct.*, **80**, 477 (1982).
4. H.F. Hameka, *J. Mol. Struct.*, **226**, 241 (1991).
5. R.A. Shaw, H. Wieser, R. Dutler and A. Rauk, *J. Am. Chem. Soc.*, **112**, 5401 (1990).
6. A.J. Barnes and H.E. Hallam, *Trans. Faraday Soc.*, **66**, 1932 (1970).
7. P. Klaboe and D.L. Powell, *J. Mol. Struct.*, **20**, 95 (1974).
8. H.M. Badawi, M.E. Abu-Zeid and Y.A. Yousef, *J. Mol. Struct.*, **240**, 225 (1990).
9. M. Perttilä, *Spectrochim. Acta, Part A*, **35A**, 37 (1979).
10. H. Hollenstein, R.W. Schär, N. Schwizgebel, G. Grassi and Hs. Günthard, *Spectrochim. Acta, Part A*, **39A**, 193 (1983).
11. J. Travert and J.C. Lavaley, *Spectrochim. Acta, Party A*, **32A**, 637 (1976).
12. M. Perttilä, *Spectrochim. Acta, Part A*, **35A**, 585 (1979).
13. J.R. Durig and R.A. Larsen, *J. Mol. Struct.*, **238**, 195 (1990).
14. H. Michelsen and P. Klaboe, *J. Mol. Struct.*, **4**, 293 (1969).
15. Y. Kobayashi, H. Takahara, H. Takahashi and H. Higasi, *J. Mol. Struct.*, **32**, 235 (1976).
16. B. Silvi and J.P. Perchard, *Spectrochim. Acta, Part A*, **32A**, 11 (1976).
17. R.A. Nyquist, *Spectrochim. Acta, Part A*, **27A**, 2513 (1971).
18. J. Travert, J.C. Lavalley and D. Chenery, *Spectrochim. Acta, Part A*, **35A**, 291 (1979).
19. S. Chakravorti, A.K. Sarkar, P.K. Mallick and S.B. Banerjee, *Indian J. Phys., B*, **56B**, 96 (1982).
20. S. Tariq, N. Ali and P.K. Verma, *Indian J. Pure Appl. Phys.*, **21**, 220 (1983).
21. S. Tariq and P.K. Verma, *Indian J. Phys., B*, **57B**, 356 (1983).
22. L. Strandman-Long and J. Murto, *Spectrochim. Acta, Part A*, **37A**, 643 (1981).
23. G. Varsányi, *Assignments for Vibrational Spectra of Seven Hundred Benzene Derivatives*, J.Wiley & Sons, New York (1974).
24. N.M.D. Brown, B.J. Meenan and G.M. Taggart, *Spectrochim. Acta, Part A*, **48A**, 939 (1992).
25. S. Chakravorti, R. De, P.K. Mallick and S.B. Banerjee, *Spectrochim. Acta, Part A*, **49A**, 543 (1993).

3.2.2 Methoxymethyl

Compounds such as XCH_2OMe (X = F, Cl, Br and I) belong to the point group C_s in which the 21 normal vibrations are divided among 13a′ and 8a″ species of vibration. When the CX stretching vibration (a′) has been replaced by a torsion (a″), the normal vibrations of —CH_2OMe are divided among 12a′ and 9a″ vibration types, which can be described as follows:

a′: ν_a'Me, ν_sCH$_2$, ν_sMe, δ_a'Me, δCH$_2$, δ_sMe, ωCH$_2$, ρ'Me, ν_aCOC, ν_sCOC and two skeletal deformations,;
a″: ν_aMe, ν_aCH$_2$, δ_aMe, τCH$_2$, ρMe, ρCH$_2$ and three torsions.

Of these, eight are ascribed to vibrations of methyl and six to vibrations of methylene, while the other seven belong to the COC antisymmetric and symmetric stretchings, skeletal deformations and torsions.

Methyl and methylene stretching vibrations

As a matter of course the absorption regions of ν_aMe (3000 $\pm$ 25), ν_a'Me (2960 $\pm$ 30) and ν_sMe (2830 $\pm$ 10 cm^{-1}) strongly overlap those of ν_aCH$_2$ (2945 $\pm$ 40) and ν_sCH$_2$ (2900 $\pm$ 60 cm^{-1}). The sharp separate peak with medium intensity of ν_sMe is a useful identification mark for methyl ethers. In general, the CH stretching frequencies of —CH$_2$OMe are ordered as follows:

$$\nu_a\text{Me} \geq \nu_a'\text{Me} \geq \nu_a\text{CH}_2 > \nu_s\text{CH}_2 \geq \nu_s\text{Me}$$

The high values for the methylene stretching vibrations within —XCH$_2$OMe (X = F, Cl, Br, I) compounds have not been taken into account here (see Section 3.1).

Methyl and methylene deformations

It is not easy to distinguish clearly the methyl and methylene deformations, which appear with medium intensity in the range 1455 $\pm$ 30 cm^{-1}, with the exception of δCH$_2$ (1496 cm^{-1}) in the spectrum of FCH$_2$OMe. Typical for —CH$_2$OMe is the high value for δ_sMe (1445 $\pm$ 15 cm^{-1}) in comparison with δ_sMe of the saturated hydrocarbons (1375 $\pm$ 15 cm^{-1}).

Methylene wagging and twisting vibrations

The methylene wag and twist mostly absorb in the regions 1350 $\pm$ 55 and 1270 $\pm$ 40 cm^{-1} respectively, with band intensities being weak to moderate. The low values for ICH$_2$OMe and BrCH$_2$OMe, in which the influence of the halogen clearly appears, have not been taken into account (see Section 3.1).

Methyl rocking vibrations

The methyl rockings are active in the ranges 1195 $\pm$ 35 and 1160 $\pm$ 30 cm^{-1} with an intensity that may vary from weak to strong. The areas overlap each other but in most cases the two rocks do not coincide at one and the same wavenumber.

COC stretching vibrations

The most characteristic normal vibration of —CH_2OMe is undoubtedly the COC antisymmetric stretching vibration, considered as νCH_2—O, which appears strongly in the range 1110 $\pm$ 70 cm^{-1}. FCH_2OMe absorbs strongly at 1175 cm^{-1} and $MeOCH_2OMe$ at 1042 cm^{-1} (the other antisymmetric stretch appears at 1111 cm^{-1}). If these values are left unconsidered, the above-mentioned range will narrow to 1120 $\pm$ 40 cm^{-1}. Vasícková *et al.* [26] find for both ν_aCOC 1150 $\pm$ 10 and 1050 $\pm$ 10 cm^{-1} in the spectra of 55 examined compounds with the protecting group $MeOCH_2O$—. The antisymmetric stretch provides a good group frequency, which is further distinguished by the great intensity with which the vibration presents itself. This does not preclude proceeding with caution, considering that the CF and S=O stretching vibrations also absorb strongly in this area.

The COC symmetrical stretch, corresponding with νMe—O, is located in the range 930 $\pm$ 40 cm^{-1}, with an intensity that might be moderate to strong but is significantly less than the intensity of the antisymmetrical stretch. High values, in the neighbourhood of 963 cm^{-1}, are revealed in the spectra of $ClCH_2OMe$ and $HOCH_2CH_2OMe$. Many compounds show this COC symmetric stretch in the range 930 $\pm$ 30 cm^{-1}, for example $MeO(CH_2)_nOMe$ (n = 3–5) [10], so that the ν_sCOC does not cut too poor a figure as a group vibration.

Methylene rocking vibration

The ρCH_2 in the spectra of RCH_2OMe compounds manifests itself in the broad region 905 $\pm$ 90 cm^{-1}, with moderate intensity. High values (995 cm^{-1}) are found in the spectra of $MeOCH_2Cl$ and $MeOCH_2F$ and the lowest value in the spectrum of $MeCH_2OMe$ (815 cm^{-1}), but the ρCH_2 of many RCH_2OMe compounds is found in the region 900 $\pm$ 50 cm^{-1}. Sometimes a difference of opinion occurs in assigning this ρCH_2. In the spectra of $HC{\equiv}CCH_2OMe$ (1006/895), $N{\equiv}CCH_2OMe$ (1016/888) and $MeCH_2OMe$ (1014/815 cm^{-1}) the highest wavenumber is assigned to the ρCH_2 and the lowest to νCC instead of the contrary.

Skeletal deformations

The two skeletal deformations are weakly to moderately active in the regions 480 $\pm$ 100 and 365 $\pm$ 100 cm^{-1}. The former represents the C—O—C deformation and the latter an external deformation.

R—CH_2OMe compounds

R = Me— [1–7], CD_3— [5], $HOCH_2$—, $MeOCH_2$— [7–9], $MeO(CH_2)_n$— (n = 2–4) [10], $MeC(=O)OCH_2$—, $MeOCH_2CH_2OCH_2$— [8], $MeSCH_2$— [9], $ClCH_2$—, $BrCH_2$— and ICH_2— [11], CF_3— [12], MeC(=O)—, MeOC(=O)—, HC≡C— [13–15], DC≡C— [13], N≡C— [15–17], MeO— [18–21], MeS— [22, 23], EtS— [22], F— [24, 28], Cl— [16, 20, 24, 25], Br— [24, 27], I— [24].

Table 3.9 Absorption regions (cm^{-1}) of the normal vibrations of —CH_2OMe

Vibration	Region	Vibration	Region
ν_aMe	3000 ± 25	ρMe	1195 ± 35
ν'_aMe	2960 ± 30	ρ'Me	1160 ± 30
$\nu_a CH_2$	2945 ± 40	ν_aCOC	1110 ± 70
$\nu_s CH_2$	2900 ± 60	ν_sCOC	930 ± 40
ν_sMe	2830 ± 10	ρCH_2	905 ± 90
δ_aMe	1465 ± 20	skeletal def.	480 ± 100
δ'_aMe	1455 ± 20	skeletal def.	365 ± 100
δCH_2	1450 ± 15	torsion	225 ± 40
δ_sMe	1445 ± 15	torsion	145 ± 35
ωCH_2	1350 ± 55	torsion	–
τCH_2	1270 ± 40		

References

1. N. Sheppard, *J. Chem. Phys.*, **17**, 79 (1949).
2. A.D.H. Clague and A. Danti, *Spectrochim. Acta, Part A*, **24A**, 439 (1968).
3. J.P. Perchard, *Spectrochim. Acta, Part A*, **26A**, 707 (1970).
4. J.P. Perchard, *J. Mol. Struct.*, **6**, 457 (1970).
5. T. Kitagawa, K. Ohno, H. Sugeta and T. Miyazawa, *Bull. Chem. Soc. Jpn.*, **45**, 969 (1972).
6. N.L. Allinger, M. Rahman and J.-H. Lii, *J. Am. Chem. Soc.*, **112**, 8293 (1990).
7. R.G. Snyder and G. Zerbi, *Spectrochim. Acta, Part A*, **23A**, 391 (1967).
8. K. Machida and T. Miyazawa, *Spectrochim. Acta*, **20**, 1868 (1964).
9. Y. Ogawa, M. Ohta, M. Sakakibara, H. Matsuura, I. Harada and T. Shimanouchi, *Bull. Chem. Soc. Jpn.*, **50**, 650 (1977).
10. H. Matsuura and H. Murata, *J. Raman Spectrosc.*, **12**, 144 (1982).
11. H. Matsuura, M. Kono, H. Iizuka, Y. Ogawa, I. Harada and T. Shimanouchi, *Bull. Chem. Soc. Jpn.*, **50**, 2272 (1977).
12. Y.-S. Li, F.O. Cox and J.R. Durig, *J. Phys. Chem.*, **91**, 1334 (1987).
13. A. Bjørseth and J. Gustavsen, *J. Mol. Struct.*, **23**, 301 (1974).
14. W.A. Seth Paul, J.P. Tollenaere, H. Meeusen and F. Höfler, *Spectrochim. Acta, Part A*, **30A**, 193 (1974).
15. S.W. Charles, F.C. Cullen, G.I.L. Jones and N.L. Owen, *J. Chem. Soc., Faraday Trans. 2*, **70**, 758 (1974).
16. R.G. Jones and W.J. Orville-Thomas, *J. Chem. Soc.*, 692 (1964).
17. R.G. Jones and W.J. Orville-Thomas, *Spectrochim. Acta*, **20**, 291 (1964).
18. K. Nukada, *Spectrochim. Acta*, **18**, 745 (1962).
19. J.E. Katon and P.D. Miller, *Appl. Spectrosc.*, **29**, 501 (1975).
20. H.R. Linton and E.R. Nixon, *Spectrochim. Acta*, **15**, 146 (1959).
21. M. Sakakibara, Y. Yonemura, H. Matsuura and H. Murata, *J. Mol. Struct.*, **66**, 333 (1980).
22. H. Matsuura, H. Murata and M. Sakakibara, *J. Mol. Struct.*, **96**, 267 (1983).
23. H. Matsuura, K. Kimura and H. Murata, *J. Mol. Struct.*, **64**, 281 (1980).
24. D.C. McKean, I. Torto and A.R. Morrisson, *J. Mol. Struct.*, **99**, 101 (1983).
25. H.F. Hameka, *J. Mol. Struct.*, **226**, 241 (1991).

26. S. Vasícková, V. Pouzar, I. Cerný, P. Drasar and M. Havel, *Coll. Czech. Chem. Commun.*, **51**, 90 (1986).
27. B.J. Van der Veken, G.A. Guirgis, J. Liu and J.R. Durig, *J. Raman Spectrosc.*, **23**, 205 (1992).
28. J.R. Durig, J. Liu, G.A. Guirgis and B.J. Van der Veken, *Struct. Chem.*, **4**, 103 (1993).

3.3 SULFUR-BONDED METHYLENE

3.3.1 Mercaptomethyl

Methanethiol [1–3, 6, 11] has 12 fundamental vibrations which, according to C_s symmetry, are distributed among 8a′ and 4a″ vibration types. In addition to the eight vibrations of methyl and a torsion one finds:

	νSH (a′)	δSH (a′)	νCS (a′)
gas (monomer)	2605	802	710
solid (dimer)	2543	802	703

A slight increase of the absorption frequency of the SH stretching vibration in the gas phase proves that thiols form hydrogen bridges, although considerably less strongly than alcohols. Ethanethiol, in the more stable *trans* conformation, possesses a plane of symmetry and therefore belongs to the point group C_s. The 21 normal vibrations are divided into 13a′ and 8a″ species of vibration. Leaving aside the eight vibrations of methyl (5a′ + 3a″) and a CC stretching vibration (a′), 7a′ and 5a″ normal vibrations remain for the —CH_2SH fragment:

a′: $\nu_s CH_2$, νSH, δCH_2, ωCH_2, δSH, νCS and δCS;
a″: $\nu_a CH_2$, τCH_2, ρCH_2 and 2 torsions.

Methylene stretching vibrations

The $\nu_a CH_2$ occurs moderately to strongly in the region 2960 ± 25 cm^{-1} with, for example $EtCH_2SH$ (2950 cm^{-1}) and also $MeCH_2SH$ and $HSCH_2CH_2SH$, which both absorb at 2970 cm^{-1}.

The $\nu_s CH_2$ vibration is moderately to strongly active in the region 2900 ± 45 cm^{-1} with $MeCH_2SH$ (2940 cm^{-1}) at the HW side; the lower limit is not unambiguously determined by compounds such as $MeNHC(=O)CH_2CH_2SH$ (2855 cm^{-1}) and $MeCH_2SH$ (2872 cm^{-1}).

These ranges are more or less 15 cm^{-1} higher than those of the methylene stretchings of CH_2OH but considerably lower than those of CH_2X (X = halogen).

The SH stretching vibration

With a range of 2570 ± 30 cm^{-1} the νSH is the most characteristic band in the spectra of thiols [34, 35]. $MeCH_2SH$ and $H_2C{=}CHCH_2SH$, measured in the gas phase, form the upper level with 2600 cm^{-1}, and they absorb in the vicinity of 2570 cm^{-1} as a liquid. As lowest value, 2545 cm^{-1} is noted in the spectra of $MeC({=}O)NHCH_2CH_2SH$, $MeNHC({=}O)CH_2SH$ and $MeSCH_2SH$. In most of the compounds this SH stretch is found in the narrow range 2565 ± 10 cm^{-1}. The intensity is mostly weak and is very weak even with butanethiol. Only for dithiols with four or fewer carbon atoms can the intensity be called moderate to strong.

Methylene deformation, wagging and twisting vibrations

The vibrations are active in the ranges:

δCH_2	ωCH_2	τCH_2
1435 ± 25	1260 ± 45	1200 ± 60 cm^{-1}

mostly as moderate to strong bands, although the twist might also turn out rather weak. The lowest value for the wag (1220 cm^{-1}) is being found in the spectra of *trans*-$HSCH_2CH_2SH$ and $MeSCH_2SH$ but most of the compounds show this wag in the range 1270 ± 30 cm^{-1}. In the spectra of $MeCH_2SH$ and $EtCH_2SH$ the twist occurs around 1250 cm^{-1}, but absorbs approximately 100 cm^{-1} lower in the spectra of $ClCH_2CH_2SH$, $BrCH_2CH_2SH$, $HSCH_2CH_2SH$, $PhCH_2SH$ and $MeSCH_2SH$.

SH deformation

The SH deformation is weakly to moderately active in the region 840 ± 55 cm^{-1} and is sensitive to conformational changes in the molecule. For $MeCH_2SH$ one finds this vibration at 870 cm^{-1} and for $HSCH_2CH_2SH$ at 890 and 800 cm^{-1}. On the other hand, *gauche*-$EtCH_2SH$, $H_2C{=}CHCH_2SH$ and $PhCH_2SH$ are responsible for the lower values around 790 cm^{-1}.

Methylene rocking vibration

The methylene rock in the $-CH_2SH$ fragment appears at relatively low wavenumbers (750 ± 35 cm^{-1}) in comparison with the rock in $-CH_2Hal$, $-CH_2O-$, $-CH_2N-$ and $-CH_2C-$ compounds, so that an assignment is possible even when other methylene groups are present. The second methylene rock assigned in the spectrum of $HSCH_2CH_2SH$ (*gauche*: 972; *trans*: 942 cm^{-1}) falls outside this area.

CS stretching vibration

The νCS is sensitive to conformation and mostly gets assigned as a moderate to strong band in the region 675 ± 45 cm^{-1}. High values (720 cm^{-1}) are reached in the spectra of *trans*-$HSCH_2CH_2SH$ [27–29] and *trans*-$EtCH_2SH$ [15], although for the first {last} compound also 693 and 655 [25] or 680 and 660 [26] {$\approx$705 cm^{-1} [12–14, 16–19]} would be. The lowest value (635 cm^{-1}) is found in the spectrum of MeNHC(=O)CH_2SH, so that the νCS concentrates itself at 670 ± 35 cm^{-1}.

CS deformation

The δ-C—S is active in the unfriendly range of 330 ± 90 cm^{-1} and, moreover, mostly at weak intensity. *Gauche*-$EtCH_2SH$ is responsible for the high value (417 cm^{-1}) whereas *trans*-$BrCH_2CH_2SH$, with 245 cm^{-1}, scores low. For most of the molecules this skeletal deformation is situated at 350 ± 50 cm^{-1} but also for this vibration *trans* and *gauche* conformers do not produce the same values.

Table 3.10 Absorption regions (cm^{-1}) of the normal vibrations of $—CH_2SH$

Vibration	C_s	Region	Vibration	C_s	Region
ν_aCH_2	a″	2960 ± 25	δSH	a′	840 ± 55
ν_sCH_2	a′	2900 ± 45	ρCH_2	a″	730 ± 35
νSH	a′	2570 ± 30	νCS	a′	675 ± 45
δCH_2	a′	1435 ± 25	δCS	a′	330 ± 90
ωCH_2	a′	1260 ± 45	torsion	a″	200 ± 50
τCH_2	a″	1200 ± 60	torsion	a″	130 ± 45

R—CH_2SH compounds

R = Me— [4–12], Et— [11–19], $HSCH_2CH_2$—, HOC(=O)CH_2— [22], MeOC(=O)CH_2—, MeNHC(=O)CH_2— and MeC(=O)$NHCH_2$— [23], $HOCH_2$—, $HSCH_2$— [24–29], $ClCH_2$— and $BrCH_2$— [20], iPr— [21], HOC(=O)— [22], MeNHC(=O)— [23], MeOC(=O)—, EtOC(=O)—, ^-O_2C—, H_2C=CH— [33], Ph— [30, 31], 2-Fu— [36], MeS— [32].

References

1. H. Siebert, *Z. Anorg. Allg. Chem.*, **271**, 65 (1952).
2. I.W. May and E.L. Pace, *Spectrochim. Acta, Part A*, **24A**, 1605 (1968).
3. O. Saur, J. Travert, J.-C. Lavalley and M. Chabanel, *C.R. Acad. Sci., Ser. C*, **291C**, 227 (1980).
4. D.W. Scott, H.L. Finke, J.P. McCullough, M.E. Gross, K.D. Williamson, G. Waddington and H.M. Huffman, *J. Am. Chem. Soc.*, **73**, 261 (1951).
5. D.W. Scott and J.P. McCullough, *J. Am. Chem. Soc.*, **80**, 3554 (1958).
6. A.J. Barnes, H.E. Hallam and J.D.R. Howells, *J. Chem. Soc.*, Faraday Trans II **68**, 737 (1972).
7. W.O. George, J.H.S. Green and D.J. Harrison, *Spectrochim. Acta, Part A*, **24A**, 367 (1968).

8. D. Smith, J.P. Devlin and D.W. Scott, *J. Mol. Spectrosc.*, **25**, 174 (1968).
9. J.R. Durig, W.E. Bucy, C.J. Wurrey and L.A. Carreira, *J. Phys. Chem.*, **79**, 988 (1975).
10. H. Wolff and J. Szydlowski, *Can. J. Chem.*, **63**, 1708 (1985).
11. R. Fausto, J.J.C. Teixeira-Dias and P.R. Carey, *J. Mol. Struct.*, **159**, 137 (1987).
12. D.W. Scott and M.Z. El-Sabban, *J. Mol. Spectrosc.*, **30**, 317 (1969).
13. K.G. Allum, J.A. Creighton, J.H.S. Green, G.J. Minkoff and L.J.S. Prince, *Spectrochim. Acta, Part A*, **24A**, 928 (1968).
14. T.H. Joo, K. Kim and M.S. Kim, *J. Phys. Chem.*, **90**, 5817 (1986).
15. R.E. Pennington, D.W. Scott, H.L. Finke, J.P. McCullough, J.F. Messerly, I.A. Hossenlopp and G. Waddington, *J. Am. Chem. Soc.*, **78**, 3266 (1956).
16. T. Torgrimsen and P. Klaboe, *Acta Chem. Scand.*, **24**, 1139 (1970).
17. M. Hayashi, Y. Shiro and H. Murata, *Bull. Chem. Soc. Jpn.*, **39**, 112 (1966).
18. G. Radinger and H. Wittek, *Z. Phys. Chem., Abt. B*, **45B**, 329 (1940).
19. I.F. Trotter and H.W. Thompson, *J. Chem. Soc.*, 481 (1946).
20. M. Hayashi, Y. Shiro, M. Murakami and H. Murata, *Bull. Chem. Soc. Jpn.*, **38**, 1740 (1965).
21. D.W. Scott, J.P. McCullough, J.F. Messerly, R.E. Pennington, I.A. Hossenlopp, H.L. Finke and G. Waddington, *J. Am. Chem. Soc.*, **80**, 55 (1958).
22. N. Saraswathi and S. Soundararajan, *J. Mol. Struct.*, **4**, 419 (1969).
23. G. Zuppiroli, C. Perchard, M.H. Baron and C. de Loze, *J. Mol. Struct.*, **69**, 1 (1980).
24. S.K. Nandy, D.K. Mukherjee, S.B. Roy and G.S. Kastha, *Indian J. Phys.*, **47**, 528 (1973).
25. D. Welti and D. Whittaker, *J. Chem. Soc.*, 4372 (1962).
26. M. Ikram and D.B. Powell, *Spectrochim. Acta, Part A*, **28A**, 59 (1972).
27. M. Ohsaku and H. Murata, *J. Mol. Struct.*, **52**, 143 (1979).
28. M. Hayashi, Y. Shiro, T. Oshima and H. Murata, *Bull. Chem. Soc. Jpn.*, **38**, 1734 (1965).
29. C.K. Kwon, K. Kim, M.S. Kim and Y.S. Lee, *J. Mol. Struct.*, **197**, 171 (1989).
30. K. Doerffel and B. Adler, *Wiss. Z. Tech. Hochsch. Chem. Leuna-Merseburg*, **10**, 7 (1968).
31. P.K. Mallick, S. Chattopadhyay and S.B. Bannerjee, *Indian J. Pure Appl. Phys.*, **11**, 609 (1973).
32. M. Ohsaku, Y. Shiro and H. Murata, *Bull. Chem. Soc. Jpn.*, **45**, 3035 (1972).
33. C.S. Hsu, *Spectrosc. Lett.*, **7**, 439 (1974).
34. N. Mori, S. Kaido, K. Suzuki, M. Nakamura and Y. Tsuzuki, *Bull. Chem. Soc. Jpn.*, **44**, 1858 (1971).
35. G.F. Bolshakov, *Izv. Vyssh. Uchebn. Zaved. Neft Gaz*, **23**, 39 (1980).
36. M. Sénéchal and P. Saumagne, *J. Chim. Phys.*, **69**, 1246 (1972).

3.3.2 Methylthiomethyl

Compounds such as XCH_2SMe (X = halogen) show a plane of symmetry in the *trans* conformation and therefore belong to the point group C_s by which the 21 normal vibrations are being divided among 13a′ and 8a″ species of vibration. Replacing the CX stretching vibration (a′) by a torsion (a″) leads to 12a′ and 9a″ vibrations for the CH_2SMe fragment:

a′: ν_a'Me, ν_sMe, ν_sCH_2, δ_a'Me, δCH_2, δ_sMe, ωCH_2, ρ'Me, ν_aCSC, ν_sCSC and two skeletal deformations;

a″: ν_aMe, ν_aCH_2, δ_aMe, τCH_2, ρMe, ρCH_2 and three torsions.

Methyl and methylene stretching vibrations

The five CH stretching vibrations are active between 3000 and 2850 cm^{-1} and it is not simple to assign the individual vibrations. Although many analysts omit further specifications, a classification of these CH stretching vibrations generally is as follows:

$$\nu_a\text{Me} \geq \nu'_a\text{Me} \geq \nu_a\text{CH}_2 > \nu_s\text{Me} \geq \nu_s\text{CH}_2$$

in which the methyl symmetric stretch appears most strongly and the other stretching vibrations exhibit moderate intensity.

Methyl and methylene deformations

Although the areas of the two methyl antisymmetric deformations partly overlap, often two bands appear in the ranges 1440 ± 15 and 1425 ± 15 cm^{-1}. The methylene scissors is weakly to moderately active at 1405 ± 30 cm^{-1}, which is 50 wavenumbers lower than the range of the CH_2 scissors in RCH_2OMe compounds. The methyl symmetric deformation (1320 ± 10 cm^{-1}) absorbs 125 cm^{-1} lower than the corresponding vibration with RCH_2OMe molecules and can be counted among the good group vibrations. Unfortunately the intensity is mostly inadequate.

Methylene wagging and twisting vibrations

The methylene wag (1250 ± 55) and twist (1200 ± 80 cm^{-1}) absorb respectively 100 and 70 wavenumbers lower than with methoxymethyl compounds. The absorption regions are delimited by the following compounds: $MeOCH_2SMe$ (ω: 1305; τ: 1280), $EtOCH_2SMe$ (ω: 1302; τ: 1273), $MeSCH_2SCH_2SMe$ (ω: 1195 and *1218*; τ: 1132 and *1165*), $MeSCH_2CH_2SMe$ (ω: *1285* and *1268*; τ: 1120 and *1195*).

Methyl and methylene rocking vibrations

The methyl rocks get assigned at 1000 ± 35 (a′) and 940 ± 30 cm^{-1} (a″). The highest value of the ρMe is located in the spectrum of $MeSCH_2CH_2SMe$ (1035 cm^{-1}) and the lowest value of the ρ'Me in that of $nPrCH_2SMe$ (916 cm^{-1}). Most of the analysed compounds show these methyl rocks in the regions 995 ± 30 and 950 ± 20 cm^{-1}. With its range of 815 ± 75 cm^{-1} and mostly a weak band intensity, the methylene rock is not an example of a good group vibration and is moreover sensitive to conformation. Not considering the values for $MeOCH_2SMe$ (888 cm^{-1}), $EtOCH_2SMe$ (877 cm^{-1}) and $MeSCH_2CH_2SMe$ (740 cm^{-1}), this vibration occurs in the area 815 ± 50 cm^{-1}.

CSC stretching vibrations

It is not always straightforward to differentiate between these a′ stretching vibrations. The two areas overlap each other and the different conformers do not show the same wavenumber. Nevertheless the CSC antisymmetric stretch (725 ± 50 cm^{-1}) is associated with νS—Me and the symmetric stretch (680 ± 45 cm^{-1}) rather with the νCH_2—S. The intensities are weak to moderate, rarely strong.

Skeletal deformations and torsions

The two skeletal deformations, of which the first is an essentially external vibration and the second the CH_2—S—Me deformation, are also sensitive to conformation and mostly result in a weak band in the regions 370 ± 50 and 250 ± 40 cm^{-1}. In all probability the torsions τMe, τSMe and τCH_2SMe will be noted at 190 ± 30, 145 ± 35 and 75 ± 30 cm^{-1}.

Table 3.11 Absorption regions (cm^{-1}) of the normal vibrations of —CH_2SMe

Vibration	Region	Vibration	Region
ν_aMe	2990 ± 10	ρMe	1000 ± 35
ν'_aMe	2970 ± 10	ρ'Me	940 ± 30
$\nu_a CH_2$	2960 ± 15	ρCH_2	815 ± 75
ν_sMe	2920 ± 10	ν_aCSC	725 ± 50
$\nu_s CH_2$	2885 ± 30	ν_sCSC	680 ± 45
δ_aMe	1440 ± 15	skeletal def.	370 ± 50
δ'_aMe	1425 ± 15	skeletal def.	250 ± 40
δCH_2	1405 ± 30	torsion	190 ± 30
δ_sMe	1320 ± 10	torsion	145 ± 35
ωCH_2	1250 ± 55	torsion	75 ± 30
τCH_2	1200 ± 80		

R—CH_2SMe compounds

R = Me— [1–10], D_3C— [9], nPr— [11], MeOCH_2— [13], MeSCH_2— [13–15], MeS$(CH_2)_n$— (n = 2–4) [16], ClCH_2— and BrCH_2— [12], iPr— [25], N≡C— [19], Ph— [17, 18], MeO— [20, 21], EtO— [21], HS— and DS— [22], MeS— [15, 23, 24], MeSCH_2S— and MeSCH_2SCH_2S— [23, 24].

References

1. M. Hayashi, T. Shimanouchi and S. Mizushima, *J. Chem. Phys.*, **26**, 608 (1957).
2. M. Ohsaku, Y. Shiro and H. Murata, *Bull. Chem. Soc. Jpn.*, **45**, 954 (1972).
3. M. Ohsaku, Y. Shiro and H. Murata, *Bull. Chem. Soc. Jpn.*, **46**, 1399 (1973).

4. D.W. Scott and J.P. McCullough, *J. Am. Chem. Soc.*, **80**, 3554 (1958).
5. D.W. Scott and M.Z. El-Sabban, *J. Mol. Spectrosc.*, **30**, 317 (1969).
6. D.W. Scott, H.L. Finke, J.P. McCullough, M.E. Gross, K.D. Williamson, G. Waddington and H.M. Huffman, *J. Am. Chem. Soc.*, **73**, 261 (1951).
7. W.O. George, J.H.S. Green and D.J. Harrison, *Spectrochim. Acta, Part A*, **24A**, 367 (1968).
8. R. Fausto, J.J.C. Teixeira-Dias and P.R. Carey, *J. Mol. Struct.*, **159**, 137 (1987).
9. M. Sakakibara, H. Matsuura, I. Harada and T. Shimanouchi, *Bull. Chem. Soc. Jpn.*, **50**, 111 (1977).
10. J.R. Durig, M.S. Rollins and H.V. Phan, *J. Mol. Struct.*, **263**, 95 (1991).
11. M. Ohta, Y. Ogawa, H. Matsuura, I. Harada and T. Shimanouchi, *Bull. Chem. Soc. Jpn.*, **50**, 380 (1977).
12. H. Matsuura, N. Miyauchi, H. Murata and M. Sakakibara, *Bull. Chem. Soc. Jpn.*, **52**, 344 (1979).
13. Y. Ogawa, M. Ohta, M. Sakakibara, H. Matsuura, I. Harada and T. Shimanouchi, *Bull. Chem. Soc. Jpn.*, **50**, 650 (1977).
14. M. Hayashi, Y. Shiro, T. Oshima and H. Murata, *Bull. Chem. Soc. Jpn.*, **39**, 118 (1966).
15. D. Welti and D. Whittaker, *J. Chem. Soc.*, 4372 (1962).
16. H. Matsuura, J. Matsumoto and H. Murata, *Spectrochim. Acta, Part A*, **36A**, 291 (1980).
17. K. Doerffel and B. Adler, *Wiss. Z. Tech.Hochsch. Chem. Leuna-Merseburg*, **10**, 7 (1968).
18. K.G. Allum, J.A. Creighton, J.H.S. Green, G.J. Minkoff and L.J.S. Prince, *Spectrochim. Acta, Part A*, **24A**, 928 (1968).
19. S.W. Charles, F.C. Cullen and N.L. Owen, *J. Mol. Struct.*, **34**, 219 (1976).
20. H. Matsuura, K. Kimura and H. Murata, *J. Mol. Struct.*, **64**, 281 (1980).
21. H. Matsuura, H. Murata and M. Sakakibara, *J. Mol. Struct.*, **96**, 267 (1983).
22. M. Ohsaku, Y. Shiro and H. Murata, *Bull. Chem. Soc. Jpn.*, **45**, 3035 (1972).
23. M. Ohsaku, *Bull. Chem. Soc. Jpn.*, **47**, 965 (1974).
24. M. Ohsaku, Y. Shiro and H. Murata, *Bull. Chem. Soc. Jpn.*, **45**, 113 (1972).
25. N. Nogami, H. Sugeta and T. Miyazawa, *Bull. Chem. Soc. Jpn.*, **48**, 2417 (1975).

3.3.3. Thiocyanatomethyl

Ethyl thiocyanate possesses 24 normal vibrations which are divided for the C_s conformation into 15a′ and 9a″ species of vibration. After subtraction of eight methyl vibrations (5a′ + 3a″) and a CC stretching vibration (a′) the remaining normal vibrations (9a′ + 6a″) have to be assigned to —CH_2SCN:

a′: $\nu_s CH_2$, $\nu C{\equiv}N$, δCH_2, ωCH_2, $\nu_a CSC$, $\nu_s CSC$, δSCN, δ—CS, δCSC;
a″: $\nu_a CH_2$, τCH_2, ρCH_2, γSCN, γ—CS and torsion.

Methylene vibrations

In the ranges of $\nu_a CH_2$ (2969 ± 20) and $\nu_s CH_2$ (2930 ± 10 cm^{-1}) the high values (3035 and 2970) of $ClCH_2SCN$ have not been listed (see Section 3.1.2).

The regions of the methylene deformations may be compared with those of other sulfur-bonded methylene compounds (see Sections 3.3.1 and 3.3.2). If the low value (1157 cm^{-1}) in the spectrum of $ClCH_2SCN$ is left aside, the methylene twist restricts itself to 1220 ± 25 cm^{-1}. The high value (868 cm^{-1}) for the ρCH_2 in the same spectrum also has not been taken into account.

The C≡N stretching vibration

The C≡N stretching vibration produces the most characteristic band in the spectra of thiocyanates. This remarkable group vibration appears at 2155 ± 10 cm^{-1}, sharp and intense, in constrast to that of isothiocyanates, which is also active in this neighbourhood but with a typical broad structure (see Section 3.4.4).

CSC stretching vibrations and skeletal deformations

The antisymmetric CSC stretch, which is associated with the S—CN stretching vibration, appears weakly to moderately at 685 ± 15 cm^{-1}, and the symmetric CSC stretch, corresponding with the CH_2—S stretching vibration, absorbs moderately to strongly in the range 635 ± 25 cm^{-1}.

The CH_2SCN group provides five skeletal deformations and a torsion, in which two R—C—S (R=C) skeletal deformations must be considered as external deformations. These skeletal deformations absorb weakly to moderately and, in the absence of a plane of symmetry, the difference between in-plane and out-of-plane disappears. The torsion in Me—CH_2SCN is found around 230 cm^{-1}; the torsions of larger molecules are expected to be much lower.

Table 3.12 Absorption regions (cm^{-1}) of the normal vibrations of —CH_2SCN

Vibration	C_s	Region	Vibration	C_s	Region
$\nu_a CH_2$	a″	2960 ± 20	$\nu_s CSC$	a′	635 ± 25
$\nu_s CH_2$	a′	2930 ± 10	δS—CN	a′	465 ± 15
νC≡N	a′	2155 ± 10	γS—CN	a″	410 ± 10
δCH_2	a′	1420 ± 20	—C—S skelet.def.	a′	320 ± 20
ωCH_2	a′	1255 ± 25	—C—S skelet.def.	a″	290 ± 20
τCH_2	a″	1200 ± 45	CSC skelet.def.	a′	150 ± 20
ρCH_2	a″	780 ± 20	torsion	a″	–
$\nu_a CSC$	a′	685 ± 15			

R—CH_2SCN molecules

R = Me— [1–5], Et— [1], nPr— [1], Ph— [6], HC≡C— and DC≡C— [9], Cl— [7, 8].

References

1. R.P. Hirschmann, R.N. Kniseley and V.A. Fassel, *Spectrochim. Acta*, **20**, 809 (1964).
2. G.A. Crowder, *J. Mol. Struct.*, **7**, 147 (1971).

3. O.H. Ellestad and T. Torgrimsen, *J. Mol. Struct.*, **12**, 79 (1972).
4. J.R. Durig, J.F. Sullivan and H.L. Heusel, *J. Phys. Chem.*, **88**, 374 (1984).
5. G.O. Braathen and A. Gatial, *Spectrochim. Acta, Part A*, **42A**, 615 (1986).
6. C.A. Sjøgren, *Acta Chem. Scand., Ser. A*, **38A**, 657 (1984).
7. G.A. Crowder, *J. Chem. Phys.*, **47**, 3080 (1967).
8. G.A. Crowder, *Tex. J. Sci.*, **26**, 565 (1975).
9. T. Midtgaard, G. Gundersen and C.J. Nielsen, *J. Mol. Struct.*, **176**, 159 (1988).

3.4 NITROGEN-BONDED METHYLENE

3.4.1 Aminomethyl

Molecules such as $MeCH_2NH_2$ possess a plane of symmetry in the *trans* conformation and belong to the point group C_s. The 24 normal vibrations must be divided among 14a′ and 10a″ types. When the eight vibrations (5a′ + 3a″) of methyl and a νCC (a′) are left out of consideration, the other 15 (8a′ + 7a″) can be assigned to the $—CH_2NH_2$ fragment:

a′: $\nu_s NH_2$, $\nu_s CH_2$, δNH_2, δCH_2, ωCH_2, νCN, ωNH_2, skeletal deformation;
a″: $\nu_a NH_2$, $\nu_a CH_2$, τCH_2, $\tau/\rho NH_2$, ρCH_2 and two torsions.

Amine stretching vibrations

The amine stretching vibrations cause two weak to moderate absorptions with approximately equal intensity. Although the NH_2 group is less subject to the formation of hydrogen bridges than the OH group, this doublet appears much sharper and at 25 cm^{-1} higher wavenumbers in dilute solutions and at 50 cm^{-1} higher wavenumbers for a gas. In the associated state $\nu_a NH_2$ occurs at 3365 ± 20 cm^{-1}. $F_3CCH_2NH_2$ absorbs at 3385 cm^{-1}, n-alkanamines at 3365 ± 5 cm^{-1} [31] and $FCH_2CH_2NH_2$ at 3345 cm^{-1}. The region 3285 ± 20 cm^{-1} is characteristic for $\nu_s NH_2$ with $PhCH_2NH_2$ at 3300 cm^{-1}, n-alkanamines at 3280 ± 5 cm^{-1} and FCH_2CH_2 at 3267 cm^{-1}. At the LW side of this band one often finds a weaker absorption at 3180 ± 20 cm^{-1}. This might be an overtone of the NH_2 scissors intensified by Fermi resonance interaction with $\nu_s NH_2$ and therefore is sometimes also assigned to this symmetric stretching [25].

Methylene stretching vibrations

With the exception of $F_3CCH_2NH_2$ (ν_a at 2987 and ν_s at 2960 cm^{-1}) the methylene antisymmetric stretching is found at 2930 ± 15 cm^{-1} and the symmetric stretching at 2870 ± 20 cm^{-1}.

Amine deformation

In the associated state the NH_2 scissoring vibration appears as a moderate to strong band in the region 1610 ± 15 cm^{-1}, a vibration which is found at 1623 cm^{-1} in the spectrum of methanamine [24]. This band is considerably broader than the sharp aromatic ring stretching modes in the neighbourhood of 1600 cm^{-1} and moves 25 cm^{-1} lower in solution. The $—CH_2NH_2$ fragment, combined with an unsaturated or aromatic part of a molecule, shows this NH_2 scissors at the lower side of the region at approximately 1600 cm^{-1}, as for instance $HC{\equiv}CCH_2NH_2$ (1597), $H_2C{=}CHCH_2NH_2$ (1604), $PhCH_2NH_2$ (1603) and 2-$FuCH_2NH_2$ (1603 cm^{-1}).

Methylene deformations

The area 1450 ± 20 cm^{-1} for the methylene deformation is determined by $H_2NCH_2CH_2NH_2$ (1469 and *1456* cm^{-1}) and $HC{\equiv}CCH_2NH_2$ (1435 cm^{-1}). Most of the $—CH_2NH_2$ compounds absorb in the region 1450 ± 10 cm^{-1} with moderate intensity.

The methylene wag absorbs with a weak to moderate intensity in the region 1360 ± 25 cm^{-1}, as defined by $F_3CCH_2NH_2$ (1384) and $PhOCH_2CH_2NH_2$ (1336 cm^{-1}).

The weak to moderate band at 1290 ± 45 cm^{-1} is mostly assigned to the methylene twist but also partly to the amine twist [3, 5, 6, 13]. The highest values are found in the spectra of RCH_2NH_2 compounds in which R is an aromatic or unsaturated fragment such as 2-$FuCH_2NH_2$ (1335), $HC{\equiv}CCH_2NH_2$ (1330), or $H_2C{=}CHCH_2NH_2$ (1328 cm^{-1}) and the lowest in that of 1—butanamine with 1245 cm^{-1}, a value which may be assigned to the amine twisting vibration as well.

Amine twisting vibration

The amine twist or rock, which is more or less connected with the CN stretch and the methylene twist, appears in the spectra of $—CH_2NH_2$ compounds in the region 1215 ± 70 cm^{-1} with a weak intensity and is not significant for identification purposes. $H_2C{=}CHCH_2NH_2$ shows this twist at 1280 cm^{-1}, $EtCH_2NH_2$ at 1275 cm^{-1} but $H_2NCH_2CH_2NH_2$ at1254 cm^{-1} [11] or at 1182 and 1148 cm^{-1} [7].

CN stretching vibration

The CN stretch, somewhat connected with the amine twist, appears as a moderate to strong band in the region 1075 ± 25 cm^{-1}, which is for the most part determined by $H_2NCH_2CH_2NH_2$ with 1096 and 1054 cm^{-1}.

Methylene rocking vibration

The methylene rock absorbs at 890 ± 55 cm^{-1} but is difficult to observe when superimposed on the wide NH_2 wag band. In contrast, high values such as 943 cm^{-1} in the spectrum of $HC{\equiv}CCH_2NH_2$ are easily assigned.

Amine wagging vibration

In the associated state, primary amines display a typical diffuse strong band in the region 845 ± 50 cm^{-1}, due to the NH_2 wagging mode. Diamines with short chains such as $H_2NCH_2CH_2NH_2$ (894 and *821*), $H_2N(CH_2)_3NH_2$ (862) and $F_3CCH_2NH_2$ (880 cm^{-1}) are responsible for the high values. The remaining primary amines produce this wagging mode at 830 ± 35 cm^{-1}; the ones with long chains are responsible for the lowest values, for example $Me(CH_2)_nNH_2$ ($n > 8$): 800 ± 5 cm^{-1}. In solution or for a gas, the hydrogen bridges disappear and the amine wag absorbs sharply at 770 ± 30 cm^{-1}. In the solid state at low temperature, however, the wag moves to higher wavenumbers (920 ± 40 cm^{-1}).

Skeletal deformation and torsions

The —C—N skeletal deformation is generally weakly active at 390 ± 75 cm^{-1}, with a high value for $H_2NCH_2CH_2NH_2$ (464 and *330*) and a low value for $H_2C{=}CDCH_2NH_2$ (315 cm^{-1}). The $—NH_2$ torsion is found at 255 ± 40 cm^{-1} and the $—CH_2NH_2$ torsion may be expected lower than 200 cm^{-1}.

Table 3.13 Absorption regions (cm^{-1}) of the normal vibrations of $—CH_2NH_2$

Vibration	C_s	Region	Vibration	C_s	Region
ν_aNH_2	a″	3365 ± 20	$\tau/\rho NH_2$	a″	1215 ± 70
ν_sNH_2	a′	3285 ± 20	νCN	a′	1075 ± 25
ν_aCH_2	a″	2930 ± 15	ρCH_2	a″	890 ± 55
ν_sCH_2	a′	2870 ± 20	ωNH_2	a′	845 ± 50
δNH_2	a′	1610 ± 15	—C—N def.	a′	390 ± 75
δCH_2	a′	1450 ± 20	torsion	a″	255 ± 40
ωCH_2	a′	1360 ± 25	torsion	a″	<200
τCH_2	a″	1290 ± 45			

The following $R—CH_2NH_2$ compounds have been taken into account:

R = Me— [1, 26], Et— [1–4], Pr— [1, 5, 6], $iPrCH_2—$, $tBuCH_2—$, $PhCH_2—$ [23], $H_2NCH_2CH_2—$ $H_2NCH_2—$ [7–11], $PhOCH_2—$ [23], 4-$MeOPhOCH_2—$ [23], $^{-}O_3SCH_2—$ [12], $FCH_2—$ [13], $ClCH_2—$ [14], $BrCH_2—$ [15], iPr—, $Ph_2CH—$ [30], tBu—, $F_3C—$ [16], $H_2C{=}CH—$ and $H_2C{=}CD—$ [17–19, 27, 28], $HC{\equiv}C—$ [20, 21], Ph— [22, 23], 2-Fu— [29].

References

1. H. Wolff and H. Ludwig, *Ber. Bunsenges. Phys. Chem.*, **71**, 1107 (1967).
2. N. Sato, Y. Hamada and M. Tsuboi, *Spectrochim. Acta, Part A*, **43A**, 943 (1987).
3. L.A.E. Batista De Carvalho, A.M. Amorim Da Costa, M.L. Duarte and J.J.C. Teixeira-Dias, *Spectrochim. Acta, Part A*, **44A**, 723 (1988).
4. J.R. Durig, W.B. Beshir, S.E. Godbey and T.J. Hizer, *J. Raman Spectrosc.*, **20**, 311 (1989).
5. J.J.C. Teixeira-Dias, L.A.E. Batista De Carvalho, A.M. Amorim Da Costa, I.M.S.Lampreia and E.F.G.Barbosa, *Spectrochim. Acta, Part A*, **42A**, 589 (1986).
6. G. Ramis and G. Busca, *J. Mol. Struct.*, **193**, 93 (1989).
7. A.-L. Borring and K. Rasmussen, *Spectrochim. Acta, Part A*, **31A**, 889 (1975).
8. A. Sabatini and S. Califano, *Spectrochim. Acta*, **16**, 677 (1960).
9. Y. Omura and T. Shimanouchi, *J. Mol. Spectrosc.*, **57**, 480 (1975).
10. L. Segal and F.V. Eggerton, *Appl. Spectrosc.*, **15**, 112 (1961).
11. M. Giorgini, M.R. Pellet, G. Paliani and R.S. Cataliotti, *J. Raman Spectrosc.*, **14**, 16 (1983).
12. K. Ohno, Y. Mandai and H. Matsuura, *J. Mol. Struct.*, **268**, 41 (1992).
13. J.A.S. Smith and V.F. Kalasinsky, *Spectrochim. Acta, Part A*, **42A**, 157 (1986).
14. M. Nakata and M. Tasumi, *Spectrochim. Acta, Part A*, **41A**, 341 (1985).
15. M. Nakata and M. Tasumi, *Spectrochim. Acta, Part A*, **41A**, 1015 (1985).
16. H. Wolff, D. Horn and H.-G. Rollar, *Spectrochim. Acta, Part A*, **29A**, 1835 (1973).
17. A.L. Verma and P. Venkateswarlu, *J. Mol. Spectrosc.*, **39**, 227 (1971).
18. B. Silvi and J.P. Perchard, *Spectrochim. Acta, Part A*, **32A**, 23 (1976).
19. J.R. Durig, J.F. Sullivan and C.M. Whang, *Spectrochim. Acta, Part A*, **41A**, 129 (1985).
20. Y. Hamada, M. Tsuboi, M. Nakata and M. Tasumi, *J. Mol. Spectrosc.*, **107**, 269 (1984).
21. N.V. Riggs, *Aust. J. Chem.*, **40**, 435 (1987).
22. S. Chattopadhyay, *Indian J. Phys.*, **41**, 759 (1967).
23. G.Varsányi, *Assignments for Vibrational Spectra of Seven Hundred Benzene Derivatives*, J.Wiley & Sons, New York (1974).
24. P. Pulay and F. Török, *J. Mol. Struct.*, **29**, 239 (1975).
25. H. Wolff, U. Schmidt and E. Wolff, *Spectrochim. Acta, Part A*, **36A**, 899 (1980).
26. Y. Hamada, K. Hashiguchi, A.Y. Hirakawa, M. Tsuboi, M. Nakata and M. Tasumi, *J. Mol. Spectrosc.*, **102**, 123 (1983).
27. Y. Hamada, M. Tsuboi, M. Nakata and M. Tasumi, *J. Mol. Spectrosc.*, **106**, 164 (1984).
28. K. Yamanouchi, T. Matsuzawa, K. Kuchitsu, Y. Hamada and M. Tsuboi, *J. Mol. Struct.*, **126**, 305 (1985).
29. H. Sénéchal and P. Saumagne, *J. Chim. Phys.*, **69**, 1246 (1972).
30. S. Chakravorti, R. De, P.K. Mallick and S.B. Banerjee, *Spectrochim. Acta, Part A*, **49A**, 543 (1993).
31. C.J. Pouchert, *The Aldrich Library of FT-IR Spectra*, Aldrich Chemical Company, first edn. (1985).

3.4.2 Ammoniomethyl

Et—NH_3^+ belongs to the point group C_s. The 27 normal vibrations are divided into 16a′ and 11a″ types. If the eight vibrations (5a′ + 3a″) of methyl and a νCC (a′) are left out of consideration, 10a′ and 8a″ vibrations remain for the —$CH_2NH_3^+$ part:

a′: $\nu'_a NH_3^+$, $\nu_s NH_3^+$, $\nu_s CH_2$, $\delta'_a NH_3^+$, $\delta_s NH_3^+$, δCH_2, ωCH_2, νC—N, $\rho' NH_3^+$, δ—C—N;
a″: $\nu_a NH_3^+$, $\nu_a CH_2$, $\delta_a NH_3^+$, τCH_2, ρNH_3^+, ρCH_2, τNH_3^+ and torsion $—CH_2NH_3^+$.

Since most of the spectra are run in an alkali halide matrix, the differentiation of types of vibration is of minor importance. Moreover, the strong ion interactions disturb the symmetry, so that the selection rules are no longer applicable. The undisturbed methanaminium ion [1–6] shows 12 ($5a_1$ + a_2 + 6e) fundamental vibrations according to C_{3v} symmetry, but 18 vibrations (11a′ + 7a″) in C_s symmetry, where the a_1 vibrations change into a′, the a_2 vibrations in a″ and the e vibrations split up into a′ and a″ species. Even the difference between a′ and a″ disappears owing to the transition into the C_1 conformation.

Typical for the amine salts and the amino acids is the very broad aminium band in the range between 3600 and 2000 cm^{-1} with superimposed on it, the NH_3^+ and CH_2 stretching vibrations, plus a series of typical absorptions due to the following interactions [6,8]:

$\approx$2700–$\approx$2600 cm^{-1}: NH^+ ... N (association)
$\approx$2600–$\approx$2500 cm^{-1}: NH^+ ... Br^- (interaction with KBr)
$\approx$2400–$\approx$2000 cm^{-1}: NH^+ ... Cl^- (HCl salt)
$\approx$2200–$\approx$2000 cm^{-1}: $\delta_a NH_3^+$ + τNH_3^+ (combination band)

Stretching vibrations

The exact wavenumbers of the stretching vibrations are difficult to locate. At best, the stretching vibrations are found as small minima on the wide aminium band. The aminium antisymmetric stretching vibrations absorb at $\approx$100 cm^{-1} higher than those of methyl and methylene, but the symmetric ones coincide with the latter. The highest values for the aminium antisymmetric stretching vibrations are found in the Raman spectrum of $MeCH_2NH_3^+$ NO_3^- (3240 and 3115 cm^{-1}) and the lowest in that of $^+H_3NCH_2CH_2NH_3^+$ (3030, 3000 and 3027, 2982 cm^{-1}). The following regions off good possibilities:

$\nu_a NH_3^+$	$\nu'_a NH_3^+$	$\nu_s NH_3^+$	$\nu_a CH_2$	$\nu_s CH_2$
3150 $\pm$ 50	3055 $\pm$ 60	2960 $\pm$ 50	2930 $\pm$ 30	2860 $\pm$ 60

Aminium deformations

The ranges of the NH_3^+ antisymmetric deformations and the CO_2^- antisymmetric stretch overlap each other in such a way as to make it difficult to assign the two vibrations unambiguously in the spectra of amino acids. Moreover, in the area 1700–1200 cm^{-1} the amino acids display a typical broad absorption on which the aminium and the methylene deformations as well as the CO_2^- stretching vibrations are superimposed. The aminium antisymmetric deformations of $^{-}O_2CCH_2CH_2NH_3^+$ are assigned at 1650 and 1630 cm^{-1} [15] or at 1565 and 1548 cm^{-1} [16]. High values for the same vibrations are assigned in the spectra of $^{-}HO_3PCH_2NH_3^+$ (1650 and 1622) and $^{-}HO_3PCH_2CH_2NH_3^+$ (1635 and 1628 cm^{-1}). The symmetric deformation attains the highest values in the spectra of $^{-}O_2CCH_2NHC(=O)CH_2NH_3^+$ (1575) and $^{-}O_2CCH_2NHC(=O)_2CH_2NH_3^+$ (1560 cm^{-1}). Most of the investigated compounds display aminium deformations in the regions:

$\delta_a NH_3^+$	$\delta'_a NH_3^+$	$\delta_s NH_3^+$
1610 ± 25	1585 ± 25	1505 ± 25

The CN stretching vibration and NH_3^+ rocking modes

Although the CN stretch and the NH_3^+ rock can couple, this stretching vibration is unambiguously assigned in the neighbourhood of 1050 cm^{-1}. For the NH_3^+ rocks, three ranges are taken into account:

ρNH_3^+	$\rho' NH_3^+/\nu CC$	$\nu CC/\rho' NH_3^+$
1150 ± 50	1070 ± 65	1015 ± 85

In the spectra of dipolar ions, the two rocking modes are assigned in the two ranges with the highest wavenumbers and an external stretching mode in the range with the lowest wavenumber. For the NH_3^+ rocks of amine salts the two regions with the lowest wavenumbers are suitable for analysis.

Skeletal deformations and torsions

With dipolar ions the absorption at 480 ± 55 cm^{-1} is assigned to the NH_3^+ twisting vibration and that at 340 ± 30 cm^{-1} to the skeletal deformation. In the spectra of amine salts, on the contrary, the band at 470 ± 20 cm^{-1} belongs to the skeletal deformation and the one at 305 ± 55 cm^{-1} to the NH_3^+ twist.

The data are collected from the spectra of following $R—CH_2NH_3^+$ compounds:

Amine salts:
R = Me— [7], Et—, $Me(CH_2)_n$ (n = 6, 8 and 10) [8], F_3C—, $^+H_3NCH_2$— [6, 9–14], $H_2C=CH$— [17], FCH_2—, $ClCH_2$— and $BrCH_2$—.

Dipolar ions:
R = $^{-}O_2CCH_2$— [15, 16], $^{-}O_3SCH_2$— [15, 18, 34], $^{-}HO_3PCH_2$— [15], $^{-}O_2C$— [18–26, 33], $^{-}O_2C$— [27], $^{-}O_2CCH_2NHC(=O)$— and $^{-}O_2CD_2NHC(=O)$— [27], $^{-}O_2CCH_2NHC(=O)_2$— [28], $^{-}O_3SCH_2CH_2NHC(=O)$— [29], $^{-}HO_3PCH_2NHC(=O)$— [30], $^{-}HO_3P$— [31, 32].

Table 3.14 Absorption regions (cm^{-1}) of the normal vibrations of —$CH_2NH_3^+$

Vibration	Amine salts	Dipolar ions	General
$\nu_a NH_3^+$	3135 ± 105		3135 ± 105
$\nu'_a NH_3^+$	3045 ± 70		3045 ± 70
$\nu_s NH_3^+$	2960 ± 50		2960 ± 50
$\nu_a CH_2$	2930 ± 30		2930 ± 30
$\nu_s CH_2$	2860 ± 60		2860 ± 60
$\delta_a NH_3^+$	1610 ± 25	1605 ± 45	1605 ± 45
$\delta'_a NH_3^+$	1585 ± 30	1575 ± 55	1575 ± 55
$\delta_s NH_3^+$	1500 ± 20	1525 ± 50	1525 ± 50
δCH_2	1465 ± 30	1455 ± 20	1465 ± 30
ωCH_2	1340 ± 25	1360 ± 30	1350 ± 40
τCH_2	1295 ± 50	1300 ± 40	1295 ± 50
νCN	1055 ± 30	1040 ± 25	1050 ± 35
ρNH_3^+	1070 ± 65	1150 ± 50	1100 ± 100
$\rho' NH_3^+$	1015 ± 85	1070 ± 65	1030 ± 105
ρCH_2	845 ± 45	865 ± 55	860 ± 60
$\tau NH_3^+/\delta CCN$	470 ± 20	480 ± 55	480 ± 55
$\delta CCN/\tau NH_3^+$	305 ± 55	340 ± 30	310 ± 60
torsion	190 ± 50	170 ± 60	175 ± 65

References

1. I.A. Oxton and O. Knop, *J. Mol. Struct.*, **37**, 59 (1977).
2. E. Castellucci, *J. Mol. Struct.*, **23**, 449 (1974).
3. A. Theorèt and C. Sandorfy, *Spectrochim. Acta, Part A*, **23A**, 519 (1967).
4. A. Cabana and C. Sandorfy, *Spectrochim. Acta*, **18**, 843 (1962).
5. E.A.V. Ebsworth and N. Sheppard, *Spectrochim. Acta*, **13**, 261 (1959).
6. J. Bellanato, *Spectrochim. Acta*, **16**, 1344 (1960).
7. T.J. O'Leary and I.W. Levin, *J. Phys. Chem.*, **88**, 4074 (1984).
8. C. Siguënza, P. Galera, E.Otero-Aenlle and P.F. González-Díaz, *Spectrochim. Acta, Part A*, **37A**, 459 (1981).
9. R.J. Mureinik and W. Scheuermann, *Spectrosc. Lett.*, **3**, 281 (1970).
10. R.J. Mureinik and W. Robb, *Spectrochim. Acta, Part A*, **24A**, 377 (1968).
11. R.D. McLachlan, *Spectrochim. Acta, Part A*, **30A**, 985 (1974).
12. R.W. Berg and K. Rasmussen, *Spectrosc. Lett.*, **4**, 288 (1971).
13. I.A. Oxton and O. Knop, *J. Mol. Struct.*, **43**, 17 (1978).
14. D.B. Powell, *Spectrochim. Acta*, **16**, 241 (1960).
15. C. Garrigou-Lagrange, *Can. J. Chem.*, **56**, 663 (1978).
16. R.S. Krishinan and R.S. Katiyar, *Bull. Chem. Soc. Jpn.*, **42**, 2098 (1969).
17. R. Rericha and P. Svoboda, *Collect. Czech. Chem. Commun.*, **41**, 1014 (1976).
18. U. Stahlberg and E. Steger, *Spectrochim. Acta, Part A*, **23A**, 475 (1967).
19. K. Machida, A. Kagayama, Y. Saito, Y. Kuroda and T. Uno, *Spectrochim. Acta, Part A*, **33A**, 569 (1977).
20. J. Herranz and J.M. Delgado, *Spectrochim. Acta, Part A*, **32A**, 821 (1976).
21. C. Destrade, C. Garrigou-Lagrange and M.-T. Forel, *J. Mol. Struct.*, **10**, 203 (1971).
22. I. Laulicht, S. Pinchas, D. Samuel and I. Wasserman, *J. Phys. Chem.*, **70**, 2719 (1966).

23. R.K. Khanna, M. Horak and E.R. Lippincott, *Spectrochim. Acta*, **22**, 1759 (1966).
24. S. Suzuki, T. Shimanouchi and M. Tsuboi, *Spectrochim. Acta*, **19**, 1195 (1963).
25. D.M. Dodd, *Spectrochim. Acta*, **15**, 1072 (1959).
26. M. Tsuboi, T. Onishi, J. Nakagawa, T. Shimanouchi and S.J. Mizushima, *Spectrochim. Acta*, **12**, 253 (1958).
27. C. Destrade, E. Dupart, M. Joussot-Dubien and C. Garrigou-Lagrange, *Can. J. Chem.*, **52**, 2590 (1974).
28. C. Destrade and C. Garrigou-Lagrange, *J. Mol. Struct.*, **31**, 301 (1976).
29. C. Garrigou-Lagrange, H. Jensen and M. Cotrait, *J. Mol. Struct.*, **36**, 275 (1977).
30. M. Cotrait, M. Avignon, J. Prigent and C. Garrigou-Lagrange, *J. Mol. Struct.*, **32**, 45 (1976).
31. C. Garrigou-Lagrange and C. Destrade, *J. Chim. Phys.*, **67**, 1646 (1970).
32. C. Garrigou-Lagrange and C. Destrade, *C.R. Acad. Sci., Ser. C*, **280**, 969 (1975).
33. S.F.A. Kettle, E. Lugwisha, J. Eckert and N.K. McGuire, *Spectrochim. Acta, Part A*, **45A**, 533 (1989).
34. K. Ohno, Y. Mandai and H. Matsuura, *J. Mol. Struct.*, **268**, 41 (1992).
35. C. Brissette and C. Sandorfy, *Can. J. Chem.*, **38**, 34 (1960).

3.4.3 Isocyanatomethyl

The most characteristic group vibrations of the isocyanates are the out-of-phase stretch (ν_aN=C=O) and the N=C=O out-of-plane deformation. The ν_aN=C=O is easily observed as an intense broad absorption in the region 2270 ± 20 cm^{-1}, in contrast with the sharp peak of the nitriles in the same area. In contrast, the in-phase stretch (ν_sN=C=O) is active in the region 1420 ± 20 cm^{-1} with a weak to moderate intensity. Both stretching vibrations give rise to a typical combination band in the neighbourhood of 3700 cm^{-1}. The characteristic broad band with medium intensity with the appearance of a 'V' is assigned to the γN=C=O.

Table 3.15 Absorption regions(cm^{-1}) of the normal vibrations of —CH_2N=C=O

Vibration	C_s	Region	Vibration	C_s	Region
$\nu_a CH_2$	a″	2960 ± 20	νCN	a′	725 ± 75
$\nu_s CH_2$	a′	2915 ± 15	γNCO	a″	605 ± 25
ν_aNCO	a′	2270 ± 20	δNCO	a′	575 ± 25
δCH_2	a′	1455 ± 25	CCN ext. sk. def.	a″	435 ± 25
ν_sNCO	a′	1420 ± 20	CCN ext. sk. def.	a′	350 ± 50
ωCH_2	a′	1325 ± 25	CNC sk. def.	a′	155 ± 25
τCH_2	a″	1260 ± 30	torsion	a″	–
ρCH_2	a″	870 ± 80			

R—CH_2N=C=O molecules:
R = Me— [1, 2], Et— [4], Pr— [5], ClCH_2—, H_2C=CH— [3], OCN$(CH_2)_5$—, Cl—.

References

1. R.P. Hirschmann, R.N. Kniseley and V.A. Fassel, *Spectrochim. Acta*, **21**, 2125 (1965).
2. J.F. Sullivan, D.T. Durig, J.R. Durig and S. Cradock, *J. Phys. Chem.*, **91**, 1770 (1987).
3. T. Torgrimsen, P. Klaboe and F. Nicolaisen, *J. Mol. Struct.*, **20**, 213 (1974).
4. R.A. Nyquist and G.L. Jewett, *Appl. Spectrosc.*, **46**, 841 (1992).
5. R.A. Nyquist, D.A. Luoma and C.L. Putzig, *Appl. Spectrosc.*, **46**, 972 (1992).

3.4.4 Isothiocyanatomethyl

In the spectra of isothiocyanates, the out-of-phase stretch (ν_aN=C=S) makes its appearance as a very strong broad absorption in the region 2105 ± 20 cm^{-1}, with a shoulder at the HW side (2160 ± 25) with an intensity which reaches that of the fundamental vibration. Often a shoulder is also observed at the LW side (2025 ± 25 cm^{-1}) . This characteristic absorption is easy to distinguish from the sharp peak of the thiocyanates (see Section 3.3.3). The in-phase stretch (ν_sN=C=S) is only weakly active at 1060 ± 35 cm^{-1} and coupled with the C—N stretching vibration. The latter may be considered as a νC—NCS and absorbs therefore at low wavenumbers (610 ± 35 cm^{-1}). In the spectrum of MeNCS the two NCS stretching vibrations are assigned at 2113 and 1090 cm^{-1} [1, 2].

Table 3.16 Absorption regions (cm^{-1}) of the normal vibrations of —CH_2N=C=S

Vibration	C_s	Region	Vibration	C_s	Region
$\nu_a CH_2$	a″	2960 ± 30	νCN	a′	610 ± 35
$\nu_s CH_2$	a′	2915 ± 15	γNCS	a″	530 ± 15
ν_aNCS	a′	2105 ± 20	CCN ext. sk. def.	a″	470 ± 30
δCH_2	a′	1445 ± 20	δNCS	a′	435 ± 15
ωCH_2	a′	1335 ± 15	CCN ext. sk. def.	a′	380 ± 30
τCH_2	a″	1260 ± 30	CNC sk. def.	a′	140 ± 40
ν_sNCS	a′	1060 ± 35	torsion	a″	–
ρCH_2	a″	865 ± 75			

R—CH_2N=C=S molecules:
R = Me— [2–4], Et—, Me_2NCH_2—, H_2C=CH— [5, 6] and Ph— [5].

References

1. J.R. Durig, J.F. Sullivan, H.L. Heusel and S. Cradock, *J. Mol. Struct.*, **100**, 241 (1983).
2. J.R. Durig, J.F. Sullivan, D.T. Durig and S. Cradock, *Can. J. Chem.*, **63**, 2000 (1985).
3. J.R. Durig, H.L. Heusel and J.F. Sullivan, *Spectrochim. Acta, Part A*, **40A**, 739 (1984).
4. R.N. Kniseley, R.P. Hirschmann and V.A. Fassel, *Spectrochim. Acta, Part A*, **23A**, 109 (1967).
5. N.S. Ham and J.B. Willis, *Spectrochim. Acta*, **16**, 279 (1960).
6. T. Torgrimsen, P. Klaboe and F. Nicolaisen, *J. Mol. Struct.*, **20**, 213 (1974).

3.5 CARBON-BONDED METHYLENE

3.5.1 Ethyl

Molecules of type X—Et (X= F, Cl, Br, I) possess 18 normal vibrations which are divided in 11a′ and 7a″ vibration types. After substitution of νCX (a′) by a torsion (a″), the 18 normal vibrations (10a′ + 8a″) for ethyl according to C_s symmetry are:

a′: ν'_aMe, ν_sMe, $\nu_s CH_2$, δ'_aMe, δCH_2, δ_sMe, ωCH_2, ρ'Me, νCC, δC—C—,
a″: ν_aMe, $\nu_a CH_2$, δ_aMe, τCH_2, ρMe, ρCH_2 and two torsions.

The two methyl rockings and the νCC are mixed vibrations. The out-of-plane ρMe may be correlated with the band at 1100 ± 95 cm^{-1}. The in-plane ρ'Me and the νCC are strongly coupled and give rise to an absorption in two regions, considered as: ρ'Me/νCC (1080 ± 80) and νCC/ρ'Me (910 ± 100 cm^{-1}).

Table 3.17 Absorption regions (cm^{-1}) of α-C saturated $—CH_2CH_3$

	Total region	Molecules absorbing at high wavenumbers	Molecules absorbing at low wavenumbers	Region of remaining molecules
ν_aMe	2980 ± 20	MeCHClEt (2997) cPrEt (2994)		2975 ± 15
ν'_aMe	2965 ± 25	Cl_3C—Et (2989)	$MeCCl_2Et$ (2944)	2965 ± 15
$\nu_a CH_2$	2935 ± 35		$ClCH_2CH_2Et$ (2900) FCH_2CH_2Et (2905)	2945 ± 25
ν_sMe	2900 ± 60	Me—Et (2962)	ICH_2Et (2840) $BrCH_2Et$ (2846)	2900 ± 40
$\nu_s CH_2$	2865 ± 25		ICH_2Et (2840) $BrCH_2Et$ (2846) XCH_2CH_2Et (2846) (X = F, Cl)	2870 ± 20
δ_aMe	1465 ± 10			1465 ± 10
δ'_aMe	1450 ± 15		$BrCH_2Et$ (1435) $ClCH_2Et$ (1442)	1455 ± 10
δCH_2	1450 ± 30	Me—Et (1476) $H_2NCH_2CH_2Et$ (1475)	$MeCCl_2Et$ (1427) Cl_3C—Et (1430)	1455 ± 15
δ_sMe	1375 ± 15	3-Br—Ph—CH_2Et (1390) FCH_2CH_2Et (1387)	EtC(=O)$NHCH_2Et$ (1369) 4-F—Ph—C(=O)CH_2Et (1370)	1379 ± 6
ωCH_2	1330 ± 35	$H_2NCH_2CH_2Et$ (1364)	MeCHClEt (1299) $cPr(Et)_2$ (1302) $H_3SiCH_2CH_2Et$ (1307)	1340 ± 20
τCH_2	1245 ± 45	MeCHClEt (1290) HC≡CCH_2CH_2Et (1290) MeC(=O)$NHCH_2Et$ (1289)	H_3SiCH_2Et (1207) D_3SiCH_2Et (1208) $MeSiH_2CH_2Et$ (1212)	1250 ± 35
ρMe	1125 ± 65	Me—Et (1190 and 1158) $BrCH_2CH$(Me)Et (1169)	$MeSiD_2CH_2Et$ (1062) $MeSiH_2CH_2Et$ (1067)	1110 ± 40
ρ'Me/ νCC	1055 ± 45	O_2NCH_2Et (1097) MeC≡CCH_2Et (1094)	ICH_2Et (1010)	1055 ± 35

(continued)

Table 3.17 (*continued*)

	Total region	Molecules absorbing at high wavenumbers	Molecules absorbing at low wavenumbers	Region of remaining molecules
νCC/ ρ'Me	940 ± 60	$BrCH_2CH(Me)Et$ (1000) D_3C—Et (999) $MeCCl_2Et$ (999)	ICH_2Et (880) $BrCH_2Et$ (886) 2-HO—Ph—C(=O)CH_2Et (880) $MeSiD_2CH_2Et$ (885)	935 ± 45
ρCH_2	780 ± 55	H_3SiCH_2Et (*tr*:830; g:824) $MeCCl_2Et$ (810)	ICH_2Et (727)	760 ± 30
skelet. def.	390 ± 100	$MeCCl_2Et$ (490) FCH_2CH_2Et (473)	ICH_2Et (290) $BrCH_2Et$ (312)	405 ± 65
torsion Me	230 ± 105	Me—Et (333 and *223*)	Cl_3C—Et (130)	225 ± 45
torsion Et	120 ± 30			120 ± 30

R—Et molecules

R = R′CH_2— (see Section 3.5.5)

R = Me— [1–7, 84], D_3C— [5, 6], EtCH(Me)— [8], $BrCH_2CH(Me)$— [9], cPr— [10], $^-O_2CCHOH$— [11], MeCHF— [12], MeCHCl— [13–15], EtCHCl— [15], MeCHBr— and MeCHI— [14], $ClCH_2CHCl$— and $BrCH_2CHBr$— [16], 1-Et—cPr— [31], tBu— [17], $MeCCl_2$— [18–22], $MeCBr_2$— [23], Cl_3C— [22, 24].

Table 3.18 Absorption regions (cm^{-1}) of α-C unsaturated —CH_2CH_3

	Total region	Molecules absorbing at high wavenumbers	Molecules absorbing at low wavenumbers	Region of remaining molecules
ν_aMe	2980 ± 20	N≡C—Et (3000)	Ph—Et (2966)	2980 ± 10
ν'_aMe	2970 ± 30	N≡C—Et (2998) HC≡C—Et (2988)		2960 ± 20
$\nu_a CH_2$	2940 ± 20	N≡C—Et (2958) H_2C=CH—Et (2952)		2935 ± 15
ν_sMe	2915 ± 45	N≡C—Et (2958) HC≡C—Et (2945)	H_2C=CBr—Et (2878)	2910 ± 30
$\nu_s CH_2$	2885 ± 45	HC≡C—Et (2925)	H_2C=CBr—Et (2848)	2885 ± 35
δ_aMe	1465 ± 10	DC≡C—Et (1471)	Et—C≡C—Et (1457 and *1460*)	1465 ± 05
δ'_aMe	1455 ± 10	HC≡C—Et (1462)	Et—C≡C—Et (1447 and *1450*)	1455 ± 05
δCH_2	1440 ± 10	H_2C=CH—Et (1450)	N≡C—Et (1431)	1440 ± 05
δ_sMe	1380 ± 10	N≡C—Et (1386)	Et—C≡C—Et (1374)	1380 ± 05
ωCH_2	1340 ± 25	2-Th—Et (1363)		1330 ± 15
τCH_2	1265 ± 10	$HOCH_2CH_2C$≡C—Et (1273)	Me—C≡C—Et (1259)	1265 ± 05
ρMe	1130 ± 50	H_2C=CH—Et (1177)	Me—C≡C—Et (1088)	1120 ± 30
ρ'Me/νCC	1085 ± 45	H_2C=CH—Et (1128)		1065 ± 25
νCC/ρ'Me	975 ± 35		2-Th—Et (946)	980 ± 30
ρCH_2	800 ± 25	2-Th—Et (821)		790 ± 10
skelet. def.	345 ± 45		H_2C=CH—Et (306)	360 ± 30
torsion Me	255 ± 35			255 ± 35
torsion Et	–			

R—Et molecules

R = $H_2C{=}CH$— [25–27], N≡C—HC=CH— [28], $H_2C{=}CCl$— and $H_2C{=}CBr$— [29], 2-Th— [30], Ph— [80], HC≡C— [32, 33, 34], DC≡C— [32], MeC≡C— [35, 36], EtC≡C— [37], $HOCH_2CH_2C{\equiv}C$—, N≡C— [34, 38].

Table 3.19 Absorption regions (cm^{-1}) of C(=O)-, O- and S- bound $—CH_2CH_3$

	C(=O)Et	OEt	SEt
ν_aMe	2985 ± 15	2985 ± 10	2985 ± 15
ν'_aMe	2970 ± 30	2965 ± 25	2965 ± 10
$\nu_a CH_2$	2930 ± 25	2940 ± 20	2940 ± 20
ν_sMe	2900 ± 40	2910 ± 30	2920 ± 25
$\nu_s CH_2$	2880 ± 60	2885 ± 30	2880 ± 30
δ_aMe	1470 ± 15	1465 ± 15	1465 ± 15
δ'_aMe	1455 ± 15	1445 ± 20	1450 ± 10
δCH_2	1425 ± 20	1475 ± 20	1430 ± 15
δ_sMe	1380 ± 15	1385 ± 15	1380 ± 10
ωCH_2	1340 ± 40	1350 ± 40	1280 ± 30
τCH_2	1265 ± 25	1285 ± 45	1250 ± 20
ρMe	1130 ± 60	1165 ± 30	1075 ± 30
ρ'Me/νCC	1050 ± 50	1120 ± 40	1035 ± 25
νCC/ρ'Me	920 ± 80	875 ± 65	975 ± 25
ρCH_2	805 ± 30	790 ± 50	765 ± 35
skelet. def.	375 ± 85	390 ± 85	350 ± 40
torsion Me	220 ± 30	240 ± 40	245 ± 35
torsion Et	130 ± 70	150 ± 50	185 ± 30

R—Et molecules

R = R′C(=O)— (see Section 7.1.6), R′O— (see Section 10.1.3), O_2NO— [82], $Cl_2P({=}O)O$— and $Cl_2P({=}S)O$— [81], H(EtO)P(=O)O— [83], R′S— (see Section 11.1.2)

Table 3.20 Absorption regions (cm^{-1}) of N-, P-, Si- and Halogen-bound $—CH_2CH_3$

	N	P	Si	Hal
ν_aMe	2980 ± 20	2985 ± 15	2975 ± 15	2990 ± 20
ν'_aMe	2965 ± 25	2975 ± 20	2965 ± 15	2965 ± 25
$\nu_a CH_2$	2940 ± 20	2940 ± 20	2950 ± 20	3010 ± 05
ν_sMe	2880 ± 40	2915 ± 25	2915 ± 25	2895 ± 25
$\nu_s CH_2$	2850 ± 40	2885 ± 25	2885 ± 15	2945 ± 25
δ_aMe	1465 ± 20	1460 ± 15	1465 ± 10	1460 ± 15
δ'_aMe	1455 ± 10	1455 ± 15	1445 ± 25	1445 ± 15
δCH_2	1450 ± 20	1415 ± 10	1415 ± 10	1455 ± 25
δ_sMe	1375 ± 15	1375 ± 15	1380 ± 05	1381 ± 11
ωCH_2	1320 ± 35	1255 ± 25	1240 ± 10	1285 ± 80
τCH_2	1265 ± 25	1240 ± 15	1225 ± 20	1195 ± 80
ρMe	1125 ± 60	1050 ± 30	1020 ± 10	1110 ± 60
ρ'Me/νCC	1055 ± 55	1035 ± 20	1020 ± 10	1045 ± 25
νCC/ρ'Me	955 ± 40	980 ± 20	970 ± 10	923 ± 50
ρCH_2	800 ± 20	770 ± 50	745 ± 30	775 ± 40
skelet. def.	400 ± 95	310 ± 30	320 ± 30	340 ± 80
torsion Me	–	210 ± 30	230 ± 35	265 ± 15
torsion Et	–	–	–	–

R—Et molecules

R = H_2N— [39], ^+H_3N— [40], EtNH— [41], MeC(=O)NH— and EtC(=O)NH— [42], MeNHC(=O)NH— [43, 44], $PhSO_2NH$— [45], Me_2N— [46], O_2N— [47, 48], OCN— [49], SCN— [50–52], CN— [53], N_3— [54], $(Et_2N)_2PN$— [55], H_2P— and D_2P— [56], Me_2P— [57], F_2P— [58], Cl_2P— [59, 60], F_2P(=O)— [61], Cl_2P(=O)— [62], F_2P(=S)— [63], Cl_2P(=S)— [64], H_3Si— and D_3Si— [65], $MeSiH_2$—, $MeSiD_2$—, $EtSiH_2$— and $EtSiD_2$— [66], Me_2SiH— and Me_2SiD— [67], F— [68, 69], Cl— [13, 70–75], Br— [69, 73, 76, 79], I— [69, 75, 77, 78].

References

1. M.A. Elyashevich and B.I. Stepanov, *Dok. Akad. Nauk. SSSR*, **32**, 481 (1941).
2. D.C. Smith, C. Yvan and J.R. Nielsen, *J. Chem. Phys.*, **18**, 706 (1950).
3. H.L. McMurry and V. Thornton, *J. Chem. Phys.*, **18**, 1515 (1950).
4. H.L. McMurry and V. Thornton, *J. Chem. Phys.*, **19**, 1014 (1951).
5. J.N. Gayles and W.T. King, *Spectrochim. Acta*, **21**, 543 (1965).
6. J.N. Gayles, W.T. King and J.H. Schachtschneider, *Spectrochim. Acta, Part A*, **23A**, 703 (1967).
7. R.L. Flurry Jr., *J. Mol. Spectrosc.*, **56**, 88 (1975).
8. G.A. Crowder and D. Hill, *J. Mol. Struct.*, **145**, 69 (1986).
9. G.A. Crowder and M.-R. Jalilian, *Spectrochim. Acta, Part A*, **34A**, 707 (1978).
10. A.B. Nease and C.J. Wurrey, *J. Raman Spectrosc.*, **9**, 107 (1980).
11. M. Morssli, G. Cassanas, L. Baroet, B. Pauvert and A. Terol, *Spectrochim. Acta, Part A*, **47A**, 529 (1991).
12. G.A. Crowder and T. Koger, *J. Mol. Struct.*, **29**, 233 (1975).
13. A.J. Barnes, M.L. Evans and H.E. Hallam, *J. Mol. Struct.*, **99**, 235 (1983).
14. E. Benedetti and P. Cecchi, *Spectrochim. Acta, Part A*, **28A**, 1007 (1972).
15. W.H. Moore and S. Krimm, *Spectrochim. Acta, Part A*, **29A**, 2025 (1973).
16. G.A. Crowder, *J. Mol. Struct.*, **100**, 415 (1983).
17. S. Konaka, J. Hirose, A. Suwa, H. Takeuchi, T. Egawa, K. Siam, J.D. Ewbank and L. Schäfer, *J. Mol. Struct.*, **244**, 1 (1991).
18. M.S. Wu, P.C. Painter and M.M. Coleman, *Spectrochim. Acta, Part A*, **35A**, 823 (1979).
19. G.A. Crowder and W.Y. Lin, *J. Mol. Struct.*, **62**, 1 (1980).
20. K. Ohno, Y. Shiro and H. Murata, *Bull. Chem. Soc. Jpn.*, **47**, 2962 (1974).
21. S.H. Cough and S. Krimm, *Spectrochim. Acta, Part A*, **46A**, 1419 (1990).
22. K. Ohno, K. Taga, I. Yoshida and H. Murata, *Spectrochim. Acta, Part A*, **36A**, 721 (1980).
23. A.O. Diallo, *Spectrochim. Acta, Part A*, **38A**, 687 (1982).
24. A. Goursot-Leray, M. Carles-Lorjou, G. Pouzard and H. Bodot, *Spectrochim. Acta, Part A*, **29A**, 1497 (1973).
25. J.R. Durig and D.A.C. Compton, *J. Phys. Chem.*, **84**, 773 (1980).
26. D.A.C. Compton and W.F. Murphy, *Spectrochim. Acta, Part A*, **41A**, 1141 (1985).
27. G. Busca, G. Ramis, V. Lorenzelli, A. Janin and J.C. Lavalley, *Spectrochim. Acta, Part A*, **43A**, 489 (1987).
28. D.A.C. Compton and W.F. Murphy, *J. Phys. Chem.*, **85**, 482 (1981).
29. G.A. Crowder and N. Smyrl, *J. Mol. Struct.*, **10**, 373 (1971).
30. J.J. Peron, P. Saumagne and J.M. Lebas, *Spectrochim. Acta, Part A*, **26A**, 1651 (1970).
31. P.M. Green, C.J. Wurrey and V.F. Kalasinsky, *Spectrochim. Acta*, **42A**, 141 (1986).
32. J.Saussey, J.Lamotte and J.C.Lavalley, *Spectrochim. Acta*, **32A**, 763 (1976).

33. G.A. Crowder and H.Fick, *J. Mol. Struct.*, **147**, 17 (1986).
34. G.A. Crowder, Spectrosc.Letters **20**, 343 (1987).
35. J.C.Lavalley, J.Saussey and J.Lamotte, *Spectrochim. Acta*, **35A**, 695 (1979).
36. G.A. Crowder and P.Blankenship, *J. Mol. Struct.*, **196**, 125 (1989).
37. G.A. Crowder and P.Blankenship, *J. Mol. Struct.*, **156**, 147 (1987).
38. P.Klaboe and J.Grundnes, *Spectrochim. Acta*, **24A**, 1905 (1968).
39. H.Wolff and H.Ludwig, Ber.Bunsenges.Phys.Chem. **71**, 1107 (1967).
40. T.J.O′Leary and I.W.Levin, *J. Phys. Chem.*, **88**, 4074 (1984).
41. A.L.Verma, *Spectrochim. Acta*, **27A**, 2433 (1971).
42. J.Jakes and S.Krimm, *Spectrochim. Acta, Part A*, **27A**, 35 (1971).
43. Y. Mido, *Spectrochim. Acta, Part A*, **28A**, 1503 (1972).
44. Y. Mido, F. Fujita, H. Matsuura and K. Machida, *Spectrochim. Acta, Part A*, **37A**, 103 (1981).
45. M. Goldstein, M.A. Russell and H.A. Willis, *Spectrochim. Acta, Part A*, **25A**, 1275 (1969).
46. J.R. Durig and F.O. Cox, *J. Mol. Struct.*, **95**, 85 (1982).
47. P. Groner, R. Meyer and H.H. Günthard, *Chem. Phys.*, **11**, 63 (1975).
48. G. Geiseler and H. Kessler, *Ber. Bunsenges Phys. Chem.*, **68**, 571 (1964).
49. R.P. Hirschmann, R.N. Kniseley and V.A. Fassel, *Spectrochim. Acta*, **21**, 2125 (1965).
50. R.N. Kniseley, R.P. Hirschmann and V.A. Fassel, *Spectrochim. Acta, Part A*, **23A**, 109 (1967).
51. J.R. Durig, H.L. Heusel, J.F. Sullivan and S. Cradock, *Spectrochim. Acta, Part A*, **40A**, 739 (1984).
52. J.R. Durig, J.F. Sullivan, D.T. Durig and S. Cradock, *Can. J. Chem.*, **63**, 2000 (1985).
53. K. Bolton, N.L. Owen and J. Sheridan, *Spectrochim. Acta, Part A*, **25A**, 1 (1969).
54. C.J. Nielsen, K. Kosa, H. Priebe and C.E. Sjøgren, *Spectrochim. Acta, Part A*, **44A**, 409 (1988).
55. G. Davidson and S. Phillips, *Spectrochim. Acta, Part A*, **35A**, 141 (1979).
56. J.R. Durig and A.W. Cox, *J. Chem. Phys.*, **63**, 2303 (1975).
57. J.R. Durig and T.J. Hizer, *J. Raman Spectrosc.*, **17**, 97 (1986).
58. J.R. Durig, J.S. Church, C.M. Wang, R.D. Johnson and B.J. Streusand, *J. Phys. Chem.*, **91**, 2769 (1987).
59. J.R. Durig, C.G. James, A.E. Stanley, T.J. Hizer and S. Cradock, *Spectrochim. Acta, Part A*, **44A**, 911 (1988).
60. A.J. Fishman, A.B. Remisov, I.Ya. Kuramshin and I.S. Pominov, *Spectrochim. Acta, Part A*, **32A**, 651 (1976).
61. J.R. Durig, T.J. Hizer and R.J. Harlan, *J. Phys. Chem.*, **96**, 541 (1992).
62. J.R. Durig, C.G. James and T.J. Hizer, *J. Raman Spectrosc.*, **21**, 155 (1990).
63. J.R. Durig, R.D. Johnson, H. Nanaie and T.J. Hizer, *J. Chem. Phys.*, **88**, 7317 (1988).
64. J.R. Durig and T.J. Hizer, *J. Raman Spectrosc.*, **18**, 415 (1987).
65. K.M. Mackay and R. Watt, *Spectrochim. Acta, Part A*, **23A**, 2761 (1967).
66. H. Matsuura, K. Ohno, T. Sato and H. Murata, *J. Mol. Struct.*, **52**, 13 (1979).
67. K. Ohno, K. Suerhiro and H. Murata, *J. Mol. Struct.*, **98**, 251 (1983).
68. G.A. Crowder and H.K. Mao, *J. Mol. Struct.*, **18**, 33 (1973).
69. D.C. Smith, R.A. Saunders, J.R. Nielsen and E.E. Ferguson, *J. Chem. Phys.*,**20**, 847 (1952).
70. R.G. Snyder and J.H. Schachtschneider, *J. Mol. Spectrosc.*, **30**, 290 (1969).
71. L.W. Daasch, C.Y. Liang and J.R. Nielsen, *J. Chem. Phys.*, **22**, 1293 (1954).
72. F.A. Miller and F.E. Kiviat, *Spectrochim. Acta, Part A*, **25A**, 1363 (1969).
73. S. Mizushima, T. Shimanouchi, I. Nakagawa and A. Miyake, *J. Chem. Phys.*, **21**, 215 (1953).

74. S. Suzuki and A.B. Dempster, *J. Mol. Struct.*, **32**, 339 (1976).
75. N.T. McDevitt, A.L. Rozek, F.F. Bentley and A.D. Davidson, *J. Chem. Phys.*, **42**, 1173 (1965).
76. R. Gaufrèzs and M. Bejaud-Bianchi, *Spectrochim. Acta, Part A*, **27A**, 2249 (1971).
77. G.A. Crowder, *J. Mol. Spectrtosc.*, **48**, 467 (1973).
78. J.R. Durig, J.W. Thompson, V.W. Thyagesan and J.D. Witt, *J. Mol. Struct.*, **24**, 41 (1975).
79. F.F. Bentley, N.T. McDevitt and A.L. Rozek, *Spectrochim. Acta*, **20**, 105 (1964).
80. G. Varsányi, *Assignments for Vibrational Spectra of Seven Hundred Benzene Derivatives*, J.Wiley & Sons New York (1974).
81. R.A. Nyquist, W.W. Muelder and M.N. Wass, *Spectrochim. Acta, Part A*, **26A**, 769 (1970).
82. J.R. Durig and T.G. Sheehan, *J. Raman Spectrosc.*, **21**, 635 (1990).
83. S.A. Katcyuba, N.I. Monakhova, L.Kh. Ashrafullina and R.R. Shagidullin, *J. Mol. Struct.*, **269**, 1 (1992).
84. P. Derreumaux, M. Dauchez and G. Vergoten, *J. Mol. Struct.*, **295**, 203 (1993).

3.5.2 Trichloroethyl

The simplest molecules with a $—CH_2CCl_3$ structure unit are XCH_2CCl_3 (X =F, Cl, Br and I). The 18 normal vibrations are distributed over 11a′ and 7a″ vibration types according to C_s symmetry. Substitution of the νCX (a′) by a torsion (a″) furnishes the 10a′ + 8a″ normal vibrations of $—CH_2CCl_3$:

a′: $\nu_s CH_2$, δCH_2, ωCH_2, νCC, ν'CCl (or $\nu'_a CCl_3$), ν''CCl (or $\nu_s CCl_3$), δ—C—C, δCCl (or $\delta_s CCl_3$), δ''CCl (or $\delta'_a CCl_3$), ρ'CCl (or $\rho' CCl_3$);
a″: $\nu_a CH_2$, τCH_2, νCCl (or $\nu_a CCl_3$), ρCH_2, δ'CCl (or $\delta_a CCl_3$), ρCCl (or ρCCl_3) and two torsions.

The trichloroethyl group, just like the trichloromethyl group, is characterized by the three strong absorptions of the CCl stretching vibrations. The normal vibrations of trichloromethyl are treated in Chapter 2 (see Section 2.2.2).

Table 3.21 Absorption regions (cm^{-1}) of the normal vibrations of $—CH_2CCl_3$

Vibration	Region	Vibration	Region
$\nu_a CH_2$	2970 ± 20	ν''CCl	545 ± 45
$\nu_s CH_2$	2935 ± 20	—CC ext. sk. def.	500 ± 70
δCH_2	1435 ± 20	δCCCl	380 ± 30
ωCH_2	1340 ± 30	δ'CCCl	355 ± 30
τCH_2	1270 ± 30	δ''CCCl	305 ± 40
νCC	1045 ± 25	ρCCl	245 ± 15
νCCl	795 ± 35	ρ'CCl	195 ± 35
ν'CCl	740 ± 60	torsion	120 ± 30
ρCH_2	765 ± 70	torsion	–

$R{-}CH_2CCl_3$ molecules
R = Me— [1, 3], MeCXH—, tBuCXH— and Me_2XC— (X = Cl, Br) [2], $ClCH_2$— [1, 2, 4], $BrCH_2$— [1, 2], Cl_2CH— [4], HO— [5, 6], ClC(=O)—O—, Cl_2PO—, $Cl_2P(=O)O$—, $Cl_3CCH_2OP(=O)O$—, Cl— [7, 8].

References

1. A. Goursot-Leray, M. Carles-Lorjou, G.Pouzard and H.Bodot, *Spectrochim. Acta, Part A*, **29A**, 1497 (1973).
2. M. Carles-Lorjou, A. Goursot-Leray and H. Bodot, *Spectrochim. Acta, Part A*, **29A**, 329 (1973).
3. K. Ohno, K. Taga, I. Yoshida and H. Murata, *Spectrochim. Acta, Part A*, **36A**, 721 (1980).
4. A.B. Dempster, K. Price and N. Sheppard, *Spectrochim. Acta, Part A*, **27A**, 1563 (1971).
5. M. Perttilä, *Spectrochim. Acta, Part A*, **35A**, 585 (1979).
6. J. Travert and J.C. Lavalley, *Spectrochim. Acta, Part A*, **32A**, 637 (1976).
7. S. Suzuki and A.B. Dempster, *J. Mol. Struct.*, **32**, 339 (1976).
8. G. Allen and H.J. Bernstein, *Can. J. Chem.*, **32**, 1124 (1954).

3.5.3 Chloroethyl

The 18 normal vibrations of chloroethyl are divided into 10a′ and 8a″ species of vibration, which are composed of 12 vibrations for methylene, a CC and a CCl stretching vibration, two skeletal deformations and two torsions:

a′: $\nu_s CH_2$ (2), δCH_2 (2), ωCH_2 (2), νCC, νCCl, δ—C—C, δC—C—Cl;
a″: $\nu_a CH_2$ (2), τCH_2 (2), ρCH_2 (2) and two torsions.

This collection contains the nine normal vibrations of $-CH_2X$ (see Section 3.1.2), the six vibrations of methylene, one CC stretching vibration, one external —C—C deformation and one $-CH_2CH_2Cl$ torsion. The four methylene stretching vibrations are often described as four individual CH stretchings. Likewise it is not always possible to assign the absorptions of the two wags and the two twists because they are coupled to one another. In this work the higher wavenumbers are assigned to the wagging modes, the lower to the twists. The CCl stretch, generally with strong band intensity, is the most characteristic band, but the notable CH_2Cl wag is also helpful for identification purposes. The $-CH_2Cl$ vibrations are especially sensitive to the conformational state of the molecule (Section 3.1.2). Self-evidently, more accurate absorption regions for the $R{-}CH_2{-}$ vibrations are available in the more relevant tables.

Table 3.22 Wavenumbers of the wagging and twisting vibrations of a few RCH_2CH_2Cl compounds

R—	ωCH_2	ωCH_2Cl	τCH_2	τCH_2Cl
HO—	1380	1300	1247	1165
MeO—	1386	1299	1255	1218
H_2N—	1385	1290	1254	1156
Cl_3C—	1342	1280	1254	1171
$ClCH_2$—	1357	1315, 1280	1258	1194, 1150
$BrCH_2$—	1355	1308	1243	1144
Me—	1339	1291	1266	1227
NC—	1333	1302	1218	1170
$H_2C{=}CH$—	1324	1260	1243	1200
Ph—	1320	1282	1242	1195
D—	1305	1263	1258	1213
Cl— (*g* and *tr*)	1315, 1304	1292, 1264	1143, 1125	1207, 1233
Br— (*g* and *tr*)	1299, 1284	1260, 1259	1127, 1111	1190, 1203

R—CH_2CH_2Cl molecules

R = D— [1], Me— [2–10], Et— [7, 9, 10], MeC(=O)CH_2—, MeOC(=O)CH_2—, EtOC(=O)CH_2—, ClC(=O)CH_2—, CH_2=CHCH_2— [11], N≡CCH_2—, PhCH_2— [12], HOCH_2—, iPrOCH_2—, PhOCH_2—, ClS(=O)$_2$$CH_2$—, Cl$CH_2$—, Br$CH_2$— and I$CH_2$ [13], MeCH(Cl)—, Cl_3C— [14–16], EtC(=O)—, ClC(=O)— [17], HOC(=O)—, H_2NC(=O)—, 4-XPhC(=O)— [18] (X = F, Cl, Br), CH_2=CH— [11, 19], HC≡C—, N≡C— [20–24], Ph— [12, 25], H_2N— [26], H_3N^+—, OCN—, HO— [27–31], DO— [27–29], MeO— [32], EtO— and ClCH_2CH_2O— [30], ClC(=O)O—, 4-BrPhO—, HS— [33], MeS— [34], EtS— and ClCH_2CH_2S— [42], NaOS(=O)$_2$—, F— [35], Cl— [36–41], Br— [36, 39].

References

1. J.S. Francisco, Z. Qingshi and J.I. Steinfeld, *Spectrochim. Acta, Part A*, **38A**, 671 (1982).
2. J.K. Brown and N. Sheppard, *Trans. Faraday Soc.*, **50**, 1164 (1954).
3. C. Komaki, I. Ichishima, K. Kutanani, T. Miyazawa, T. Shimanouchi and S.Mizushima, *Bull. Chem. Soc. Jpn.*, **28**, 330 (1955).
4. V.A. Pozdyshev, Yu.A. Pentin and V.M. Tatevskii, *Opt. Spectrosc.*, **3**, 211 (1957).
5. N.T. McDevitt, A.L. Rozek, F.F. Bentley and A.D. Davidson, *J. Chem. Phys.*, **42**, 1173 (1965).
6. K. Radcliffe and J.L. Wood, *Trans. Faraday Soc.*, **62**, 1678 (1966).
7. R.G. Snyder and J.H. Schachtschneider, *J. Mol. Spectrosc.*, **30**, 290 (1969).
8. K. Tanabe and S. Saëki, *J. Mol. Struct.*, **27**, 79 (1975).
9. A.J. Barnes, M.L. Evans and H.E. Hallam, *J. Mol. Struct.*, **99**, 235 (1983).
10. Y. Ogawa, S. Imazeki, H. Yamaguchi, H. Matsuura, I. Harada and T. Shimanouchi, *Bull. Chem. Soc. Jpn.*, **51**, 748 (1978).

Table 3.23 Absorption regions (cm^{-1}) of the normal vibrations of $—CH_2CH_2Cl$

	Total region	Molecules absorbing at high wavenumbers	Molecules absorbing at low wavenumbers	Region of most molecules
ν_aCH_2Cl	3010 ± 20	$N{\equiv}CCH_2CH_2Cl$ (3030) $ClC(=O)CH_2CH_2Cl$ (3030)	$MeOCH_2CH_2Cl$ (2990)	3010 ± 10
ν_aCH_2	2975 ± 35	$BrCH_2CH_2Cl$ (3010) $ClCH_2CH_2Cl$ (3005)		2965 ± 25
ν_sCH_2Cl	2960 ± 20		$EtCH_2CH_2Cl$ (2945)	2965 ± 15
ν_sCH_2	2910 ± 50	*g*-$N{\equiv}CCH_2CH_2Cl$ (2960)	$EtCH_2CH_2Cl$ (2861)	2905 ± 35
δCH_2	1450 ± 20	$EtCH_2CH_2Cl$ (1469) $MeCH_2CH_2Cl$ (1468)	*g*-$ClCH_2CH_2Cl$ (1430) *g*-$BrCH_2CH_2Cl$ (1430)	1450 ± 15
δCH_2Cl	1430 ± 20		4-F—$PhC(=O)CH_2CH_2Cl$ (1411) $EtC(=O)CH_2CH_2Cl$ (1412) $N{\equiv}CCH_2CH_2Cl$ (1416)	1435 ± 15
ωCH_2	1335 ± 55	$MeOCH_2CH_2Cl$ (1386) $H_2NCH_2CH_2Cl$ (1385)	$BrCH_2CH_2Cl$ (*tr*:1284; *g*:1299) $ClCH_2CH_2Cl$ (*tr*:1304; *g*:1315)	1350 ± 30
ωCH_2Cl	1280 ± 35		$PhCH_2CH_2Cl$ (1245)	1285 ± 30
τCH_2	1205 ± 95	$H_2NCH_2CH_2Cl$ (1290)	$BrCH_2CH_2Cl$ (*tr*:1111; *g*:1127) $ClCH_2CH_2Cl$ (*tr*:1125; *g*:1143)	1240 ± 40
τCH_2Cl	1195 ± 55	$HC{\equiv}CCH_2CH_2Cl$ (1247)	$BrCH_2CH_2CH_2Cl$ (1144)	1190 ± 45
νCC	1050 ± 60	$EtCH_2CH_2Cl$ (1110)	$Cl_3CCH_2CH_2Cl$ (990)	1050 ± 50
ρCH_2Cl	910 ± 80	$HC{\equiv}CCH_2CH_2Cl$ (989) *tr*-$ClCH_2CH_2Cl$ (982)	4-X—$PhC(=O)CH_2CH_2Cl$ (830) (X = F, Cl, Br)	905 ± 55
ρCH_2	795 ± 90	*g*-$ClCH_2CH_2Cl$ (880) *tr*-$N{\equiv}CCH_2CH_2Cl$ (878)	$Cl_3CCH_2CH_2Cl$ (705)	800 ± 60
νCCl	700 ± 60	*tr*-$N{\equiv}CCH_2CH_2Cl$ (755) *tr*-$ClCH_2CH_2Cl$ (754, *709*)	$ClCH_2CH_2CH_2Cl$ (679, 641) *tr*-DCH_2CH_2Cl (642)	705 ± 50
δC—C—	440 ± 120	$N{\equiv}CCH_2CH_2Cl$ (*tr*:494; *g*:565)	$H_2C{=}CHCH_2CH_2Cl$ (329)	400 ± 70
δC—C—Cl	295 ± 70	$ClCH_2CH_2CH_2Cl$ (362, 348)	*tr*-$ClCH_2CH_2Cl$ (300, 225)	280 ± 50
torsion CH_2Cl	155 ± 50	*g*-$N{\equiv}CCH_2CH_2Cl$ (202)	*g*-$BrCH_2CH_2Cl$ (107)	155 ± 35
torsion	–			

11. G.A. Crowder, *J. Mol. Struct.*, **10**, 290 (1971).
12. J.E. Saunders, J.J. Lucier and J.N. Willis Jr, *Spectrochim. Acta, Part A*, **24A**, 2027 (1968).
13. J.A. Thørbjornsrud, O.H. Ellestad, P. Klaboe and T. Torgrimsen, *J. Mol. Struct.*, **15**, 61 (1973).
14. A. Goursot-Leray, M. Carles-Lorjou, G. ouzard and H. Bodot, *Spectrochim. Acta, Part A*, **29A**, 1497 (1973).
15. M. Carles-Lorjou, A. Goursot-Leray and H. Bodot, *Spectrochim. Acta, Part A*, **29A**, 329 (1973).
16. A.B. Dempster, K. Price and N. Sheppard, *Spectrochim. Acta, Part A*, **27A**, 1563 (1971).
17. J. Som, D. Bhaumik, D.K. Mukherjee and G.S. Kastha, *Indian J. Pure Appl. Phys.*, **12**, 149 (1974).
18. W.A. Seth Paul, B. Van der Veken and M.A. Herman, *Can. J. Spectrosc.*, **27**, 21 (1982).
19. G.A. Crowder and N. Smyrl, *J. Mol. Struct.*, **8**, 255 (1971).
20. K. Tanabe, *J. Mol. Struct.*, **25**, 259 (1975).
21. E. Wyn-Jones and W.J. Orville-Thomas, *J. Chem. Soc.*, 101 (1966).
22. M.F. El Bermani and N. Jonathan, *J. Chem. Soc.*, 1712 (1968).
23. P. Klaboe and J. Grundnes, *Spectrochim. Acta, Part A*, **24A**, 1905 (1968).
24. T. Fujiyama, *Bull. Chem. Soc. Jpn.,*, **44**, 3317 (1971).
25. A.M. North, R.A. Pethrik and A.D. Wilson, *Spectrochim. Acta, Part A*, **30A**, 1317 (1974).
26. M. Nakata and M. Tasumi, *Spectrochim. Acta, Part A*, **41A**, 341 (1985).
27. E. Wyn-Jones and W.J. Orville-Thomas, *J. Mol. Struct.*,**1**, 79 (1967).
28. G. Davidovics, J. Pourcin, M. Carles, L. Pizzala and H. Bodot, *J. Mol. Struct.*, **99**, 165 (1983).
29. M. Perttilä, J. Murto and L. Halonen, *Spectrochim. Acta, Part A*, **34A**, 469 (1978).
30. H.F. Hameka, *J. Mol. Struct.*, **226**, 241 (1991).
31. P. Buckley, P.A. Giguère and M. Schneider, *Can. J. Chem.*, **47**, 901 (1969).
32. H. Matsuura, M. Kono, H. Iizuka, Y. Ogawa, I. Harada and T. Shimanouchi, *Bull. Chem. Soc. Jpn.*, **50**, 2272 (1977).
33. M. Hayashi, Y. Shiro, M. Murakami and H. Murata, *Bull. Chem. Soc. Jpn.*, **38**, 1740 (1965).
34. H. Matsuura, N. Miyauchi, H. Murata and M. Sakakibara, *Bull. Chem. Soc. Jpn.*, **52**, 344 (1979).
35. J.R. Durig, J. Liu and T.S. Little, *J. Phys. Chem.*, **95**, 4664 (1991).
36. K. Tanabe, *Spectrochim. Acta, Part A*, **28A**, 407 (1972).
37. Y. Duchesne and A. Van De Vorst, *J. Mol. Struct.*, **2**, 47 (1968).
38. K. Tanabe, *Spectrochim. Acta, Part A*, **30A**, 1901 (1974).
39. K. Tanabe, J. Hiraishi and T. Tamura, *J. Mol. Struct.*, **33**, 19 (1976).
40. S. Suzuki and A.B. Dempster, *J. Mol. Struct.*, **32**, 339 (1976).
41. S. Mizushima, T. Shimanouchi, I. Nakagawa and A. Miyake, *J. Chem. Phys.*, **21**, 215 (1953).
42. S.D. Christesen, *J. Raman Spectrosc.*, **22**, 459 (1991).

3.5.4 Hydroxyethyl

Molecules of type XCH_2CH_2OH (X = F, Cl, Br and I) in the *trans* conformation belong to the point group C_s. The 21 normal vibrations are divided over 13a′ and 8a″ species of vibration. Substitution of νCX (a′) by a torsion (a″) gives the 12a′ + 9a″ vibrations of hydroxyethyl.

a′: νOH, $\nu_s CH_2$ (2), δCH_2 (2), δOH, ωCH_2 (2), νCO, νCC, δCCO, δ—C—C;
a″: $\nu_a CH_2$ (2), τCH_2 (2), ρCH_2 (2), γOH and two torsions.

The 12 normal vibrations of the associated hydroxymethyl have already been studied (see Section 3.2.1): νOH, $\nu_a CH_2OH$, $\nu_s CH_2OH$, δCH_2OH, δOH, ωCH_2OH, τCH_2OH, νCO, ρCH_2OH, γOH, δ—CO and torsion —CH_2OH.

For the six vibrations of methylene ($\nu_a CH_2$, $\nu_s CH_2$, δCH_2, ωCH_2, τCH_2 and ρCH_2), more appropriate tables with absorption regions of R—CH_2— compounds are available.

The remaining normal vibrations are: a CC stretching vibration, an external —C—C skeletal deformation and a hydroxyethyl torsion.

The CC stretching vibration, which is also referred to as C—C—O in-phase stretching, exhibits a weak to medium band at 1015 ± 35 cm^{-1}, and is coupled with the stronger C—O stretching or C—C—O out-of-phase stretching mode. The skeletal —C—C deformation is an external vibration which, for the investigated compounds, appears weakly to moderately in the region 340 ± 60 cm^{-1}. The free OH torsion of alcohols in the unbonded state may be correlated with a weak absorption in the neighbourhood of 300 cm^{-1}.

Table 3.24 Absorption regions (cm^{-1}) of associated —CH_2CH_2OH

Vibration	Region	Vibration	Region
νOH	3300 ± 120	τCH_2	1210 ± 80
$\nu_a CH_2$	2980 ± 45	νCO	1065 ± 25
$\nu_a CH_2OH$	2950 ± 30	νCC	1015 ± 35
$\nu_s CH_2$	2920 ± 45	ρCH_2OH	900 ± 60
$\nu_s CH_2OH$	2885 ± 45	ρCH_2	820 ± 70
δCH_2OH	1450 ± 25	γOH	630 ± 60
δCH_2	1435 ± 25	CCO sk. def.	475 ± 75
δOH	1405 ± 35	—CC ext. sk. def.	340 ± 60
ωCH_2OH	1360 ± 30	torsion	–
ωCH_2	1260 ± 90	torsion	–
τCH_2OH	1240 ± 60		

R—CH_2CH_2OH molecules

R = Me— [1], Et— [2], MeC(=O)CH_2—, H_2NCH_2—, PhOCH_2—, ClCH_2—, BrCH_2—, $^-O_2CCH_2$— [3], H_2C=CH—, H_2C=C(Me)—, EtCH=CH—, 2-Th—, 4-MeOPh—, HC≡C— [4], N≡C—, EtC≡C—, H_2N—, ^+H_3N—, MeNH—, EtNH—, H_2NNH—, HOCH_2CH_2NHC(=S)C(=S)NH— [17], O_2N—, HO— [5, 6, 7], MeO—, EtO—, HS—, MeS—, F— [8–12], Cl— [9, 10, 13–15], Br— [9, 14, 16], I— [9, 16].

References

1. K. Fukushima and B.J. Zwolinski, *J. Mol. Spectrosc.*, **26**, 368 (1968).
2. G.A. Crowder and M.J. Townsend, *J. Mol. Struct.*, **42**, 27 (1977).
3. M. Morssli, G. Cassanas, L. Bardet, B. Pauvart and A. Terol, *Spectrochim. Acta*, **47A**, 529 (1991).
4. G.A. Crowder and E.W. Loya, *J. Mol. Struct.*, **62**, 297 (1980).
5. H. Matsuura and T. Miyazawa, *Bull. Chem. Soc. Jpn.*, **40**, 85 (1967).
6. W. Sawodny, K.K. Niedenzu and J.W. Dawson, *Spectrochim. Acta, Part A*, **23A**, 799 (1967).
7. H. Matsuura, M. Hiraishi and T. Miyazawa, *Spectrochim. Acta, Part A*, **28A**, 2299 (1972).
8. O. Schrems and W.A.P. Luck, *J. Mol. Struct.*, **80**, 477 (1982).
9. E. Wyn-Jones and W.J. Orville-Thomas, *J. Mol. Struct.*, **1**, 79 (1967).
10. G. Davidovics, J. Pourcin, M. Carles, L. Pizzala and H. Bodot, *J. Mol. Struct.*, **99**, 165 (1983).
11. M. Perttilä, J. Murto, A. Kivinen and K. Turunen, *Spectrochim. Acta, Part A*, **34A**, 9 (1978).
12. G. Davidovics, J. Pourcin, M. Monnier, P. Verlaque and H. Bodot, *J. Mol. Struct.*, **116**, 39 (1984).
13. H.F. Hameka, *J. Mol. Struct.*, **226**, 241 (1991).
14. P. Buckley, P.A. Giguère and M. Schneider, *Can. J. Chem.*, **47**, 901 (1969).
15. M. Perttilä, J. Murto and L. Halonen, *Spectrochim. Acta, Part A*, **34A**, 469 (1978).
16. L. Homanen, *Spectrochim. Acta, Part A*, **39A**, 77 (1983).
17. P. Geboes, H. Hofmans, H.O. Desseyn, R. Dommisse, A.T.H. Lenstra, S.B. Sanni, M.M. Smits and P.T. Beurskens, *Spectrochim. Acta, Part A*, **43A**, 35 (1987).

3.5.5 n-Propyl

The simplest compounds with a n-propyl unit in the structure are $MeCH_2CH_2X$ (X = halogen). The $3N - 6 = 27$ normal vibrations are divided into $16a' + 11a''$ types of vibration. Substitution of the CX stretching vibration (a′) by a torsion (a″) yields the 27 normal vibrations ($15a' + 12a''$) for n-propyl:

a′: ν_a'Me, ν_sMe, $\nu_s CH_2$ (2), δ_a'Me, δCH_2 (2), δ_sMe, ωCH_2 (2), ρ'Me, νCC (2), skeletal deformations (2);
a″: ν_aMe, $\nu_a CH_2$ (2), δ_aMe, τCH_2 (2), ρMe, ρCH_2 (2) and torsions (3).

Consequently, n-propyl has nine vibrations more with respect to ethyl: $\nu_a CH_2$, $\nu_s CH_2$, δCH_2, ωCH_2, τCH_2, ρCH_2, νCC, δ—C—C and torsion —CH_2CH_2Me.

The four CH antisymmetric stretchings in propyl exhibit moderate to strong absorptions between 2990 and 2900 cm^{-1}, which may fuse together into one band. The three CH symmetric stretchings are moderately to strongly active, coincident or not, between 2940 and 2840 cm^{-1}. The methylene scissors and the methyl antisymmetric deformations also tend to amalgamate, but the methyl symmetric deformation appears totally free at about 1375 cm^{-1}. The methylene wags and twists give separate absorptions but for lack of data in relation to the band structure, the wags are assigned at higher wavenumbers than the twists. The notation ρ'Me/νCH_2—Me and νCH_2—Me/ρ'Me illustrates that the two vibrations are strongly coupled. The absorption frequency of the methylene rockings depends largely upon the surroundings. Ciampelli and Tosi found values around 739 cm^{-1} for the ρCH_2 in saturated hydrocarbons with a propyl unit [51].

Table 3.25 Absorption regions (cm^{-1}) of $—CH_2CH_2CH_3$

	Total region	Molecules absorbing at high wavenumbers	Molecules absorbing at low wavenumbers	Region of most molecules
ν_aMe	2975 ± 15	nPrCl (2990)		2970 ± 10
ν'_aMe	2965 ± 15	nPrCl (2978) $nPrCH_2Cl$ (2974)		2960 ± 10
$\nu_a CH_2Et$	2965 ± 40	nPrCl (3005) nPrBr (3003) nPrF (2995)	nPrSH (2925) nPrOH (2929) $nPrNH_2$ (2933)	2960 ± 25
$\nu_a CH_2Me$	2935 ± 35		$nPrCH_2Cl$ (2900) $nPrCH_2F$ (2905)	2945 ± 25
$\nu_s CH_2Et$	2905 ± 55	nPrCl (2958) nPrF (2944)	$nPrNH_2$ (2850) nPrSH (2855)	2900 ± 40
ν_sMe	2890 ± 50	4-X—Ph—C(=O)nPr (2934) (X = F, Cl, Br, OH)	nPrI (2840) nPrBr (2846)	2890 ± 40
$\nu_s CH_2Me$	2865 ± 25	nPrCl (2887) nPrF (2883)	nPrI (2840) nPrBr (2846) $nPrCH_2X$ (2846) (X = F, Cl)	2865 ± 15
δ_aMe	1465 ± 10			1465 ± 10
δ'_aMe	1450 ± 15		nPrBr (1435) nPrCl (1442)	1455 ± 10
δCH_2Me	1460 ± 20	$nPrCH_2NH_2$ (1475) $nPrNH_2$ (1470) $nPrCH_2Cl$ and nPrCl (1468)	$nPrCH_2SiH_3$ (1443)	1455 ± 10
δCH_2Et	1430 ± 30	$nPrNH_2$ (1456) $nPrND_2$ (1455)	4-X—Ph—C(=O)nPr (1405) $nPrSiH_3$ (1411) (X = F, Cl, Br) $nPrCH_2SiH_3$ (1415) $nPrSiH_2Me$ (1416)	1435 ± 15
δ_sMe	1375 ± 15	3-Br—Ph—nPr (1390) $nPrCH_2F$ (1387)	EtC(=O)NHnPr (1369) 4-F—Ph—C(=0)nPr (1370)	1379 ± 06
ωCH_2Me	1335 ± 30	$nPrCH_2NH_2$ (1364)	$nPrCH_2SiH_3$ (1307)	1340 ± 20
ωCH_2Et	1270 ± 85	$nPrNH_2$ (1353) nPrOH (1340)	nPrI (tr:1186; *g*:1202) nPrBr (tr:1228; *g*:1233)	1290 ± 40
τCH_2Me	1245 ± 45	$nPrCH_2C{\equiv}CH$ (1290) MeC(=O)NHnPr (1289)	$nPrSiH_3$ (1207) $nPrSiD_3$ (1208) $nPrSiH_2Me$ (1212)	1250 ± 35
τCH_2Et	1215 ± 65	$nPrNH_2$ (1274)	nPrI (1158)	1225 ± 35

(*continued*)

Table 3.25 (*continued*)

	Total region	Molecules absorbing at high wavenumbers	Molecules absorbing at low wavenumbers	Region of most molecules
ρMe	1105 ± 45	nPrC≡CMe (1143)	$nPrSiD_2Me$ (1062) $nPrSiH_2Me$ (1067)	1105 ± 35
ρ'Me/	1055 ± 45	$nPrNO_2$ (1097)	nPrI (1010)	1055 ± 35
νCH_2—Me		nPrC≡CMe (1094)		
νCH_2—Et	995 ± 65	$nPrCH_2NH_2$ (1060) $nPrCH_2Cl$ (1051)	EtC(=O)NHnPr (931) nPrC≡N (943)	995 ± 45
νCH_2—Me/	925 ± 45	$nPrCH_2NH_2$ (970)	nPrI (880) and nPrBr (886)	925 ± 35
ρ'Me		$nPrCH_2F$ (970)	2-HO—Ph—C(=O)nPr (880) $nPrSiD_2Me$ (885)	
ρCH_2Et	810 ± 85	nPrSH (tr:890; *g*:881)	$nPrSiH_3$ (tr:728; *g*:787) nPrSH (tr:736; *g*:780) nPrSCN (782)	835 ± 45
ρCH_2Me	780 ± 55	$nPrSiH_3$ (tr:830; *g*:824)	nPrI (727)	760 ± 30
C—C—C skelet.def.	380 ± 95	$nPrCH_2F$ (473) nPrF (465)	nPrI (290) nPrBr (312)	400 ± 60
—C—C ext. sk. def.	270 ± 95		nPrC≡CH (175) nPrC≡N and $nPrSiD_3$ (188)	280 ± 85
torsion Me	225 ± 45	nPrC(=O)H (270)	nPrCl (187)	220 ± 30
torsion Et	120 ± 30			120 ± 30
torsion nPr	–	–		

R—nPr molecules

R = Me— [52, 53], $tBuCH_2$— [1], $Cl_3CC(=O)CH_2$— [2], $HC\equiv CCH_2$— [3, 4], $N\equiv CCH_2$— [5], H_2NCH_2— [6–8], $MeNHC(=O)NHCH_2$— [9], $HOCH_2$— [10], $iPrSCH_2$— [11], FCH_2— [12], $ClCH_2$— [13, 14], $BrCH_2$— [14, 15], ICH_2—[14, 16], H_3SiCH_2— and D_3SiCH_2— [17], HC(=O)— [18], ClC(=O)— and BrC(=O)— [19], 4-XPhC(=O)— (X = F, Cl and Br) [20], $3\text{-}O_2NPhC(=O)$—, 2-HOPhC(=O)— [21, 23], 4-HOPh—C(=O)— [23], Ph—, 3-BrPh— [22], HC≡C— [24, 25], MeC≡C— [26], N≡C— [27–29], H_2N— and D_2N— [6, 30–32], MeC(=O)NH— and EtC(=O)NH— [33], MeNHC(=O)NH— [9], O_2N— [34], HO— [35], $Cl_3C(=O)O$— [2], HS— [36–39], EtS— [40], iPrS— [11], NCS— [41], F— [42], Cl— [13, 14, 43–48, 50], Br— [14, 15, 43–47, 49], I— [14, 16, 43, 45, 46], H_3Si— and D_3Si— [17], $MeSiH_2$— and $MeSiD_2$— [17].

References

1. G.A. Crowder and R.M.P. Jaiswal, *J. Mol. Struct.*, **102**, 145 (1983).
2. Y. Mido, N. Komatsu, J. Morcillo and M.V. Garcia, *J. Mol. Struct.*, **172**, 49 (1988).

3. R.F. Kendall, *Spectrochim. Acta, Part A*, **24A**, 1839 (1968).
4. G.A. Crowder, *J. Mol. Struct.*, **172**, 151 (1988).
5. G.A. Crowder, *J. Mol. Struct.*, **200**, 235 (1989).
6. H. Wolff and H. Ludwig, *Ber. Bunsenges. Phys. Chem.*, **71**, 1107 (1967).
7. J.J.C. Teixeira-Dias, L.A.E. Batista De Carvalho, A.M. Amorim Da Costa, I.M.S. Lampreia and E.F.G. Barbosa, *Spectrochim. Acta, Part A*, **42A**, 589 (1986).
8. G. Ramis and G. Busca, *J. Mol. Struct.*, **193**, 93 (1989).
9. Y. Mido, F. Fujita, H. Matsuura and K. Machida, *Spectrochim. Acta, Part A*, **37A**, 103 (1981).
10. G.A. Crowder and M.J. Townsend, *J. Mol. Struct.*, **42**, 27 (1977).
11. M. Ohsaku, H. Murata and Y. Shiro, *Spectrochim. Acta, Part A*, **33A**, 467 (1977).
12. G.A. Crowder and H.K. Mao, *J. Mol. Struct.*, **23**, 161 (1974).
13. A.J. Barnes, M.L. Evans and H.E. Hallam, *J. Mol. Struct.*, **99**, 235 (1983).
14. Y. Ogawa, S. Imazeki, H. Yamaguchi, H. Matsuura, I. Harada and T. Shimanouchi, *Bull. Chem. Soc. Jpn.,*. **51**, 748 (1978).
15. G.A. Crowder and M.R. Jalilian, *Can. J. Spectrosc.*, **22**, 1 (1977).
16. G.A. Crowder and S. Ali, *J. Mol. Struct.*, **25**, 377 (1975).
17. H. Murata, H. Matsuura, K. Ohno and T. Sato, *J. Mol. Struct.*, **52**, 1 and 13 (1979).
18. G. Sbrana and V. Schettino, *J. Mol. Spectrosc.*, **33**, 100 (1970).
19. J. Som. D. Bhaumik, D.K. Mukherjee and G.S. Kastha, *Indian J. Pure Appl. Phys.*, **12**, 149 (1974).
20. W.A. Seth Paul and J. Meeuwesen, *Bull. Soc. Chim. Belg.*, **90**, 127 (1981).
21. W.A.L.K. Al-Rashid and M.F. El-Bermani, *Spectrochim. Acta, Part A*, **47A**, 35 (1991).
22. S. Chattopadhyay, L. Chakravorti and G.S. Kastha, *Indian J. Pure Appl. Phys.*, **25**, 456 (1987).
23. G. Varsányi, *Assignments for Vibrational Spectra of Seven Hundred Benzene Derivatives*, J.Wiley & Sons, New York (1974).
24. G.A. Crowder and H. Fick, *J. Mol. Struct.*, **147**, 17 (1986).
25. G.A. Crowder, *Spectrochim. Acta, Part A*, **42A**, 941 (1986).
26. G.A. Crowder and P. Blankenship, *J. Mol. Struct.*, **196**, 125 (1989).
27. S.W. Charles, F.C. Cullen and N.L. Owen, *J. Mol. Struct.*, **34**, 219 (1976).
28. G.A. Crowder, *J. Mol. Struct.*, **158**, 229 (1987).
29. T. Fujiyama, *Bull. Chem. Soc. Jpn.*, **44**, 3317 (1971).
30. L.A.E. Batista de Carvalho, A.M. Amorim da Costa, M.L. Duarte and J.J.C. Teixeira-Dias, *Spectrochim. Acta, Part A*, **44A**, 723 (1988).
31. N. Sato, Y. Hamada and M. Tsuboi, *Spectrochim. Acta, Part A*, **43A**, 943 (1987).
32. J.R. Durig, W.B. Beshir, S.E. Godbey and T.J. Hizer, *J. Raman Spectrosc.*, **20**, 311 (1989).
33. J. Jakes and S. Krimm, *Spectrochim. Acta, Part A*, **27A**, 35 (1971).
34. G. Geiseler and H. Kessler, *Ber. Bunsenges Phys. Chem.*, **68**, 571 (1964).
35. K. Fukushima and B.J. Zwolinski, *J. Mol. Spectrosc.*, **26**, 368 (1968).
36. T. Torgrimsen and P. Klaboe, *Acta Chem. Scand., Ser. A*, **24**, 1139 (1970).
37. D.W. Scott and M.Z. El-Sabban, *J. Mol. Spectrosc.*, **30**, 317 (1969).
38. K.G. Allum, J.A. Creighton, J.H.S. Green, G.J. Minkoff and L.J.S. Prince, *Spectrochim. Acta, Part A*, **24A**, 928 (1968).
39. M. Hayashi, Y. Shiro and H. Murata, *Bull. Chem. Soc. Jpn.*, **39**, 112 (1966).
40. M. Otha, Y. Ogawa, H. Matsuura, I. Harada and T. Shimanouchi, *Bull. Chem. Soc. Jpn.*, **50**, 380 (1977).
41. R.P. Hirschmann, R.N. Kniseley and V.A. Fassel, *Spectrochim. Acta*, **20**, 809 (1964).
42. G.A. Crowder and H.K. Mao, *J. Mol. Struct.*, **18**, 33 (1973).
43. J.K. Brown and N. Sheppard, *Trans. Faraday Soc.*, **50**, 1164 (1954).

44. C. Kombaki, I. Ichishima, K. Kutanani, T. Miyazawa, T. Shimanouchi and S. Mizushima, *Bull. Chem. Soc. Jpn.*, **28**, 330 (1955).
45. K. Tanabe and S. Saëki, *J. Mol. Struct.*, **27**, 79 (1975).
46. N.T. McDevitt, A.L. Rozek, F.F. Bentley and A.D. Davidson, *J. Chem. Phys.*, **42**, 1173 (1965).
47. K. Radcliffe and J.L. Wood, *Trans. Faraday Soc.*, **62**, 1678 (1966).
48. R.G. Snyder and J.H. Schachtschneider, *J. Mol. Spectrosc.*, **30**, 290 (1969).
49. M. Hayashi, K. Ohno and H. Murata, *Bull. Chem. Soc. Jpn.*, **46**, 2332 (1973).
50. V.A. Pozdyshev, Yu.A. Pentin and V.M. Tatevskii, *Opt. Spectrosc.*, **3**, 211 (1957).
51. F. Ciampelli and C. Tosi, *Spectrochim. Acta, Part A*, **24A**, 2158 (1968).
52. J.R. Durig, A. Wang, W. Beshir and T.S. Little, *J. Raman Spectrosc.*, **22**, 683 (1991).
53. P. Derreumaux, M. Dauchez and G. Vergoten, *J. Mol. Struct.*, **295**, 203 (1993).

3.5.6 Propynyl

The $XCH_2C{\equiv}CH$ compounds (X = F, Cl, Br, I) belong to the point group C_s and the $3N - 6 = 15$ normal vibrations are divided into 10a′ and 5a″ types of vibration. Substitution of νCX by a torsion leads to the 15 normal vibrations (9a′ + 6a″) of $-CH_2C{\equiv}CH$:

a′: νCH, $\nu_s CH_2$, $\nu C{\equiv}C$, δCH_2, ωCH_2, νC—C, δCH, δ—C—C, δC—C≡C;
a″: $\nu_a CH_2$, τCH_2, ρCH_2, γCH, γC—C≡C and torsion.

The CH stretching vibration

The ≡CH stretch, an excellent group vibration, is characterized by the very sharp and strong absorption in the range $3315 \pm 25\ cm^{-1}$; this is the highest CH stretching vibration, since the sp^3- and sp^2-hybridized CH bonds show the CH stretching at lower frequencies. The band of the NH stretching vibration in secondary amides (Section 7.3) and amines (Section 9.2), also occurring in this region, is broader and not so strong. Nyquist found values of $3332 \pm 8\ cm^{-1}$ for a series of 1-alkynes with the basic structure $Me(CH_2)_nC{\equiv}CH$ (n = 1–14) [32].

Methylene stretching vibrations

The methylene antisymmetric stretch appears weakly to moderately at 2970 $\pm$ 40 cm^{-1}. High values are found in the spectra of $ClC({=}O)OCH_2C{\equiv}CH$ (3010), $ICH_2C{\equiv}CH$ (3008) and $BrCH_2C{\equiv}CH$ (3006 cm^{-1}) and low values in those of $H_2NCH_2C{\equiv}CH$ (2930), $HC{\equiv}CCH_2CH_2C{\equiv}CH$ (2938 and 2945) and $MeCH_2C{\equiv}CH$ (2939 cm^{-1}). There is a great change to find the $\nu_a CH_2$ in the region $2970 \pm 30\ cm^{-1}$.

The methylene symmetric stretch is assigned in the range $2935 \pm 40\ cm^{-1}$. If the contribution of the absorption at $\approx$2875 cm^{-1} to this $\nu_s CH_2$ is taken into account, the region increases to $2915 \pm 60\ cm^{-1}$. $BrCH_2C{\equiv}CH$ (2975) and $ClCH_2C{\equiv}CH$ (2968 cm^{-1}) give high values. In the spectra of $MeOCH_2CH_2C{\equiv}CH$ (ν_s 2935 and 2855) and $HOCH_2CH_2C{\equiv}CH$ (ν_s 2928 and 2890 cm^{-1}) the lower value may be assigned to ν_sCH_2O and the higher to ν_sCH_2C, but $H_2NCH_2C{\equiv}CH$ and $HOCH_2C{\equiv}CH$ exhibit a medium band at 2855 and 2873 cm^{-1} respectively which is assigned to the ν_sCH_2.

Although the regions of ν_aCH_2 and ν_sCH_2 overlap each other, the two vibrations appear at different wavenumbers.

The C≡C stretching vibration

With a range of 2125 ± 25 cm^{-1} and a weak to moderate band intensity, the C≡C stretching vibration takes a unique place as a good group vibration. Because of the influence of the halogen, $FCH_2C{\equiv}CH$ (2150) and $ClCH_2C{\equiv}CH$ (2147) absorb at high values. This band shifts to lower wavenumbers in the spectra of $MeOCH_2C{\equiv}CH$ and $EtOCH_2C{\equiv}CH$ (2115) and $H_2NCH_2C{\equiv}CH$, $MeHNCH_2C{\equiv}CH$ and $Me_2NCH_2C{\equiv}CH$ (2100 cm^{-1}). The C≡C stretching frequencies of ten 1-alkynes in the gas phase are reported as being in the range 2147 ± 15 cm^{-1} [32].

Methylene deformations

The methylene scissoring vibration is moderately active in the region 1440 ± 25 cm^{-1}, in which other CH_2 scissors also absorb.

The methylene wag is assigned in a range (1305 ± 75 cm^{-1}) in which also overtones of δ ≡CH and γ ≡CH give rise to a weak but broad typical band. The HW sidc of this region is limited by $FCH_2C{\equiv}CH$ (1375), $Cl_2P({=}S)OCH_2C{\equiv}CH$ (1367) and $ClC({=}O)OCH_2C{\equiv}CH$ (1366 cm^{-1}) and the LW side by $NCSCH_2C{\equiv}CH$ with 1235 for the anti conformer and 1247 cm^{-1} for the *gauche*. The low values observed in the spectra of $ICH_2C{\equiv}CH$ (1160) (Section 3.1.4) and $BrCH_2C{\equiv}CH$ (1218) (Section 3.1.3) are outside the above—mentioned region.

The twist is assigned at 1255 ± 80 cm^{-1}, with the highest values for $H_2NCH_2C{\equiv}CH$ (1335), $MeOCH_2C{\equiv}CH$ (1284) and $ClC({=}O)OCH_2C{\equiv}CH$ (1270 cm^{-1}) and the lowest for $NCSCH_2C{\equiv}CH$ (1179) and $ClCH_2C{\equiv}CH$ (1179 cm^{-1}). The lower wavenumbers from $ICH_2C{\equiv}CH$ (1116) and $BrCH_2C{\equiv}CH$ (1152 cm^{-1}) are not taken into account.

The C—C stretch and methylene rock

Based on spectra of analogous compounds, it appears that the methylene rock absorbs at lower wavenumbers than the C—C stretching vibration. Nevertheless some spectroscopists assign higher wavenumbers to the rock. Table 3.26 shows that the spectroscopists are not always unanimous in assigning the C—C stretch and the methylene rock.

Table 3.26

R—$CH_2C{\equiv}CH$	νCC	ρCH_2	ref.
HC≡CCH_2—	1021	950	[7]
$HOCH_2$—	1020	950	[8]
Me—	1008	782	[2]
H_2C=C=CH—	1005	890	[10]
ClC(=O)O—	992	937	[21]
Br—	961	866	[24, 25]
Cl—	960	908	[25, 26]
MeO—	938	899	[16]
Me—	840	782	[1]
F—	938	1018	[15, 23, 24]
EtO—	920	1016	[19]
MeO—	893	1007	[17, 18]
HO—	917	980	[15]
DO—	907	971	[15]

In this text the ρCH_2 is assigned at a lower wavenumber (865 $\pm$ 85) than that of the C—C stretching vibration (930 $\pm$ 90). Nyquist and Potts assign the νC—C in a few $RCH_2C{\equiv}CH$ compounds at 930 $\pm$ 30 cm^{-1} [31].

CH deformations

The in-plane and out-of-plane ≡CH deformations are moderately to strongly active in the regions 660 $\pm$ 30 and 635 $\pm$ 15 cm^{-1}. In addition to the νCH and νC≡C, these deformations take a special place as group vibrations of —$CH_2C{\equiv}CH$. Often the two deformations coincide and give rise to one broad band because there is only a slight difference in energy between in-plane and out-of-plane deformations in molecules with a straight chain. In the spectra of $FCH_2C{\equiv}CH$ (675 and 635) and $ClC(=O)OCH_2C{\equiv}CH$ (689 and 645 cm^{-1}) the δ ≡CH and γ ≡CH are clearly separated, but in those of $MeCH_2C{\equiv}CH$ (634 and 630) and $EtCH_2C{\equiv}CH$ (634 and 629 cm^{-1}) both deformations absorb closely together because of the small difference in electronegativity. Higher 1-alkynes exhibit one broad band which for $Me(CH_2)_nC{\equiv}CH$ (n = 1–14) is situated at 631 $\pm$ 3 cm^{-1} [32]. Overtones of these deformations give a weak but broad absorption in the region of the CH_2 wag and twist. Such a typical band is usually seen in the spectra of 1-alkynes at 1250 $\pm$ 10 cm^{-1} [31].

Skeletal deformations

The vibrational analysis reveals that the external skeletal R—C—C deformation is in the range 505 $\pm$ 70 cm^{-1}, in which the low values for $BrCH_2C{\equiv}CH$ (399) and $ICH_2C{\equiv}CH$ (364 cm^{-1}) are not included. The utility of the out-of-plane and in-plane C—C≡C skeletal deformations, which are weakly active in the respective ranges 330 ± 35 and 195 ± 45 cm^{-1}, is very limited for identification purposes.

Table 3.27 Absorption regions (cm^{-1}) of the normal vibrations of $—CH_2C≡CH$

Vibration	Region	Vibration	Region
νCH	3315 ± 25	ρCH_2	865 ± 85
$\nu_a CH_2$	2970 ± 40	δCH	660 ± 30
$\nu_s CH_2$	2915 ± 60	γCH	635 ± 15
νC≡C	2125 ± 25	—C—C ext. sk. def.	505 ± 70
δCH_2	1440 ± 25	γC—C≡C	330 ± 35
ωCH_2	1305 ± 75	δC—C≡C	195 ± 45
τCH_2	1255 ± 80	torsion	–
νC—C	930 ± 90		

The following $R—CH_2C≡CH$ molecules are taken into account:

R = D— [30], Me— [1–3], CD_3— [1], Et— [3–5], nPr— [5, 6], nBu— and nPent— [5], $HC≡CCH_2$— [7], $HOCH_2$— [8], $MeOCH_2$—, iPr— [9], $H_2C=C=CH$— [10], H_2N— [11, 12], O_2N— [11], MeHN—, Me_2N—, N_3— [13], HO— [14, 15], DO— [15], MeO— [16–18], EtO— [19], HC(=O)O— [20], ClC(=O)O— [21], $Cl_2P(=S)O$— [21, 23], NCS— [22], F— [15, 23, 24], Cl— [24–27], Br— [24, 25, 28, 29], I— [24].

References

1. J. Saussey, J. Lamotte and J.C. Lavalley, *Spectrochim. Acta, Part A*, **32A**, 763 (1976).
2. G.A. Crowder, *Spectrosc. Lett.*, **20**, 343 (1987).
3. G.A. Crowder and H. Fick, *J. Mol. Struct.*,**147**, 17 (1986).
4. G.A. Crowder, *Spectrochim. Acta, Part A*, **42A**, 941 (1986).
5. R.F. Kendall, *Spectrochim. Acta, Part A*, **24A**, 1839 (1968).
6. G.A. Crowder, *J. Mol. Struct.*, **172**, 151 (1988).
7. D.L. Powell, P. Klaboe, A. Phongsatha, B.N. Cyvin, S.J. Cyvin and H. Hopf, *J. Mol. Struct.*,**41**, 203 (1978).
8. G.A. Crowder and E.W. Loya, *J. Mol. Struct.*, **62**, 297 (1980).
9. G.A. Crowder, *J. Mol. Struct.*, **193**, 307 (1989).
10. P. Klaboe, A. Phongsatha, B.N. Cyvin, S.J. Cyvin and H. Hopf, *J. Mol. Struct.*, **43**, 1 (1978).
11. Y. Hamada, M. Tsuboi, M. Nakata and M. Tasumi, *J. Mol. Spectrosc.*, **107**, 269 (1984).
12. N.V. Riggs, *Aust. J. Chem.*, **40**, 435 (1987).
13. J. Almlof, G.O. Braathen, P. Klaboe, C.J. Nielsen and H. Priebe, *J. Mol. Struct.*, **160**, 1 (1987).
14. J. Travert, J.C. Lavalley and D. Chenery, *Spectrochim. Acta, Part A*, **35A**, 291 (1971).
15. R.A. Nyquist, *Spectrochim. Acta, Part A*, **27A**, 2513 (1971).

16. A. Bjørseth and J. Gustavsen, *J. Mol. Struct.*, **23**, 301 (1974).
17. W.A. Seth Paul, J.P. Tollenaere, H. Meeusen and F. Höfler, *Spectrochim. Acta, Part A*, **30A**, 193 (1974).
18. S.W. Charles, F.C. Cullen, G.I.L. Jones and N.L. Owen, *J. Chem. Soc., Faraday Trans. 2*, **70**, 758 (1974).
19. S.W. Charles, F.C. Cullen and N.L. Owen, *J. Chem. Soc. Faraday Trans. 2*, **72**, 351 (1976).
20. G.I.L. Jones, D.G. Lister and N.L. Owen, *J. Chem. Soc. Faraday Trans. 2*, **71**, 1330 (1975).
21. R.A. Nyquist, *Spectrochim. Acta, Part A*, **28A**, 285 (1972).
22. T. Midtgaard, G. Gundersen and C.J. Nielsen, *J. Mol. Struct.*, **176**, 159 (1988).
23. R.A. Nyquist and W.W. Muelder, *J. Mol. Struct.*, **2**, 465 (1968).
24. J.C. Evans and R.A. Nyquist, *Spectrochim. Acta*, **19**, 1153 (1963).
25. R.A. Nyquist, A.L. Johnson and Y.S. Lo, *Spectrochim. Acta*, **21**, 77 (1965).
26. R.A. Nyquist, T.L. Reder, G.R. Ward and G.J. Kallos, *Spectrochim. Acta, Part A*, **27A**, 541 (1971).
27. E. Hirota and Y. Morino, *Bull. Chem. Soc. Jpn.*, **34**, 341 (1961).
28. E. Kikuchi, E. Hirota and Y. Morino, *Bull. Chem. Soc. Jpn.*,. **34**, 348 (1961).
29. R.A. Nyquist, T.L. Reder, F.F. Stec and G.J. Kallos, *Spectrochim. Acta, Part A*, **27A**, 897 (1971).
30. H. Priebe, C.J. Nielsen and P. Klaboe, *Spectrochim. Acta, Part A*, **36A**, 1017 (1980).
31. R.A. Nyquist and W.J. Potts, *Spectrochim. Acta*, **16**, 419 (1960).
32. R.A. Nyquist, *Appl. Spectrosc.*, **39**, 1088 (1985).

3.5.7 Cyanomethyl

The simplest compounds that contain the —$CH_2C{\equiv}N$ structure unit are $XCH_2C{\equiv}N$ (X = F, Cl, Br, I). The 12 normal vibrations are composed of 8a′ and 4a″ species of vibration. Substitution of νCX (a′) by a torsion (a″) gives the 12 vibrations of —$CH_2C{\equiv}N$:

a′: $\nu_s CH_2$, $\nu C{\equiv}N$, δCH_2, ωCH_2, νC—C, δ—C—C, δC—C≡N;
a″: $\nu_a CH_2$, τCH_2, ρCH_2, γC—C≡N and torsion.

Methylene stretching vibrations

The $\nu_a CH_2$ is found in the range 2980 ± 50 cm^{-1}. The highest wavenumbers (3020 ± 10 cm^{-1}) are due to $XCH_2C{\equiv}N$ (X = F, Cl, Br, I) (see Section 3.1), followed by *trans*-$N{\equiv}CCH_2CH_2C{\equiv}N$ (2998) and $N_3CH_2C{\equiv}N$ (2996 cm^{-1}). The lowest wavenumbers have been traced in the spectra of $H_2C{=}CHCH_2C{\equiv}N$ (2936) and $H_2C{=}C(Me)CH_2C{\equiv}N$ (2947 cm^{-1}). Most of the investigated molecules were found to give this antisymmetric stretching in the region 2970 ± 30 cm^{-1}. The symmetric counterpart appears in the range 2915 ± 65 cm^{-1} with compounds such as $XCH_2C{\equiv}N$ (X = F, Cl, Br) (2973 ± 4 cm^{-1}) at the HW side and $MeSCH_2C{\equiv}N$ (2857), $BrCH_2CH_2C{\equiv}N$ (2857), $nPrCH_2C{\equiv}N$ (2875) and $EtCH_2C{\equiv}N$ (2885 cm^{-1}) at the LW side. The remaining molecules show the $\nu_s CH_2$ at 2930 ± 40 cm^{-1}.

The C≡N stretching vibration

The —CH_2C≡N group is best recognized by the C≡N stretching vibration in the narrow region 2260 ± 15 cm^{-1}. The intensity of the sharp peak is moderate to strong with the exception of the νC≡N in XCH_2C≡N, for which X presents a strongly negative group (e.g. FCH_2C≡N). The C≡C stretching vibration absorbs upwards of 100 cm^{-1} lower with a much weaker intensity (Section 3.5.6). The highest values for the C≡N stretching vibration have been attributed in the spectra of N≡CCH_2C≡N (2275), H_2NC(=O)CH_2C≡N (2271) and HO(O=)CCH_2C≡N (2271 cm^{-1}), the lowest in those of CD_3CH_2C≡N, $cPrCH_2C$≡N, $MeSCH_2C$≡N and ICH_2C≡N in the neighbourhood of 2245 cm^{-1}. The νC≡N in H_2C=C(Me)CH_2C≡N and H_2C=$CHCH_2C$≡N is situated at 2247 cm^{-1}. The lower absorption band (2220 cm^{-1}) in the spectrum of H_2C=C(Me)CH_2C≡N [30] comes probably from the conjugated C≡N of the isomers [29]. Conjugated organonitriles, however, exhibit νC≡N at lower frequencies (2225 ± 10) than non-conjugated aliphatic nitriles (2245 ± 15 cm^{-1}) [42, 43].

Methylene deformations

The methylene scissors absorbs in the range 1425 ± 30 cm^{-1} with a moderate to strong intensity. This range is limited by *trans*-N≡CCH_2CH_2C≡N (1455 and *1445*) and FCH_2C≡N (1453 cm^{-1}) at the HW side and by N≡CCH_2C≡N (1395), $MeSCH_2C$≡N (1400), ICH_2C≡N (1408) and $BrCH_2C$≡N (1410 cm^{-1}) at the LW side. The methylene scissors is usually observed at 1430 ± 20 cm^{-1}.

The methylene wag is weakly to moderately active in the region 1295 ± 65 cm^{-1}, in which the extreme values from FCH_2C≡N (1381), ICH_2C≡N (1155) and $BrCH_2C$≡N (1220 cm^{-1}) are not included (Section 3.1). Disregarding also the values 1359, 1353 and 1237 cm^{-1} in the spectra of N≡CCH_2CH_2C≡N, $MeOCH_2C$≡N and $MeSCH_2C$≡N, most of the R—CH_2C≡N compounds absorb in the region 1295 ± 45 cm^{-1}.

Disregarding the low values of $BrCH_2C$≡N (1155) and ICH_2C≡N (1100 cm^{-1}), the methylene twist is assigned in the region 1230 ± 55 cm^{-1}. High values originate from the spectra of $MeOCH_2C$≡N (1285) and $MeCH_2C$≡N (1272 cm^{-1}) and low values from those of N≡CCH_2CH_2C≡N (1178 and *1232*), $MeSCH_2C$≡N (1184) and $ClCH_2C$≡N (1185 cm^{-1}). Usually the twist is assigned at 1230 ± 40 cm^{-1}, well separated from the wag.

C—C stretching and methylene rocking vibration

It is no easy matter to establish the absorption regions of the C—C stretch and the CH_2 rock from published data, probably because of their coupling one to the other. With reservations, it is assumed that the C—C stretch absorbs in the region 955 ±

65 cm^{-1} and the CH_2 rock at 840 ± 80 cm^{-1}. In the examples in Table 3.28 the maximum and minimum values are included.

Table 3.28

Compound	$\nu C—C$	ρCH_2
$FCH_2C\equiv N$	1019	911
trans-$N\equiv CCH_2CH_2C\equiv N$	1005, 951	917, 762
gauche-$N\equiv CCH_2CH_2C\equiv N$	975, 963	818, 816
$MeCH_2C\equiv N$	1004	784
$MeSCH_2C\equiv N$	965	844
$cPrCH_2C\equiv N$	960	830
$H_2C=CHCH_2C\equiv N$	936	865
$N\equiv CCH_2C\equiv N$	936	893
$ClCH_2CH_2C\equiv N$	919	885
$BrCH_2CH_2C\equiv N$	896	800

Skeletal deformations

Setting aside the low values from halogen-bonded $—CH_2C\equiv N$ (Section 3.1), the external —C—C skeletal deformation, strongly coupled with $\delta C—C\equiv N$, is assigned in the range 535 ± 70 cm^{-1}. In the spectrum of *gauche*-$N\equiv CCH_2CH_2C\equiv N$ the absorptions at 604 and 480 cm^{-1} are attributed to this skeletal deformation. The out-of-plane $C—C\equiv N$ skeletal deformation is weakly active at 365 ± 25 cm^{-1} and the in-plane $C—C\equiv N$ deformation at 225 ± 65 cm^{-1}.

Table 3.29 Absorption regions (cm^{-1}) of the normal vibrations of $—CH_2C\equiv N$

Vibration	Region	Vibration	Region
$\nu_a CH_2$	2980 ± 50	$\nu C—C$	955 ± 65
$\nu_s CH_2$	2915 ± 65	ρCH_2	840 ± 80
$\nu C\equiv N$	2260 ± 15	—C—C ext. sk. def.	535 ± 70
δCH_2	1425 ± 30	$\gamma C—C\equiv N$	365 ± 25
ωCH_2	1295 ± 65	$\delta C—C\equiv N$	225 ± 65
τCH_2	1230 ± 55	torsion	–

$R—CH_2C\equiv N$ molecules

R = Me— [1–7, 41], CD_3— [3], Et— [7–10], nPr— [7, 11], cPr— [12], $N\equiv CCH_2CH_2$— [6], $N\equiv CCH_2$— [6, 13, 14], $CNCH_2$— [15], $ClCH_2$— [4, 9, 16–18], $BrCH_2$— [4, 9], HO(O=)C— [19], NaO(O=)C— [19, 20],

$H_2N(O{=})C$— [21], iBuO(O=)C—, EtO(O=)CHN(O=)C—, $H_2N(S{=})C$— [22], $H_2C{=}CH$— [23–26], $H_2C{=}C(Me)$— [26–30], N≡C— [6, 31, 32], Ph— [44], 4-ClPh— [33], 4-MeOPh—, 2-Th—, N_3— [34], MeO— [35, 36], MeS— [10], F—, Cl— and I— [37–39], Br— [37–40].

References

1. G.A. Crowder, *Spectrosc. Lett.*, **20**, 343 (1987).
2. G.A. Crowder, *Spectrochim. Acta, Part A*, **42A**, 1229 (1986).
3. H.H. Heise, F. Winther and H. Lutz, *J. Mol. Spectrosc.*, **90**, 531 (1981).
4. P. Klaboe and J. Grundnes, *Spectrochim. Acta, Part A*, **24A**, 1905 (1968).
5. N.E. Duncan and G.J. Janz, *J. Chem. Phys.*, **23**, 434 (1955).
6. R. Yamadera and S.Krimm, *Spectrochim. Acta, Part A*, **24A**, 1677 (1968).
7. J.J. Lucier, E.C. Tuazon and F.F. Bentley, *Spectrochim. Acta, Part A*, **24A**, 771 (1968).
8. G.A. Crowder, *J. Mol. Struct.*, **158**, 229 (1987).
9. T. Fujiyama, *Bull. Chem. Soc. Jpn.*, **44**, 3317 (1971).
10. S.W. Charles, F.C. Cullen and N.L. Owen, *J. Mol. Struct.*, **34**, 219 (1976).
11. G.A. Crowder, *J. Mol. Struct.*, **200**, 235 (1989).
12. C.J. Wurrey, Y.Y. Yeh, M.D. Weakly and V.F. Kalasinsky, *J. Raman Spectrosc.*, **15**, 179 (1984).
13. W.E. Fitzgerald and G.J. Janz, *J. Mol. Spectrosc.*, **1**, 49 (1957).
14. T. Fujiyama, K. Tokumaru and T. Shimanouchi, *Spectrochim. Acta*, **20**, 415 (1964).
15. G. Schrumpf and S. Martin, *J. Mol. Struct.*, **101**, 57 (1983).
16. E. Wyn-Jones and W.J. Orville-Thomas, *J. Chem. Soc.*, 101 (1966).
17. M.F. El-Bermani and N. Jonathan, *J. Chem. Soc.*, 1712 (1968).
18. K. Tanabe, *J. Mol. Struct.*, **25**, 259 (1975).
19. D. Sinha and J.E. Katon, *Appl. Spectrosc.*, **26**, 599 (1972).
20. E. Spinner, *J. Chem. Soc.*, 4217 (1964).
21. L. Van Haverbeke and M.A. Herman, *Spectrochim. Acta, Part A*, **31A**, 959 (1975).
22. A. Ray and D.N. Sathyanarayana, *Bull. Chem. Soc. Jpn.*, **46**, 1969 (1973).
23. D.A.C. Compton and W.F. Murphy, *Spectrochim. Acta, Part A*, **41A**, 1141 (1985).
24. A.L. Verma, *J. Mol. Spectrosc.*, **39**, 247 (1971).
25. G.H. Griffith, L.A. Harrah, J.W. Clark and J.R. Durig, *J. Mol. Struct.*, **4**, 255 (1969).
26. D.A.C. Compton, S.C. Hsi and H.H. Mantsch, *J. Phys. Chem.*, **85**, 3721 (1981).
27. D.A.C. Compton, S.C. Hsi, H.H. Mantsch and W.F. Murphy, *J. Raman Spectrosc.*, **13**, 30 (1982).
28. S.H. Schei, *Spectrochim. Acta, Part A*, **39A**, 327 (1983).
29. D.A.C. Compton, W.F. Murphy and H.H. Mantsch, *Spectrochim. Acta, Part A*, **37A**, 453 (1981).
30. A.O. Diallo, *Spectrochim. Acta, Part A*, **35A**, 1189 (1979).
31. T. Fujiyama and T. Shimanouchi, *Spectrochim. Acta*, **20**, 829 (1964).
32. B.J. Van der Veken and H.O. Desseyn, *J. Mol. Struct.*, **23**, 427 (1974).
33. S. Chakravorti, A.K. Sarkar, P.K. Mallick and S.B. Banerjee, *Indian J. Phys.*, **56B**, 96 (1982).
34. P. Klaboe, K. Kosa, C.J. Nielsen, H. Priebe and S.H. Schei, *J. Mol. Struct.*, **160**, 245 (1987).
35. S.W. Charles, F.C. Cullen, G.I.L. Jones and N.L. Owen, *J. Chem. Soc. Faraday Trans. 2*, **70**, 758 1974)
36. R.G. Jones and W.J. Orville-Thomas, *J. Chem. Soc.*, 692 (1964).

37. R.G. Jones and W.J. Orville-Thomas, *J. Chem. Soc.*, 4623 (1965).
38. J.R. Durig and D.W. Wertz, *Spectrochim. Acta, Part A*, **24A**, 21 (1968).
39. G.A. Crowder, *Mol. Phys.*, **23**, 707 (1972).
40. F. Watari and K. Aida, *Spectrochim. Acta, Part A*, **23A**, 2951 (1967).
41. C.J. Wurrey, W.E. Bucy and J.R. Durig, *J. Phys. Chem.*, **80**, 1129 (1976).
42. R.A. Nyquist, *Appl. Spectrosc.*, **41**, 904 (1987).
43. R.E. Kitson and N.E. Griffith, *Analyt. Chem.*, **24**, 334 (1952).
44. G. Varsányi, *Assignments for Vibrational Spectra of Seven Hundred Benzene Derivatives*, J.Wiley & Sons, New York (1974).

4

Normal Vibrations and Absorption Regions of CHX_2

4.1 DIHALOGENOMETHYL

The $—CHX_2$ group (X = F, Cl, Br, I), just like the CX_3 group, has nine normal vibrations:

νCH, δCH, ωCH, $\nu_a CX_2$, $\nu_s CX_2$, ωCX_2, τCX_2, δCX_2 and torsion.

It is quite possible that rotational isomers give rise to more than one band for the same vibration. Molecules such as $MeCHCl_2$, *trans*-$F_2CHCHCl_2$ and *trans*-$ClCH_2CHCl_2$ show a plane of symmetry and belong to the point group C_s. In this case eleven (a′) of the eighteen vibrations cause a change in dipole moment in this plane and seven (a″) cause a dipole change in a direction perpendicular to the plane of symmetry. The 5a′ and 4a″ vibrations of the $—CHCl_2$ group in 1,1-dichloroethane are listed in Table 4.1.

Table 4.1 Vibrations of $MeCHCl_2$

a′		a″	
νCH	3008	ωCH	1230
δCH	1280	$\nu_a CCl_2$	691
$\nu_s CCl_2$	647	τCCl_2	318
ωCCl_2	405	torsion	293
δCCl_2	274		

Gauche-$F_2CHCHCl_2$ and *gauche*-$ClCH_2CHCl_2$ have no plane of symmetry and belong to the point group C_1. The difference between a′ and a″ disappears and the literature does not agree in assigning the —CHX_2 deformations in terms of in-plane bending, wag, twist or rock.

4.1.1 Difluoromethyl

The CH vibrations

The vibrational analysis of difluoromethyl compounds reveals the CH stretching vibration at 2990 ± 15 cm^{-1}, or mostly even at 2995 ± 10 cm^{-1} if the value 2995 cm^{-1} [2] is preferred to 2975 cm^{-1} [1] in the spectrum of 1,1-difluoroethane.

The two CH deformations are moderately to strongly active, one at 1395 ± 50 and the other at 1275 ± 70 cm^{-1}. They are sensitive to conformation but the wavenumbers of the *trans* and the *gauche* conformer rarely differ by more than 40 cm^{-1}. By analogy with C_s molecules, the first is considered as the δCH (a′) and occurs normally at 1370 ± 25 cm^{-1}, neglecting the wavenumber 1443 cm^{-1} in the Raman spectrum of *trans*-F_2CHCHF_2. The second deformation is then the CH wagging vibration (a″) occurring in the range 1295 ± 50 cm^{-1}, except for the low value (1212 cm^{-1}) in the spectrum of $ClCF_2CHF_2$.

The CF_2 stretching vibrations

The CF_2 antisymmetric stretch gives rise to a strong band in the range 1155 ± 50 cm^{-1}, which diminishes to 1140 ± 30 cm^{-1} if the values 1205, 1203 and 1104, in the spectra of respectively *gauche*-F_2CHCHF_2, $BrCF_2CHF_2$ and $Cl_2C{=}C(Cl)CHF_2$, are ignored but without rejection.

The CF_2 symmetric stretch appears strongly at 1090 ± 35, or at 1080 ± 25 cm^{-1} without the highest values of 1125 cm^{-1} in the spectrum of *trans*-F_2CHCHF_2 and 1118 cm^{-1} in that of $MeCHF_2$.

The CF_2 stretching vibrations, although sensitive to the conformational state of the molecule, merit the title 'group vibration' on account of their intensity.

The CF_2 skeletal deformations

The CF_2 wag is sensitive to rotational isomerism, yielding band separations of up to 100 wavenumbers and more. The highest wags have been assigned in the spectra of *gauche*-F_2CHCHF_2 (780 and *598*), *gauche*-Cl_2CHCHF_2 (765) and $HOCH_2CHF_2$ (749 cm^{-1}), the lowest in those of $MeCHF_2$ (568), *trans*-F_2CHCHF_2 (*625 and* 542) and $HO(O{=})CCHF_2$ (574 cm^{-1}). In many cases the CF_2 wag is observed at 630 ± 30 cm^{-1} as a moderate to strong band.

The CF_2 in-plane deformation absorbs mostly at $525 \pm 50\ cm^{-1}$, with a moderate to strong intensity, if the low values for *trans*-F_2CHCHF_2 (*479* and 417) and $HOCH_2CHF_2$ ($429\ cm^{-1}$) are ignored.

The third skeletal deformation, the CF_2 twisting vibration, absorbs weakly to moderately in the range $260 \pm 60\ cm^{-1}$ if the high value $383\ cm^{-1}$ from the spectrum of $MeCHF_2$ is not taken into account. The δCF_2 and τCF_2 absorptions are virtually insensitive to conformation.

Torsion

With the exception of the methyl torsion in $MeCHF_2$ ($220\ cm^{-1}$), many —CHF_2 torsions are assigned at $105 \pm 35\ cm^{-1}$.

Table 4.2 Absorption region (cm^{-1}) of the normal vibrations of —CHF_2

Vibration	C_s	Region	Vibration	C_s	Region
νCH	a′	2990 ± 15	ωCF_2	a′	660 ± 120
δCH	a′	1395 ± 50	δCF_2	a′	495 ± 80
ωCH	a″	1275 ± 70	τCF_2	a″	290 ± 95
$\nu_a CF_2$	a″	1155 ± 50	torsion	a″	145 ± 75
$\nu_s CF_2$	a′	1090 ± 35			

R—CHF_2 molecules

R = Me— [1, 2], $HOCH_2$— [3], FCH_2— [4], F_2CH— [4, 5], Cl_2CH— [6], F_3C— [7], $ClCF_2$— and $BrCF_2$— [8], $Cl_2C{=}CCl$— [9], HOC(=O)— [10], $H_2NC(=O)$— [11].

4.1.2 Dichloromethyl

The CH vibrations

With the exception of the low values for *gauche*-$HC(=O)CHCl_2$ (2970) and $Cl_2CHCHCl_2$ (*3012* and $2980\ cm^{-1}$), the CH stretching vibration is observed at $3000 \pm 15\ cm^{-1}$.

The two CH deformations are sensitive to the conformational state of the molecule. The band with the highest wavenumber is often assigned to the δCH in the region $1255 \pm 55\ cm^{-1}$, with *gauche*-$MeOC(=O)CHCl_2$ (1310) and *trans*-$HC(=O)CHCl_2$ ($1200\ cm^{-1}$) as extremities. The absorption region of the CH wag ($1215 \pm 35\ cm^{-1}$) is limited by $FC(=O)CHCl_2$ (1248) and $Cl_2CHCHClCHCl_2$ (*1229* and $1184\ cm^{-1}$).

CCl_2 stretching vibrations

The CCl_2 stretching vibrations provide two moderate to strong, but conformation-sensitive, absorptions. The $\nu_a CCl_2$ is observed in the region 750 $\pm$ 90 cm^{-1} with at the HW side, $HOC(=O)CHCl_2$ (840 [30, 35] or 817 [32]) and at the LW side, $MeCHCl_2$ (691 [12, 13] or 707 [14]) and CD_3CHCl_2 (660 cm^{-1}). For most of the $R{-}CHCl_2$ molecules this absorption lies in the narrower region 765 $\pm$ 65 cm^{-1}. The CCl_2 symmetric stretching vibration absorbs in the extensive region 680 $\pm$ 100 cm^{-1}, which diminishes to 705 $\pm$ 65 cm^{-1} if some high values in the spectra of $Me_2NC(=O)CHCl_2$ (780), $FC(=O)CHCl_2$ (778) and $NaOC(=O)CHCl_2$ (777) and low values in those of aromatic-bonded dichloromethyl (605 $\pm$ 25 cm^{-1}), $MeCHCl_2$ (604) and $HC(=O)CHCl_2$ (*trans*: 610; *gauche*: 630 cm^{-1}) are not taken into account.

CCl_2 skeletal deformations

The three skeletal deformations in 1,1-dichloroethane, which belongs to the point group C_s, are arranged in order of decreasing wavenumber as follows: $\omega CCl_2(a')$ $> \tau CCl_2(a'') > \delta CCl_2(a')$. In more complex molecules such as $R{-}C(=O)CHCl_2$ the deformation with the lowest wavenumber is described as the CCl_2 rock or twist (a'').

High values for ωCCl_2 have been assigned in the spectra of $Cl_2CHCHCl_2$ (546 and *353*) and $ClCH_2CHCl_2$ (525 cm^{-1}) and low values in those of $ClC(=O)CHCl_2$ (262) and $MeOC(=O)CHCl_2$ (262 cm^{-1}), in which the dichloromethyl group is attached to a carbonyl group.

For the second skeletal deformation, $Cl_2CHCHCl_2$ and $ClCH_2CHCl_2$ also score at the HW side with 333 cm^{-1} and $ClC(=O)CHCl_2$ at the LW side with 176 cm^{-1}.

For the third skeletal deformation, the lowest wavenumbers have been observed in the spectra of $BrC(=O)CHCl_2$ (151) and $ClC(=O)CHCl_2$ (166 cm^{-1}) and are assigned to the CCl_2 rock (a'').

The CCl_2 skeletal deformations are sensitive to rotational isomerism. For the wag, the band separations between *trans* and *gauche* conformer can reach 100 wavenumbers, and those for the in-plane deformation and rock rarely above 50 wavenumbers.

Torsion

The absorption region of the torsion (165 $\pm$ 130) narrows to 110 $\pm$ 75 cm^{-1} if the values 293 for $MeCHCl_2$ and 274 cm^{-1} for CD_3CHCl_2 are not taken into account.

Table 4.3 Absorption regions (cm^{-1}) of the normal vibrations of $—CHCl_2$

	C_s	α—saturated	keto bonded	aromatic
νCH	a′	2995 ± 20	2990 ± 25	≈3005
δCH	a′	1255 ± 55	1255 ± 55	1275 ± 25
ωCH	a″	1215 ± 35	1220 ± 30	1210 ± 10
$\nu_a CCl_2$	a″	745 ± 85	775 ± 65	725 ± 45
$\nu_s CCl_2$	a′	685 ± 85	695 ± 85	605 ± 25
ω or δCCl_2	a′	435 ± 115	340 ± 80	385 ± 25
δ or ωCCl_2	a′	285 ± 50	225 ± 50	≈300
ρ or τCCl_2	a″	225 ± 60	190 ± 40	≈280
torsion	a″	170 ± 125	105 ± 70	≈115

$R—CHCl_2$ molecules

R = Me— [12–14], CD_3— [14], $HOCH_2$— [3], Cl_2CHCH_2— [15, 16], $ClCH_2$— [13, 17–19], F_2CH— [6], Cl_2CH— [13, 20, 21], $ClCH_2CHCl$—, $Cl_2CHCHCl$— and $ClCH_2CCl_2$— [15], Cl_3CCH_2—, MeCHCl—, Cl_3CCHCl—, $MeCCl_2$— and Cl_3CCCl_2— [22], F_3C— [23], Cl_3C— [13], $MeOCF_2$— [24], Ph— [25, 26], 2-, 3- and 4-FPh— [27], HC(=O)— [28], FC(=O)— [31, 46], ClC(=O)— [29–33], BrC(=O)— [31], MeC(=O)— [34], MeOC(=O)— [30, 32, 36], HOC(=O)— [30, 32, 35], NaOC(=O)— [32, 38], HSC(=O)— [37], H_2NC(=O)— [11], MeHNC(=O)— and Me_2NC(=O)— [32], SiX_3— (X = H, D, F, Cl) [39].

4.1.3 Dibromomethyl

The CH vibrations

The absorption region of νCH (3005 ± 20 cm^{-1}) is bounded by $MeCHBr_2$ (3023) and $Br_2CHCHBr_2$ (2985 cm^{-1}).

Disregarding the low δCH (1143 cm^{-1}) for *trans*-$Br_2CHCHBr_2$ and the high CH waggings (1183 and 1172 cm^{-1}) in the spectra of $MeCHBr_2$ and *trans*-$BrCH_2CHBr_2$, the two absorption regions do not overlap one another: δCH: 1235 ± 45; ωCH: 1135 ± 20 cm^{-1}.

CBr stretching vibrations

Disregarding the low value (620 cm^{-1}) in the spectrum of $MeCHBr_2$, the $\nu_a CBr_2$ occurs at 685 ± 45 cm^{-1}. The $\nu_s CBr_2$ usually absorbs at 575 ± 50 cm^{-1} with the exception of one of the symmetric stretchings of *trans*-$Br_2CHCHBr_2$ (637 and *586* cm^{-1}). Both stretching vibrations exhibit strong bands but they are sensitive to conformation.

CBr_2 skeletal deformations

The first CBr_2 deformation or ωCBr_2 is often located in the region 305 ± 95 cm^{-1} with the exception of *gauche*-$Br_2CHCHBr_2$ (450 and *217* cm^{-1}).

The second deformation, absorbing in the range 205 ± 45 cm^{-1}, is often described as δCBr_2. The value 275 cm^{-1} in the spectrum of $MeCHBr_2$, however, is assigned to the CBr_2 twist (a″) and is outside the above-mentioned region.

The third deformation, with the lowest wavenumber (155 ± 45 cm^{-1}), is assigned to τCBr_2, but in the spectrum of $MeCHBr_2$ (172 cm^{-1}) to δCBr_2.

Table 4.4 Absorption regions (cm^{-1}) of the normal vibrations of —$CHBr_2$

Vibration	C_s	Region	Vibration	C_s	Region
νCH	a′	3005 ± 20	ωCBr_2	a′	330 ± 120
δCH	a′	1210 ± 70	δCBr_2	a′	220 ± 55
ωCH	a″	1150 ± 35	τCBr_2	a″	155 ± 45
$\nu_a CBr_2$	a″	675 ± 55	torsion	a″	100 ± 65
$\nu_s CBr_2$	a′	585 ± 60			

R—$CHBr_2$ molecules

R = Me— [40], $BrCH_2$— [41, 42], Br_2CH— [20, 43], N≡C— [44], ClC(=O)— and HSC(=O)— [45], NaOC(=O)— [38].

References

1. G.A. Guirgis and G.A. Crowder, *J. Fluorine Chem.*, **25**, 405 (1984).
2. D.C. Smith, R.A. Saunders, J.R. Nielsen and E.E. Ferguson, *J. Chem. Phys.*, **20**, 847 (1952).
3. M. Perttilä, *Spectrochim. Acta, Part A*, **35A**, 37 (1979).
4. V.F. Kalasinsky, H.V. Anjaria and T.S. Little, *J. Phys. Chem.*, **86**, 1351 (1982).
5. P. Klaboe and J.R. Nielsen, *J. Chem. Phys.*, **32**, 899 (1960).
6. V.B. Kartha, S.B. Kartha and N.A. Narasimham, *Proc. Indian Acad. Sci.*, **65A**, 1 (1967).
7. J.R. Nielsen, H.H. Claassen and N.B. Moran, *J. Chem. Phys.*, **23**, 329 (1955).
8. P. Klaboe and J.R. Nielsen, *J. Chem. Phys.*, **34**, 1819 (1961).
9. P. Klaboe, G. Neerland and S.H. Schei, *Spectrochim. Acta, Part A*, **38A**, 1025 (1982).
10. J.R. Barcelo and C. Otero, *Spectrochim. Acta*, **18**, 1231 (1962).
11. D. Troitino, E. Sanchez de la Blanca and M.V. Garcia, *Spectrochim. Acta, Part A*, **46A**, 1281 (1990).
12. L.W. Daasch, C.Y. Liang and J.R. Nielsen, *J. Chem. Phys.*, **22**, 1293 (1954).
13. S. Suzuki and A.B. Dempster, *J. Mol. Struct.*, **32**, 339 (1976).
14. D.C. McKean, J.C. Lavalley, O. Sau, H.G.M. Edwards and V. Fawcett, *Spectrochim. Acta, Part A*, **33A**, 913 (1977).
15. A.B. Dempster, K. Price and N. Sheppard, *Spectrochim. Acta, Part A*, **27A**, 1579 (1971).
16. M. Braathen, D.H. Christensen, P. Klaboe, R. Seip and R. Stølevik, *Acta Chem. Scand., Ser. A*, **33A**, 437 (1979).
17. S.D. Christian, J. Grundnes, P. Klaboe, C.J. Nielsen and T. Woldbaek, *J. Mol. Struct.*, **34**, 33 (1976).

18. K. Kuratani and S.-I. Mizushima, *J. Chem. Phys.*, **22**, 1403 (1954).
19. R.H. Harrison and K.A. Kobe, *J. Chem. Phys.*, **26**, 1411 (1957).
20. G.W. Chantry, H.A. Gebbie, P.R. Griffiths and R.F. Flake, *Spectrochim. Acta*, **22**, 125 (1966).
21. K. Naito, I. Nakagawa, K. Kuratani, I. Ichishima and S.-I. Mizushima, *J. Chem. Phys.*, **23**, 1907 (1955).
22. A.B. Dempster, K. Price and N. Sheppard, *Spectrochim. Acta, Part A*, **27A**, 1563 (1971).
23. J.R. Nielsen, C.Y. Liang and D.C. Smith, *J. Chem. Phys.*, **21**, 1060 (1953).
24. Y.S. Li and J.R. Durig, *J. Mol. Struct.*, **81**, 181 (1982).
25. P.J.A. Ribeiro-Claro and J.J.C. Teixeira-Dias, *J. Raman Spectrosc.*, **15**, 224 (1984).
26. P.J.A. Ribeiro-Claro, A.M. D'A Rocha Gonsalves and J.J.C. Teixeira-Dias, *Spectrochim. Acta, Part A*, **41A**, 1055 (1985).
27. S. Tariq and P.K. Verma, *Spectrochim. Acta, Part A*, **39A**, 1027 (1983).
28. G. Lucazeau and A. Novak, *J. Mol. Struct.*, **5**, 85 (1970).
29. A. Miyaka, I. Nakagawa, T. Miyazawa, I. Ichishima, T. Shimanouchi and S. Mizushima, *Spectrochim. Acta*, **13**, 161 (1958).
30. R. Fausto and J.J.C. Teixeira-Dias, *J. Mol. Struct.*, **144**, 241 (1986).
31. A.J. Woodward and N. Jonathan, *J. Phys. Chem.*, **74**, 798 (1970).
32. J.E. Katon, T.H. Stout and G.G. Hess, *Appl. Spectrosc.*, **40**, 1 (1986).
33. J.R. Durig, M.M. Bergana and H.V. Phan, *J. Mol. Struct.*, **242**, 179 (1991).
34. J.R. Durig, J.A. Hardin and C.L. Tolley, *J. Mol. Struct.*, **224**, 323 (1990).
35. L.M. Babkow, V.V. Vashchinskaya, M.A. Kovner, G.A. Puchkovskaya and Yu.Ya. Fialkov, *Spectrochim. Acta, Part A*, **32A**, 1379 (1976).
36. Y. Mido and H. Hashimoto, *J. Mol. Struct.*, **131**, 71 (1985).
37. H.S. Randhawa and W. Walter, *J. Mol. Struct.*, **38**, 89 (1977).
39. I.V. Kochikov, G.M. Kuramshina, S.V. Syn'ko and Yu.A. Pentin, *J. Mol. Struct.*, **172**, 299 (1988).
40. J.R. Durig, A.E. Sloan, J.W. Thompson and J.D. Witt, *J. Chem. Phys.*, **60**, 2260 (1974).
41. S. Suzuki and G. Vergoten, *Spectrochim. Acta, Part A*, **37A**, 37 (1981).
42. T. Torgrimsen and P. Klaboe, *Acta Chem. Scand.*, **24**, 1145 (1970).
43. G.L. Carlson, W.G. Fateley and J. Hiraishi, *J. Mol. Struct.*, **6**, 101 (1970).
44. F. Watari and K. Aida, *Spectrochim. Acta, Part A*, **23A**, 2951 (1967).
45. H.S. Randhawa, *Indian J. Chem.*, **19A**, 152 (1980).
46. J.R. Durig, M.M. Bergana and H.V. Phan, *J. Raman Spectrosc.*, **22**, 141 (1991).

4.2 ISOPROPYL

Molecules of type Me_2CHX (X = F, Cl, Br, I) have a plane of symmetry, which encloses the halogen, the central carbon and the secondary hydrogen. They belong to the point group C_s. The 27 normal vibrations are distributed among 15a′ and 12a″ species of vibration. Substitution of the νCX (a′) by a torsion (a″) results in 27 normal vibrations for the isopropyl group, which in the case of C_s symmetry are divided in 14a′ and 13a″ types of vibration:

a′: ν_aMe(2), ν_sMe, νCH, δ_aMe(2), δ_sMe, δCH, ρMe(2), $\nu_s CC_2$, δCC_2, skeletal deformation and torsion Me;

a″: ν_aMe(2), ν_sMe, δ_aMe(2), δ_sMe, γCH, $\nu_a CC_2$, ρMe(2), skeletal deformation, torsion Me and torsion CC_2.

The methyl stretching vibrations, as well as the methyl deformations and rocks, can be in-phase or out-of-phase movements, leading to a′ or a″ vibrations.

The CH stretching vibrations

The seven CH stretching vibrations absorb with moderate to strong intensity, but seldom give separate bands. The in-phase and out-of-phase vibrations often coincide. The four methyl antisymmetric stretchings are found between 3005 and 2935 cm^{-1}, but usually lower than 3000 cm^{-1} except for iPrF (3005), iPrCl (3005), iPrBr (3003)) and $iPrNO_2$ (3003 cm^{-1}). Both methyl symmetric stretching vibrations absorb between 2940 and 2860 cm^{-1}, or even between 2920 and 2860 cm^{-1} with the exception of iPrBr (2931), iPrNCO (2929), iPrCl (2927), iPrC(=O)F (2927) and iPrNCS (2924 cm^{-1}). The CH stretching vibration probably occurs at 2930 ± 25 cm^{-1}.

The CH deformations

The four methyl antisymmetric deformations are active between 1485 and 1430 cm^{-1} with moderate intensity. The a′ and a″ vibrations often coincide, principally those with the highest wavenumber. The HW side is limited by iPrC(=O)OMe (1483), $iPrC(=O)OCD_3$ (1482), iPr—S—iPr (1483) and iPrC(=O)Me (1482 cm^{-1}) and the LW side by iPrI (1430 cm^{-1}). The remaining methyl antisymmetric deformations have been observed at 1460 ± 20 cm^{-1}.

Setting aside the high values for the highest δ_sMe in the spectra of $iPrNO_2$ (1400), iPrC(=O)F (1399), iPrC(=O)H (1395), $iPrSiCl_3$ (1393), iPrOC(=O)F (1392) and iPrC≡N (1391 cm^{-1}) and the low values for the lowest δ_sMe in the spectra of iPrF (1354) and iPrCH(Me)iPr (1351 and *1366* cm^{-1}), most of the methyl symmetric stretching vibrations have been observed between 1390 and 1360 cm^{-1}. The isopropyl group is best recognized by the two moderate to strong absorptions with about equal intensity of δ_sMe, often named as a ‘symmetrical doublet’.

With the exception of Si- and P- isopropyl (1290 ± 15 cm^{-1}) the a″ CH deformation (mentioned as γCH but also as δCH) is assigned at 1320 ± 30 cm^{-1}. The a′ δCH appears usually in the region 1265 ± 35 cm^{-1}, but in the spectrum of iPrI at 1210 cm^{-1}.

Skeletal stretching vibrations and methyl rocking vibrations

In the range 1190–765 cm^{-1} four methyl rocks and two CC_2 skeletal stretchings are observed, well separated from each other and mostly with a weak to moderate intensity.

The highest in-phase methyl rock at 1170 ± 20 cm^{-1} is a good group vibration. In the spectra of $iPrNH_2$ (1177), $iPrPF_2$ (1168) and $iPrPCl_2$ (1165) and iPrSH (1162 cm^{-1}), however, this absorption is assigned to the $\nu_a CC_2$.

The CC_2 antisymmetric stretching vibration is usually seen at 1115 ± 45 cm^{-1}, a wavenumber that in the above-mentioned cases is assigned to a methyl rocking vibration. Disregarding the values 1159 and *1123* in the spectrum of iPrCH(Me)iPr, 1080 in that of 2-HO—Ph—iPr and 1085 ± 15 cm^{-1} in those of Si- and P-bonded iPr, this region is reduced to 1125 ± 25 cm^{-1}.

The second in-phase methyl rock is usually located in the region 1080 ± 40 cm^{-1}, with the exception of iPrSCN (1130), iPrCH(Me)iPr (1038 and *1076*) and $iPrNH_2$ (1036 cm^{-1}). In the spectra of iPrC(=O)X (X = H, Cl, F, OMe) this absorption is strongly influenced by the $\nu_a CC_2$.

The highest out-of-phase methyl rock may be assigned at 970 ± 30 cm^{-1}. A shift to higher frequencies occurs when the methyl group is attached to Si or P (1010 ± 15 cm^{-1}) and a small shift to lower frequencies in the spectra of iPrI (937), iPrOMe (938) and iPrSCN (938 cm^{-1}).

The second out-of-phase methyl rock can be found in the range 930 ± 20 cm^{-1} if some higher values in the spectra of Si— and P—bonded isopropyl (945 ± 30 cm^{-1}) and in that of iPrC(=O)Me (952 cm^{-1}) are not taken into account.

The CC_2 symmetric stretching vibration is reported in the extensive region 835 $\pm$ 70 cm^{-1}, which is reduced to 855 ± 40 cm^{-1} by neglecting the extreme values in the spectra of $iPrCH_2OC(=O)CH_2C{\equiv}N$ (905), iPrNCO (903), $iPrC{\equiv}N$ (769), iPrC(=O)F (780) and iPrOMe (800 cm^{-1}).

Skeletal deformations

The CC_2 in-plane deformation provides a weak band in the range 450 ± 65 cm^{-1} but most of these skeletal bendings have been observed at 440 ± 40 cm^{-1}. The following compounds are responsible for the wide spread: $iPrC{\equiv}N$ (510), iPrC(=O)F (486), iPr—CH=N—iPr (483 and *408*), iPrI (398), iPrSH (396), $iPrNO_2$ (387) and iPrC(=O)D (385 cm^{-1}).

The vibrational analysis of isopropyl compounds reveals another two skeletal deformations: a C—C—R in-plane deformation or CC_2 wag (a′) and a C—C—R out-of-plane deformation or CC_2 twist (rock). These more external deformations are often weakly active at 360 ± 50 and 320 ± 45 cm^{-1} with the exception of iPrBr (290 and 282) and iPrI (270 and 250 cm^{-1}), which are influenced by the heavy halogen.

The following R—iPr compounds have been taken into account:

R = $ClMe_2CCH_2$— and $ClMe_2CH_2CH_2$— [1], $MeCHClCH_2$— and $MeCHBrCH_2$— [2], $PhCH_2$— [58], $HC{\equiv}CCH_2$— [3], $MeNHC(=O)NHCH_2$—, $EtNHC(=O)NHCH_2$—, $nPrNHC(=O)NHCH_2$—, $nBuNHC(=O)NHCH_2$— and $iBuNHC(=O)NHCH_2$— [4], $cHexNHC(=O)NHCH_2$— [5], $N{\equiv}CCH_2C(=O)OCH_2$—, $MeSCH_2$— [6], $ClCH_2$— [7–10], $BrCH_2$— [9–11], iPrCH(Me)— and EtCHMeCHMe— [12], cPr— [13], $ClMe_2C$— [1], Ph— [58], 2-, 3- and 4-MePh— [58], 2-, 3- and 4-iPrPh— [58], 2-, 3- and 4-HOPh— [58], HC(=O)— [14,

Table 4.5 Absorption regions (cm^{-1}) of the normal vibrations of iPr

adjacent to		Saturated C	Aromatic	C=O	O and S	N	Si and P	Halogen and C≡N	Total
ν_aMe	a′	2980 ± 15	2965 ± 10	2985 ± 15	2985 ± 15	2985 ± 20		2985 ± 20	2980 ± 25
ν_aMe	a″	2980 ± 15	2965 ± 10	2985 ± 15	2985 ± 15	2980 ± 20		2980 ± 20	2975 ± 25
ν_aMe	a′	2960 ± 20	2960 ± 10	2965 ± 15	2970 ± 15	2965 ± 20		2960 ± 25	2960 ± 25
ν_aMe	a″	2960 ± 20	2960 ± 10	2960 ± 15	2960 ± 20	2960 ± 20		2955 ± 20	2960 ± 25
νCH	a′	2930 ± 20	2930 ± 10	2930 ± 20	2935 ± 15	2940 ± 15		2935 ± 15	2935 ± 25
ν_sMe	a′	2890 ± 20	2900 ± 15	2905 ± 20	2895 ± 25	2905 ± 25		2910 ± 25	2905 ± 35
ν_sMe	a″	2880 ± 15	2875 ± 15	2885 ± 10	2885 ± 15	2880 ± 15		2885 ± 15	2880 ± 20
δ_aMe	a′	1470 ± 10	1465 ± 10	1475 ± 10	1470 ± 15	1470 ± 10	1470 ± 10	1470 ± 10	1470 ± 15
δ_aMe	a″	1465 ± 10	1465 ± 10	1470 ± 10	1465 ± 15	1470 ± 10	1470 ± 10	1470 ± 10	1465 ± 15
δ_aMe	a′	1455 ± 10	1450 ± 10	1460 ± 10	1455 ± 15	1460 ± 10	1460 ± 10	1460 ± 15	1455 ± 20
δ_aMe	a″	1450 ± 10	1450 ± 10	1450 ± 10	1455 ± 15	1450 ± 10	1450 ± 10	1445 ± 15	1450 ± 20
δ_sMe	a′	1380 ± 10	1380 ± 10	1390 ± 10	1385 ± 10	1385 ± 15	1385 ± 10	1385 ± 10	1385 ± 15
δ_sMe	a″	1365 ± 15	1365 ± 05	1370 ± 10	1370 ± 10	1370 ± 10	1370 ± 10	1365 ± 15	1365 ± 15
γCH	a″	1325 ± 25		1325 ± 30	1320 ± 30	1325 ± 25	1290 ± 15	1330 ± 10	1315 ± 40
δCH	a′	1285 ± 45		1265 ± 35	1285 ± 45	1285 ± 45	1245 ± 10	1255 ± 45	1270 ± 60
ρMe	a′	1170 ± 20		1175 ± 15	1170 ± 20	1170 ± 15	1165 ± 10	1165 ± 15	1170 ± 20
$\nu_a CC_2$	a″	1130 ± 30	1110 ± 30	1130 ± 20	1125 ± 25	1135 ± 15	1085 ± 15	1125 ± 20	1115 ± 45
ρMe	a′	1065 ± 30	1060 ± 20	1090 ± 20	1080 ± 35	1085 ± 50	1075 ± 15	1080 ± 40	1085 ± 50
ρMe	a″	960 ± 20		980 ± 15	965 ± 30	970 ± 30	1010 ± 15	955 ± 25	980 ± 45
ρMe	a″	925 ± 20		935 ± 20	930 ± 15	925 ± 15	945 ± 30	925 ± 15	940 ± 35
$\nu_s CC_2$	a′	855 ± 45	850 ± 45	865 ± 40	850 ± 50	855 ± 50	885 ± 10	825 ± 60	835 ± 70
δCC_2	a′	440 ± 25		440 ± 50	435 ± 40	435 ± 50	420 ± 10	455 ± 60	450 ± 65
skeletal def.	a′	375 ± 35	340 ± 25	350 ± 20	370 ± 40	380 ± 30	345 ± 25	320 ± 50	340 ± 70
skeletal def.	a″	315 ± 35		310 ± 30	320 ± 25	330 ± 40	310 ± 20	300 ± 50	310 ± 60
Me torsion	a″	240 ± 25		245 ± 25	245 ± 25	260 ± 20	225 ± 25	245 ± 30	240 ± 40
Me torsion	a′	225 ± 25		195 ± 35		230 ± 20	195 ± 35	215 ± 35	205 ± 45
iPr torsion	a″	135 ± 40		≈120	≈150	≈130	≈140	≈130	135 ± 40

15], DC(=O)— [14], MeC(=O)— [16], 2-HOPhC(=O)— [17], FC(=O)— [18], ClC(=O)— [19], MeOC(=O)— and CD_3OC(=O)— [20], HO— [21, 53], MeO— [22, 23], EtO— [22], iPrO— [22, 23], FC(=O)O— [24], O=N—O— [25, 60], O_2NO— [60], HS— [26–28], DS— [28], MeS— [26, 29, 30], EtS— [26, 31, 32], nPrS— and nBuS— [31], iPrS— [26, 31], iPrSS— [26], NCS— [33, 34], H_2N— [35, 36], D_2N— [35], MeNHC(=O)NH—, EtNHC(=O)NH—, nPrNHC(=O)NH—, nBuNHC(=O)NH— and iPrNHC(=O)NH— [4], cHexNHC(=O)NH— [5], R—SO_2NH— [37], iPr—CH=N— and iPr—CD=N— [36], OCN— [38, 39], SCN— [40–42], O_2N— [43], Me_3Si—, $ClMe_2Si$—, Cl_2MeSi— and Cl_3Si— [46], H_2P— and D_2P— [59], F_2P— [44], Cl_2P— [45], F— [47–50], Cl— [51–53, 61], Br— [52–54], I— [52, 53], N≡C— [52, 53, 55–57].

References

1. G.A. Crowder and M.T. Richardson, *Spectrochim. Acta, Part A*, **38A**, 1123 (1982).
2. G.A. Crowder and R.M.P. Jaiswal, *J. Mol. Struct.*, **99**, 93 (1983).
3. G.A. Crowder, *J. Mol. Struct.*, **193**, 307 (1989).
4. Y. Mido, *Spectrochim. Acta, Part A*, **28A**, 1503 (1972).
5. Y. Mido, *Spectrochim. Acta, Part A*, **32A**, 1105 (1976).
6. N. Nogami, H. Sugeta and T. Miyzawa, *Bull. Chem. Soc. Jpn.*, **48**, 2417 (1975).
7. G.A. Crowder and W.-L. Lin, *J. Mol. Struct.*, **64**, 193 (1980).
8. A. J.Barnes, M.L. Evans and H.E. Hallam, *J. Mol. Struct.*, **99**, 235 (1983).
9. J.R. Durig, J.F. Sullivan and S.E. Godbey, *J. Mol. Struct.*, **146**, 213 (1986).
10. N.T. Devitt, A.L. Rozek, F.F. Bentley and A.D. Davidson, *J. Chem. Phys.*, **42**, 1173 (1965).
11. G.A. Crowder and M.-R. Jalilian, *Spectrochim. Acta, Part A*, **34A**, 707 (1978).
12. G.A. Crowder and L. Gross, *J. Mol. Struct.*, **118**, 135 (1984).
13. A.B. Nease and C.J. Wurrey, *J. Raman Spectrosc.*, **9**, 107 (1980).
14. A. Piart-Goypiron, M.H. Baron, J. Belloc, M.J. Coulange and H. Zine, *Spectrochim. Acta, Part A*, **47A**, 363 (1991).
15. J.R. Durig, G.A. Guirgis, W.E. Brewer and T.S. Little, *J. Mol. Struct.*, **248**, 49 (1991).
16. T. Sakurai, M. Ishiyama, H. Takeuchi, K. Takeshita, K. Fukushi and S. Konaka, *J. Mol. Struct.*, **213**, 245 (1989).
17. W.A.L.K. Al-Rashid and M.F. El-Bermani, *Spectrochim. Acta, Part A*, **47A**, 35 (1991).
18. J.R. Durig, G.A. Guirgis, W.E. Brewer and G. Baranovic, *J. Phys. Chem.*, **96**, 7547 (1992).
19. G.A. Guirgis, H.V. Phan and J.R. Durig, *J. Mol. Struct.*, **266**, 265 (1992).
20. R.M. Moravie and J. Corset, *J. Mol. Struct.*, **24**, 91 (1975).
21. L.I. Lafer, V.I. Yakerson and G.A. Kogan, *Bull. Sci. Acad. U.S.S.R.* (*Div. Chem. Sci.*), **8**, 1717 (1969).
22. R.G. Snyder and G. Zerbi, *Spectrochim. Acta, Part A*, **23A**, 391 (1967).
23. A.D.H. Clague and A. Danti, *Spectrochim. Acta, Part A*, **24A**, 439 (1968).
24. B.J. Van der Veken and H.H. Liefooghe, *J. Mol. Struct.*, **247**, 257 (1991).
25. B.J. Van der Veken and R. Maas, *J. Mol. Struct.*, **200**, 413 (1989).
26. D.W. Scott and J.P. McCullough, *J. Am. Chem. Soc.*, **80**, 3554 (1958).
27. D. Smith and J.P. Devlin, *J. Mol. Spectrosc.*, **25**, 174 (1968).

28. J.R. Durig, G.A. Guirgis and D.A.C. Compton, *J. Phys. Chem.*, **84**, 3547 (1980).
29. M. Oshaku, Y. Shiro and H. Murata, *Bull. Chem. Soc. Jpn.*, **45**, 3480 (1972).
30. J.P. McCullough, H.L. Finke, J.F. Messerly, R.E. Pennington, I.A. Hossenlopp and G. Waddington, *J. Am. Chem. Soc.*, **77**, 6119 (1955).
31. M. Ohsaku, H. Murata and Y. Shiro, *Spectrochim. Acta, Part A*, **33A**, 467 (1977).
32. M. Sakakibara, I. Harada, H. Matsuura and T. Shimanouchi, *J. Mol. Struct.*,**49**, 29 (1978).
33. R.P. Hirschmann, R.N. Kniseley and V.A. Fassel, *Spectrochim. Acta*, **20**, 816 (1964).
34. G.A. Crowder, *J. Mol. Struct.*, **7**, 147 (1971).
35. J.R. Durig, G.A. Guirgis and D.A.C. Compton, *J. Phys. Chem.*, **83**, 1313 (1979).
36. A. Piart-Goypiron, M.H. Baron, H. Zine, J. Belloc and M.J. Coulange, *Spectrochim. Acta, Part A*, **49A**, 103 (1993).
37. M. Goldstein, M.A. Russel and H.A. Willis, *Spectrochim. Acta, Part A*, **25A**, 1275 (1969).
38. J.R. Durig, K.J. Kanes and J.F. Sullivan, *J. Mol. Struct.*, **99**, 61 (1983).
39. R.P. Hirschmann, R.N. Kniseley and V.A. Fassel, *Spectrochim. Acta*, **21**, 2125 (1965).
40. R.N. Kniseley, R.P. Hirschmann and V.A. Fassel, *Spectrochim. Acta, Part A*, **23A**, 109 (1967).
41. J.R. Durig, J.F. Sullivan, T.S. Little and S. Cradock, *J. Mol. Struct.*, **118**, 103 (1984).
42. J.R. Durig, J.F. Sullivan, D.T. Durig and S. Cradock, *Can. J. Chem.*, **63**, 2000 (1985).
43. J.R. Durig, J.A. Smooter Smith, Y.S. Li and F.M. Wasacz, *J. Mol. Struct.*, **99**, 45 (1983).
44. J.R. Durig, M.-S. Cheng, Y.S. Li, P. Grower and A.E. Stanley, *J. Phys. Chem.*, **93**, 3492 (1989).
45. J.R. Durig, M.-S. Cheng, M.E. Harris and T.J. Hizer, *J. Mol. Struct.*, **192**, 47 (1989).
46. K. Ohno, K. Taga, I. Yoshida and H. Murata, *Spectrochim. Acta, Part A*, **35A**, 883 (1979).
47. J.H. Griffiths, N.L. Owen and J. Sheridan, *J. Chem. Soc. Faraday Trans. 2*, **9**, 1359 (1973).
48. G.A. Crowder and T. Koger, *J. Mol. Struct.*, **23**, 311 (1974).
49. G.A. Crowder and T. Koger, *J. Mol. Struct.*, **29**, 233 (1975).
50. J. Gustavsen and P. Klaboe, *Spectrochim. Acta, Part A*, **32A**, 755 (1976).
51. C.G. Opaskar and S. Krimm, *Spectrochim. Acta, Part A*, **23A**, 2261 (1967).
52. P. Klaboe, *Spectrochim. Acta, Part A*, **26A**, 87 (1970).
53. J.R. Durig, C.M. Player Jr., Y.S. Li, J. Bragin and C.W. Hawley, *J. Chem. Phys.*, **57**, 4544 (1972).
54. P. Klaboe, A. Linde and B.N. Cyvin, *Spectrochim. Acta, Part A*, **30A**, 1513 (1974).
55. R. Yamadara and S. Krimm, *Spectrochim. Acta, Part A*, **24A**, 1677 (1968).
56. B.N. Cyvin and S.J. Cyvin, *Acta Chem. Scand.*, **26**, 3943 (1972).
57. D.A.C. Compton and W.F. Murphy, *Spectrochim. Acta, Part A*, **41A**, 1141 (1985).
58. G. Varsányi, *Assignments for Vibrational Spectra of Seven Hundred Benzene Derivatives*, J.Wiley & Sons, New York (1974).
59. J.R. Durig and A.W. Cox Jr., *J. Phys. Chem.*, **80**, 2493 (1976).
60. R. Maas, Thesis, UIA, Antwerp, 1992.
61. J.F. Sullivan, A. Wang, M.-S. Cheng and J.R. Durig, *Can. J. Chem.*, **69**, 1845 (1991).

5

Normal Vibrations and Absorption Regions of CHX

5.1 HALOGENOMETHYLENE

The 27 normal vibrations of Me—CHX—Me are divided, according to C_s symmetry, into 15a′ vibrations, of which nine are for the methyl group, one CC_2 symmetric stretch and one CC_2 deformation, and 12a″ vibrations: nine methyl vibrations and one CC_2 antisymmetric stretch. Six fundamental vibrations (4a′ + 2a″) belong to the —CHX— group:

a′: νCH, νCX, δCH, δCX;
a″: ωCH and ωCX.
although the CX vibrations are influenced by the skeleton of the compound. In less symmetrical molecules the differentiation between a′ and a″ disappears.

5.1.1 Fluoromethylene

The absorption regions of CHF are deduced from 2-fluoropropane-d_0, -d_3 and -d_6 and 2-fluorobutane and therefore not entirely representative but only indicative. The CH vibrations are active in the regions of those of the CHCl group but the CF vibrations absorb at considerably higher frequencies.

R—CHF—R′ molecules

R	R′
Me—	—Me [1–4], —Et [3], CD_3— [49];
CD_3—	—CD_3 [4].

5.1.2 Chloromethylene

The CH vibrations

Other methyl and/or methylene vibrations sometimes cover the CH stretching vibration in the range 2940 ± 40 cm^{-1}. The weak to moderate band in the region 1330 ± 50 cm^{-1} is due to the CH out-of-plane deformation. The limits are found in the spectra of $Me(CHCl)_2Me$ (*tr*: *1357* and 1289; *g*: 1379 and *1303*) and in that of $MeCHClCH_2CHClMe$ (1379 and *1325* cm^{-1}). This deformation is often observed at 1340 ± 25 cm^{-1}. The CH in-plane deformation exhibits a moderate to strong band in the range 1245 ± 45 cm^{-1}.

CCl vibrations

The most characteristic absorption is the moderate to strong band at 650 ± 60 cm^{-1}, due to the C—Cl stretch. The conformers of 2,3-dichlorobutane provide the limits: 706 and 597 cm^{-1}. The first CCl deformation, attributed to the δCCl, is weakly to moderately active at 345 ± 55 cm^{-1}. The second is assigned to ωCCl or γCCl and absorbs weakly to moderately at 280 ± 50 cm^{-1}.

R—CHCl—R′ molecules

R	R′
Me—	—Me [5–8], —Et [8–11], —iBu [12], —CH_2Cl [13–15], —CH_2CCl_3 [16], —$CH_2CHClMe$ [10], —$CHCl_2$ [13], —CHClMe [17, 18], —CHClCHClMe [18], —CCl_2Me [19], —C(=O)Cl, —C(=O)OMe [20], —C(=O)OEt [21], —CH=CH_2 [22, 48];
CD_3—	—CD_3 [7];
Et—	—Et [10], —CH_2Cl [15], —CH_2CCl_3 [16];
$ClCH_2$—	—CH_2Cl [23, 24], —$CHCl_2$ [23], —CCl_3 [13], —$CHClCH_2Cl$ [25];
Cl_2CH—	—$CHCl_2$ [23], —CCl_3 [13].

5.1.3 Bromomethylene

The CH vibrations

The CH stretching vibration provides a weak to moderate band in the range 2940 ± 40 cm^{-1}, in which also methyl and methylene are active. The CH out-of-plane deformation gives rise to a weak to moderate band at 1340 ± 35 cm^{-1} and the CH in-plane deformation absorbs moderately to strongly at 1240 ± 40 cm^{-1}.

The CBr vibrations

The CBr stretching vibration occurring in the region 545 ± 75 cm^{-1} provides the most characteristic absorption. The intensity varies from moderate to strong and the band is generally sensitive to conformation. N≡CCHBrC(=O)NH_2 (620) absorbs at the HW side and MeCHBrC(=O)OEt (475 cm^{-1}) at the LW side. Most of the compounds show this νCBr at 550 ± 50 cm^{-1}. The in-plane and out-of-plane CBr deformations are observed respectively in the regions 300 ± 60 and 230 ± 65 cm^{-1}.

R—CHBr—R′ molecules

R	R′
Me—	—Me [6-8, 26], —Et [8, 9], —iBu [12], —CH_2Cl [14], —CH_2Br [14, 15, 27], —CH_2CCl_3 [16], —CHBrMe [28], —C(=O)Cl, —C(=O)Br, —C(=O)OMe [20], —C(=O)OEt [21];
CD_3—	—CD_3 [7, 26];
Et—	—CH_2Br [15];
tBu—	—CH_2CCl_3 [16];
$BrCH_2$—	—CH_2Br [24];
N≡C—	—C(=O)NH_2 [29].

5.2 CYANOMETHYLENE

The 30 normal vibrations of a simple compound such as MeCH(CN)Me are divided over 17a′ and 13a″ types of vibration. After deducting 11a′ and 10a″ vibrations, nine vibrations for the —CH(CN)— fragment remain:

a′: νCH, νC≡N, δCH, νC—CN, δ-C—CN, δC—C≡N;
a″: ωCH, γ-C—CN and γC—C≡N.

The CH vibrations

The CH vibrations cover almost the same regions as those of the R—CHX—R′ (X = F, Cl, Br) compounds: νCH 2940 ± 40, ωCH 1345 ± 30 and δCH 1250 ± 50 cm^{-1}.

The C≡N stretching vibration

The C≡N stretch (2250 ± 15 cm^{-1}) is the most characteristic band and a good group vibration, but the intensity leaves much to be desired.

Skeletal vibrations

The skeletal vibrations are coupled with other vibrations, and the number of investigated molecules is too small to deduce useful absorption regions. The following regions are only indicative:

νC—CN	δC—C—CN	γC—C—CN	γC—C≡N	δC—C≡N
810 ± 80	495 ± 45	380 ± 35	325 ± 45	175 ± 45

R—CH(CN)—R′ molecules

R	R′
Me—	—Me [6, 7, 30–32], —CH=CH_2 [32];
Cl$(CH_2)_3$—	—Ph—4-Br;
Ph—	—CH(CN)Ph [33];
Br—	—C(=O)NH_2 [29].

5.3 HYDROXYMETHYLENE

MeCH(OH)Me exhibits in the sterically favoured C_s structure 17a′ and 13a′′ vibrations. The methyl groups are responsible for 9a′ and 9a′′ vibrations and the CC_2 skeleton takes 2a′ and 1a′′ vibrations for itself. The following vibrations belong to the —CH(OH)— unit:

a′: νOH, νCH, δOH, δCH, νCO, δCO;
a′′: ωCH, γOH and γCO.

The OH vibrations

The OH stretching vibrations in associated molecules show a broad, strong band in the region 3370 ± 30 cm^{-1}. In dilute solutions the free OH absorbs sharply in the neighbourhood of 3600 cm^{-1}. The associated OH in-plane deformation gives rise to a weak but broad absorption in the range 1400 ± 30 cm^{-1} and is coupled to ωCH. In dilute solutions this band shifts to 1280 ± 30 cm^{-1} [44–47]. The out-of-plane deformation is assigned in the region 630 ± 30 cm^{-1} as a broad band with moderate intensity. In the unbonded state this broad band disappears and returns as a free torsion at 280 ± 50 cm^{-1} [41].

The CH vibrations

The weak bands of the CH vibrations occur in nearly the same regions as the CH vibrations of methyl and methylene: νCH 2940 ± 40, ωCH 1365 ± 35 and δCH 1320 ± 30 cm^{-1}.

The CO vibrations

After the OH stretching vibration, the CO stretch is the most characteristic absorption of the —CH(OH)— fragment. This vibration absorbs in secondary alcohols in the region 1110 ± 30 cm^{-1} with a moderate to strong intensity, that is, higher than in primary alcohols (1045 ± 45) [43]. This band often shows multiple minima because of the coupling with the CC_2 stretching vibration and the methyl rock. On dilution, the association with the OH deformation is removed and this band shifts to lower wavenumbers. In unsaturated secondary alcohols of type R—CH(OH)—CH=, the νCO absorbs at 1050 ± 30 cm^{-1}, in the middle of the region of primary alcohols. The CO in-plane deformation appears weakly at about 470 ± 30 cm^{-1} and the out-of-plane counterpart is a little stronger in the region 360 ± 30 cm^{-1}.

R—CH(OH)—R′ molecules

R	R′
Me—	—Me [34], —Et, —tBu, —CF_3 [35, 36], —cPr [37], —$CH_2CO_2^-$ [38];
Et—	—CO_2^- [38];
F_3C—	—CF_3 [39–41], —CCl_3 [42];
Ph—	—CH(Me)NH_2 [50].

Table 5.1 Absorption regions (cm^{-1}) of the normal vibrations of R—CHX—R′ (X = F, Cl, Br, CN, OH)

Vibration	C_s	—CHF—	—CHCl—	—CHBr—	—CH(CN)—	—CH(OH)—
νOH	a′					3370 ± 30
νCH	a′	2930 ± 10	2940 ± 40	2940 ± 40	2940 ± 40	2940 ± 40
νC≡N	a′				2250 ± 15	
δOH	a′					1400 ± 30
ωCH	a″	1340 ± 15	1330 ± 50	1340 ± 35	1345 ± 30	1365 ± 35
δCH	a′	1260 ± 20	1245 ± 45	1240 ± 40	1250 ± 50	1320 ± 30
νCO	a′					1110 ± 30
νCF	a′	955 ± 15				
νC—CN	a′				810 ± 80	
νCCl	a′		650 ± 60			
γOH	a″					630 ± 30
νCBr	a′			545 ± 75		
δC—C—CN	a′				495 ± 45	
δCO	a′					470 ± 30
δCF	a′	420 ± 20				
γC—C—CN	a″				380 ± 35	
γCO	a″					360 ± 30
δCCl	a′		345 ± 55			
γC—C≡N	a″				325 ± 45	
δCBr	a′			300 ± 60		
ωCF	a″	370 ± 20				
ωCCl	a″		280 ± 50			
ωCBr	a″			230 ± 65		
δC—C≡N	a′				175 ± 45	

References

1. J.H. Griffiths, N.L. Owen and J. Sheridan, *J. Chem. Soc. Faraday Trans. 2*, **69**, 1359 (1973).
2. G.A. Crowder and T. Koger, *J. Mol. Struct.*, **23**, 311 (1974).
3. G.A. Crowder and T. Koger, *J. Mol. Struct.*, **29**, 233 (1975).
4. J. Gustavsen and P. Klaboe, *Spectrochim. Acta, Part A*, **32A**, 755 (1976).
5. C.G. Opaskar and S. Krimm, *Spectrochim. Acta, Part A*, **23A**, 2261 (1967).
6. P. Klaboe, *Spectrochim. Acta, Part A*, **26A**, 87 (1970).
7. J.R. Durig, C.M. Player, Y.S. Li, J. Bragin and C.W. Hawley, *J. Chem. Phys.*, **57**, 4544 (1972).
8. N.T. McDevitt, A.L. Rozek, F.F. Bentley and A.D. Davidson, *J. Chem. Phys.*, **42**, 1173 (1965).
9. E. Benedetti and P. Cecchi, *Spectrochim. Acta, Part A*, **28A**, 1007 (1972).
10. W.H. Moore and S. Krimm, *Spectrochim. Acta, Part A*, **29A**, 2025 (1973).
11. A.J. Barnes, M.L. Evans and H.E. Hallam, *J. Mol. Struct.*, **99**, 235 (1983).
12. G.A. Crowder and R.M.P. Jaiswal, *J. Mol. Struct.*, **99**, 93 (1983).
13. A.B. Dempster, K. Price and N. Sheppard, *Spectrochim. Acta, Part A*, **27A**, 1563 (1971).
14. J. Thorbjørnsrud, O.H. Ellestad, P. Klaboe and T. Torgrimsen, *J. Mol. Struct.*, **15**, 45 (1973).
15. G.A. Crowder, *J. Mol. Struct.*, **100**, 415 (1983).
16. M. Carles-Lorjou, A. Goursot-Leray, H. Bodot and R. Gaufrès, *Spectrochim. Acta, Part A*, **29A**, 329 (1973).
17. X. Jing and S. Krimm, *Spectrochim. Acta, Part A*, **39A**, 251 (1983).
18. S.H. Chough and S. Krimm, *Spectrochim. Acta, Part A*, **46A**, 1405 (1990).
19. S.H. Chough and S. Krimm, *Spectrochim. Acta, Part A*, **46A**, 1419 (1990).
20. R. Das and S.K. Nandy, *Indian J. Phys.*, **52B**, 85 (1977).
21. N. Dube and R. Prasad, *Spectrochim. Acta*, **43A**, 83 (1987).
22. N. Som and G.S. Kastha, *Indian J. Phys.*, **51B**, 77 (1977).
23. A.B. Dempster, K. Price and N. Sheppard, *Spectrochim. Acta, Part A*, **27A**, 1579 (1971).
24. J. Thorbjørnsrud, O.H. Ellestad, P. Klaboe, T.T. Torgrimsen and D.H. Christensen, *J. Mol. Struct.*, **17**, 5 (1973).
25. A.B. Dempster, K. Price and N. Sheppard, *Spectrochim. Acta, Part A*, **31A**, 331 (1975).
26. P. Klaboe, A. Linde and B.N. Cyvin, *Spectrochim. Acta, Part A*, **30A**, 1513 (1974).
27. J. Som and G.S. Kastha, *Indian J. Phys.*, **47**, 494 (1973).
28. K. Imura, *Bull. Chem. Soc. Jpn.*, **42**, 3135 (1969).
29. L. Van Haverbeke and M.A. Herman, *Spectrochim. Acta, Part A*, **31A**, 959 (1975).
30. R. Yamadera and S. Krimm, *Spectrochim. Acta, Part A*, **24A**, 1677 (1968).
31. B.N. Cyvin and S.J. Cyvin, *Acta Chem. Scand.*, **26**, 3943 (1972).
32. D.A.C. Compton and W.F. Murphy, *Spectrochim. Acta, Part A*, **41A**, 1141 (1985).
33. A.M. North, R.A. Pethrick and A.D. Wilson, *Spectrochim. Acta, Part A*, **30A**, 1317 (1974).
34. L.I. Lafer, V.I. Yakerson and G.A. Kogan, *Bull. Sci. Acad. U.S.S.R. (Div. Chem. Sci.)* **8**, 1717 (1969).
35. J. Murto, A. Kivinen, K. Edelmann and E. Hassinen, *Spectrochim. Acta, Part A*, **31A**, 479 (1975).
36. J.R. Durig, F.O. Cox, P. Grower and B.J. Van der Veken, *J. Phys. Chem.*, **91**, 3211 (1987).
37. P. Klaboe and D.L. Powell, *J. Mol. Struct.*, **20**, 95 (1974).
38. M. Morssli, G. Cassanas, L. Bardet, B. Pauvert and A. Terol, *Spectrochim. Acta, Part A*, **47A**, 529 (1991).
39. S.J. Cyvin, J. Brunvoll and M. Perttilä, *J. Mol. Struct.*, **17**, 17 (1973).

40. J. Murto, A. Kivinen, R. Viitala and J. Hyömäki, *Spectrochim. Acta, Part A*, **29A**, 1121 (1973).
41. J.R. Durig, R.A. Larsen, F.O. Cox and B.J. Van der Veken, *J. Mol. Struct.*, **172**, 183 (1988).
42. J. Murto, A. Kivinen and P. Saarinen, *Acta Chem. Scand., Ser. A*, **30A**, 448 (1976).
43. H.H. Zeiss and M. Tsutsui, *J. Am. Chem. Soc.*, **75**, 897 (1953).
44. A.V. Stuart and G.B.B.M. Sutherland, *J. Chem. Phys.*, **24**, 559 (1956).
45. S. Krimm, C.Y. Liang and G.B.B.M. Sutherland, *J. Chem. Phys.*, **25**, 778 (1956).
46. P. Tarte and R. Duponthière, *J. Chem. Phys.*, **26**, 962 (1957).
47. P. Tarte and R. Duponthière, *Bull. Soc. Chim. Bel.*, **66**, 525 (1957).
48. S.H. Schei and P. Klaboe, *Acta Chem. Scand., Ser. A*, **37A**, 315 (1983).
49. J.R. Durig, H. Nanaie and G.A. Guirgis, *J. Raman Spectrosc.*, **22**, 155 (1991).
50. Y.-S. Li, A.S. Lee and Yu. Wang, *J. Raman Spectrosc.*, **22**, 191 (1991).

6

Normal Vibrations and Absorption Regions of CX_2

Just like CH_2, the CX_2 fragment displays six fundamental vibrations. R—CX_2—R′ compounds in which R is identical to R′ can belong to the point group C_{2v}. In this case the six normal vibrations are divided into $2a_1 + a_2 + 2b_1 + b_2$ species in which b_1 and b_2 can be exchanged according to the choice of the symmetry plane. For compounds in which R differs from R′ the a_1 and b_2 vibrations change into a′ and the a_2 and b_1 vibrations change into a″ if the molecule belongs to the point group C_s. The following vibrations are considered as inherent in the —CX_2— structure unit:

$\nu_a CX_2$ (b_1)(a″), $\nu_s CX_2$ (a_1)(a′), δCX_2 (a_1)(a′), ωCX_2 (b_2)(a′), ρCX_2 (b_1)(a″), τCX_2 (a_2)(a″).

6.1 DIFLUOROMETHYLENE

The CF_2 stretching vibrations

The —CF_2— fragment gives rise to two strong bands: $\nu_a CF_2$ (1195 ± 85) and $\nu_s CF_2$ (1130 ± 70 cm^{-1}). The broad region of the antisymmetric stretch is due to the low values in the spectra of CF_2I_2 (1110), CF_2Br_2 (1153), CF_2Cl_2 (1167) and F_3CCF_2I (1161 cm^{-1}). The molecules CF_2I_2 (1067) and $ClCF_2CFCl_2$ (1087 cm^{-1}) are responsible for the lowest values of the symmetric stretch. For most of the difluoromethylene compounds the regions are more attractive: $\nu_a CF_2$ at 1225 ± 50 and $\nu_s CF_2$ at 1150 ± 50 cm^{-1}.

The CF_2 deformations

The —CF_2— scissoring vibration is situated in the broad region 520 ± 155 cm^{-1} as a moderate to strong band. The molecules N≡C(CF_2)C≡N (*trans*: 650; *gauche*:

671 cm^{-1}) and $ClCF_2NO$ (644 cm^{-1}) take care of the upper limit, and F_3CCF_2X (X = Br and I) and CF_2Br_2 with 367 cm^{-1} keep the lower limit. There is a considerable chance of finding these CF_2 scissors in the range 510 ± 70 cm^{-1}. The coupled vibrations CF_2 wag, CF_2 rock and CF_2 twist are classified in three large regions which are reduced to:

ωCF_2: 415 ± 85; ρCF_2: 375 ± 65 and τCF_2: 295 ± 65 cm^{-1}

if the extreme values in the spectra of the compounds in Table 6.1 are not taken into account.

Table 6.1 CF_2 deformations

	ωCF_2	ρCF_2	τCF_2
trans-$MeCF_2CH_2Cl$	515	*428*	*358*
gauche-$MeCF_2CH_2Cl$	508	*428*	*325*
$MeOCF_2CHCl_2$	526	*375*	*313*
gauche-$N{\equiv}CCF_2CF_2C{\equiv}N$	523 and *484*	469 and *364*	*280* and *238*
CF_3CF_2Cl	*362*	*316*	183
CF_3CF_2Br	*332*	298	154
CF_3CF_2I	300	262	133

$R{-}CF_2{-}R'$ molecules

R	R'
Me—	—Me [1, 2], $—CH_2Cl$ [3], —Cl and —Br [4];
$(CF_2CH_2)_n$	$(CH_2CF_2)_n$ [5];
F_3C—	$—CF_3$ [6, 7], —Cl, —Br and —I [8], —C(=O)OH [9], —C(=O)OR [10] (R = Me, Et, nPr, iPr, nBu, iBu and nPent);
N≡C—	—C≡N [12], $—CF_2C{\equiv}N$ [13];
MeO—	$—CHCl_2$ [11];
Cl—	$—CFCl_2$ [14], N=O [15], —Cl [16–18];
Br—	—Br [17];
I—	—I [19].

6.2 DICHLOROMETHYLENE

The CCl_2 stretching vibrations

Although the absorption regions overlap each other, the $—CCl_2—$ stretching vibrations are observed as moderate to strong separate bands. The large region of the antisymmetric stretching vibration is produced by compounds such as $FCCl_2F$ (920) and $F_3CCCl_2CF_3$ (912 cm^{-1}) on the one side and $MeCCl_2CHCl_2$ (640 or 696), *trans*-$MeCCl_2Et$ (645) and $D_3CCCl_2CD_3$ (600 cm^{-1}) on the other side. Extreme values for the symmetric stretch are found in the spectra of $O_2NCCl_2NO_2$ (762), *gauche*-$MeCCl_2CCl_2Me$ (722 and *565*), *trans*-$MeCCl_2Et$ (545) and $D_3CCl_2CD_3$ (526 cm^{-1}). Most of the R—CCl_2—R′ compounds display these stretching vibrations in the regions $\nu_a CCl_2$: 755 ± 100 and $\nu_s CCl_2$: 620 ± 70 cm^{-1}.

The CCl_2 deformations

The CCl_2 deformations are observed in four regions which overlap each other. Disregarding the high values in the spectra of CCl_2F_2 (wag: 465 and rock: 435), N≡$CCCl_2C$(=O)NH_2 (wag: 494 and twist: 376), $FCCl_2CF_2Cl$ (wag: 440) and N≡$CCCl_2C$≡N (wag: 440 cm^{-1}) and the values assigned for *gauche*-$MeCCl_2CCl_2Me$ (wag: 428 and *365*; rock: 288 and 249 and twist: 245 and 238 cm^{-1}), the regions become much more attractive:

ωCCl_2	ρCCl_2	τCCl_2	δCCl_2
380 ± 40	340 ± 40	300 ± 40	250 ± 40

R—CCl_2—R′ molecules

R	R′
Me—	—Me [20–25], —Et [22, 25–28], —$CHCl_2$ [29], —CCl_2Me [30], —F [31];
D_3C—	—CD_3 [21];
$ClCH_2$—	—CH_2Cl [32, 33], —$CHCl_2$ [32];
Cl_2CH—	—CCl_3 [29];
F_3C—	—CF_3 [6];
N≡C—	—C≡N [34, 35], —C(=O)NH_2 [36];
O_2N—	—NO_2 [37];
F—	—CF_2Cl [14], —N=O [15], —F [16–18];
Br—	—N=O [15].

6.3 DIBROMOMETHYLENE

The CBr_2 stretching vibrations

The —CBr_2— stretching vibrations appear moderately to strongly in the regions: $\nu_a CBr_2$ 650 ± 70 and $\nu_s CBr_2$ 530 ± 50 cm^{-1}. The high values assigned in the spectra of CF_2Br_2 (ν_a: 831 and ν_s: 623) and $CF_3CBr_2CF_3$ (ν_a: 850 cm^{-1}) are outside the above—mentioned regions.

The CBr_2 deformations

Ignoring, without rejecting, the high values for ωCBr_2 in the spectrum of N≡$CCBr_2C$(=O)NH_2 (447) and for δCBr_2 in that of $MeCBr_2Et$ (258 cm^{-1}), the CBr_2 deformations are expected in the regions:

ωCBr_2	ρCBr_2	τCBr_2	δCBr_2
360 ± 40	320 ± 30	250 ± 40	180 ± 30

R—CBr_2—R′ molecules

R	R′
Me—	—Me [20, 24], —Et[39], —CBr_2Me [38];
F_3C—	—CF_3 [6];
N≡C—	—C≡N [34, 35], —C(=O)NH_2 [36];
F—	—F [17];
Cl—	—NO [15].

Table 6.2 Absorption regions (cm^{-1}) of the normal vibrations of —CX_2— (X = F, Cl and Br)

Vibration	—CF_2—	—CCl_2—	—CBr_2—
$\nu_a CF_2$	1195 ± 85		
$\nu_s CF_2$	1130 ± 70		
$\nu_a CCl_2$		760 ± 160	
$\nu_a CBr_2$			650 ± 70
$\nu_s CCl_2$		645 ± 120	
$\nu_s CBr_2$			530 ± 50
δCF_2	520 ± 155		
ωCF_2	415 ± 115		
ωCCl_2		418 ± 80	
ωCBr_2			385 ± 65
ρCF_2	365 ± 105		
ρCCl_2		340 ± 95	
ρCBr_2			320 ± 30
τCCl_2		305 ± 75	
δCCl_2		250 ± 40	
τCBr_2			250 ± 40
τCF_2	245 ± 115		
δCBr_2			205 ± 55

References

1. G.A. Crowder and D. Jackson, *Spectrochim. Acta, Part A*, **27A**, 2505 (1971).
2. J.R. Durig, G.A. Guirgis and Y.S. Li, *J. Chem. Phys.*, **74**, 5946 (1981).
3. G.A. Crowder, *J. Mol. Struct.*, **15**, 356 (1973).
4. J.R. Durig, S.M. Craven, C.W. Hawley and J. Bragin, *J. Chem. Phys.*, **57**, 131 (1972).
5. G. Cortili and G. Zerbi, *Spectrochim. Acta, Part A*, **23A**, 285 (1967).
6. H. Bürger and G. Pawelke, *Spectrochim. Acta, Part A*, **35A**, 525 (1979).
7. E.L. Pace, A.C. Plaush and H.V. Samuelson, *Spectrochim. Acta*, **22**, 993 (1966).
8. O. Risgin and R.C. Taylor, *Spectrochim. Acta*, **15**, 1036 (1959).
9. G.A. Crowder, *J. Fluorine Chem.*, **1**, 385 (1971/72).

10. G.A. Crowder, *J. Fluorine Chem.*, **2**, 217 (1972/73).
11. Y.S. Li and J.R. Durig, *J. Mol. Struct.*, **81**, 181 (1982).
12. H.O. Desseyn, B.J. Van der Veken, L. Van Haverbeke and M.A. Herman, *J. Mol. Struct.*, **13**, 227 (1972).
13. J.E. Gustavsen, P. Klaboe, C.J. Nielsen and D.L. Powell, *Spectrochim. Acta, Part A*, **35A**, 109 (1979).
14. P. Klaboe and J.R. Nielsen, *J. Mol. Spectrosc.*, **6**, 379 (1961).
15. N.P. Ernsting and J. Pfab, *Spectrochim. Acta, Part A*, **36A**, 75 (1980).
16. S. Giorgianni, A. Gambi, L. Franco and S. Ghersetti, *J. Mol. Struct.*, **75**, 389 (1979).
17. L.H. Ngai and R.H. Mann, *J. Mol. Spectrosc.*, **38**, 322 (1971).
18. H.H. Claassen, *J. Chem. Phys.*, **22**, 50 (1954).
19. I. McAlpine and H. Sutcliffe, *Spectrochim. Acta, Part A*, **25A**, 1723 (1969).
20. M.C. Tobin, *J. Am. Chem. Soc.*, **75**, 1788 (1953).
21. G.A. Crowder, *Spectrochim. Acta, Part A*, **42A**, 1079 (1986).
22. M.S. Wu, P.C. Painter and M.M. Coleman, *Spectrochim. Acta, Part A*, **35A**, 823 (1979).
23. J.H.S. Green and D.J. Harrison, *Spectrochim. Acta, Part A*, **27A**, 1217 (1971).
24. P. Klaboe, *Spectrochim. Acta, Part A*, **26A**, 977 (1970).
25. S.H. Chough and S. Krimm, *Spectrochim. Acta, Part A*, **46A**, 1419 (1990).
26. K. Ohno, Y. Shiro and H. Murata, *Bull. Chem. Soc. Jpn.*, **47**, 2962 (1974).
27. G.A. Crowder and W.-Y. Lin, *J. Mol. Struct.*, **62**, 1 (1980).
28. K. Ohno, K. Taga, I. Yoshida and H. Murata, *Spectrochim. Acta, Part A*, **36A**, 721 (1980).
29. A.B. Dempster, K. Price and N. Sheppard, *Spectrochim. Acta, Part A*, **27A**, 1563 (1971).
30. A.O. Diallo, *Spectrochim. Acta, Part A*, **35A**, 597 (1979).
31. J.R. Durig, C.J. Wurrey, W.E. Bucy and A.E. Sloan, *Spectrochim. Acta, Part A*, **32A**, 175 (1976).
32. A.B. Dempster, K. Price and N. Sheppard, *Spectrochim. Acta, Part A*, **27A**, 1579 (1971).
33. D.L. Powell, P. Klaboe, K. Saebø and G.A. Crowder, *J. Mol. Struct.*, **98**, 55 (1983).
34. L. Van Haverbeke, H.O. Desseyn, B.J. Van der Veken, *Bull. Soc. Chim. Bel.*, **82**, 133 (1973).
35. S.Bjørklund and E. Augdahl, *Spectrochim. Acta, Part A*, **32A**, 1021 (1976).
36. L. Van Haverbeke and M.A. Herman, *Spectrochim. Acta, Part A*, **31A**, 959 (1975).
37. A.O. Diallo, *Spectrochim. Acta, Part A*, **27A**, 239 (1971).
38. A.O. Diallo, *Spectrochim. Acta, Part A*, **32A**, 295 (1976).
39. A.O. Diallo, *Spectrochim. Acta, Part A*, **38A**, 687 (1982).

7

Normal Vibrations and Absorption Regions of C(=X)Y

7.1 CARBONYL COMPOUNDS

7.1.1 Formyl

The —C(=O)H group possesses six normal vibrations which are described as follows:

νCH, νC=O, δCH, γCH/C=O, δC=O and torsion.

The CH stretching vibration

The highest CH stretching vibrations are found in the spectra of formates (2935 ± 45 cm^{-1}). Formyl fluoride scores highest with 2980 cm^{-1}. For $ClCH_2OC(=O)H$ in the crystalline state the value is 2980 cm^{-1} but in the liquid state 2968 cm^{-1} [109, 111, 112]; formic acid as a dimer gives 2957 [93, 95, 96] and 2949 [96] and as a monomer 2942 cm^{-1} [93, 94, 96]. Aldehydes, thioformates and formamides absorb in the region 2850 ± 45 cm^{-1} with high values for 2-$F_3CPhC(=O)H$ (2895) and $H_2NC(=O)H$ (2882) and low values for MeCH=CHC(=O)H (2805) and 2-MePhC(=O)H (2808 cm^{-1}). The unambiguous assignment of this weak to moderate band in aldehydes is often disturbed by the overtone of the CH in-plane deformation (2745 ± 45 cm^{-1}), which increases in intensity by Fermi resonance. In α-halogen-substituted aldehydes, in which the halogen slightly lowers the absorption frequency of δCH, the overtone does not cause Fermi resonance, for example $Cl_3CC(=O)H$ and $Br_3CC(=O)H$. This typical Fermi doublet, in which both absorptions are in essence implicated in the CH stretching vibration, is very useful in identifying aldehydes [138, 139].

	saturated aldehyde	unsaturated aldehyde	aromatic aldehyde
νCH	2840 ± 30	2830 ± 30	2850 ± 45
2 × δCH	2720 ± 20	2720 ± 20	2755 ± 35

The C=O stretching vibration

Logically the C=O stretching vibration provides the most characteristic band of the C(=O)H group, which absorbs very strongly in the region 1745 ± 95 cm^{-1}. In formates the large region (1765 ± 75) is produced by F—C(=O)H, which displays the νC=O at 1837 cm^{-1} (Section 7.1.2). Formic acid and its esters (Section 7.1.7) absorb 1730 ± 40 cm^{-1}. Saturated aldehydes absorb strongly in the region 1755 ± 35 cm^{-1} with at the HW side the α-halogenated aldehydes such as F_3CC(=O)H (1788) and Cl_3CC(=O)H (1758) and at the LW side *trans*-cPrC(=O)H (1700 cm^{-1}). For Me(CH_2)$_n$C(=O)H (n = 0–12) the region narrows to 1730 ± 5 cm^{-1} [141]. Most of the investigators [38, 41–46, 49, 50] find 1745 and 1732 cm^{-1} bands in the vapour state spectrum of H(O=)CC(=O)H while Durig *et al.* [39] find 1729 and 1707 cm^{-1} in the spectrum of the crystals. Aromatic aldehydes, formamides (Section 7.2) and thioformates show the νC=O in the range 1690 ± 30 cm^{-1}. Most of the benzaldehydes absorb in the neighbourhood of 1700 cm^{-1}. High values are found in the spectrum of 4-HO(O=)CPhC(=O)H (1719 with 1688 for the acid) and in the spectra of pyridinecarboxaldehydes (≈1712 cm^{-1}), and low values in those of 4-Me_2NPhC(=O)H (1661) and 2-, 3- and 4-HOPhC(=O)H (≈1665 cm^{-1}). MeHNC(=O)H absorbs at 1720 cm^{-1} as a gas [125] or in a N_2 matrix [124] but at 1666 cm^{-1} as a liquid. Unsaturated aldehydes exhibit the C=O band in the region 1685 ± 35 cm^{-1}. The highest values (≈1720 cm^{-1}) are found in the vapour phase spectra of the compounds H_2C=CHC(=O)H, MeCH=CHC(=O)H and H_2C=C(Me)C(=O)H, which absorb in the vicinity of 1695 cm^{-1} as a liquid. For most of the unsaturated aldehydes the C=O stretching vibration occurs at 1675 ± 25 cm^{-1}. The low values (≈1605) in the spectra of salts of β-ketoaldehydes with the formula R—C(=O)CH_2C(=O)H (R = Me, Et, iPr, tBu), in which the C=O bond is weakened by the contribution of the enol tautomer [12], are not taken into account.

The CH in-plane deformation

The CH in-plane deformation or CH rocking vibration mostly appears in the region 1360 ± 55 cm^{-1} as a weak to moderate band. On the HW side of this region one finds the absorptions of a few benzaldehydes such as 2-F_3CPhC(=O)H (1415), 3-MePhC(=O)H (1408) and 3-FPhC(=O)H (1406 cm^{-1}). In the spectra of 4-X—substituted benzaldehydes Nyquist *et al.* [140] assign the δCH at 1387 ± 7 cm^{-1}, with overtones near 2775 cm^{-1}. The lowest values appear in the spectra of H(O=)CC(=O)H (1312 and 1338), D(O=)CC(=O)H (1335) and HSC(=O)H

(1339 cm^{-1}). The value 1504 cm^{-1} in the vapour phase spectrum of formaldehyde falls outside the above-mentioned region. Most of the aldehydes show this δCH at 1370 $\pm$ 25 cm^{-1}. The overtones should be expected near 2745 $\pm$ 45 cm^{-1}.

The CH/C=O wagging vibration

The CH/C=O wag is assigned in the large region 885 $\pm$ 185 cm^{-1} with a weak to moderate intensity. The broadness of the region stems from the fact that the absorption at high {low} frequencies has mostly the character of a CH {C=O} wag. The ν_6 (ωCH_2: b_1) at 1167 cm^{-1} in the spectrum of formaldehyde is not taken into account. In MeC(=O)H the CH wag (764 cm^{-1}) is coupled to the methyl rock (1102 cm^{-1}). In the spectrum of EtC(=O)H the CH wag is assigned at 896 cm^{-1}. A compound such as H(O=)CC(=O)H displays the CH wagging vibrations at 1048 and 801 cm^{-1}. High wags are found in the formates with 1050 cm^{-1} for HC(=O)OH and values in the neighbourhood of 1040 cm^{-1} for the esters of formic acid. In the spectrum of ethyl formate Dahlqvist and Euranto [109] assign the shoulder (1015 cm^{-1}) next to the ν_sCOC to the ωCH, whereas Charles *et al.* [110] prefer the weak band at 920 cm^{-1}. Aromatic aldehydes show the CH/C=O wag probably in the region 775 $\pm$ 55 cm^{-1}, although in the spectra of some benzaldehydes this vibration is assigned in the neighbourhood of 1000 cm^{-1} [69, 70].

C=O in-plane deformation

The C=O in-plane deformation can be found in the extensive region 550 $\pm$ 220 cm^{-1} as a weak to moderate absorption. Mostly the δC=O appears at 615 $\pm$ 155 cm^{-1} if the low value of a δC=O in the spectrum of H(O=)CC(=O)H (*551* and 339 as a gas and *551* and 388 cm^{-1} in the solid phase) is not taken into account. Saturated aldehydes [135] show the δC=O in the region 565 $\pm$ 100 cm^{-1} with EtC(=O)H (660), nPrC(=O)H (661) and $ClCH_2C$(=O)H (463 cm^{-1}) as examples. The range of unsaturated aldehydes (640 $\pm$ 100 cm^{-1}) is in good agreement with that of aromatic aldehydes (645 $\pm$ 55 cm^{-1}). In the spectra of XC≡CC(=O)H (X = Cl, Br, I, H, D) the δC=O is assigned respectively at 738, 691, 670, 615 and 609 cm^{-1} and in the spectrum of MeCH=CHC(=O)H at 542 cm^{-1}. The highest values are found in the spectra of formates and formamides: MeOC(=O)H (767), HC≡CCH_2OC(=O)H (765) and MeHNC(=O)H (770 cm^{-1}).

Torsion

The weak absorption in the region 125 $\pm$ 65 cm^{-1} is assigned to the —C(=O)H torsion.

Table 7.1 Absorption regions (cm^{-1}) of the normal vibrations of —C(=O)H

Vibration	saturated	C=O bonded	unsaturated	aromatic	formate	thioformate	formamide
νCH	2840 ± 30	2860 ± 30	2830 ± 30	2850 ± 45	2935 ± 45	2840 ± 10	2865 ± 25
νC=O	1755 ± 35	1730 ± 25	1685 ± 35	1690 ± 30	1765 ± 75	1680 ± 20	1690 ± 30
δCH	1375 ± 25	1340 ± 30	1380 ± 25	1390 ± 25	1365 ± 15	1340 ± 10	1385 ± 15
ωCH/C=O	865 ± 125	925 ± 125	860 ± 160	865 ± 145	990 ± 70	935 ± 20	960 ± 95
δC=O	565 ± 100	465 ± 135	640 ± 100	645 ± 55	720 ± 50	–	665 ± 105
torsion	110 ± 50	150 ± 40	140 ± 40	130 ± 40	–	–	–

R—C(=O)H molecules

R = H— [1, 2], Me— [1, 3–5, 136], CD_3— [1, 5], CH_2D— [5], Et— [6–9], $MeCD_2$— [8], nPr—[7], $PhCH_2$—, $HOCH_2$— [10, 11], R′C(=O)CH_2— (R′= Me, Et, iPr, tBu) [12], FCH_2— [13], $ClCH_2$— [14, 15], Cl_2CH— [16], cPr— [17], iPr— [18, 19], tBu— [20, 21], CF_3— [22–28], CCl_3— [29–31], CBr_3— [29, 30, 32], H(O=)C— [33–51], D(O=)C— [38, 41, 44, 45, 47], HO(O=)C— [52, 53], H_2C=CH— [43–48, 54], MeCH=CH— [54–56], H_2C=C(Me)— [56, 57], Me—$(CH=CH)_2$—, Me—$(CH=CH)_3$— and Me$(CH=CH)_2$C(Me)=CH— [58], HOCH=CH— [46, 60], H_2NCH=CH— and D_2NCH=CH— [59], HOCH=C(Br)— [60, 61], HC≡C— [62–65], DC≡C— [62–64], MeC≡C— [66], N≡C— [65], XC≡C— (X = Cl, Br, I) [64, 67], Ph— [68–72], 2-, 3- and 4-MePh— [70], 4-EtPh—, 2-, 3- and 4-F_3CPh— [73], 4-tBuPh—, 4-N≡CPh— [74], 4-HO(O=)CPh— [74], 2-HOPh— [75, 76], 3-HOPh— [92], 4-HOPh— [77, 92], 2-, 3- and 4-MeOPh— [78, 79], 2- and 4-EtOPh— [74], 2-, 3- and 4-O_2NPh— [77, 80, 92], 4-Me_2NPh— [77, 81], 4-MeSPh—, 2-, 3- and 4-FPh— [70, 71], 2-, 3- and 4-ClPh— [70, 71, 82], 2-, 3- and 4-BrPh— [70], 2,4-Me_2Ph— and 2,5-Me_2Ph— [83], 2,3-$(HO)_2Ph$— and 3,4-$(HO)_2Ph$— [84], 2-HO—3-MeOPh—, 3-HO—4-MeOPh— and 4-HO—3-MeOPh— [85], 2,3-$(MeO)_2Ph$—, 2,4-$(MeO)_2Ph$— and 3,4-$(MeO)_2Ph$— [86], 2,4-Cl_2Ph—, 3,4-Cl_2Ph— and 2,6—Cl_2Ph— [87], 2,4,6—$(MeO)_3Ph$—, 2,4,5-$(MeO)_3Ph$— and 2,3,4-$(MeO)_3Ph$— [88], 2-, 3- and 4-Py— [89], 2-Fu— [90, 91], 2-Th—, HO— [93–98], DO— [93, 97–99], MeO— [100–109], CD_3O— [105–107, 109], EtO— [109, 110], $ClCH_2O$— [109, 111–113], $ClCD_2O$— [109, 111], H_2C=CHO— [114], HC≡CCH_2O— [109, 115], HS— [116, 137], MeS— [117], H_2N— [118–123], D_2N— [118, 120, 123], MeNH— [123–125], MeND— [123, 126], H_2NHNC(=S)NH— [127], Me_2N— [125, 128–132], F— [133, 134], Cl— [134].

References

1. P. Cossee and J.H. Schachtschneider, *J. Chem. Phys.*, **44**, 97 (1966).
2. J.W.C. Johns and W.B. Olsen, *J. Mol. Spectrosc.*, **39**, 479 (1971).
3. K.S. Pitzer and W. Weltner, *J. Am. Chem. Soc.*, **71**, 2842 (1949).
4. J.C. Evans and H.J. Bernstein, *Can. J. Chem.*, **34**, 1083 (1956).
5. H. Hollenstein and H.H. Günthard, *Spectrochim. Acta, Part A*, **27A**, 2027 (1971).
6. E.F. Worden Jr., *Spectrochim. Acta*, **18**, 1121 (1962).
7. G. Sbrana and V. Schettino, *J. Mol. Spectrosc.*, **33**, 100 (1970).
8. S.G. Frankiss and W. Kynaston, *Spectrochim. Acta, Part A*, **28A**, 2149 (1972).
9. P. Van Nuffel, L. Van den Enden, C. Van Alsenoy and H.J. Geise, *J. Mol. Struct.*, **116**, 99 (1984).
10. H. Michelsen and P. Klaboe, *J. Mol. Struct.*, **4**, 293 (1969).
11. Y. Kobayashi, H. Takahara, H. Takahashi and K. Higasi, *J. Mol. Struct.*, **32**, 235 (1976).

12. J. Terpinski, *Spectrochim. Acta, Part A*, **36A**, 621 (1980).
13. H.V. Phan and J.R. Durig, *J. Mol. Struct.*, **209**, 333 (1990).
14. G. Lucazeau and A. Novak, *J. Chim. Phys. Phys.-chim. Biol.*, **67**, 1614 (1970).
15. S. Dyngeseth, H. Schei and K. Hagen, *J. Mol. Struct.*, **102**, 45 (1983).
16. G. Lucazeau and A. Novak, *J. Mol. Struct.*, **5**, 85 (1970).
17. J.R. Durig and T.S. Little, *Croat. Chem. Acta*, **61**, 529 (1988).
18. A. Piart-Goypiron, M.H. Baron, J. Belloc, M.J. Coulange and H. Zine, *Spectrochim. Acta, Part A*, **47A**, 363 (1991).
19. J.R. Durig, G.A. Guirgis, W.E. Brewer and T.S. Little, *J. Mol. Struct.*, **248**, 49 (1991).
20. G.A. Crowder, *J. Chem. Soc. Perkin Trans. 2*, 1241 (1973).
21. J.R. Durig, S.M. Craven, J.H. Mulligan and C.W. Hawley, *J. Chem. Phys.*, **58**, 1281 (1973).
22. R.E. Dodd, H.L. Roberts and L.A. Woodward, *J. Chem. Soc.*, 2783 (1957).
23. C.V. Berney, *Spectrochim. Acta, Part A*, **25A**, 793 (1969).
24. G.A. Crowder, *J. Fluorine Chem.*, **2**, 107 (1972/73).
25. R.L. Redington, *Spectrochim. Acta, Part A*, **31A**, 1699 (1975).
26. E. Ottavianelli, E.A. Castro and A.H. Jubert, *J. Mol. Struct.*, **254**, 279 (1992).
27. J.S. Francisco and I.H. Williams, *Spectrochim. Acta, Part A*, **48A**, 1115 (1992).
28. J.R. Durig, G.A. Guirgis and B.J. Van der Veken, *J. Raman Spectrosc.*, **18**, 549 (1987).
29. G. Lucazeau and A. Novak, *Spectrochim. Acta, Part A*, **25A**, 1615 (1969).
30. G. Hagen, *Acta Chem. Scand.*, **25**, 813 (1971).
31. J.R. Durig and W.J. Natter, *J. Raman Spectrosc.*, **11**, 32 (1981).
32. M.I. Suero, F. Marquez and M.J. Martin-Delgado, *Spectrosc. Lett.*, **23**, 771 (1990).
33. A.R.H. Cole and H.W. Thompson, *Proc. R. Soc. London, Ser. A*, **200A**, 10 (1949).
34. A.R.H. Cole, *Aust. J. Sci.*, **25**, 225 (1962).
35. W.G. Fateley, R.K. Harris, F.A. Miller and R.E. Witkowsky, *Spectrochim. Acta*, **21**, 231 (1965).
36. F.D. Verderame, E. Castellucci and S. Califano, *J. Chem. Phys.*, **52**, 719 (1970).
37. J.R. Durig and S.E. Hannum, *J. Cryst. Mol. Struct.*, **1**, 131 (1971).
38. A.R.H. Cole and G.A. Osborne, *Spectrochim. Acta, Part A*, **27A**, 2461 (1971).
39. J.R. Durig, S.C. Brown and S.A. Hannum, *J. Chem. Phys.*, **54**, 4428 (1971).
40. A.R.H. Cole and J.R. Durig, *J. Raman Spectrosc.*, **4**, 31 (1975).
41. C. Cossart-Magos, *Spectrochim. Acta, Part A*, **34A**, 415 (1978).
42. R. Naaman, D.M. Lubman and R.N. Zare, *J. Mol. Struct.*, **59**, 225 (1980).
43. R.K. Harris, *Spectrochim. Acta*, **20**, 1129 (1964).
44. Yu. N. Panchenko, P. Pulay and F. Török, *J. Mol. Struct.*, **34**, 283 (1976).
45. H.J. Oelichman, D. Bougeard and B. Schrader, *J. Mol. Struct.*, **77**, 149 (1981).
46. Z. Smith, E.B. Wilson and R.W. Duerst, *Spectrochim. Acta, Part A*, **39A**, 1117 (1983).
47. P. Pulay, G. Fogarasi, G. Pongor, J.E. Boggs and A. Vargha, *J. Am. Chem. Soc.*, **105**, 7037 (1983).
48. R.K. Harris and R.E. Witkowsky, *Spectrochim. Acta*, **20**, 1651 (1964).
49. C.Cossart-Magos, A. Frad and A. Tramer, *Spectrochim. Acta, Part A*, **34A**, 195 (1978).
50. G.E. Scuseria and H.F. Schaefer, *J. Am. Chem. Soc.*, **111**, 7761 (1989).
51. G.R. Demaré, *J. Mol. Struct.*, **253**, 199 (1992).
52. G. Fleury and V. Tabacik, *J. Mol. Struct.*, **10**, 359 (1971).
53. G. Fleury and V. Tabacik, *J. Mol. Struct.*, **12**, 156 (1972).
54. A. Bowles, W.O. George and W.F. Maddams, *J. Chem. Soc.*, B, 817 (1969).
55. J.R. Durig, S.C. Brown, V.F. Kalasinski and W.O. George, *Spectrochim. Acta, Part A*, **32A**, 807 (1976).
56. H.J. Oelichmann, D. Bougeard and B. Schrader, *J. Mol. Struct.*, **77**, 179 (1981).
57. J.R. Durig, J. Qui, B. Dehoff and T.S. Little, *Spectrochim. Acta, Part A*, **42A**, 89 (1986).

58. M.M.A. Aly, M.H. Baron, J. Favrot, J. Belloc and M. Revault, *Can. J. Chem.*, **63**, 1587 (1985).
59. J. Terpinsky and J. Dabrowski, *J. Mol. Struct.*, **4**, 285 (1969).
60. P. Piaggio, M. Rui and G. Dellepiane, *J. Mol. Struct.*, **75**, 171 (1981).
61. W.O. George and V.G. Mansell, *Spectrochim. Acta, Part A*, **24A**, 154 (1968).
62. G.W. King and D. Moule, *Spectrochim. Acta*, **17**, 286 (1961).
63. J.C.D. Brand and D.G. Williamson, *Discuss. Faraday Soc.*, **35**, 184 (1963).
64. P. Klaboe and G.Kremer, *Spectrochim. Acta, Part A*, **33A**, 947 (1977).
65. W.J. Balfour, S.G. Fougere and D. Klapstein, *Spectrochim. Acta, Part A*, **47A**, 1127 (1991).
66. J.C.D. Brand and R.A. Powell, *J. Mol. Spectrosc.*, **43**, 342 (1972).
67. E. Lagset, P. Klaboe, E. Kloster-Jensen, S.J. Cyvin and F.M. Nicolaisen, *Spectrochim. Acta, Part A*, **29A**, 17 (1973).
68. C. Garrigou-Lagrange, N. Claverie, J.-M. Lebas and M.-L. Josien, *J. Chim. Phys.*, **59**, 559 (1962).
69. R. Zwarich, J. Smolarek and L. Goodman, *J. Mol. Spectrosc.*, **38**, 336 (1971).
70. J.H.S. Green and D.J. Harrison, *Spectrochim. Acta, Part A*, **32A**, 1265 (1976).
71. M.K. Haque and S.N. Thakur, *J. Mol. Struct.*, **57**, 163 (1979).
72. S. Chattopadhyay and J. Jha, *Indian J. Phys.*, **42**, 610 (1968).
73. R.Y. Yadav and I.S. Singh, *Indian J. Phys.*, **58B**, 556 (1984).
74. P. Venkoji, *Spectrochim. Acta, Part A*, **42A**, 1301 (1986).
75. A.P. Upadhyay and K.N. Upadhyay, *Indian J. Phys.*, **55B**, 232 (1981).
76. M.M. Radhi and M.F. El-Bermani, *Spectrochim. Acta, Part A*, **46A**, 33 (1990).
77. G.E. Campagnaro and J.L. Wood, *J. Mol. Struct.*, **6**, 117 (1970).
78. M.P. Srivastava, O.N. Singh and I.S. Singh, *Curr. Sci.*, **37**, 100 (1968).
79. C.P.D. Dwivedi, *Indian J. Pure Appl. Phys.*, **6**, 440 (1968).
80. S. Mohan and A.R. Prabakaran, *J. Raman Spectrosc.*, **20**, 263 (1989).
81. J.G. Rosencrance and P.W. Jagodzinsky, *Spectrochim. Acta, Part A*, **42A**, 869 (1986).
82. S.H.W. Hankin, O.S. Khalil and L. Goodman, *J. Mol. Spectrosc.*, **72**, 383 (1978).
83. P. Venkoji, *Proc. Indian Acad. Sci. (Chem. Sci.)*, **93**, 105 (1984).
84. P. Venkoji, *Indian J. Pure Appl. Phys.*, **24**, 166 (1986).
85. S.P. Gupta, C. Gupta, S. Sharma and R.K. Goel, *Indian J. Pure Appl. Phys.*, **24**, 111 (1986).
86. S.J. Singh and R. Singh, *Indian J. Pure Appl. Phys.*, **16**, 939 (1978).
87. H.S. Singh and N.K. Sanyal, *Indian J. Pure Appl. Phys.*, **10**, 545 (1972).
88. P. Venkoji, *Acta Chim. Acad. Sci. Hung.*, **117**, 163 (1984).
89. J.H.S. Green and D.J. Harrison, *Spectrochim. Acta, Part A*, **33A**, 75 (1977).
90. J. Bánki, F. Billes, M. Gál, A. Grofcsik, G. Jalsovszky and L. Sztraka, *J. Mol. Struct.*, **142**, 351 (1986).
91. J. Bánki, F. Billes, M. Gál, A. Grofcsik, G. Jalsovszky and L. Sztraka, *Acta Chim. Hung.*, **123**, 115 (1986).
92. G. Varsányi, *Assignments for Vibrational Spectra of Seven Hundred Benzene Derivatives*, J.Wiley & Sons, New York (1974).
93. R.C. Millikan and K.S. Pitzer, *J. Am. Chem. Soc.*, **80**, 3515 (1958).
94. J.E.D. Davies, *J. Mol. Struct.*, **9**, 483 (1971).
95. I. Alfheim, G. Hagen and S.J. Cyvin, *J. Mol. Struct.*, **8**, 159 (1971).
96. J.E. Bertie and K.H. Michaelian, *J. Chem. Phys.*, **76**, 886 (1982).
97. I. Yokoyama, Y. Miwa and K. Machida, *J. Phys. Chem.*, **95**, 9740 (1991).
98. I. Yokoyama, Y. Miwa and K. Machida, *J. Am. Chem. Soc.*, **113**, 6458 (1991).
99. J.E. Bertie, K.H. Michaelian, H.H. Eysel and D. Hager, *J. Chem. Phys.*, **85**, 4779 (1986).

100. A. Hadni, J. Deschamps and M.L. Josien, *C.R. Acad. Sci.*, **242**, 1014 (1956).
101. J.K. Wilmshurst, *J. Mol. Spectrosc.*, **1**, 201 (1957).
102. H. Susi and J.R. Scherer, *Spectrochim. Acta, Part A*, **25A**, 1243 (1969).
103. M. Matzke, O. Chacón and C. Andrade, *J. Mol. Struct.*, **9**, 255 (1971).
104. J. Derouault, J. Le Calve and M.T. Forel, *Spectrochim. Acta, Part A*, **28A**, 359 (1972).
105. W.C. Harris, D.A. Coe and W.O. George, *Spectrochim. Acta, Part A*, **32A**, 1 (1976).
106. E.B. Marmar, C. Pouchan, A. Dargelos and M. Chaillet, *J. Mol. Struct.*, **57**, 189 (1979).
107. H. Susi and T. Zell, *Spectrochim. Acta*, **19**, 1933 (1963).
108. R.M. Moravie and J. Corset, *J. Mol. Struct.*, **30**, 113 (1976).
109. M.G. Dahlqvist and K. Euranto, *Spectrochim. Acta, Part A*, **34A**, 863 (1978).
110. S.W. Charles, G.I.L. Jones, N.L. Owen, S.J. Cyvin and B.N. Cyvin, *J. Mol. Struct.*, **16**, 225 (1973).
111. M.G. Dahlqvist, *Spectrochim. Acta, Part A*, **36A**, 37 (1980).
112. F. Daeyaert and B.J. Van der Veken, *J. Mol. Struct.*,.**213**, 97 (1989).
113. M. Räsänen, H. Kunttu, J. Murto and M. Dahlqvist, *J. Mol. Struct.*, **159**, 65 (1987).
114. W. Pyckhout, C. Alsenoy, H.J. Heise, B.J. Van der Veken, P. Coppens and M. Traetteberg, *J. Mol. Struct.*, **147**, 85 (1986).
115. G.I.L. Jones, D.G. Lister and N.L. Owen, *J. Chem. Soc. Faraday Trans. 2*, **71**, 1330 (1975).
116. H.S. Randhawa and C.N.R. Rao, *J. Mol. Struct.*, **21**, 123 (1974).
117. G.I.L. Jones, D.G. Lister, N.L. Owen, M.C.L. Gerry and P. Palmier, *J. Mol. Spectrosc.*, **60**, 348 (1976).
118. I. Suzuki, *Bull. Chem. Soc. Jpn.*, **33**, 1359 (1960).
119. N. Jonathan, *J. Mol. Spectrosc.*, **6**, 205 (1961).
120. K. Itoh and T. Shimanouchi, *J. Mol. Spectrosc.*, **42**, 86 (1972).
121. M. Räsänen, *J. Mol. Struct.*, **101**, 275 (1983).
122. N. Østergard, P.L. Christiansen and O.F. Nielsen, *J. Mol. Struct.*, **235**, 423 (1991).
123. A. Balázs, *Acta Chim. Acad. Sci. Hung.*, **108**, 265 (1981).
124. S. Ataka, H. Takeuchi and M. Tasumi, *J. Mol. Struct.*, **113**, 147 (1984).
125. R.L. Jones, *J. Mol. Spectrosc.*, **11**, 411 (1963).
126. I. Suzuki, *Bull. Chem. Soc. Jpn.*, **35**, 540 (1962).
127. S.K. Sinha, S. Ram and O.P. Lamba, *Spectrochim. Acta, Part A*, **44A**, 713 (1988).
128. G. Kaufmann and M.J.F. Leroy, *Bull. Soc. Chim. Fr.*, 402 (1967).
129. G. Durgaprasad, D.N. Sathyanarayana and C.C. Patel, *Bull. Chem. Soc. Jpn.*, **44**, 316 (1971).
130. T.C. Jao, I. Scott and D. Steele, *J. Mol. Spectrosc.*, **92**, 1 (1982).
131. Z. Mielke, H. Ratajczak, M. Wiewiorowski, A.J. Barnes and S.J. Mitson, *Spectrochim. Acta, Part A*, **42A**, 63 (1986).
132. D. Steele and A. Quatermain, *Spectrochim. Acta, Part A*, **43A**, 781 (1987).
133. M. Mizuno and S. Saëki, *Spectrochim. Acta, Part A*, **34A**, 407 (1978).
134. I.C. Hisatsune and J. Heicklen, *Can. J. Spectrosc.*, **18**, 77 (1973).
135. J.V. Pustinger Jr., J.E. Katon, F.F. Bentley, *Appl. Spectrosc.*, **18**, 36 (1964).
136. C.O. Della Védova and O. Sala, *J. Raman Spectrosc.*, **22**, 505 (1991).
137. C.O. Della Védova, *J. Raman Spectrosc.*, **22**, 291 (1991).
138. S. Pinchas, *Anal. Chem.*, **27**, 1 (1955).
139. S. Pinchas, *Anal. Chem.*, **29**, 334 (1957).
140. R.A. Nyquist, S.E. Settineri and D.A. Luoma, *Appl. Spectrosc.*, **46**, 293 (1992).
141. C.J. Pouchert, *The Aldrich Library of FT-IR Spectra*, Aldrich Chemical Company, 1st edn., (1985).

7.1.2 Fluoroformyl (fluorocarbonyl)

The —C(=O)F group provides six normal vibrations of which only the C=O stretching vibration leads to a good group vibration:

νC=O, νCF, δC=O/CF, $\gamma(\omega)$C=O/F, $\delta(\rho)$CF/C=O and torsion.

The C=O stretching vibration

FC(=O)F shows the C=O stretch at 1928 cm^{-1} (gas) or 1909 cm^{-1} (liquid) and F(O=)COOC(=O)F shows ν_aC=O at 1927 and ν_sC=O at 1902 cm^{-1}, but the other R—C(=O)F compounds absorb strongly in the region 1845 ± 55 cm^{-1}. These high values are influenced by the electronegativity of the fluorine atom, which confers a largely double bonded character to the C=O bond and causes the frequency to deviate significantly from the mean C=O frequency (≈1700 cm^{-1}). For $CF_3C(=O)F$ and $CF_3OC(=O)F$ this value reaches 1900 cm^{-1}. At the lower side there is absorption by MeSSC(=O)F at 1793 cm^{-1} as a liquid and 1829 cm^{-1} as a gas and also by the α—unsaturated acid fluorides with, as examples, H_2C=C(Me)C(=O)F (solid: 1790; liquid: 1804; gas: 1826), H_2C=CHC(=O)F (S-*trans*: 1803; S-*cis*: 1813 cm^{-1}) and PhC(=O)F (1810 cm^{-1}).

The C—F stretching vibration

The C—F stretching vibration appears moderately to strongly in the extensive region 1150 ± 140 cm^{-1}. As the CF stretch couples readily to the R—C stretching vibration, the former is sometimes described as an R—C—F [7] out-of-phase stretch, absorbing in the above-mentioned region, and the latter as an in-phase R—C—F stretch absorbing at lower frequencies. The origin of the latter absorption can be for instance a C—C stretching vibration which occurs near 830 cm^{-1} or an O—C stretch near 900 cm^{-1}. In F_2C=O the terms out-of-phase νCF (1249) and in-phase νCF (965 cm^{-1}) are clearly relevant. In acid fluorides this νCF occurs at higher wavenumbers (1175 ± 115 cm^{-1}) than in fluoroformates (1075 ± 65 cm^{-1}) and in sulfur-bonded R—C(=O)F compounds (1070 ± 30 cm^{-1}). F(O=)CC(=O)F gives rise to a νCF at 1290 and at 1122 cm^{-1} but for most of the acid fluorides the range is reduced to 1150 ± 85 cm^{-1}. Low C—F stretching vibrations are assigned in the spectra of $Me(CD_3)HCOC(=O)F$ (1010), iPrOC(=O)F (1015) and MeOC(=O)F (1054 cm^{-1}).

The C=O/CF deformations

The C=O/CF deformations give rise to three absorption regions usually assigned to the in-plane deformation or δC=O/CF (680 ± 110), the out-of-plane deformation or $\gamma(\omega)$C=O/CF (545 ± 125) and the rocking vibration or $\rho(\delta)$CF/C=O (415 ±

155 cm^{-1}). The impracticably wide ranges are caused by the fluoroformates, which absorb at ≈100 cm^{-1} higher wavenumbers than other R—C(=O)F compounds. The three deformations of MeOC(=O)F, $ClCH_2COC(=O)F$ and iPrOC(=O)F in the gas phase are reported in the vicinity of 790 (γC=O/CF), 670 (δC=O/CF) and 550 cm^{-1} (ρCF/C=O). In fluoroformates the higher wavenumber (775 ± 15 cm^{-1}) is assigned to the out-of-plane deformation but, as these vibrations are mixed, it is not at all obvious if a deformation occurs in-plane or out-of-plane or if a deformation possesses more the character of a C=O or a CF deformation.

Table 7.2 Absorption regions (cm^{-1}) of the normal vibrations of —C(=O)F

Vibration	α-saturated	C=O bonded	α-unsaturated	fluoroformates	sulfur-bonded
νC=O	1845 ± 55	1870 ± 30	1825 ± 35	1855 ± 45	1820 ± 30
νCF	1150 ± 85	1195 ± 95	1155 ± 70	1075 ± 65	1070 ± 30
δC=O/CF	670 ± 100	625 ± 55	655 ± 75	770 ± 20	650 ± 20
γC=O/CF	510 ± 90	485 ± 35	545 ± 65	650 ± 20	535 ± 45
ρCF/C=O	420 ± 80	345 ± 85	470 ± 80	540 ± 30	420 ± 80
torsion	110 ± 65	70 ± 35	85 ± 35	105 ± 20	–

R—C(=O)F molecules

R= H— and D— [1–3, 53], Me— [4–8], CD_3— [4, 5, 7], Et— [9, 10, 54], FH_2C— [11–13], ClH_2C— [14, 15], BrH_2C— [16], Cl_2HC— [17, 58], cPr— [18], cBu— [55], iPr— [19], CF_3— [20–25], F(O=)C— [26–28], Cl(O=)C— [26, 29], H_2C=CH— [30, 56], H_2C=C(Me)— [31, 32], H_2C=CF— [33], F_2C=CF— [34], HC≡C— [35–37], DC≡C— [35, 36], MeC≡C— [38], N≡C— [37], Ph— [39, 40], 4-MePh— [40], MeO— [41, 42], CD_3O— [42], EtO— [43], ClH_2CO— and ClD_2CO— [44], iPrO— [45], Me_2DCO— and $Me(CD_3)HCO$— [45], CF_3O— [46], CH_2=CHO— [57], F(O=)COO— [59], MeS— [47], MeSS— [48], Cl(O=)CSS— [49], F(O=)CSS— [50], F— [51], Cl— [51, 52].

References

1. R.F. Stratton and A.H. Nielsen, *J. Mol. Spectrosc.*, **4**, 373 (1960).
2. I.C. Hisatsune and J. Heicklen, *Can. J. Spectrosc.*, **18**, 77 (1973).
3. M. Mizuno and S. Saëki, *Spectrochim. Acta, Part A*, **34A**, 407 (1978).
4. J. Overend, R.A. Nyquist, J.C. Evans and W.J. Potts, *Spectrochim. Acta*, **17**, 1205 (1961).
5. J.C. Evans and J. Overend, *Spectrochim. Acta*, **19**, 701 (1963).
6. J.A. Ramsay and J.A. Ladd, *J. Chem. Soc.*, **B**, 118 (1968).
7. C.V. Berney and A.D. Cormier, *Spectrochim. Acta, part A*, **28A**, 1813 (1972).
8. S. Tsuchiya, *J. Mol. Struct.*, **22**, 77 (1974).
9. S.G. Frankiss and W. Kynaston, *Spectrochim. Acta, Part A*, **31A**, 661 (1975).
10. J.R. Durig, G.A. Guirgis and H.V. Phan, *J. Raman Spectrosc.*, **21**, 359 (1990).

11. R. Fausto, J.J.C. Teixeira-Dias and M.N. Ramos, *Spectrochim. Acta, Part A*, **44A**, 47 (1988).
12. J.R. Durig, H.V. Phan, J.A. Hardin, R.J. Berry and T.S. Little, *J. Mol. Struct.*, **198**, 365 (1989).
13. A.Y. Khan and N. Jonathan, *J. Chem. Phys.*, **52**, 147 (1970).
14. J.R. Durig, W. Zhao, D. Lewis and T.S. Little, *J. Chem. Phys.*, **89**, 1285 (1988).
15. A.Y. Khan and N. Jonathan, *J. Chem. Phys.*, **50**, 1801 (1969).
16. J.R. Durig, H.V. Phan and T.S. Little, *J. Mol. Struct.*, **212**, 187 (1989).
17. A.J. Woodward and N. Jonathan, *J. Phys. Chem.*, **74**, 798 (1970).
18. J.R. Durig, H.D. Bist and T.S. Little, *J. Chem. Phys.*, **77**, 4884 (1982).
19. J.R. Durig, G.A. Guirgis, W.E. Brewer and G. Baranovic, *J. Phys. Chem.*, **96**, 7547 (1992).
20. K.R. Loos and R.C. Lord, *Spectrochim. Acta*, **21**, 119 (1965).
21. C.V. Berney, *Spectrochim. Acta, Part A*, **27A**, 663 (1971).
22. K.O. Christe and D. Naumann, *Spectrochim. Acta, Part A*, **29A**, 2017 (1973).
23. R.L. Redington, *Spectrochim. Acta, Part A*, **31A**, 1699 (1975).
24. E. Ottavianelli, E.A. Castro and A.H. Jubert, *J. Mol. Struct.*, **254**, 279 (1992).
25. J.S. Francisco and I.H. Williams, *Spectrochim. Acta, Part A*, **48A**, 1115 (1992).
26. J. Goubeau and M. Adelhelm, *Spectrochim. Acta, Part A*, **28A**, 2471 (1972).
27. J.R. Durig, S.C. Brown and S.E. Hannum, *J. Chem. Phys.*, **54**, 4428 (1971).
28. J.G. Contraras and J.O. Machuca, *An. Quim.*, **77**, 370 (1981).
29. J.R. Durig and M.E. Harris, *J. Mol. Struct.*, **81**, 195 (1982).
30. J.R. Durig, R.J. Berry and P. Groner, *J. Chem. Phys.*, **87**, 6303 (1987).
31. J.R. Durig, P.A. Brletic and J.S. Church, *J. Chem. Phys.*, **76**, 1723 (1982).
32. B.C. Laskowski, R.L. Jaffe and A. Komornicki, *J. Chem. Phys.*, **82**, 5089 (1985).
33. J.R. Durig, A.-Y. Wang, T.S. Little, P.A. Brletic and J.R. Bucenell, *J. Chem. Phys.*, **91**, 7361 (1989).
34. G.A. Crowder, *J. Mol. Struct.*, **16**, 161 (1973).
35. W.J. Balfour, D. Klapstein and S. Visaisouk, *Spectrochim. Acta, Part A*, **31A**, 1085 (1975).
36. W.J. Balfour and M.K. Phibbs, *Spectrochim. Acta, part A*, **35A**, 385 (1979).
37. W.J. Balfour, S.G. Fougere and D. Klapstein, *Spectrochim. Acta, Part A*, **47A**, 1127 (1991).
38. W.J. Balfour, K. Beveridge and J.C.M. Zwinkels, *Spectrochim. Acta, Part A*, **35A**, 163 (1979).
39. J.H.S. Green and D.J. Harrison, *Spectrochim. Acta, Part A*, **33A**, 583 (1977).
40. G. Varsányi, *Assignments for Vibrational Spectra of Seven Hundred Benzene Derivatives*, J.Wiley & Sons, New York (1974).
41. G. Williams and N.L. Owen, *Trans. Faraday Soc.*, **67**, 950 (1971).
42. J.R. Durig, T.S. Little and C.L. Tolley, *Spectrochim. Acta, Part A*, **45A**, 567 (1989).
43. S.W. Charles, G.I.L. Jones, N.L. Owen, S.J. Cyvin and B.N. Cyvin, *J. Mol. Struct.*, **26**, 249 (1975).
44. F. Daeyaert and B.J. Van der Veken, *Spectrochim. Acta, Part A*, **45A**, 993 (1989).
45. B.J. Van der Veken and H.H. Liefooghe, *J. Mol. Struct.*, **247**, 257 (1991).
46. E.L. Varetti, P.J. Aymonino, *J. Mol. Struct.*, **1**, 39 (1967).
47. C.O. Della Védova, *J. Raman Spectrosc.*, **20**, 483 (1989).
48. C.O. Della Védova, *Spectrochim. Acta, Part A*, **47A**, 1619 (1991).
49. C.O. Della Védova, *J. Raman Spectrosc.*, **20**, 581 (1989).
50. S.E. Ulic, C.O. Della Védova and P.J. Aymonino, *J. Raman Spectrosc.*, **20**, 655 (1989).
51. A.H. Nielsen, T.G. Burke, P.J.H. Woltz and E.A. Jones, *J. Chem. Phys.*, **20**, 596 (1952).
52. R.A. Nyquist, *Spectrochim. Acta, Part A*, **28A**, 285 (1972).

53. G. Yarwood, H. Niki and P.D. Maker, *J. Phys. Chem.*, **95**, 4773 (1991).
54. G.A. Guirgis, B.A. Barton Jr. and J.R. Durig, *J. Chem. Phys.*, **79**, 5918 (1983).
55. J.R. Durig, H.M. Badawi, H.D. Bist and T.S. Little, *J. Chem. Phys.*, **85**, 5446 (1986).
56. J.R. Durig, J.S. Church and D.A.C. Compton. *J. Chem. Phys.*, **71**, 1175 (1979).
57. J.R. Durig, J. Lin and B.J. Van der Veken, *J. Raman Spectrosc.*, **23**, 287 (1992).
58. J.R. Durig, M.M. Bergana and H.V. Phan, *J. Raman Spectrosc.*, **22**, 141 (1991).
59. C.O. Della Védova and H.G. Mack, *J. Mol. Struct.*, **274**, 25 (1992).

7.1.3 Chloroformyl (chlorocarbonyl)

The six —C(=O)Cl normal vibrations are usually assigned as follows:

νC=O, νC—Cl, γC=O/CCl, δC=O/CCl, δCCl/C=O and torsion.

The C=O stretching vibration

The C=O stretch absorbs strongly in the region 1790 ± 55 cm^{-1}, with the exception of FC(=O)Cl (1868 cm^{-1}). More than one conformation leads to several bands in the above-mentioned absorption region. The highest C=O stretching vibrations can be found in the spectra of Cl(O=)CC(=O)Cl (1845 and 1774), ClC(=O)Cl (1827) and tBuC(=O)Cl (1825 and 1778 cm^{-1}) and the lowest in the spectra of carbamoyl chlorides R′R″NC(=O)Cl (R′, R″ = H, Me, Et) (≈1740), PhCH=CHC(=O)Cl (1753), 2-ThC(=O)Cl (1753), 3-O_2NPhC(=O)Cl (1755), 3-ClPhC(=O)Cl (1758) and Me—C≡CC(=O)Cl (1758 cm^{-1}). The majority of the investigated molecules were found to give this ν-O in the region 1790 ± 30 cm^{-1}, which is reduced to 1800 ± 15 cm^{-1} for R′CH_2C(=O)Cl compounds and even to 1802 ± 2 cm^{-1} for Me$(CH_2)_n$C(=O)Cl (n = 2–16). In the spectra of aromatic acid chlorides the νC=O (1775 ± 20 cm^{-1}) is often accompanied by a smaller band (1740 ± 10 cm^{-1}). The higher frequency band has the larger intensity and is called the carbonyl stretching absorption band, and the lower frequency band is called the first overtone of a complex planar mode involving stretching of the phenyl—C(=O)—Cl bonds in the neighbourhood of 865 cm^{-1} [84].

The C—Cl stretching vibration

The C—Cl stretch 'wanders' in the extensive region 745 ± 185 cm^{-1}. Although the intensity of the band is moderate to strong, it is not a good group vibration. In addition, this νC—Cl is sensitive to conformation [83] and the assignments in literature are often controversial. The highest values are observed in the spectra of aromatic acid chlorides (865 ± 65 cm^{-1}) with 3-O_2NPhC(=O)Cl (930) and 3-ClPhC(=O)Cl (909 cm^{-1}) as maxima, but most of the benzoyl chlorides show a strong band in the region 855 ± 45 cm^{-1}. This complex vibration is due to the coupling between the C—Cl stretching vibration and the phenyl—C stretch in which the νC—Cl furnishes the highest contribution [57, 63, 65, 84]. For benzoyl chloride this vibration is assigned at 875 cm^{-1} although Condit *et al.* [60] prefer 673 cm^{-1} and assign the strong band at 875 cm^{-1} to an in-plane substituent sensitive ring deformation. To all probability the chloroformates also absorb at high wavenumbers (790 $\pm$ 60), while values in the vicinity of 690 cm^{-1} are assigned to the γC=O [48–51]. Hory [85] on the other hand puts this νC—Cl at 691 $\pm$ 3 cm^{-1} in a series of chloroformates. The lowest C—Cl stretching vibrations are reported in the spectra of $CD_3C(=O)Cl$ (563), *trans*-$CH_2DC(=O)Cl$ (565), *gauche*-$Cl_2CHC(=O)Cl$ (578) and MeC(=O)Cl (595 cm^{-1}). For most of the acid chlorides the C—Cl stretching vibration is situated at 750 $\pm$ 150 cm^{-1}. Neglecting the above-mentioned values and the high value for $CBrF_2C(=O)Cl$ (830), the region for the α-saturated compounds narrows to 705 $\pm$ 75 cm^{-1}. EtC(=O)Cl absorbs at 693 cm^{-1} and the remaining $Me(CH_2)_nC(=O)Cl$ compounds (n = 2–16) at $\approx$680 cm^{-1}. In the spectrum of ClC(=O)Cl ν_1 and ν_4 occur respectively at 575 and 849 cm^{-1}.

The C=O deformations

The C=O/C—Cl deformations, absorbing weakly to moderately in the regions 575 $\pm$ 115 and 465 $\pm$ 75 cm, are mainly C=O deformations. The highest wavenumber is often assigned to the out-of-plane deformation but, for lack of a plane of symmetry, it is not always clear if a vibration occurs in-plane or out-of-plane. The chloroformates alkyl—OC(=O)Cl (alkyl = Me, Et, nPr, iPr, nBu and iBu) ($\approx$690), $HC\equiv CCH_2OC(=O)Cl$ (689) and PhOC(=O)Cl (684 cm^{-1}) show a sharp absorption in the neighbourhood of 690 cm^{-1}, assigned to the γC=O [48–51] or to the νC—Cl [85]. Low values for the γC=O are found in the spectra of Cl(O=)CCH=CHC(=O)Cl (497 and 460) and $H_2C=CHC(=O)Cl$ (495 cm^{-1}). The very low value of 440 cm^{-1} for ClC(=O)Cl is outside this region.

The domain of the δC=O (465 $\pm$ 75 cm^{-1}) is delimited by 4-BrPhC(=O)Cl (536), 2,4-$Cl_2PhC(=O)Cl$ (535) and 3,4-$Cl_2PhC(=O)Cl$ (532 cm^{-1}) on the one side and by Cl(O=)CCH=CHC(=O)Cl (393 and 430), MeO(O=)CCH=CHC(=O)Cl (410) and *cis*-cPrC(=O)Cl (415 cm^{-1}) on the other side. The low value for ClC(=O)Cl (297 cm^{-1}) is not taken into account.

The C—Cl deformation

The C—Cl/C=O in-plane deformation is generally assigned in the range 325 $\pm$ 90 cm^{-1} with compounds such as FC(=O)Cl (415), $HC\equiv CC(=O)Cl$ (414), MeOC(=O)Cl (413) and *trans*-$Cl_2CHC(=O)Cl$ (410 cm^{-1}) at the HW side and Cl(O=)CC(=O)Cl (290 and 225), $Cl(O=)CCH_2CH_2C(=O)Cl$ (260 and 230), *trans,trans*-Cl(O=)CCH=CHC(=O)Cl (232 and 241), ClC(=O)Cl (240) and $CH_2=C(Me)C(=O)Cl$ (249 cm^{-1}) at the LW side. Especially in the case of low wavenumbers the C—Cl group makes the highest contribution to this deformation.

Table 7.3 Absorption region (cm^{-1}) of the normal vibrations of —C(=O)Cl

Vibration	α-saturated	C=O bonded	chloroformates	aromatic	unsaturated	carbamoyl chlorides
νC=O	1800 ± 30	1810 ± 35	1780 ± 20	1775 ± 20	1760 ± 25	1740 ± 05
νCCl	670 ± 110	690 ± 70	790 ± 60	865 ± 65	700 ± 100	640 ± 40
γC=O/CCl	575 ± 95	630 ± 50	675 ± 15	620 ± 50	560 ± 100	615 ± 45
δC=O/CCl	465 ± 55	440 ± 50	500 ± 40	480 ± 60	440 ± 50	460 ± 20
δCCl/C=O	320 ± 90	315 ± 90	365 ± 50	330 ± 50	320 ± 95	390 ± 20
torsion	90 ± 60	–	–	–	–	–

R—C(=O)Cl molecules

R= H— [1, 2], Me— [3–6], CD_3— [3, 5], CH_2D— [3], Et— [7, 8], nPr— [9], $PhCH_2CH_2$—, $ClCH_2CH_2$— [9], Cl(O=)CCH_2CH_2— [10], $PhCH_2$—, FCH_2— [11–15], $ClCH_2$— [16–20, 89], $BrCH_2$— [16, 21], Cl_2CH— [22–25], Br_2CH— [26], iPr— [27], cPr— [28–30], cBu— [31], tBu—, CF_3— [32–37], CCl_3— [24], CBr_3— [38], $CBrF_2$— [86], F(O=)C— [39, 40], Cl(O=)C— [41–43], Me(O=)C— [44], MeO(O=)C— [45, 46], MeO— [4, 48, 49], CD_3O— [48], EtO— [47], nPrO—, nBuO—, iPrO—, iBuO—, $ClCH_2O$— and $ClCD_2O$— [50, 51], H_2C=$CHCH_2O$—, HC≡CCH_2O— [49], $PhCH_2O$—, PhO—, MeS— [4, 52, 53], F(O=)CSS— [54], Cl(O=)CSS— [88], Ph— [55–62, 90], C_6D_5— [60], 2-, 3- and 4-XPh— (X = F, Cl, Br, I, Me, OMe and NO_2 [63], F_3C [87]), 4-XPh— (X = Cl, Br, OMe, NMe_2 and NO_2 [64, 65], N≡C, nPr, nBu, tBu), 2,4-Cl_2Ph—, 3,4-Cl_2Ph— and 3,5-Cl_2Ph— [66, 67], 2-Fu— [68], 2-Th—, H_2C=CH— [69–72], MeCH=CH— and Me_2C=CH— [73], Cl(O=)CCH=CH— [74], MeO(O=)CCH=CH— [46, 75], XCH=CH— (X = Cl, Br, I) [76], H_2C=C(Me)— [77], PhCH=CH—, HC≡C— [78–80], DC≡C— [80], MeC≡C— [78], MeHN— and MeDN— [81], Me_2N—, Et_2N—, F— [49, 82], Cl— [82].

References

1. I.C. Hisatsune and J. Heicklen, *Can. J. Spectrosc.*, **18**, 77 (1973).
2. G. Yarwood, H. Niki and P.D. Maker, *J. Phys. Chem.*, **95**, 4773 (1991).
3. J. Overend, R.A. Nyquist, J.C. Evans and W.J. Potts, *Spectrochim. Acta*, **17**, 1205 (1961).
4. J.C. Evans and J. Overend, *Spectrochim. Acta*, **19**, 701 (1963).
5. J.A. Ramsay and J.A. Ladd, *J. Chem. Soc.*, **B**, 118 (1968).
6. R. Fausto and J.J.C. Teixeira-Dias, *J. Mol. Struct.*, **144**, 215 (1986).
7. S.G. Frankiss and W. Kynaston, *Spectrochim. Acta, Part A*, **31A**, 661 (1975).
8. S. Dyngeseth, S.H. Schei and K. Hagen, *J. Mol. Struct.*, **116**, 257 (1984).
9. J. Som, D. Bhaumik, D.K. Mukherjee and G.S. Kastha, *Indian J. Pure Appl. Phys.*, **12**, 149 (1974).
10. J.E. Katon and S.R. Lobo, *J. Mol. Struct.*, **127**, 229 (1985).
11. J.E.F. Jenkins and J.A. Ladd, *J. Chem. Soc.*, **B**, 1237 (1968).

12. A.Y. Khan and N. Jonathan, *J. Chem. Phys.*, **52**, 147 (1970).
13. J.R. Durig, H.V. Phan, J.A. Hardin and T.S. Little, *J. Chem. Phys.*, **90**, 6840 (1989).
14. M. Monnier, G. Davidovics and A. Allouche, *J. Mol. Struct.*, **243**, 13 (1991).
15. A.A. El-Bindary, A. Horn, P. Klaboe, C.J. Nielsen and F.I.M. Taha, *J. Mol. Struct.*, **273**, 27 (1992).
16. I. Nakagawa, I. Ichishima, K. Kuratani, T. Miyazawa, T. Shimanouchi and S. Mizushima, *J. Chem. Phys.*, **20**, 1720 (1952).
17. A.Y. Khan and N. Jonathan, *J. Chem. Phys.*, **50**, 1801 (1969).
18. K. Tanabe and S. Saëki, *Spectrochim. Acta, Part A*, **28A**, 1083 (1972).
19. R. Fausto and J.J.C. Teixeira-Dias, *J. Mol. Struct.*, **144**, 225 (1986).
20. G. Davidovics, A. Allouche and M. Monnier, *J. Mol. Struct.*, **243**, 1 (1991).
21. J.R. Durig, H.V. Phan and T.S. Little, *J. Mol. Struct.*, **212**, 187 (1989).
22. A. Miyaka, I. Nakagawa, T. Miyazawa, I. Ichishima, T. Shimanouchi and S. Mizushima, *Spectrochim. Acta*, **13**, 161 (1958).
23. A.J. Woodward and N. Jonathan, *J. Phys. Chem.*, **74**, 798 (1970).
24. R. Fausto and J.J.C. Teixeira-Dias, *J. Mol. Struct.*, **144**, 241 (1986).
25. J.R. Durig, M.M. Bergana and H.V. Phan, *J. Mol. Struct.*, **242**, 179 (1991).
26. H.S. Randhawa, *Indian J. Chem., Sect. A*, **19A**, 152 (1980).
27. G.A. Guirgis, H.V. Phan and J.R. Durig, *J. Mol. Struct.*, **266**, 265 (1992).
28. J.E. Katon, W.R. Feairheller Jr. and J.T. Miller Jr., *J. Chem. Phys.*, **49**, 823 (1968).
29. J.R. Durig, H.D. Bist, S.V. Saari, J.A.S. Smith and T.S. Little, *J. Mol. Struct.*, **99**, 217 (1983).
30. J.R. Durig, A. Wang and T.S. Little, *J. Mol. Struct.*, **269**, 285 (1992).
31. K. Hanai and J.E. Katon, *J. Mol. Struct.*, **70**, 127 (1981).
32. C.V. Berney, *Spectrochim. Acta*, **20**, 1437 (1964).
33. G.A. Crowder, *Appl. Spectrosc.*, **27**, 440 (1973).
34. R.L. Redington, *Spectrochim. Acta, Part A*, **31A**, 1699 (1975).
35. C.V. Berney and A.D. Cormier, *Spectrochim. Acta, Part A*, **33A**, 929 (1977).
36. E. Ottavianelli, E.A. Castro and A.H. Jubert, *J. Mol. Struct.*, **254**, 279 (1992).
37. J.S. Francisco and I.H. Williams, *Spectrochim. Acta, Part A*, **48A**, 1115 (1992).
38. H.S. Randhawa and W. Walter, *J. Mol. Struct.*, **35**, 303 (1976).
39. J. Goubeau and M. Adelhelm, *Spectrochim. Acta, Part A*, **28A**, 2471 (1972).
40. J.R. Durig and M.E. Harris, *J. Mol. Struct.*, **81**, 195 (1982).
41. J.R. Durig and S.E. Hannum, *J. Chem. Phys.*, **52**, 6089 (1970).
42. J.R. Durig, S.C. Brown and S.E. Hannum, *J. Chem. Phys.*, **54**, 4428 (1971).
43. J.G. Contreras and J.O. Machuca, *An. Quim.*, **77**, 370 (1981).
44. W.J. Ray and J.E. Katon, *Spectrochim. Acta, Part A*, **36A**, 793 (1980).
45. S.W. Charles, G.I.L. Jones, N.L. Owen and L.A. West, *J. Mol. Struct.*, **32**, 111 (1976).
46. J.E. Katon and P.-H. Chu, *J. Mol. Struct.*, **78**, 141 (1982).
47. S.W. Charles, G.I.L. Jones, N.L. Owen, S.J. Cyvin and B.N. Cyvin, *J. Mol. Struct.*, **16**, 225 (1973).
48. J.R. Durig and M.G. Griffin, *J. Mol. Spectrosc.*, **64**, 252 (1977).
49. R.A. Nyquist, *Spectrochim. Acta, Part A*, **28A**, 285 (1972).
50. F. Daeyaert, Thesis, UIA, Antwerp, 1988.
51. F. Daeyaert and B.J. Van der Veken, *J. Mol. Struct.*, **198**, 239 (1989).
52. T. Miyazawa and K.S. Pitzer, *J. Chem. Phys.*, **30**, 1076 (1959).
53. R.A. Nyquist, *J. Mol. Struct.*, **1**, 1 (1967).
54. C.O. Della Védova, *J. Raman Spectrosc.*, **20**, 581 (1989).
55. C. Garrigou-Lagrange, N. Claverie, J.M. Lebas and M.L. Josien, *J. Chim. Phys.*, **59**, 559 (1962).
56. S. Yoshida, *Chem. Pharm. Bull.*, **10**, 450 (1962).

57. P. Delorme, V. Lorenzelli and A. Alemagna, *J. Chim. Phys.*, **62**, 3 (1965).
58. S. Chattopadhyay and J. Jha, *Indian J. Phys.*, **42**, 610 (1968).
59. S.R. Singh, B.B. Lal, I.S. Singh and M.P. Srivastava, *Indian J. Pure Appl. Phys.*, **8**, 116 (1970).
60. D. Condit, S.M. Craven and J.E. Katon, *Appl. Spectrosc.*, **28**, 420 (1974).
61. J.H.S. Green and D.J. Harrison, *Spectrochim. Acta, Part A*, **33A**, 583 (1977).
62. R.A.Yadav, *Spectrochim. Acta*, **49A**, 891 (1993).
63. H.N. Al-Jallo and M.G. Jalhoom, *Spectrochim. Acta, Part A*, **28A**, 1655 (1972).
64. C.N.R. Rao and R. Venkataraghavan, *Spectrochim. Acta*, **18**, 273 (1962).
65. E. Ortiz, J.F. Bertran and L. Ballester, *Spectrochim. Acta, Part A*, **27A**, 1713 (1971).
66. U.C. Joshi, R.N. Singh and S.N. Sharma, *Spectrochim. Acta, Part A*, **38A**, 205 (1982).
67. U.C. Joshi, M. Joshi, R.N. Singh and S.N. Sharma, *Indian J. Phys.*, **55**B, 220 (1981).
68. G. Cassanas-Fabre and L. Bardet, *J. Mol. Struct.*, **25**, 281 (1975).
69. J.E. Katon and W.R. Feairheller Jr., *J. Chem. Phys.*, **47**, 1248 (1967).
70. R.L. Redington and J.R. Kennedy, *Spectrochim. Acta, Part A*, **30A**, 2197 (1974).
71. D.A.C. Compton, W.O. George, J.E. Goodfield and W.F. Maddams, *Spectrochim. Acta, Part A*, **37A**, 147 (1981).
72. J.R. Durig, R.J. Berry and P. Groner, *J. Chem. Phys.*, **87**, 6303 (1987).
73. R.K. Gupta, R. Prasad and H.L. Bhatnagar, *Spectrochim. Acta, Part A*, **45A**, 595 (1989).
74. J.M. Landry and J.E. Katon, *Spectrochim. Acta, Part A*, **40A**, 871 (1984).
75. J.E. Katon and P.-H. Chu, *J. Mol. Struct.*, **82**, 61 (1982).
76. K. Kamieńska-Trela, H. Barańska and A. Labudzińska, *J. Mol. Struct.*, **54**, 59 (1979).
77. J.R. Durig, P.A. Brletic, Y.S. Li, A.-Y. Wang and T.S. Little, *J. Mol. Struct.*, **223**, 291 (1990).
78. E. Augdahl, E. Kloster-Jensen and A. Rogstad, *Spectrochim. Acta, Part A*, **30A**, 399 (1974).
79. W.J. Balfour, R.H. Mitchell and S. Visaisouk, *Spectrochim. Acta, Part A*, **31A**, 967 (1975).
80. W.J. Balfour and M.K. Phibbs, *Spectrochim. Acta, Part A*, **35A**, 385 (1979).
81. W. Buder and A. Schmidt, *Spectrochim. Acta, Part A*, **29A**, 1419 (1973).
82. A.H. Nielsen, T.G. Burke, P.J.H. Woltz and E.A. Jones, *J. Chem. Phys.*, **20**, 596 (1952).
83. J.E. Katon and W.R. Feairheller Jr., *J. Chem. Phys.*, **44**, 144 (1966).
84. R.A. Nyquist, *Appl. Spectrosc.*, **40**, 79 (1986).
85. H.A. Ory, *Spectrochim. Acta*, **16**, 1488 (1960).
86. T.A. Mohamed, H.D. Stidham, G.A. Guirgis, H.V. Phan and J.R. Durig, *J. Raman Spectrosc.*, **24**, 1 (1993).
87. R. Shanker, R.A. Yadav, I.S. Singh and O.N. Singh, *J. Raman Spectrosc.*, **23**, 141 (1992).
88. S.E. Ulic, P.J. Aymonino and C.O. Della Védova, *J. Raman Spectrosc.*, **22**, 675 (1991).
89. A.A. El-Bindary, P. Klaboe and C.J. Nielsen, *Acta Chem. Scand.*, **45**, 877 (1991).
90. G. Varsányi and S. Szoke, *Vibrational Spectra of Benzene Derivatives*, Academic Press, New York (1969), pp. 361, 370.

7.1.4 Bromoformyl (bromocarbonyl)

Just like —C(=O)Cl, the —C(=O)Br furnishes six normal vibrations:

νC=O, νCBr, γC=O/CBr, δC=O/CBr, δCBr/C=O and torsion. Only the νC=O is a good group vibration.

The C=O stretching vibration

The C=O stretching vibration provides a strong band in the region 1780 ± 50 cm^{-1}. The α-saturated —C(=O)Br compounds absorb at higher wavenumbers (1800 ± 30) than the α-unsaturated ones (1765 ± 30 cm^{-1}). The highest values are observed in the spectra of $F_3CC(=O)Br$ (1826), $CD_3C(=O)Br$ (1825) and EtC(=O)Br (1824 cm^{-1}) and the lowest in those of MeNHC(=O)Br (1738), $H_2C=C(Me)C(=O)Br$ (1750), Br(O=)CC(=O)Br (*1792* and 1752) and PhC(=O)Br (1765 cm^{-1}). The region 1790 ± 30 cm^{-1} is appropriate to locate a C(=O)Br group.

The C—Br stretching vibration

The C—Br stretching vibration is observed in the extensive region 685 ± 165 cm^{-1}, which is made so extensive by the strong and broad absorption at 851 cm^{-1} in the spectrum of benzoyl bromide; this band is assigned to the νC—Br by Green and Harrison [29]. Condit *et al.* [30], however, attribute the absorption at 657 cm^{-1} to the νC—Br and that at 851 cm^{-1} to an in-plane ring mode. Low νC—Br frequencies are found in the spectra of $CD_3C(=O)Br$ (525) and MeNHC(=O)Br (545 cm^{-1}) but generally the C—Br stretching vibration is active in the region 655 ± 90 cm^{-1}. For MeC(=O)Br this stretching mode is assigned at 570 cm^{-1}.

The C=O deformations

The C=O/CBr deformation with the highest wavenumber (520 ± 160 cm^{-1}) is usually assigned to the out-of-plane deformation so far as there is a plane of symmetry. The HW side of this region is limited by 678 cm^{-1} from $F_3CC(=O)Br$ and 629 cm^{-1} from PhC(=O)Br. The lowest values are from Br(O=)CC(=O)Br (405 and 362) and *trans*-$ClCH_2C(=O)Br$ (398 cm^{-1}). A narrower region such as 500 ± 60 cm^{-1} must be acceptable for this out-of-plane deformation.

The region 400 ± 90 cm^{-1} is often assigned to the C=O/CBr in-plane deformation with 486 cm^{-1} for PhC(=O)Br and 476 cm^{-1} for EtC(=O)Br. The lowest values are from $BrCH_2C(=O)Br$ with 310 cm^{-1} for the *gauche* conformer and 359 cm^{-1} for the *trans* conformer.

The C—Br deformation

The CBr/C=O in-plane deformation or CBr/C=O rock is active in the region 260 ± 85 cm^{-1}. The extreme values are found in the spectra of $H_2C=CHC(=O)Br$ (*s—trans:* 345; *s—cis:* 327 cm^{-1}), $H_2C=C(Me)C(=O)Br$ (332) and Br(O=)CC(=O)Br (190 and 176 cm^{-1}). The region 260 ± 60 cm^{-1} is a good approximation for this CBr deformation.

Table 7.4 Absorption regions (cm^{-1}) of the normal vibrations of —C(=O)Br

Vibration	Region	Vibration	Region
νC=O	1780 ± 50	δC=O/CBr	400 ± 90
νCBr	685 ± 165	δCBr/C=O	260 ± 85
γC=O/CBr	520 ± 160	torsion	90 ± 60

R—C(=O)Br molecules

R = H— [1], Me— [2–6], CD_3— [5], Et— [7], FCH_2— [8–10], $ClCH_2$— [11, 17], $BrCH_2$— [12–14], Cl_2CH— [15], MeCHBr—, cPr— [16], cBu— [26], F_3C— [18, 19], Me_2CBr—, MeC(=O)— [20], BrC(=O)— [21–23], H_2C=CH— [24], H_2C=C(Me)— [25], ClCH=CH— and BrCH=CH— [31], Ph— [29, 30], 4-MePh— [28], MeNH— [27].

References

1. G. Yarwood, H. Niki and P.D. Maker, *J. Phys. Chem.*, **95**, 4773 (1991).
2. J.C. Evans and H.J. Bernstein, *Can. J. Chem.*, **34**, 1083 (1956).
3. J. Overend, R.A. Nyquist, J.C. Evans and W.J. Potts, *Spectrochim. Acta*, **17**, 1205 (1961).
4. J.C. Evans and J. Overend, *Spectrochim. Acta*, **19**, 701 (1963).
5. L.C. Hall and J. Overend, *Spectrochim. Acta, Part A*, **23A**, 2535 (1967).
6. J.A. Ramsay and J.A. Ladd, *J. Chem. Soc.*, **B**, 118 (1968).
7. S.G. Frankliss and W. Kynaston, *Spectrochim. Acta, Part A*, **31A**, 661 (1975).
8. J.E.F. Jenkins and J.A. Ladd, *J. Chem. Soc.*, **B**, 1237 (1968).
9. A.Y. Khan and N. Jonathan, *J. Chem. Phys.*, **52**, 147 (1970).
10. A.A. El-Bindary, A. Horn, P. Klaboe, C.J. Nielsen and F.I.M. Taha, *J. Mol. Struct.*, **273**, 27 (1992).
11. A.Y. Khan and N. Jonathan, *J. Chem. Phys.*, **50**, 1801 (1969).
12. K. Tanabe and S. Saëki, *Spectrochim. Acta, Part A*, **28A**, 1083 (1972).
13. J.R. Durig, H.V. Phan and T.S. Little, *J. Mol. Struct.*, **212**, 187 (1989).
14. I. Nakagawa, I. Ichishima, K. Kuratani, T. Miyazawa, T. Shimanouchi and S. Mizushima, *J. Chem. Phys.*, **20**, 1720 (1952).
15. A.J. Woodward and J. Jonathan, *J. Phys. Chem.*, **74**, 798 (1970).
16. J.E. Katon, W.R. Feairheller Jr. and J.T. Miller Jr., *J. Chem. Phys.*, **49**, 823 (1968).
17. A.A. El-Bindary, P. Klaboe and C.J. Nielsen, *Acta Chem. Scand.*, **45**, 877 (1991).
18. C.V. Berney, *Spectrochim. Acta*, **20**, 1437 (1964).
19. C.V. Berney and A.D. Cormier, *Spectrochim. Acta, Part A*, **33A**, 929 (1977).
20. W.J. Ray and J.E. Katon, *Spectrochim. Acta, Part A*, **36A**, 793 (1980).
21. J.R. Durig, S.C. Brown and S.E. Hannum, *J. Chem. Phys.*, **54**, 4428 (1971).
22. J.R. Durig, S.E. Hannum and F.G. Baglin, *J. Chem. Phys.*, **54**, 2367 (1971).
23. J.G. Contreras and J.O. Machuca, *An. Quim.*, **77**, 370 (1981).
24. J.R. Durig, R.J. Berry and P. Groner, *J. Chem. Phys.*, **87**, 6303 (1987).
25. J.R. Durig, W. Zhao, R.J. Berry and T.S. Little, *J. Mol. Struct.*, **212**, 169 (1989).
26. K. Hanai and J.E. Katon, *J. Mol. Struct.*, **70**, 127 (1981).
27. W. Buder and A. Schmidt, *Spectrochim. Acta, Part A*, **29A**, 1419 (1973).

28. G. Varsányi, *Assignments for Vibrational Spectra of Seven Hundred Benzene Derivatives*, John Wiley & Sons, New York (1974).
29. J.H.S. Green and D.J. Harrison, *Spectrochim. Acta, Part A*, **33A**, 583 (1977).
30. D. Condit, S.M. Craven and J.E. Katon, *Appl. Spectrosc.*, **28**, 420 (1974).
31. K. Kamiénska-Trela, H. Barańska and A. Kabudzińska, *J. Mol. Struct.*, **54**, 59 (1979).

7.1.5 Acetyl

The $—C(=O)CH_3$ structure unit gives rise to $3N - 6 = 15$ normal vibrations of which nine are inherent to the methyl group:

ν_aMe, ν_a'Me, ν_sMe, νC=O, δ_aMe, δ_a'Me, δ_sMe, ρMe, ρ'Me, νCC, δC=O, γC=O, δ—C—C and two torsions.

Methyl stretching vibrations

Typical for acetyl compounds is the weak intensity of the methyl stretching vibrations (Section 2.1.6).

ν_aMe	ν_a'Me	ν_sMe
3005 ± 40	2975 ± 45	2905 ± 65 cm^{-1}

The C=O stretching vibration

The highest values for the νC=O are furnished by acetyl halides XC(=O)Me (X = F, Cl, Br, I) (1840 ± 30), acetic anhydride MeC(=O)OC(=O)Me (1827 and 1755) and acetylhypochlorite ClOC(=O)Me (1818 cm^{-1}). Setting aside these high values, the C=O stretching vibration exhibits a strong band at 1710 ± 70 cm^{-1}. With the exception of F_3CC(=O)Me (1780 cm^{-1}]) the α-saturated methyl ketones show the νC=O in the region 1720 ± 25 cm^{-1}. At the HW side Cl_2CHC(=O)Me (1743) and FCH_2C(=O)Me (1740 cm^{-1}) clearly show the influence of the halogen. At the LW side cPrC(=O)Me absorbs at 1697 cm^{-1}. Most of the saturated ketones are active at 1715 ± 15 cm^{-1} and even at 1715 ± 5 cm^{-1} for R—C(=O)Me compounds if R contains only methyl or methylene, branched or not. The thioanhydride MeC(=O)SC(=O)Me absorbs at 1769 and 1712 cm^{-1} but the thiol acetates do so in the region 1695 ± 15 cm^{-1} [80]. A hydrogen bridge in aromatic methyl ketones produces a supplementary lowering of the absorption frequency of the C=O stretching vibration: 2-HOPhC(=O)Me shows the νC=O at 1643 and 4-HOPhC(=O)Me at 1646 cm^{-1}. The very low values for the enol form [13] of MeC(=O)CH_2C(=O)Me (1620) and MeC(=S)CH_2C(=O)Me (1610 cm^{-1}) are outside the above-mentioned absorption region of the C=O stretching vibration.

Methyl deformations

As contrasted with the weak absorptions of the methyl stretching vibrations, the methyl symmetric deformation absorbs moderately to strongly.

δ_aMe	δ'_aMe	δ_sMe
1445 ± 35	1430 ± 40	1365 ± 25 cm^{-1}

Methyl rocking vibrations and C—C stretch

In most of the compounds the C—C stretching vibrations are coupled to the methyl rock (Section 2.1.6)

ρMe	ρ'Me	νC—C
1085 ± 70	985 ± 85	905 ± 95

The C=O in-plane deformation

With the exception of XC(=O)Me (X = H, Cl, Br, I), the C=O deformation in the region 580 ± 115 cm^{-1} is described as the in-plane C=O deformation. The CH/C=O deformation in HC(=O)Me at 764 cm^{-1} falls outside this region, but DC(=O)Me reveals this vibration at 668 cm^{-1}, which demonstrates the influence of the light hydrogen atom. The highest wavenumbers are those for ClNHC(=O)Me (693), *s—cis* H_2C=CHC(=O)Me (690) and MeC(=O)NHC(=O)Me (648 and *560* cm^{-1}) and the lowest emerge in the spectra of IC(=O)Me (465), cPrC(=O)Me (487) and HSC(=O)Me (525 cm^{-1}). The remaining values are situated in the region 585 ± 55 cm^{-1} with 530 cm^{-1}, for 2-propanone.

The C=O out-of-plane deformation

The C=O out-of-plane deformation occurs variably in the region 505 ± 115 cm^{-1}. High values originate from the spectra of EtNHC(=O)Me and nBuNHC(=O)Me (620), D_3SiOC(=O)Me (618 cm^{-1}, coincident with δC=O) and methyl and ethyl acetate with 607 cm^{-1}. The lowest values come from HOC(=O)C(=O)Me (394), MeOC(=O)C(=O)Me (403), FCH_2C(=O)Me (404) and 3-PyC(=O)Me (405 cm^{-1}). The remaining out-of-plane deformations are observed in the range 500 ± 100 cm^{-1}.

Skeletal deformation and torsions

The —C(=O)—Me skeletal deformation is weakly to moderately active in the region 345 ± 130 cm^{-1}. The highest values have been attributed in the spectra of FC(=O)Me (473), Me_2NC(=O)Me (473), EtOC(=O)Me, $ClCH_2OC$(=O)Me and H_2C=CHOC(=O)Me with 462 cm^{-1} and nBuNHC(=O)Me with 450 cm^{-1}. At the lower limit a few RC(=O)C(=O)Me compounds absorbs, such as those with R = MeO (121), Cl (230) and HO (258 cm^{-1}). The remaining skeletal deformations are situated at 355 ± 95 cm^{-1}.

The methyl torsion may be expected at 190 ± 80 cm^{-1} and the acetyl torsion at 100 ± 60 cm^{-1}.

Table 7.5 Absorption regions (cm^{-1}) of the normal vibrations of —C(=O)Me

Vibration	α-saturated	α-unsaturated	aromatic	C=O bonded
ν_aMe	3005 ± 40	3000 ± 30	3010 ± 10	3020 ± 15
ν'_aMe	2990 ± 30	2960 ± 30	2975 ± 25	2990 ± 10
ν_sMe	2905 ± 65	2900 ± 50	2925 ± 15	2930 ± 10
νC=O	1735 ± 45	1675 ± 35	1680 ± 30	1720 ± 20
δ_aMe	1440 ± 25	1425 ± 15	1445 ± 25	1425 ± 15
δ'_aMe	1425 ± 15	1415 ± 25	1445 ± 25	1410 ± 20
δ_sMe	1365 ± 25	1355 ± 10	1360 ± 15	1350 ± 10
ρMe	1085 ± 70	1060 ± 40	1070 ± 25	1095 ± 45
ρ'Me	985 ± 85	1000 ± 25	1020 ± 20	1015 ± 10
νC—C	895 ± 80	895 ± 85	955 ± 25	900 ± 80
δC=O	555 ± 70	615 ± 80	605 ± 20	595 ± 55
γC=O	460 ± 50	455 ± 45	445 ± 45	475 ± 85
δ-C—C	375 ± 50	410 ± 25	375 ± 45	310 ± 95
torsion Me	200 ± 70	–	205 ± 20	175 ± 45
torsion C(=O)Me	100 ± 60	–	≈150	≈130
Vibration	**N-bonded**	**O-bonded**	**S-bonded**	**Halogen-bonded**
ν_aMe	2990 ± 20	3010 ± 30	3000 ± 10	3030 ± 15
ν'_aMe	2965 ± 35	2970 ± 30	2990 ± 10	3010 ± 10
ν_sMe	2900 ± 45	2910 ± 40	2920 ± 10	2945 ± 25
νC=O	1690 ± 45	1750 ± 20	1725 ± 45	1840 ± 30
δ_aMe	1450 ± 30	1445 ± 20	1435 ± 15	1430 ± 05
δ'_aMe	1440 ± 20	1435 ± 15	1420 ± 10	1425 ± 10
δ_sMe	1365 ± 10	1370 ± 20	1355 ± 10	1365 ± 15
ρMe	1080 ± 50	1050 ± 30	1120 ± 20	1080 ± 30
ρ'Me	995 ± 55	975 ± 45	1000 ± 65	1035 ± 35
νC—C	915 ± 85	860 ± 50	960 ± 30	895 ± 65
δC=O	625 ± 70	620 ± 30	575 ± 55	535 ± 70
γC=O	530 ± 90	600 ± 20	480 ± 50	500 ± 70
δ-C—C	420 ± 55	415 ± 50	390 ± 50	375 ± 100
torsion Me	–	160 ± 50	–	170 ± 40
torsion C(=O)Me	–	–	–	–

R—C(=O)Me compounds

R = H— [1–5], D— [1, 2, 4], Me— [3, 6–10], Et— [82], nPr—, nBu—, nPent—, $HOCH_2$—, $Me_2C(OH)CH_2$—, $MeC(=O)CH_2$— [11–13],

MeC(=S)CH_2— [13], FCH_2— [14–16], ClCH_2— [16–18], BrCH_2— [16], Cl_2CH— [19], iPr— [20], cPr— [21, 22], F_3C— [23], H_2C=CH— [24–27], MeCH=CH— [24], N=N=CH— and N=N=CD— [28], N=N=C(Me)— [29], HS(Me)C=CH— [30], HC≡C— [31], Ph— [32–35], 2-, 3- and 4-XPh— (X = Me [35], F_3C, H_2N [35], O_2N [35], HO [35, 36], MeO [35[, F, Cl [35, 37], Br [35]), 2- and 3-MeC(=O)Ph—, 3- and 4-EtPh—, 3- and 4-N≡CPh— [35], 3-FS(=O)$_2$Ph—, 2-, 3- and 4-Py— [38], 2-Th—, 2-Fu— [81], 2-Pyr— [39], HOC(=O)— [40–43], MeOC(=O)— [41, 44], NaOC(=O)— [41, 45], ClC(=O)— and BrC(=O)— [41], H_2N— [46–48], MeND— [49–53], MeC(=O)ND— [54, 55], Me_2N— [56–61], Cl_2N— [62], N—Im—, R′NH— (see Section 9.3), HO— [48, 63–69], DO— [64, 65, 67], R′O— (see Section 10.2.3), HS— and DS— [70], MeS— [71], MeC(=O)S— [72], F— [73–77], Cl— [69, 73–75], Br— [2, 73–75, 78], I— [75, 77, 79].

References

1. K.S. Pitzer and W. Weltner, *J. Am. Chem. Soc.*, **71**, 2842 (1949).
2. J.C. Evans and H.J. Bernstein, *Can. J. Chem.*, **34**, 1083 (1956).
3. P. Cosse and J.H. Schachtschneider, *J. Chem. Phys.*, **44**, 97 (1966).
4. H. Hollenstein and H.H. Günthard, *Spectrochim. Acta, Part A*, **27A**, 2027 (1971).
5. C.O. Della Védova and O. Sala, *J. Raman Spectrosc.*, **22**, 505 (1991).
6. T. Miyazawa, *Nippon Kagaku Zasshi*, **74**, 915 (1953).
7. J. Overend and J.R. Scherer, *Spectrochim. Acta*, **16**, 773 (1960).
8. P. Mirone and P. Chiorboli, *Ann. Chim. (Rome)*, **50**, 1095 (1960).
9. P. Mirone, *Spectrochim. Acta*, **20**, 1646 (1964).
10. G. Dellepiane and J. Overend, *Spectrochim. Acta*, **22**, 593 (1966).
11. E.E. Ernstbrunner, *J. Chem. Soc. A*, 1558 (1970).
12. P.C. Metha, S.S.L. Surana and S.P. Tandon, *Can. J. Spectrosc.*, **18**, 56 (1973).
13. Z. Jablonski, I. Rychlowska-Himmel and M. Dyrek, *Spectrochim. Acta, Part A*, **35A**, 1297 (1979).
14. G.A. Crowder and P. Pruettiangkura, *J. Mol. Struct.*, **15**, 197 (1973).
15. J.R. Durig, J.A. Hardin, H.V. Phan and T.S. Little, *Spectrochim. Acta, Part A*, **45A**, 1239 (1989).
16. J.R. Durig, J. Lin and H.V. Phan, *J. Raman Spectrosc.*, **23**, 253 (1992).
17. K. Tanabe and S. Saëki, *J. Mol. Struct.*, **25**, 243 (1975).
18. J.R. Durig, J. Lin, C.L. Tolley and T.S. Little, *Spectrochim. Acta, part A*, **47A**, 105 (1991).
19. J.R. Durig, J.A. Hardin and C.L. Tolley, *J. Mol. Struct.*, **224**, 323 (1990).
20. T. Sakurai, M. Ishiyama, H. Takeuchi, K. Takeshita, K. Fukushi and S. Konaka, *J. Mol. Struct.*, **213**, 245 (1989).
21. D.L. Powell, P. Klaboe and D.H. Christensen, *J. Mol. Struct.*, **15**, 77 (1973).
22. J.R. Durig, H.D. Bist and T.S. Little, *J. Mol. Struct.*, **116**, 345 (1984).
23. J.R. Durig and J.S. Church, *Spectrochim. Acta, Part A*, **36A**, 957 (1980).
24. A. Bowles, W.O. George and W.F. Maddams, *J. Chem. Soc.*, **B**, 810 (1969).
25. H.J. Oelichmann, D. Bougeard and B. Schrader, *J. Mol. Struct.*, **77**, 179 (1981).
26. J.R. Durig and T.S. Little, *J. Chem. Phys.*, **75**, 3660 (1981).
27. J. De Smedt, F. Van Houtegem, C. Van Alsenoy, H.J. Geise, B.J. Van der Veken and B. Coppens, *J. Mol. Struct.*, **195**, 227 (1989).

28. A. Poletti, G. Paliani, M.G. Giorgini and R. Cataliotti, *Spectrochim. Acta, Part A*, **31A**, 1869 (1975).
29. G. Davidovics, F. Debu, C. Marfisi, M. Monnier, J.P. Aycard, J. Pourcin and H. Bodot, *J. Mol. Struct.*, **147**, 29 (1986).
30. O. Siiman, J. Fresco and H.B. Gray, *J. Am. Chem. Soc.*, **96**, 2347 (1974).
31. G.A. Crowder, *Spectrochim. Acta, Part A*, **29A**, 1885 (1973).
32. W.D. Mross and G. Zundel, *Spectrochim. Acta, Part A*, **26A**, 1097 (1970).
33. J.H.S. Green and D.J. Harrison, *Spectrochim. Acta, Part A*, **33A**, 583 (1977).
34. A. Gambi, S. Giorgianni, A. Passerini, R. Visinoni and S. Ghersetti, *Spectrochim. Acta, Part A*, **36A**, 871 (1980).
35. G. Varsányi, *Assignments for Vibrational Spectra of Seven Hundred Benzene Derivatives*, J.Wiley & Sons, New York (1974).
36. W.A.L.K. Al-Rashid and M.F. El-Bermani, *Spectrochim. Acta, Part A*, **47A**, 35 (1991).
37. A. Gambi, S. Giorgiani, A. Passerini and R. Visinoni, *Spectrochim. Acta, Part A*, **38A**, 871 (1982).
38. K.G. Medhi, *Indian J. Phys.*, **51A**, 399 (1977).
39. J.C. Viljoen and A.M. Heyns, *Spectrosc. Lett.*, **20**, 765 (1987).
40. H. Hollenstein, F. Akermann and H.H. Günthard, *Spectrochim. Acta, Part A*, **34A**, 1041 (1978).
41. W.J. Ray and J.E. Katon, *Spectrochim. Acta, Part A*, **36A**, 793 (1980).
42. W.J. Ray, J.E. Katon and D.B. Phillips, *J. Mol. Struct.*, **74**, 75 (1981).
43. J. Murto, T. Raaska, H. Kunttu and M. Räsänen, *J. Mol. Struct.*, **200**, 93 (1989).
44. J.K. Wilmshurst and J.F. Horwood, *Aust. J. Chem.*, **24**, 1183 (1971).
45. J.E. Katon and D.T. Covington, *Spectrosc. Lett.*, **12**, 761 (1979).
46. N. Jonathan, *J. Mol. Spectrosc.*, **6**, 205 (1961).
47. T. Uno, K. Machida and Y. Saito, *Bull. Chem. Soc. Jpn.*, **42**, 900 (1969).
48. S.T. King, *Spectrochim. Acta, Part A*, **28A**, 165 (1972).
49. B. Schneider, A. Horeni, H. Picová and J. Honzl, *Collect Czech. Chem. Commun.*, **30**, 2196 (1965).
50. H. Pivcová, B. Schneider and J. Stokr, *Collect Czech. Chem. Commun.*, **30**, 2215 (1965).
51. A. Warshel, M. Levitt and S. Lifson, *J. Mol. Spectrosc.*, **33**, 84 (1970).
52. J. Jakes and S. Krimm, *Spectrochim. Acta, Part A*, **27A**, 19 (1971).
53. M. Ray-Lafon, M.T. Forel and C. Garrigou-Lagrange, *Spectrochim. Acta, Part A*, **29A**, 471 (1973).
54. Y. Kuroda, Y. Saito, K. Machida and T. Uno, *Spectrochim. Acta, Part A*, **27A**, 1481 (1971).
55. Y. Kuroda, Y. Saito, K. Machida and T. Uno, *Spectrochim. Acta, Part A*, **29A**, 411 (1973).
56. R.L. Jones, *J. Mol. Spectrosc.*, **11**, 411 (1963).
57. C. Garrigou-Lagrange, C. De Loze, P. Bacelon, P. Combelas and J. Dagaut, *J. Chim. Phys.*, **67**, 1936 (1970).
58. G. Durgaprasad, D.N. Sathyanarayana, C.C. Patel, H.S. Randhawa, A. Goel and C.N.R. Rao, *Spectrochim. Acta, Part A*, **28A**, 2311 (1972).
59. Z. Mielke and A.J. Barnes, *J. Chem. Soc. Faraday Trans. 2*, **82**, 437 (1986).
60. A.M. Dwivedi, S. Krimm and S. Mierson, *Spectrochim. Acta, Part A*, **45A**, 271 (1989).
61. V.V. Chalapathi and K.V. Ramiah, *Proc. Indian Acad. Sci.*, **68A**, 105 (1968).
62. J.E. Devia and J.C. Carter, *Spectrochim. Acta, Part A*, **29A**, 613 (1973).
63. J.K. Wilmshurst, *J. Chem. Phys.*, **25**, 478, 1171 (1956).
64. M. Haurie and A. Novak, *Spectrochim. Acta*, **21**, 1217 (1965).
65. D. Clague and A. Novak, *J. Mol. Struct.*, **5**, 149 (1970).
66. J.L. Derissen, *J. Mol. Struct.*, **7**, 67 (1971).

67. P.F. Krause, J.E. Katon, J.M. Rogers and D.B. Phillips, *Appl. Spectrosc.*, **31**, 110 (1977).
68. H. Hollenstein and H.H. Günthard, *J. Mol. Spectrosc.*, **84**, 457 (1980).
69. R. Fausto and J.J.C. Teixeira-Dias, *J. Mol. Struct.*, **144**, 215 (1986).
70. H.S. Randhawa, W. Walter and C.O. Meese, *J. Mol. Struct.*, **37**, 187 (1977).
71. A. Smolders, G. Maes and T. Zeegers-Huyskens, *J. Mol. Struct.*, **172**, 23 (1988).
72. B. Fortunato, M.G. Giorgini and P. Mirone, *J. Mol. Struct.*, **25**, 237 (1975).
73. J. Overend, R.A. Nyquist, J.C. Evans and W.J. Potts, *Spectrochim. Acta*, **17**, 1205 (1961).
74. J.C. Evans and J.Overend, *Spectrochim. Acta*, **19**, 701 (1963).
75. J.A. Ramsay and J.A. Ladd, *J. Chem. Soc. B*, 118 (1968).
76. C.V. Berney and A.D. Cormier, *Spectrochim. Acta, Part A*, **28A**, 1813 (1972).
77. S. Tsuchiya, *J. Mol. Struct.*, **22**, 77 (1974).
78. L.C. Hall and J. Overend, *Spectrochim. Acta, Part A*, **23A**, 2535 (1967).
79. F.N. Nicolaisen and J.S. Hansen, *Acta Chem. Scand., Ser. A*, **38A**, 453 (1984).
80. R.A. Nyquist and W.J. Potts, *Spectrochim. Acta*, **15**, 514 (1959).
81. M. Sénéchal and P. Saumagne, *J. Chim. Phys.*, **69**, 1246 (1972).
82. J.R. Durig, F.S. Feng, A. Wang and H.V. Phan, *Can. J. Chem.*, **69**, 1827 (1991).

7.1.6 Propionyl (propanoyl)

The —C(=O)CH_2CH_3 structure unit possesses 24 normal vibrations, of which 18 are derived from the ethyl group. The remaining six are described as follows:

νC=O, νC(=O)—C, δC=O, γC=O, δ-C(=O)—C and torsion.

Evidently the concepts in-plane and out-of-plane lose their meaning if the molecule has no plane of symmetry. In this case the C=O in-plane and out-of-plane deformations are skeletal deformations also, partly dependent on the α-atom.

Methyl and methylene stretching vibrations

The five CH stretching vibrations are situated between 3000 and 2820 cm^{-1}. The ν_aMe in XC(=O)Et compounds (X = F, Cl, Br) absorb in the neighbourhood of 3000 cm^{-1} and the lowest ν_sCH_2 is assigned in the spectrum of EtC(=O)NHC(=O)Et (2820 cm^{-1}). Disregarding these extreme values, the CH stretching vibrations are observed between 2990 and 2870 cm^{-1}:

ν_aMe	ν'_aMe	ν_aCH_2	ν_sMe	ν_sCH_2
2980 ± 10	2965 ± 25	2930 ± 25	2915 ± 25	2905 ± 35

The C=O stretching vibration

With the exception of FC(=O)Et, the C=O stretching vibration provides a strong absorption in the region 1740 ± 90 cm^{-1}. The highest wavenumbers are due to XC(=O)Et (X = F, Cl, Br) with, respectively, 1862, 1830 and 1815 cms^{-1}, which clearly demonstrates the influence of the halogen. Likewise in the HW region one finds the esters of propanoic acid: EtC(=O)OMe (1744), EtC(=O)OEt (1740) and EtC(=O)OiPr (1734 cm^{-1}). Aliphatic ketones with the formula R—C(=O)Et in which R = Et, nPr, iPr, nBu, iBu, nPent, $BrCH_2$, $ClCH_2CH_2$, MeC(=O) and $PhCH_2$ absorb in the narrow region 1715 $\pm$ 2 cm^{-1}. Conjugation results in a considerable lowering of the wavenumber of νC=O, as in CH_2=CHC(=O)Et (1685) and PhC(=O)Et (1688 cm^{-1}), but halogen substituted phenyl ethanones absorb somewhat higher: 4-ClPhC(=O)Et (1691) and 3, 4-Cl_2PhC(=O)Et (1696 cm^{-1}). In delimiting the absorption region of νC=O, the additional lowering of the wavenumber by an intramolecular hydrogen bridge is not taken into account, for example: 2-HOPhC(=O)Et (1642 cm^{-1}). The propionamides also absorb in the LW region: EtC(=O)NH_2 (1660), EtC(=O)NHMe (1653), EtC(=O)NHEt (1650), EtC(=O)NHPr (1653) and EtC(=O)NMe_2 (1651 cm^{-1}).

Methyl and methylene deformations

The methyl deformations are found in the same regions as those of α-saturated ethyl; the methylene scissors absorbs at lower wavenumbers. The intensity of the bands is moderate to strong.

δ_aMe	δ'_aMe	δCH_2	δ_sMe
1470 $\pm$ 15	1455 $\pm$ 15	1425 $\pm$ 20	1380 $\pm$ 15 cm^{-1}

The methylene wag and twist give rise to weak or moderate bands in the regions 1340 $\pm$ 40 and 1265 $\pm$ 25 cm^{-1}.

Methyl rocks and C—C stretching vibrations

In the spectra of R—C(=O)Et compounds the C—C stretching vibrations and the methyl rocks are responsible for four absorption regions. These vibrations are coupled in such a way to make it difficult to determine which vibration has the greatest contribution in a distinct absorption. Sometimes the νC(=O)—C, with the highest wavenumber, is called the antisymmetric stretch and the νC—Me, with the lowest wavenumber, the symmetric stretch [13, 16].

ρMe	ρ'Me	νC(=O)—C	νC—Me
1130 $\pm$ 60	1050 $\pm$ 50	1015 $\pm$ 60	920 $\pm$ 80 cm^{-1}

Methylene rocking vibration

The methylene rock provides a weak to moderate band at 805 $\pm$ 30 cm^{-1}, a range comparable with that of the α-unsaturated ethyl compounds. X—C(=O)Et (X = Cl, Br) shows this ρCH_2 at the LW side ($\approx$784 cm^{-1}) and DC(=O)Et at the HW side at 833 cm^{-1}.

The C=O deformations

The absorption in the range 585 ± 115 cm^{-1} is usually assigned to the C=O in-plane deformation. High values originate from the spectra of nPrNHC(=O)Et (700) and EtNHC(=O)Et (659 cm^{-1}) and low values from those of BrC(=O)Et (470) and ClC(=O)Et (505 cm^{-1}), in which the influence of the halogen catches the eye. Most of the C=O in-plane deformations have been observed at 620 ± 65 cm^{-1}.

The C=O out-of-plane deformation appears in the range 520 ± 90 cm^{-1}. The HW side of this region is delimited by 608 cm^{-1} from the spectrum of EtC(=O)NDC(=O)Et (with the second γC=O at 545) and by 430 cm^{-1} in that of 2-ThC(=O)Et. The remaining C=O out-of-plane deformations absorb in the region 520 ± 70 cm^{-1}.

Skeletal deformations

The vibrational analysis of R—C(=O)Et compounds reveals two skeletal deformations: a deformation with the ethyl group and an external deformation. The lower limits are determined by the skeletal deformations of BrC(=O)Et at 296 and 206 cm^{-1}.

δC(=O)—C—C	δR—C(=O)—C
375 ± 85	260 ± 55

Table 7.6 Absorption regions (cm^{-1}) of the normal vibrations of —C(=O)Et

Vibration	Region	Vibration	Region
ν_aMe	2985 ± 15	ρMe	1130 ± 60
ν'_aMe	2970 ± 30	ρ'Me	1050 ± 50
$\nu_a CH_2$	2930 ± 25	νC(=O)—C	1015 ± 60
ν_sMe	2900 ± 40	νC—Me	920 ± 80
$\nu_s CH_2$	2880 ± 60	ρCH_2	805 ± 30
νC=O	1740 ± 90	δC=O	585 ± 115
δ_aMe	1470 ± 15	γC=O	520 ± 90
δ'_aMe	1455 ± 15	δC(=O)—Et	375 ± 85
δCH_2	1425 ± 20	δ-C(=O)—C	260 ± 55
δ_sMe	1380 ± 15	torsion Me	220 ± 30
ωCH_2	1340 ± 40	torsion Et	130 ± 70
τCH_2	1265 ± 25	torsion C(=O)Et	–

The following R—C(=O)Et compounds are taken into account:

R = H— [1–3], D— [2], Me— [19], Et— [4], nPr—, nBu—, nPent—, $BrCH_2$—, $ClCH_2CH_2$—, $PhCH_2$—, iPr—, MeC(=O)—, CH_2=CH—,

Ph—, 4-MePh—, 2- and 4-HOPh— [9, 20], 4-MeOPh—, 4-PhCH$_2$OPh—, 4-FPh— [6], 3- and 4-ClPh— [6], 4-BrPh— [6], 3, 4-Cl$_2$Ph—, 2-Th— [5], MeNH—, EtNH— and nPrNH— [7], EtC(=O)NH— and EtC(=O)ND— [8], HO— [10–12], DO— [10, 12], MeO— [13, 14], CD$_3$O— [13], EtO—, NaO— [18], F— [15, 16], Cl— [15, 17], Br— [15].

References

1. G. Sbrana and V. Schettino, *J. Mol. Spectrosc.*, **33**, 100 (1970).
2. S.G. Frankiss and W. Kynaston, *Spectrochim. Acta, Part A*, **28A**, 2149 (1972).
3. P. Van Nuffel, L. Van den Enden, C. Van Alsenoy and H.J. Geise, *J. Mol. Struct.*, **116**, 99 (1984).
4. Z. Buric and P.J. Krueger, *Spectrochim. Acta, Part A*, **30A**, 2069 (1974).
5. J.J. Peron, P. Saumange and J.M. Lebas, *Spectrochim. Acta, Part A*, **26A**, 1651 (1970).
6. W.A. Seth Paul and J. Meeuwesen, *Bull. Soc. Chim. Belg.*, **90**, 127 (1981).
7. J. Jakes and S. Krimm, *Spectrochim. Acta, Part A*, **27A**, 35 (1971).
8. Y. Kuroda, K. Machida and T. Uno, *Spectrochim. Acta, Part A*, **30A**, 47 (1974).
9. W.A.L.K. Al-Rashid and M.F. El-Bermani, *Spectrochim. Acta, Part A*, **47A**, 35 (1991).
10. Y. Mikawa, J.W. Brasch and R.J. Jakobsen, *J. Mol. Struct.*, **3**, 103 (1969).
11. R.J. Jakobsen, Y. Mikawa, J.R. Allkins and G.L. Carlson, *J. Mol. Struct.*, **10**, 300 (1971).
12. J. Umemura, *J. Mol. Struct.*, **36**, 35 (1977).
13. R.M. Moravie and J. Corset, *J. Mol. Struct.*, **24**, 91 (1975).
14. R.M. Moravie and J. Corset, *J. Mol. Struct.*, **30**, 113 (1976).
15. G. Frankiss and W. Kynaston, *Spectrochim. Acta, Part A*, **31A**, 661 (1975).
16. J.R. Durig, G.A. Guirgis and H.V. Phan, *J. Raman Spectrosc.*, **21**, 359 (1990).
17. S. Dyngeseth, S.H. Schei and K. Hagen, *J. Mol. Struct.*, **116**, 257 (1984).
18. E. Spinner, P. Yang, P.T.T. Wong and H.H. Mantsch, *Aust. J. Chem.*, **39**, 475 (1986).
19. J.R. Durig, F.S. Feng, A. Wang and H.V. Phan, *Can. J. Chem.*, **69**, 1827 (1991).
20. G. Varsányi, *Assignments for Vibrational Spectra of Seven Hundred Benzene Derivatives*, J.Wiley & Sons, New York (1974).

7.1.7 Carboxyl

The nine normal vibrations of the —C(=O)OH group are described as follows:

νOH, νC=O, δOH, νC—O, γ OH, δC=O, γ C=O, δ-C—O and torsion.

Carboxylic acids are best characterized by the OH stretch, the C=O stretch and the OH out-of-plane deformation and even by the C—O stretch and the OH in-plane deformation.

The OH stretching vibration

The very broad band with a maximum absorption at 3050 ± 150 cm^{-1}, characteristic of carboxylic acids, is due to the associated OH ...O stretching vibration of mainly cyclic dimers. The typical shoulders on the low-frequency wing are assigned to overtones and combinations of δOH and νC—O, enhanced by Fermi resonance. In saturated carboxylic acids this νOH...O appears in the vicinity of 3050 cm^{-1}. The exact wavenumber of this broad band is difficult to detect because of the CH stretching bands superimposed on it. $F_3CC(=O)OH$ and $Cl_3CC(=O)OH$ absorb at ≈100 cm^{-1} higher and benzoic acids at approximately 100 cm^{-1} lower. Even in dilute solutions this band does not disappear completely, and the absorption in the neighbourhood of 3500 cm^{-1} is due to the free OH stretching vibration.

The C=O stretching vibration

The C=O stretching vibration in the spectra of carboxylic acids gives rise to a strong band in the region 1725 ± 65 cm^{-1}. In the vapour state the monomer absorbs at a wavenumber 50 cm^{-1} higher. The extensive region of the saturated carboxylic acids (1735 ± 50 cm^{-1}) is attributable to the higher values in the spectra of α-fluoro and α-chloro acids and to the lower values, even to 1685 cm^{-1}, in the spectra of dicarboxylic acids. The following series illustrates the influence of the halogen: $F_3CC(=O)OH$ (1785), $F_2CHC(=O)OH$ (1766), $FCH_2C(=O)OH$ (1730), $Cl_3CC(=O)OH$ (1751), $Cl_2CHC(=O)OH$ (1740) and $ClCH_2C(=O)OH$ (1723 cm^{-1}). In most of the saturated carboxylic acids the νC=O is found at 1720 ± 20 cm^{-1}, a region that narrows to 1710 ± 5 cm^{-1} for the series $Me(CH_2)_nC(=O)OH$ (n = 0–16) [99]. Benzoic acids absorb at 1680 ± 20 cm^{-1}, but the νC=O of the three isomeric pyridinecarboxylic acids fall outside this region (≈1715 cm^{-1}) [97].

The OH in-plane deformation

The OH...O in-plane deformation, coupled to the C—O stretching vibration, is easy to observe as a broad band with medium intensity, occurring in the region 1395 ± 55 cm^{-1}. The sharp peaks of the CH deformations superimposed on it make it somewhat difficult to detect the exact wavenumber of this OH deformation. The highest wavenumbers have been traced in the spectra of $F_3CC(=O)OH$ (1450) and $F_2CHC(=O)OH$ (1445) and the lowest in those of HC(=O)OH (1340) and HO(O=)CC(=O)OH (1345 and *1395* cm^{-1}). Most of the carboxylic acids show the δOH...O at 1415 ± 25 cm^{-1}. The free δOH of the monomer in the vapour state occurs at wavenumbers approximately 50 cm^{-1} lower.

The C—O stretching vibration

The C(=O)—O stretching vibration, coupled to the OH in-plane deformation, exhibits a moderate to strong band in the region 1250 ± 80 cm^{-1}. High values appear in the spectra of 2-PyrC(=O)OH (1329) and 4-BrPhC(=O)OH (1323) and low values in those of $N\equiv CCH_2C(=O)OH$ (1173) and HO(O=)CC(=O)OH

(1173 and *1230* cm^{-1}). The remaining carboxylic acids display this vibration in the region 1260 ± 60 cm^{-1}. Hexanedioic acid gives both stretchings at 1300 and 1280 cm^{-1}. In the series Me(CH$_2$)$_n$C(=O)OH (n = 0–16) this region is 1285 ± 10 cm^{-1}. Monomer acids in the vapour state absorb ≈100 cm^{-1} lower.

The OH out-of-plane deformation

The γOH...O or out-of-plane OH...O wag exhibits a moderate band in the shape of a V in the region 905 ± 65 cm^{-1} [96]. The highest values come from H$_2$NC(=O)C(=O)OH (970), HO(O=)CCHMeC(=O)OH (970 and *930*) and HO(O=)CC(=O)CH$_2$C(=O)OH (960 and *940* cm^{-1}). α-Halogen compounds such as ClCH$_2$C(=O)OH (843) and Cl$_3$CC(=O)OH (849 cm^{-1}) are responsible for the lowest wavenumbers. Usually the γOH...O in carboxylic acids occurs at 920 ± 30 cm^{-1} and in the series Me(CH$_2$)$_n$C(=O)OH (n = 0–16) at 935 ± 5 cm^{-1}. In the unbound state the γOH disappears to make place for the free torsion at lower wavenumbers (≈650 cm^{-1}).

The C=O in-plane deformation

The C=O in-plane deformation is weakly to moderately active in the region 725 ± 95 cm^{-1}. The highest values are observed in the spectra of 2-XPhC(=O)OH (X = F, Cl, Br) (≈820 cm^{-1}), 2-MePhC(=O)OH (809) and benzoic acid (802 cm^{-1}). Acetic acid (630), ICH$_2$C(=O)OH (635), EtC(=O)OH (640), BrCH$_2$CH$_2$C(=O)OH (640) and FCH$_2$C(=O)OH (646 cm^{-1}) score at low wave-numbers, but formic acid and the remaining n-alkanoic acids display this δC=O in the vicinity of 670 cm^{-1}. Most carboxylic acids show the C=O in-plane deformation at 715 ± 65 cm^{-1} with the benzoic acids to the HW side of this region. These values are in good agreement with those of methyl and ethyl esters (715 ± 115 cm^{-1}), but they are significantly higher than the corresponding values in RC(=O)X (X = Me or Et) compounds (580 ± 115 cm^{-1}).

The C=O out-of-plane deformation

The C=O out-of-plane deformation absorbs weakly to moderately in the extensive region 595 ± 120 cm^{-1}. The highest wavenumbers are 715, 703, 702, 698 and 693 cm^{-1} respectively from the spectra of 4-XPhC(=O)OH (X = O$_2$N, F$_3$C, Et, H$_2$N and OH) and the lowest come from H$_2$NC(=O)C(=O)OH (480) and ICH$_2$C(=O)OH (485 cm^{-1}). For acetic acid and propanoic acid the wavenumber is 600 cm^{-1} and for the remaining n-alkanoic acids wavenumbers are in the neighbourhood of 630 cm^{-1}. Most carboxylic acids display the γC=O in a range (595 ± 85) which is in the vicinity of that of methyl and ethyl esters (635 ± 130) but higher than that of methyl and ethyl ketones (505 ± 115 cm^{-1}).

The C—O deformation

In the —C(=O)O deformation or —C(=O)O rock, the accent lies much more on the —C—O deformation than on the C=O deformation. This band has a weak to medium intensity and makes its appearance in the extensive region 445 ± 120 cm^{-1}. The highest wavenumbers are assigned in the spectra of substituted benzoic acids (530 ± 35 cm^{-1}). Low values are attributed in the spectra of $H_2NC(=O)C(=O)OH$ (326), HC(=O)C(=O)OH (365) and $HO(O=)CCH_2CH_2C(=O)OH$ (388 and *545* cm^{-1}). This region agrees with that of the corresponding vibration in the spectra of methyl and ethyl esters (435 ± 95 cm^{-1}).

Table 7.7 Normal vibrations and absorption regions (cm^{-1}) of —C(=O)OH

Vibration	α-saturated	C=O bonded	α-unsaturated	aromatic
νOH...O	3050 ± 50	3150 ± 50	3000 ± 50	2950 ± 50
νC=O	1735 ± 50	1730 ± 30	1690 ± 20	1680 ± 20
δOH...O	1395 ± 55	1395 ± 55	1400 ± 40	1415 ± 25
νC—O	1245 ± 75	1220 ± 50	1270 ± 50	1300 ± 30
γ OH...O	905 ± 65	905 ± 65	900 ± 50	900 ± 50
δC=O	705 ± 75	705 ± 75	725 ± 75	770 ± 50
γ C=O	580 ± 100	565 ± 85	590 ± 60	660 ± 55
ρC(=O)—O	465 ± 80	395 ± 70	490 ± 50	530 ± 35
torsion	–	–	–	–

R—C(=O)OH compounds

R = H— [1–12], D— [1–5], Me— [10, 11, 13–18], CD_3— [14, 17], Et— [19–21], $MeCD_2$— [21[, CD_3CD_2— [19], $Me(CH_2)_n$— (n= 2–16), $KO(O=)CC(Me)_2CH_2CH_2$— [22], $HO(O=)CCH_2CH_2$— [23, 24], $HO(O=)C(CH_2)_4$— [25], $HSCH_2CH_2$— [26], $ClCH_2CH_2$—, $BrCH_2CH_2$—, $iPrCH_2$—, $HO(O=)CCH_2$— [27], $KO(O=)CCH_2$— [28], $HO(O=)CC(=O)CH_2$— [31], $PhCH_2$—, $N{\equiv}CCH_2$— [29], $HOCH_2$—, $HOCD_2$— and HOCHD— [30], $HSCH_2$—, $HO(O=)CCH_2SSCH_2$—, $HO(O=)CCH(OH)CH_2$— and $HO(O=)CCH(SH)CH_2$— [26], FCH_2— [32], $ClCH_2$— [33–36], $BrCH_2$— [37], ICH_2— [38], HO(O=)CCH(Me)— [39], iPr—, cPr— [40–42], cBu— [43, 44], cPent— [45], cHex—, F_2CH— [32], Cl_2CH— [46, 47], FClCH— and FClCD— [48], tBu— [49], $HO(O=)CC(Me)_2$— $DO(O=)CC(Me)_2$— and $KO(O=)CC(Me)_2$— [50], F_3C— [11, 32, 51–58], Cl_3C— [46, 59, 60], F_3CCF_2— [61], H(O=)C— [62, 63], MeC(=O)— [64–67], $CH_2DC(=O)$—, $CHD_2C(=O)$— and $CD_3C(=O)$— [64], HO(O=)C— [68, 69], $H_2NC(=O)$— [70–73], $MeO(O=)CCH_2C(=O)$— [74], $H_2C=CH$— [75–77], $H_2C=C(Me)$—, MeCH=CH—, HO(O=)CCH=CH— and HO(O=)CCD=CD—

[78], KO(O=)CCH=CH— [79], MeO(O=)CCH=CH— [80], HO(O=)CCH=CH—CH=CH— [81], HC≡C— [82], Me(CH_2)$_4$C≡C—, 2-ThC≡C—, HO(O=)CC≡C—, Ph— [83–85, 95], 2-XPh— (X = Me [85, 90], HO(O=)C [87, 88, 90, 95], H_2N [90], MeHN [86], Me_2N [86], O_2N [95], F [85], Cl [85, 90], Br [85]), 3-XPh— (X = Me [85, 91], HO(O=)C [88, 91], HO [91, 95], MeO [91], H_2N [91], ^+H_3N [89], O_2N [95], F [85], Cl [85, 91, 95], Br [85], FS(=O)$_2$), 4-XPh— (X = Me [85, 92], Et, F_3C, HO(O=)C [88, 92, 93, 95], H(O=)C, H_2N [92, 95], Me_2N [94], O_2N [95], HO [92, 95], MeO [92], EtO, F [85, 95], Cl [85, 92, 95], Br [85, 95], I, MeS(=O)$_2$, FS(=O)$_2$, H_2NS(=O)$_2$), 2-, 3- and 4-Py— [97], 2-Fu— [98], 2-Th, 2-Pyr—.

References

1. J.K. Wilmshurst, *J. Chem. Phys.*, **25**, 478 (1956).
2. J.E. Bertie and K.H. Michaelian, *J. Chem. Phys.*, **76**, 886 (1982).
3. J.E. Bertie, K.H. Michaelian, H.H. Eysel and D. Hager, *J. Chem. Phys.*, **85**, 4779 (1986).
4. I. Yokoyama, Y. Miwa and K. Machida, *J. Phys. Chem.*, **95**, 9740 (1991).
5. I. Yokoyama, Y. Miwa and K. Machida, *J. Am. Chem. Soc.*, **113**, 6458 (1991).
6. W.J. Orville-Thomas, *Discuss. Faraday Soc.*, **9**, 339 (1950).
7. I. Alfheim, G. Hagen and S.J. Cyvin, *J. Mol. Struct.*, **8**, 159 (1971).
8. J.E.D. Davies, *J. Mol. Struct.*, **9**, 483 (1971).
9. Y.-T. Chang, Y. Yamaguchi, W.H. Miller and H.F. Schaefer, *J. Am. Chem. Soc.*, **109**, 7245 (1987).
10. H. Hollenstein and H.H. Günthard, *J. Mol. Spectrosc.*, **84**, 457 (1980).
11. D. Clague and A. Novak, *J. Mol. Struct.*, **5**, 149 (1970).
12. J. Nieminen, M. Räsänen and J. Murto, *J. Phys. Chem.*, **96**, 5303 (1992).
13. J.K. Wilmshurst, *J. Chem. Phys.*, **25**, 1171 (1956).
14. M. Haurie and A. Novak, *Spectrochim. Acta*, **21**, 1217 (1965).
15. J.L. Derissen, *J. Mol. Struct.*, **7**, 67 (1971).
16. S.T. King, *Spectrochim. Acta, Part A*, **28A**, 165 (1972).
17. P.F. Krause, J.E. Katon, J.M. Rogers and D.B. Phillips, *Appl. Spectrosc.*, **31**, 110 (1977).
18. R. Fausto and J.J.C. Teixeira-Dias, *J. Mol. Struct.*, **144**, 215 (1986).
19. Y. Mikawa, J.W. Brasch and R.J. Jakobsen, *J. Mol. Struct.*, **3**, 103 (1969).
20. R.J. Jakobsen, Y. Mikawa, J.R. Allkins and G.L. Carlson, *J. Mol. Struct.*, **10**, 300 (1971).
21. J. Umemura, *J. Mol. Struct.*, **36**, 35 (1977).
22. S. Yolou, J.J. Delarbre and L. Maury, *J. Raman Spectrosc.*, **23**, 501 (1992).
23. M. Suzuki and T. Shimanouchi, *J. Mol. Spectrosc.*, **28**, 394 (1968).
24. J.E. Katon and S.R. Lobo, *J. Mol. Struct.*, **127**, 229 (1985).
25. M. Suzuki and T. Shimanouchi, *J. Mol. Spectrosc.*, **29**, 415 (1969).
26. N. Saraswathi and S. Soundarajan, *J. Mol. Struct.*, **4**, 419 (1969).
27. D. Bougeard, J. De Villepin and A. Novak, *Spectrochim. Acta, Part A*, **44A**, 1281 (1988).
28. L. Angeloni, M.P. Marzocchi, S. Detoni, D. Hadzi, B. Orel and G. Sbrana, *Spectrochim. Acta, Part A*, **34A**, 253 (1978).
29. D. Sinha and J.E. Katon, *Appl. Spectrosc.*, **26**, 599 (1972).
30. H. Hollenstein, R.W. Schar, N. Schwizgebel, G. Grassi and H.H. Günthard, *Spectrochim. Acta, Part A*, **39A**, 193 (1983).
31. D.W. Schiering and J.E. Katon, *J. Mol. Struct.*, **144**, 71 (1986).
32. J.R. Barcelo and C. Otero, *Spectrochim. Acta*, **18**, 1231 (1962).

33. J.R. Barcelo, M.P. Jorge and C. Otero, *J. Chem. Phys.*, **28**, 1230 (1958).
34. D. Sinha, J.E. Katon and R.J. Jakobsen, *J. Mol. Struct.*, **20**, 381 (1975).
35. D. Sinha, J.E. Katon and R.J. Jakobsen, *J. Mol. Struct.*, **24**, 279 (1975).
36. R. Fausto and J.J.C. Teixeira-Dias, *J. Mol. Struct.*, **144**, 225 (1986).
37. J.E. Katon and R.L. Kleinlein, *Spectrochim. Acta, Part A*, **29A**, 791 (1973).
38. J.E. Katon and T.P. Carll, *J. Mol. Struct.*, **7**, 391 (1971).
39. L. Maury, J.L. Delarbre and L. Bardet, *Spectrochim. Acta, Part A*, **41A**, 1477 (1985).
40. D.L. Powell and P. Klaboe, *J. Mol. Struct.*, **15**, 217 (1973).
41. J. Maillols, *J. Mol. Struct.*, **14**, 171 (1972).
42. V. Tabacik and J. Maillols, *Spectrochim. Acta, Part A*, **34A**, 315 (1978).
43. J.E. Katon, R.O. Carter and W. Yellin, *J. Mol. Struct.*, **11**, 347 (1972).
44. L. Bardet, J. Maillols, R. Granger and E. Fabregue, *J. Mol. Struct.*, **10**, 343 (1971).
45. L. Bardet, G. Cassanas-Fabre and E. Bourret, *J. Mol. Struct.*, **28**, 45 (1975).
46. R. Fausto and J.J.C. Teixeira-Dias, *J. Mol. Struct.*, **144**, 241 (1986).
47. L.M. Babkow, V.V. Vashchinskaya, M.A. Kovner, G.A. Puchkovskaya and Yu.Ya. Fialkov, *Spectrochim. Acta, Part A*, **32A**, 1379 (1976).
48. J. Calienni, J.B. Trager, M.A. Davies, U. Gunnia and M. Diem, *J. Phys. Chem.*, **93**, 5049 (1989).
49. W. Longueville, H. Fontaine and G. Vergoten, *J. Raman Spectrosc.*, **13**, 213 (1982).
50. J.L. Delarbre, L. Maury and L. Bardet, *J. Raman Spectrosc.*, **13**, 1 (1982).
51. N. Fuson, M.L. Josien, E.A. Jones and J.R. Lawson, *J. Chem. Phys.*, **20**, 1627 (1952).
52. M.L. Josien, N. Fuson, J.R. Lawson and E.A. Jones, *C.R. Acad. Sci.*, **234**, 1163 (1952).
53. N. Fuson and M.L. Josien, *J. Opt. Soc. Am.*, **43**, 1102 (1953).
54. R.E. Kagarise, *J. Chem. Phys.*, **27**, 519 (1957).
55. T.S.S.R. Murty and K.S. Pitzer, *J. Phys. Chem.*, **73**, 1426 (1969).
56. R.L. Redington and K.C. Lin, *Spectrochim. Acta, Part A*, **27A**, 2445 (1971).
57. C.V. Berney, *J. Am. Chem. Soc.*, **95**, 708 (1973).
58. R.L. Redington, *Spectrochim. Acta, Part A*, **31A**, 1699 (1975).
59. M.D.P. Jorge and J.R. Barcelo, *An. R. Soc. Esp. Fis. Quim., Ser. B*, **53B**, 339 (1957).
60. J. Adams and H. Kim, *Spectrochim. Acta, Part A*, **29A**, 675 (1973).
61. G.A. Crowder, *J. Fluorine Chem.*, **1**, 385 (1971/72).
62. G. Fleury and V. Tabacik, *J. Mol. Struct.*, **10**, 359 (1971).
63. G. Fleury and V. Tabacik, *J. Mol. Struct.*, **12**, 156 (1972).
64. H. Hollenstein, F. Akermann and H.H. Günthard, *Spectrochim. Acta, Part A*, **34A**, 1041 (1978).
65. W.J. Ray and J.E. Katon, *Spectrochim. Acta, Part A*, **36A**, 793 (1980).
66. W.J. Ray, J.E. Katon and D.B. Phillips, *J. Mol. Struct.*, **74**, 75 (1981).
67. J. Murto, T. Raaska, H. Kunttu and M. Räsänen, *J. Mol. Struct.*, **200**, 93 (1989).
68. R.L. Redington and T.E. Redington, *J. Mol. Struct.*, **48**, 165 (1978).
69. T.A. Shippey, *J. Mol. Struct.*, **65**, 71 (1980).
70. H.O. Desseyn, F.K. Vansant and B.J. Van der Veken, *Spectrochim. Acta, Part A*, **31A**, 625 (1975).
71. F. Wallace and E. Wagner, *Spectrochim. Acta, Part A*, **34A**, 589 (1978).
72. G.N.R. Tripathi and J.E. Katon, *J. Mol. Struct.*, **54**, 19 (1979).
73. G.N.R. Tripathi and J.E. Katon, *Spectrochim. Acta, Part A*, **35A**, 401 (1979).
74. D.W. Schiering and J.E. Katon, *J. Mol. Struct.*, **144**, 25 (1986).
75. W.R. Feairheller and J.E. Katon, *Spectrochim. Acta, Part A*, **23A**, 2225 (1967).
76. P.F. Krause, J.E. Katon and K.K. Smith, *Spectrochim. Acta, Part A*, **32A**, 960 (1976).
77. S.W. Charles, F.C. Cullen, N.L. Owen and G.A. Williams, *J. Mol. Struct.*, **157**, 17 (1987).
78. J. Maillols, L. Bardet and L. Maury, *J. Mol. Struct.*, **30**, 57 (1976).

79. F. Avbelj, B. Orel, M. Klanjsek and D. Hadzi, *Spectrochim. Acta, Part A*, **41A**, 75 (1985).
80. J.E. Katon and P.-H. Chu, *J. Mol. Struct.*, **82**, 61 (1982).
81. P. Sohár and G. Varsányi, *J. Mol. Struct.*, **1**, 437 (1967/68).
82. J.E. Katon and N.T. McDevitt, *Spectrochim. Acta*, **21**, 1717 (1965).
83. W. Lewandowski, *J. Mol. Struct.*, **101**, 93 (1983).
84. Y. Kim and K. Machida, *Spectrochim. Acta, Part A*, **42A**, 881 (1986).
85. J.H.S. Green, *Spectrochim. Acta, Part A*, **33A**, 575 (1977).
86. A. Tramer, *J. Mol. Struct.*, **4**, 313 (1969).
87. L. Colombo, V. Volovsek and M. Le Postollec, *J. Raman Spectrosc.*, **15**, 252 (1984).
88. J.F. Arenas and J.I. Marcos, *Spectrochim. Acta, Part A*, **36A**, 1075 (1980).
89. L. Gopal, C.I. Jose and A.B. Biswas, *Spectrochim. Acta, Part A*, **23A**, 513 (1967).
90. E. Sanchez de la Blanca, J.L. Nunez and P. Martinez, *J. Mol. Struct.*, **142**, 45 (1986).
91. E. Sanchez de la Blanca, J.L. Nunez and P. Martinez, *An. Quim.*, **82**, 490 (1986).
92. E. Sanchez de la Blanca, J.L. Nunez and P. Martinez, *An. Quim.*, **82**, 480 (1986).
93. G.N.R. Tripathi and S.J. Sheng, *J. Mol. Struct.*, **57**, 21 (1979).
94. J.G. Rosencrance and P.W. Jagodzinsky, *Spectrochim. Acta, Part A*, **42A**, 869 (1986).
95. G. Varsányi, *Assignments for Vibrational Spectra of Seven Hundred Benzene Derivatives*, J.Wiley & Sons, New York (1974).
96. I. Fischmeister, *Spectrochim. Acta*, **20**, 1071 (1964).
97. S. Chattopadhyay and S.K. Brahma, *Spectrochim. Acta, Part A*, **49A**, 589 (1993).
98. M. Sénéchal and P. Saumagne, *J. Chim. Phys.*, **69**, 1246 (1972).
99. C.J. Pouchert, *The Aldrich Library of FT-IR Spectra*, Aldrich Chemical Company, 1st edn. (1985).

7.1.8 Methoxycarbonyl

The —C(=O)OMe group is responsible for eighteen normal vibrations, of which twelve are due to the —OMe and the remaining six to the —C(=O)O structure unit:

ν_aMe (2), ν_sMe, νC=O, δ_aMe (2), δ_sMe, νC(=O)—O, ρMe (2), νO—C, δC=O, γ C=O, δ-C(=O)—O, δC—O—C and three torsions.

Methyl stretching vibrations

In methyl esters the methyl antisymmetric stretching vibrations are regularly seen above 3000 cm^{-1} (Section 2.1.7). Halogen-substituted acids score highest: FC(=O)OMe (3047), ClC(=O)OMe (3044) and Cl_3CC(=O)OMe (3040 cm^{-1}).

The C=O stretching vibration

The most characteristic band of esters arises from the C=O stretching vibration occurring at 1750 $\pm$ 50 cm^{-1} with a strong to very strong intensity. The exceptionally high wavenumbers for FC(=O)OMe (1854) and MeOC(=O)OC(=O)OMe (1830 and *1765* cm^{-1}) and the very low wavenumbers for KOC(=O)OMe (1680) and 2-HOPhC(=O)OMe (1679 cm^{-1}, intramolecular H-bridge) are outside this region. Most methyl esters of α-saturated carboxylic acids absorb at 1745 ± 15 cm^{-1} if the influence of the halogen is not too substantial, such as in FC(=O)OMe (1854), ClC(=O)OMe (1787), F_3CC(=O)OMe (1793) and Cl_3CC(=O)OMe (1771 cm^{-1}). For Me$(CH_2)_n$C(=O)OMe (n = 1–20), this region narrows to 1743 ± 2 cm^{-1} [68]. Methyl esters of α-unsaturated and aromatic carboxylic acids show the νC=O at 1725 ± 20 cm^{-1}. Seth Paul and Van Duyse [64] identified the region 1730 ± 15 cm^{-1} for mono- and di-substituted methyl benzoates and Nyquist [65] proposed 1733 ± 5 cm^{-1} for *o*-phtalic esters.

Methyl deformations

More often than not both methyl antisymmetric deformations occur at the same wavenumber (1460 ± 25 cm^{-1}). The high wavenumber of the methyl symmetric deformation (1435 ± 15 cm^{-1}), displayed at the LW side of the ν_aMe, is remarkable. The low wavenumbers for δ_sMe in the spectra of H_2NC(=O)OMe (1369) and Cl_2NC(=O)OMe (1386 cm^{-1}) are outside the above—mentioned region.

The C(=O)—O stretching vibration

This vibration, often considered as the C—O antisymmetric stretch, appears strongly at 1255 ± 60 cm^{-1}, a region in good agreement with that of the νC—O in carboxylic acids (1250 ± 80 cm^{-1}). The highest wavenumbers are furnished by such compounds as XC(=O)CH=CHC(=O)OMe (X = Cl, HO, DO and NaO) (≈1315 cm^{-1}) followed by KOC(=O)OMe (1310), H_2C=C(Me)C(=O)OMe (1307) and MeO(O=)COC(=O)OMe (1303 cm^{-1}). The lowest values come from H_2NC(=O)OMe (1195) and Cl_2CHC(=O)OMe (1196 cm^{-1}). The remaining compounds show the νC(=O)—O at 1245 ± 45 cm^{-1}, with methyl formate (1215) and methyl acetate (1246 cm^{-1}) as examples. In addition to the νC=O, this vibration is characteristic for esters.

Methyl rocks

With the exception of Cl_2NC(=O)OMe (1071 cm^{-1}), the ρMe has been observed at 1185 ± 35 cm^{-1}, often as a shoulder on the low-frequency wing of the νC(=O)—O absorption. In iPrC(=O)OMe this ρMe (1194) is next to the νC(=O)—O (1202) but in tBuC(=O)OMe both vibrations coincide at 1193 cm^{-1}.

Disregarding the low values of 1075 and 1050 cm^{-1} respectively originating from the spectra of H_2NC(=O)OMe and Cl_2C(=O)OMe [55], the ρ'Me absorbs at 1155 ± 35 cm^{-1}.

The O—C stretching vibration

The O—Me stretch, coupled with the methyl rock, appears in the wide region 975 ± 125 cm^{-1} with an intensity varying from weak to strong. This vibration is often called the symmetric COC stretching vibration. The high values originated from cPrC(=O)OMe (1098), EtC(=O)OMe and iPrC(=O)OMe (1091), KOC(=O)OMe (1080) and PhNHC(=O)OMe (1068 cm^{-1}) appear probably under the influence of the methyl rock. The lower wavenumbers from the compounds H_2C=CHC(=O)OMe (853), MeO(O=)CC(=O)OMe (859), D_3CC(=O)OMe (863), DC(=O)OMe (878) and H_2NC(=O)OMe (880 cm^{-1}) come more in the neighbourhood of the free ν_sCOC. Most Me esters display the νO—Me at 980 ± 80 cm^{-1} but it is not a good group vibration.

The C=O deformations

More than once it happens that both deformations are assigned at the same wavenumber.

The δC=O has been traced in the region 715 ± 115 cm^{-1}, limited by 830 cm^{-1} from Cl_2NC(=O)OMe followed by 826 cm^{-1} from KOC(=O)OMe and 797 cm^{-1} from MeOC(=O)OMe on the one side and by 600 cm^{-1} in the spectrum of D_3CC(=O)OMe on the other side. Generally this vibration appears moderately at 710 ± 80 cm^{-1}. Fluoro compounds such as FC(=O)OMe (789) and F_3CC(=O)OMe (780) score higher than esters of *n*-alkanoic acids such as MeC(=O)OMe (640) and EtC(=O)OMe (671 cm^{-1}). The absorption at 690 cm^{-1} in the spectrum of ClC(=O)OMe is assigned to the γC=O (Section 7.1.3).

The γC=O absorption varies within the large region 635 ± 130 cm^{-1} with at the HW side DC(=O)OMe (765), Cl_2NC(=O)OMe (735), Me_2C=CHC(=O)OMe (735) and $(CD_3)_2C$=CHC(=O)OMe (715 cm^{-1}), and at the LW side H_2C=C(Me)C(=O)OMe (511), MeOC(=O)CH_2C(=O)OMe (534 and 517), H_2C=CHC(=O)OMe (530) and ClC(=O)OMe (540 cm^{-1}). The remaining RC(=O)OMe molecules display the γC=O in the more convenient region 625 ± 75 cm^{-1}, just like the ethyl esters. Influenced by the hydrogen, the γCH/C=O in methyl formate absorbs at a higher wavenumber (Section 7.1.1).

The C(=O)—O deformation

The —C(=O)—O deformation or —C(=O)—O rock absorbs weakly to moderately in the region 435 ± 95 cm^{-1}. The compounds D_3CC(=O)OMe, H_2NC(=O)OMe, MeOC(=O)CH=CHC(=O)OMe and MeOC(=O)OMe define the upper limit with 521 ± 3 cm^{-1} and RC≡CC(=O)OMe molecules (R = MeC(=O), Me and H) occupy the lower part of the region (≈340 cm^{-1}). For most of the methyl esters the region 445 ± 60 cm^{-1} gives satisfaction. Methyl acetate absorbs at 435 cm^{-1} and methyl propanoate at 440 cm^{-1}. Saunders *et al.* [66] found 445 ± 10 cm^{-1} for a series of fourteen aliphatic methyl esters.

Skeletal C—O—C deformation

The skeletal C—O—C deformation appears in the region 320 ± 70 cm^{-1}. The molecules MeCH=CHC(=O)OMe (385) and H_2C=C(Me)C(=O)OMe

(381) exhibit this band at high wavenumbers and MeC≡CC(=O)OMe (252), D_3CC(=O)OMe (283) and ClC(=O)OMe (284 cm^{-1}) at low wavenumbers. MeOC(=O)OMe gives both skeletal deformations at 372 and 257 cm^{-1}. Disregarding these extreme values, most of the methyl esters give this skeletal deformation at 325 ± 40 cm^{-1}, examples being methyl formate (330) and methyl acetate (318 cm^{-1}). Saunders *et al.* [66] found 330 ± 15 cm^{-1} for a series of aliphatic methyl esters.

Table 7.8 Absorption regions (cm^{-1}) of the normal vibrations of —C(=O)OMe

Vibration	Region	Vibration	Region
ν_aMe	3020 ± 30	ρ'Me	1155 ± 35
ν_a'Me	2990 ± 40	νO—C	975 ± 125
ν_sMe	2920 ± 80	δC=O	715 ± 115
νC=O	1750 ± 50	γC=O	635 ± 130
δ_aMe	1460 ± 25	δ-C(=O)O	435 ± 95
δ_a'Me	1450 ± 15	δCOC	320 ± 70
δ_sMe	1435 ± 15	torsion Me	225 ± 65
νC(=O)O	1255 ± 60	torsion OMe	–
ρMe	1185 ± 35	torsion C(=O)OMe	–

R—C(=O)OMe molecules

R = H— [1–10], D— [1–3], Me— [7–15], CD_3— [9–11, 15], Et— [8, 25], $HOCH_2$— and $DOCH_2$— [16], N≡CCH_2— [17], $PhCH_2$—, $ClCH_2$— [18–20], MeOC(=O)C(=O)CH_2— [21], cPr— [22], iPr— [8, 25], Cl_2CH— [23, 24], tBu— [8], F_3C— [8, 26–28], Cl_3C— [24], MeC(=O)— [29, 30], MeOC(=O)CH_2C(=O)— [21], MeNHC(=O)— and MeNDC(=O)— [31], MeOC(=O)— [7, 32, 33], ClC(=O)— [34, 35], H_2C=CH— [8, 36–40], MeCH=CH— [38, 40], Me_2C=CH— and $(CD_3)_2$C=CH— [41], H_2C=C(Me)— [42], MeOC(=O)CH=CH— [35, 43, 44], HOC(=O)CH=CH—, DOC(=O)CH=CH— and NaOC(=O)CH=CH— [45], ClC(=O)CH=CH— [35, 45], PhCH=CH— [46], HC≡C— [47–49], MeC≡C—, MeOC(=O)C≡C— [48], N≡C— [49], Ph— [50–53], 3-MePh— [50], 4-MeO(O=)CPh— [52], 2-H_2NS$(=O)_2$Ph—, 2-, 3- and 4-HOPh— [50, 54], 3-O_2NPh— [50], 3-ClPh—, 2-Fu— [67], H_2N— [55], PhNH—, Cl_2N— [55], MeO— [56, 57], MeOC(=O)O— [58], KO— [59], F— [49, 60], Cl— [61–63].

References

1. H. Susi and T. Zell, *Spectrochim. Acta*, **19**, 1933 (1963).
2. W.C. Harris, D.A. Coe and W.O. George, *Spectrochim. Acta, Part A*, **32A**, 1 (1976).

3. M.G. Dahlqvist and K. Euranto, *Spectrochim. Acta, Part A*, **34A**, 863 (1978).
4. H. Susi and J.R. Scherer, *Spectrochim. Acta, Part A*, **25A**, 1243 (1969).
5. E.B. Marmar, C. Pouchan, A. Dargelos and M. Chaillet, *J. Mol. Struct.*, **57**, 189 (1979).
6. A. Hadni, J. Deschamps and M.L. Josien, *C.R. Acad. Sci.*, **242**, 1014 (1956).
7. P. Matzke, O. Chacon and C. Andrade, *J. Mol. Struct.*, **9**, 255 (1971).
8. R.M. Moravie and J. Corset, *J. Mol. Struct.*, **30**, 113 (1976).
9. J. Derouault, J. Le Calve and M.-T. Forel, *Spectrochim. Acta, Part A*, **28A**, 359 (1972).
10. J.K. Wilmshurst, *J. Mol. Struct.*, **1**, 201 (1957).
11. W.O. George, T.E. Houston and W.C. Harris, *Spectrochim. Acta, Part A*, **30A**, 1035 (1974).
12. H. Hollenstein and H.H. Günthard, *J. Mol. Spectrosc.*, **84**, 457 (1980).
13. R. Fausto and J.J.C. Teixeira-Dias, *J. Mol. Struct.*, **144**, 215 (1986).
14. D. Steele and A. Muller, *J. Phys. Chem.*, **95**, 6163 (1991).
15. J. Dybal and S. Krimm, *J. Mol. Struct.*, **189**, 383 (1988).
16. H. Hollenstein, R.W. Schar, N. Schwizgebel, G. Grassi and H.H. Günthard, *Spectrochim. Acta, Part A*, **39A**, 193 (1983).
17. S.W. Charles, G.I.L. Jones and N.L. Owen, *J. Chem. Soc. Faraday Trans. 2*, **69**, 1454 (1973).
18. J.E. Katon and D. Sinha, *Spectrochim. Acta, Part A*, **33A**, 45 (1977).
19. Y. Mido and M. Hashimoto, *J. Mol. Struct.*, **129**, 253 (1985).
20. R. Fausto and J.J.C. Teixeira-Dias, *J. Mol. Struct.*, **144**, 225 (1986).
21. D.W. Schiering and J.E. Katon, *Spectrochim. Acta*, **42A**, 487 (1986).
22. D.L.Powell, P.Klaboe and D.H.Christensen, *J. Mol. Struct.*, **15**, 77 (1973).
23. Y. Mido and M. Hashimoto, *J. Mol. Struct.*, **131**, 71 (1985).
24. R. Fausto and J.J.C. Teixeira-Dias, *J. Mol. Struct.*, **144**, 241 (1986).
25. R.M. Moravie and J. Corset, *J. Mol. Struct.*, **24**, 91 (1975).
26. A.G. Robiette and J.C. Thompson, *Spectrochim. Acta*, **21**, 2023 (1965).
27. G.A. Crowder and D. Jackson, *Spectrochim. Acta, Part A*, **27A**, 1873 (1971).
28. G.A. Crowder, *Spectrochim. Acta, Part A*, **28A**, 1625 (1972).
29. W.J. Ray and J.E. Katon, *Spectrochim. Acta, Part A*, **36A**, 793 (1980).
30. J.K. Wilmshurst and J.F. Horwood, *Aust. J. Chem.*, **24**, 1183 (1971).
31. H.O. Desseyn, B.J. Van der Veken and M.A. Herman, *Bull. Soc. Chim. Belg.*, **84**, 1057 (1975).
32. J.K. Wilmshurst and J.F. Horwood, *J. Mol. Spectrosc.*, **21**, 48 (1966).
33. J.R. Durig, S.C. Brown and S.E. Hannum, *J. Chem. Phys.*, **54**, 4428 (1971).
34. S.W. Charles, G.I.L. Jones, N.L. Owen and L.A. West, *J. Mol. Struct.*, **32**, 111 (1976).
35. J.E. Katon and P.H. Chu, *J. Mol. Struct.*, **78**, 141 (1982).
36. W.L. Walton and R.B. Hughes, *J. Am. Chem. Soc.*, **79**, 3985 (1957).
37. W.R. Feairheller and J.E. Katon, *J. Mol. Struct.*, **1**, 239 (1967).
38. A. Bowles, W.O. George and D. Cunliffe-Jones, *J. Chem. Soc. B*, 1070 (1970).
39. W.O. George, D.V. Hassid, W.C. Harris and W.F. Maddams, *J. Chem. Soc. Perkin Trans. 2*, 392 (1975).
40. P. Carmona and J. Moreno, *J. Mol. Struct.*, **82**, 177 (1982).
41. J.M.M. Droog and W.M.A. Smit, *Spectrochim. Acta, Part A*, **33A**, 745 (1977).
42. T.R. Manley and C.G. Martin, *Spectrochim. Acta, Part A*, **32A**, 357 (1976).
43. D.A.C. Compton, W.O. George and A.J. Porter, *J. Chem. Soc. Perkin Trans. 2*, 400 (1975).
44. C. Téllez, R. Knudsen and O. Sala, *J. Mol. Struct.*, **67**, 189 (1980).
45. J.E. Katon and P.-H. Chu, *J. Mol. Struct.*, **82**, 61 (1982).
46. M. Dulce, G. Faria, J.J.C. Teixeira-Dias and R. Fausto, *J. Raman Spectrosc.*, **22**, 519 (1991).

47. G. Williams and N.L. Owen, *Trans. Faraday Soc.*, **67**, 950 (1971).
48. J.E. Katon and T.B. BenKinney, *Spectrochim. Acta, Part A*, **39A**, 877 (1983).
49. G. Williams and N.L. Owen, *Trans. Faraday Soc.*, **67**, 950 (1971).
50. G. Varsányi, *Assignments for Vibrational Spectra of Seven Hundred Benzene Derivatives*, J.Wiley and Sons, N.Y. (1974).
51. S. Pinchas, D. Samuel and M. Weiss-Broday, *J. Chem. Soc.*, 2382 (1961).
52. F.J. Boerio and S.K. Bahl, *Spectrochim. Acta, Part A*, **32A**, 987 (1976).
53. J.H.S. Green and D.J. Harrison, *Spectrochim. Acta, Part A*, **33A**, 583 (1977).
54. M.M. Radhi and M.F. El-Bermani, *Spectrochim. Acta, Part A*, **46A**, 33 (1990).
55. J.C. Carter and J.E. Devia, *Spectrochim. Acta, Part A*, **29A**, 623 (1973).
56. B. Collingwood, H. Lee and J.K. Wilmshurst, *Aust. J. Chem.*, **19**, 1637 (1966).
57. J.E. Katon and M.D. Cohen, *Can. J. Chem.*, **53**, 1378 (1975).
58. P.L. Liang and J.E. Katon, *J. Mol. Struct.*, **172**, 113 (1988).
59. R. Mattes and K. Scholten, *Spectrochim. Acta, Part A*, **31A**, 1307 (1975).
60. J.R. Durig, T.S. Little and C.L. Tolley, *Spectrochim. Acta, Part A*, **45A**, 567 (1989).
61. R.A. Nyquist and W.J. Potts, *Spectrochim. Acta*, **17**, 679 (1961).
62. J.C. Evans and J. Overend, *Spectrochim. Acta*, **19**, 701 (1963).
63. R.A. Nyquist, *Spectrochim. Acta, Part A*, **28A**, 285 (1972).
64. W.A. Seth Paul and A. Van Duyse, *Spectrochim. Acta, Part A*, **28A**, 211 (1972).
65. R.A. Nyquist, *Appl. Spectrosc.*, **26**, 81 (1972).
66. J.E. Saunders, J.J. Lucier and F.F. Bentley, *Appl. Spectrosc.*, **22**, 697 (1968).
67. M. Sénéchal and P. Saumagne, *J. Chim. Phys.*, **69**, 1246 (1972).
68. C.J. Pouchert, *The Aldrich Library of FT-IR Spectra*, Aldrich Chemical Company, 1st edn. (1985).

7.1.9 Ethoxycarbonyl

The 18 vibrations of ethyl (see Section 3.5.1), together with νC=O, νC(=O)—O, νO—C, δC=O, γC=O, δ-C(=O)—O, δC—O—C and two torsions, make up the 27 vibrations of —C(=O)OEt.

Methyl and methylene stretching vibrations

The five ethyl CH stretching vibrations absorb weakly to moderately between 2995 and 2860 cm^{-1}. The antisymmetric stretchings are active between 2995 and 2930 cm^{-1} and the symmetric counterparts between 2930 and 2860 cm^{-1}.

C=O stretching vibration

The wide range (1750 $\pm$ 80 cm^{-1}) in which the νC=O in ethyl esters is active is due to the esters of halogenated acids such as FC(=O)OEt (1829), ClC(=O)OEt (1779), X_3CC(=O)OEt (1775 $\pm$ 15) and X_2CHC(=O)OEt (1760 $\pm$ 15 cm^{-1}) (X = F, Cl, Br). The majority of ethyl esters of α-saturated acids were found to give this νC=O in the region 1740 $\pm$ 15 cm^{-1}. This band shifts to lower frequencies in esters of α-unsaturated and aromatic acids (1715 $\pm$ 15 cm^{-1}). Compounds with an intramolecular hydrogen bridge such as 2- and 4-HOPhC(=O)OEt (1674) absorb at lower wavenumbers, just as KOC(=O)OEt (1670 cm^{-1}).

Methyl and methylene deformations

The high absorption region of the methylene scissors (1475 ± 15 cm^{-1}), comparable with that of the ethyl ethers (Section 10.1.3), is remarkable. The methyl antisymmetric deformations, which often occur at the same wavenumber (1455 ± 20 cm^{-1}), absorb at the low-frequency side of the δCH_2 region. In contrast to methyl esters, the methyl symmetric deformation occurs in the 'normal' region 1385 ± 15 cm^{-1}. In ethyl acetate this δ_sMe appears weakly at 1392 cm^{-1}. The stronger absorption at 1374 cm^{-1} is due to the δ_sMe of the MeC(=O) fragment.

Methylene wagging and twisting vibration

Ethyl esters show the methylene wag at higher wavenumbers (1360 ± 25 cm^{-1}) than the wag in hydrocarbons. This band, occurring sometimes as a shoulder on the δ_sMe absorption, has medium intensity. The CH deformation of —C(=O)H, iPr and cPr also appears in this neighbourhood. Ethyl ethers absorb in the region 1345 ± 35 cm^{-1}. In ethyl acetate the shoulder at 1361 cm^{-1} on the 1374 cm^{-1} band is assigned to this wag.

The methylene twist is weakly to moderately active in the range 1285 ± 45 cm^{-1}. The absorption at 1301 cm^{-1} in the spectrum of ethyl acetate is probably due to this CH_2 twist.

The C(=O)—O stretching vibration

The C(=O)—O stretching vibration provides a moderate to strong band in the region 1245 ± 65 cm^{-1} and is, together with the νC=O, of practical use in elucidating the ester structure. The upper limit is provided by 2-PyC(=O)OEt (1307) and F_2CHC(=O)OEt (1305 cm^{-1}) and the lower limit by ClC(=O)OEt and Me$(CH_2)_8$C(=O)OEt with 1178 cm^{-1}, followed by MeCH=CHC(=O)OEt with 1184 cm^{-1}. Ethyl formate absorbs at 1189 and ethyl acetate at 1241 cm^{-1}.

Methyl rocking vibrations and O—C/C—C stretching vibrations

Usually the highest absorption region (1165 ± 30 cm^{-1}) is assigned to the methyl rock, but the literature does not agree in assigning the other regions. The lowest region (890 ± 50 cm^{-1}) is derived from the νC—C [3, 4, 6, 9, 11, 15] or νC—O [1, 2, 16] or ρMe [10]. The question is that these vibrations are coupled in such a way as to make it difficult to determine which vibration provides the greatest part in a distinct absorption [7, 8, 13].

The highest values for the ρMe are found in the spectra of EtC(=O)OEt (1191) and PhC(=O)OEt (1176, although 1125 cm^{-1} is preferable) and the lowest in those of H_2C=C(CN)C(=O)OEt (1145) and HC≡CC(=O)OEt (1148 cm^{-1}). The remaining ethyl esters absorb at 1160 $\pm$ 15 cm^{-1}. This ρMe in ethyl acetate appears at 1160 cm^{-1}.

The compound ClC(=O)OEt displays the ρ'Me at 1142 cm^{-1}. FCH_2C(=O)OEt absorbs at 1083 cm^{-1} and ethyl acetate weakly at 1098 cm^{-1}.

The ethyl esters exhibit a moderate to strong band in the region 1060 $\pm$ 40 cm^{-1} due to a skeletal stretching vibration with a contribution from the C—O and the C—C bonds. Ethyl acetate has an absorption at 1048 cm^{-1} which is assigned to the methyl rocking vibration of the acetyl fragment and to the C—O stretching vibration. In the spectra of some RC(=O)OEt compounds (R = H, F, Cl, N≡C and HC≡C) a band with medium to strong intensity in the above-mentioned region is attributed to the C—C stretching vibration [1, 2, 16], but for most of the ethyl esters the absorption in this region is assigned to the νC—O.

The range 890 $\pm$ 50 cm^{-1} is attributed to the ethyl C—C stretching vibration which is under the influence of the νC—O. Ethyl acetate absorbs weakly at 847 cm^{-1}, a value which also may be assigned to the νC—C of the acetyl fragment. This vibration is sometimes called the ν_sCOC, and in ethyl ethers is active at 875 $\pm$ 65 cm^{-1}.

Methylene rocking vibration

The methylene rock is active in the region 800 $\pm$ 25 cm^{-1}. Ethyl acetate absorbs at 786 cm^{-1}, ethyl chloroacetate at 805 cm^{-1} and ethyl trichloroacetate at 803 cm^{-1}.

The C=O deformations

Ignoring (without rejecting) the high values of 920 cm^{-1} in the spectrum of HC(=O)OEt [1] (Section 7.1.1) and 821 cm^{-1} in that of KOC(=O)OEt, the C=O deformations group themselves in two regions: 690 $\pm$ 65 and 625 $\pm$ 75 cm^{-1}. In this work the higher wavenumber is assigned to the C=O in-plane deformation and the lower to the C=O out-of-plane deformation. Ethyl acetate absorbs near 634 cm^{-1} (δC=O) and 607 cm^{-1} (γC=O), ethyl formate at 920 cm^{-1} (γCH/C=O) and 672 cm^{-1} (δC=O).

Skeletal deformations

The —C(=O)—O rocking vibration is weakly to moderately active in the region 430 $\pm$ 55 cm^{-1}, in good agreement with that of —C(=O)OH and —C(=O)OMe. The wavenumber 579 cm^{-1} in the spectrum of KOC(=O)OEt is outside this region. Saunders *et al.* [22] found a band at 465 $\pm$ 20 cm^{-1} for ten aliphatic ethyl esters.

The skeletal O—C—C deformation occurs at 350 ± 45 cm^{-1} with a weak to moderate intensity. Saunders *et al.* gave 380 ± 15 cm^{-1} for ethyl esters. Ethyl acetate shows this vibration at 378 cm^{-1}.

The skeletal C—O—C deformation is assigned at 310 ± 60 cm^{-1}, a region comparable with that of methyl esters (320 ± 70 cm^{-1}). In RC(=O)OEt compounds (R = H, Me, Cl, CN) this skeletal deformation is assigned respectively at 325, 314, 292 and 254 cm^{-1}. For a series of aliphatic ethyl esters Saunders *et al.* [22] note an absorption at 330 ± 20 cm^{-1}, which may be correlated with the C—O—C skeletal deformation.

Table 7.9 Absorption regions (cm^{-1}) of the normal vibrations of —C(=O)OEt

Vibration	Region	Vibration	Region
ν_aMe	2985 ± 10	ρ'Me	1115 ± 35
ν'_aMe	2975 ± 15	νO—C	1060 ± 40
$\nu_a CH_2$	2945 ± 15	νC—C	890 ± 50
ν_sMe	2910 ± 20	ρCH_2	800 ± 25
$\nu_s CH_2$	2885 ± 25	δC=O	690 ± 65
νC=O	1750 ± 80	γC=O	625 ± 75
δCH_2	1475 ± 15	$\delta(\rho)$—C(=O)—O	430 ± 55
δ_aMe	1460 ± 15	δO—C—C	350 ± 45
δ'_aMe	1450 ± 15	δC—O—C	310 ± 60
δ_sMe	1385 ± 15	torsion Me	245 ± 35
ωCH_2	1360 ± 25	torsion Et	160 ± 40
τCH_2	1285 ± 45	torsion OEt	–
νC(=O)—O	1245 ± 65	torsion C(=O)OEt	–
ρMe	1165 ± 30		

R—C(=O)OEt molecules

R = H— [1, 2], Me— [3–5], CD_3— [4], Et—, nPr—, $Me(CH_2)_8$—, $ClCH_2CH_2$—, FCH_2— [3], $ClCH_2$— [3, 6, 7], $BrCH_2$— [3, 6], ICH_2—, N≡CCH_2— [8], MeC(=O)CH_2—, EtC(=O)CH_2—, EtOC(=O)CH_2—, cPr—, F_2CH—, Cl_2CH—, Br_2CH— and ClFCH— [9], MeCH(Cl)— and MeCH(Br)— [10], F_3C—, Cl_3C— [11], MeC(=O)—, EtOC(=O)—, H_2NC(=O)— [12], H_2C=CH— [13], MeCH=CH— [13], H_2C=C(CH_2Br)—, H_2C=C(CN)— [14], Me_2C=CH—, EtO(O=)CCH=CH— [15], HC≡C— [16], N≡C— [1], Ph— [17, 18], 4-EtO(O=)CPh— [18], 2-H_2NPh—, 3-O_2NPh— and 2- and 4-HOPh— [20], 2-, 3- and 4-Py—, 2-Th— [19], 2-Fu— [23], H_2N—, EtO—, KO— [21], F— [16], Cl— [1].

References

1. S.W. Charles, G.I.L. Jones, N.L. Owen, S.J. Cyvin and B.N. Cyvin, *J. Mol. Struct.*, **16**, 225 (1973).
2. M.G. Dahlqvist and K. Euranto, *Spectrochim. Acta, Part A*, **34A**, 863 (1978).

3. M.A. Raso, M.V. Garcia and J. Morcillo, *J. Mol. Struct.*, **115**, 449 (1984).
4. Y. Mido, H. Shiomi, H. Matsuura, M.A. Raso, M.V. Garcia and J. Morcillo, *J. Mol. Struct.*, **176**, 253 (1988).
5. T.-K. Ha, C. Pal and P.N. Ghosh, *Spectrochim. Acta, Part A*, **48A**, 1083 (1992).
6. N. Dube, P. Porwal and P. Prasad, *J. Raman Spectrosc.*, **19**, 189 (1988).
7. Y. Mido, N. Kakizawa, H. Matsuura, M.A. Raso, M.V. Garcia and J. Morcillo, *J. Mol. Struct.*, **220**, 169 (1990).
8. S.W. Charles, G.I.L. Jones and N.L. Owen, *J. Chem. Soc. Faraday Trans. 2*, **69**, 1454 (1973).
9. M.A. Raso, M.V. Garcia and J. Morcillo, *J. Mol. Struct.*, **142**, 41 (1986).
10. N. Dube and R. Prasad, *Spectrochim. Acta, Part A*, **43A**, 83 (1987).
11. Y. Mido, T. Kawashita, K. Suzuki, J. Morcillo and M.V. Garcia, *J. Mol. Struct.*, **162**, 169 (1987).
12. H.O. Desseyn, F.K. Vansant and B.J. Van der Veken, *Spectrochim. Acta, Part A*, **31A**, 625 (1975).
13. A.J. Bowles, W.O. George and D.B. Cunliffe-Jones, *J. Chem. Soc. B*, 1070 (1970).
14. S. Reynolds, D.P. Oxley and R.G. Pritchard, *Spectrochim. Acta, Part A*, **38A**, 103 (1982).
15. D.A.C. Compton, W.O. George and A.J. Porter, *J. Chem. Soc. Perkin Trans. 2*, 400 (1975).
16. S.W. Charles, G.I.L. Jones, N.L. Owen and L.A. West, *J. Mol. Struct.*, **26**, 249 (1975).
17. J.H.S. Green and D.J. Harrison, *Spectrochim. Acta, Part A*, **33A**, 583 (1977).
18. F.J. Boerio and S.K. Bahl, *Spectrochim. Acta, Part A*, **32A**, 987 (1976).
19. J.J. Peron, P. Saumagne and J.M. Lebas, *Spectrochim. Acta, Part A*, **26A**, 1651 (1970).
20. G. Varsányi, *Assignments for Vibrational Spectra of Seven Hundred Benzene Derivatives*, J.Wiley & Sons, New York (1974).
21. R. Mattes and K. Scholten, *Spectrochim. Acta, Part A*, **31A**, 1307 (1975).
22. J.E. Saunders, J.J. Lucier and F.F. Bentley, *Appl. Spectrosc.*, **22**, 697 (1968).
23. M. Sénéchal and P. Saumagne, *J. Chim. Phys.*, **69**, 1246 (1972).

7.2 AMINO(THIO)CARBONYL COMPOUNDS

$HC(=X)NH_2$ (X = O or S, formamide or thioformamide) is the simplest molecule with a $-C(=X)NH_2$ structure unit. The compound gives $3N - 6 = 12$ normal vibrations (9a′ + 3a″) of which three (2a′ + a″) are considered for the external CH vibrations by name νCH (a′), δCH (a′) and ωCH (a″) (see Section 7.1.1). The remaining nine are derived from the $-C(=X)NH_2$ fragment. A compound such as $MeC(=X)NH_2$ has $3N - 6 = 21$ normal vibrations of which nine are methyl vibrations. If one torsion is replaced by a C—C stretching vibration, the remaining twelve vibrations are due to the $-C(=X)NH_2$ group. To the nine normal vibrations of formamide or thioformamide are added two skeletal deformations and one torsion:

a′: $\nu_a NH_2$, $\nu_s NH_2$, νC=X, δNH_2, νC—N, ρNH_2, δC=X, δ—C—N;
a″: τNH_2, ωNH_2, γC=X and torsion.

Some of the above-mentioned normal vibrations are mixed in such a way as to make them difficult to describe. The skeletal (C=X) deformation, an in-plane deformation in essence, is found in the literature as δCX, δCCX, δNCX or ρNCX. The NH_2 wag is also named NH_2/CX wag, πCX or πNCX. In addition, the above-mentioned vibrations are difficult to distinguish from the external R—C(=X) deformations, the wavenumbers of which depend on the —C(=X)NH_2 fragment as well as on the R-substituent. Sometimes these mixed vibrations are described in terms of amide vibrations [21]:

amide I:	νC=X	amide V:	ωNH_2/CX
amide II:	δNH_2	amide VI:	γC=X
amide III:	νC—N	amide VII:	τNH_2/CX
amide IV:	δC=X		

In the solid or liquid state the NH_2 and C=O vibrations are associated to some degree and give rise to broad bands.

7.2.1 Carbamoyl (aminocarbonyl)

Formamide possesses twelve normal vibrations (9a′ + 3a″) of which nine are assigned to —C(=O)NH_2:

$\nu_a NH_2$	a′	3330	ρNH_2	a′	1090
$\nu_s NH_2$	a′	3200	τNH_2	a″	750
νC=O	a′	1687	ωNH_2	a″	657
δNH_2	a′	1611	δC=O	a′	608
νC—N	a′	1309			

and three to the —C(=O)H fragment (Section 7.1.1):

νCH	a′	2882	instead of a torsion
δCH	a′	1391	instead of in-plane deformation
ωCH/C=O	a″	1050	instead of γC=O

The NH_2 stretching vibrations

The NH_2 antisymmetric stretching vibration in the associated state gives rise to a strong broadish band in the region 3390 ± 60 cm^{-1}. High values originate from the spectra of MeHNC(=O)NH_2 (3450), H_2NC(=O)NH_2 (3441), MeOC(=O)NH_2 (3434) and NaOC(=O)C(=O)NH_2 (3420 cm^{-1}) and low values from those of HC(=O)NH_2 and DC(=O)NH_2 (3330 cm^{-1}). Most primary amides show the $\nu_a NH_2$ bond at 3370 ± 30 cm^{-1}. In dilute solutions the free $\nu_a NH_2$ appears sharply and strongly at 3530 ± 30 cm^{-1}.

The NH_2 symmetric stretch in the associated state makes its appearance in the region 3210 ± 60 cm^{-1} with a somewhat weaker intensity than that of its antisymmetric counterpart. Compounds absorbing at the HW side of this region are the α-halogen-substituted amides (3250 ± 10), MeOC(=O)NH_2 (3262) and

N≡CC(=O)NH_2 (3260 cm^{-1}) and at the LW side MeSC(=O)NH_2 (3150) and H_2NC(=O)C(=O)NH_2 (3153 and 3185 cm^{-1}). For most of the primary amides this region can be reduced to 3190 ± 30 cm^{-1}. In dilute solution the free $\nu_s NH_2$ occurs at 3420 ± 20 cm^{-1}.

The intensity of both NH_2 stretching vibrations is greater than that of the aliphatic amines and reaches that of the benzenamines. Overtones of the amide II absorption intensified by Fermi resonance may accompany these bands. If both hydrogen atoms are unbonded or equally bonded, $\nu_a = 0.89\ \nu_s + 484$ [56]. In dilute solutions the difference in wavenumber between the to NH_2 stretching vibrations approaches 115 cm^{-1}.

The C=O stretching vibration

The C=O stretching vibration (amide I) in the associated state appears strongly at 1680 ± 40 cm^{-1}. The highest values are observed in the spectra of amides in which the α-carbon is halogenated, such as Cl_2(CN)CC(=O)NH_2 (1713), F_3CC(=O)NH_2 (1710) and Cl_3CC(=O)NH_2 (1695 cm^{-1}). Low wavenumbers are found in the spectra of MeSC(=O)NH_2 (1640) and in those of aromatic amides with, as extremes, 2-ClPhC(=O)NH_2 (1642) and 3-ThC(=O)NH_2 (1647 cm^{-1}). The majority of the investigated molecules were found to give this νC=O in the region 1670 ± 20 cm^{-1}. The absorption occurs in the high-frequency wing of the amide II band and is sometimes partly merged with it. In dilute solutions the amide I band is shifted to a 40 cm^{-1} higher wavenumber and that of the amide II band to a 20 cm^{-1} lower wavenumber so that the bands are clearly separated from each other. If in solution two absorptions for the amide I band are observed, that with the higher {lower} wavenumber is assigned to the free {associated} C=O stretching vibration. Rotational isomers also show more than one C=O band.

The NH_2 deformation

The NH_2 deformation (amide II) appears in the region 1610 ± 30 cm^{-1}. The highest values are furnished by F_3CC(=O)NH_2 and MeC(=O)NHC(=O)NH_2 (1640 cm^{-1}) and the lowest by KOC(=O)C(=O)NH_2 (1580), NaOC(=O)C(=O)NH_2 (1584), MeHNC(=O)NH_2 (1585) and MeHNC(=S)C(=O)NH_2 (1585 cm^{-1}). Most primary amides display the amide II band at 1610 ± 20 cm^{-1}. In the solid or liquid state this band appears sometimes as a shoulder with a moderate to strong intensity on the νC=O absorption. In dilute solutions the NH_2 scissors absorbs with a 30–50% lower intensity than that of the C=O stretch, and is well separated from it.

The C—N stretching vibration

The C—N stretching vibration (amide III) is only weakly to moderately active in the region 1385 ± 85 cm^{-1} and is therefore difficult to detect. Formamide gives this amide III band at 1309 cm^{-1}, acetamide at 1398 cm^{-1}, propanamide and butanamide at about 1420 cm^{-1}. Urea and $F_3CC(=O)NH_2$ score high (1465 cm^{-1}). Most absorptions come within the region 1390 ± 40 cm^{-1}, in which also the methyl or methylene deformations are active. This vibration is coupled to the R—C(=O) stretch and in some degree also to the NH_2 deformation. In dilute solutions a small shift to higher wavenumbers occurs.

The NH_2 rocking, twisting and wagging vibration

The in-plane NH_2 rock absorbs weakly to moderately in the region 1125 ± 45 cm^{-1}. In the spectrum of formamide this vibration is assigned at 1090 cm^{-1} in the liquid state and at 1150 cm^{-1} in the vapour phase. The NH_2 rock in acetamide, propanamide and butanamide occurs respectively at 1152, 1142 and 1144 cm^{-1}. High values are found in the spectra of $ICH_2C(=O)NH_2$ (1170 a′; τCH_2: 1150 cm^{-1} a″) and $MeHNC(=O)NH_2$ (1169 cm^{-1}). Low values in the neighbourhood of 1080 cm^{-1} are assigned in the spectra of $HOC(=O)C(=O)NH_2$, $KOC(=O)C(=O)NH_2$ and $R_2NC(=O)C(=O)NH_2$. Most rocks are situated at 1130 ± 30 cm^{-1}.

The amide VII band, absorbing in the region 765 ± 55 cm^{-1}, is mainly assigned to the NH_2 out-of-plane twist and is also under the influence of the C=O out-of-plane deformation. High wavenumbers in the neighbourhood of 815 ± 5 cm^{-1} are assigned in the spectra of $HOC(=O)C(=O)NH_2$, $EtOC(=O)C(=O)NH_2$, $MeC(=O)NHC(=O)NH_2$ and $XCH_2C(=O)NH_2$ (X = F, Cl, I). The weak broad band at 810 cm^{-1} in the spectrum of acetamide is attributed to this twist. To the LW side of the above-mentioned region urea absorbs at 717 and *789* cm^{-1}. Most of the primary amides give an often broadish band with a weak to moderate intensity in the range 770 ± 30 cm^{-1}. In the vapour state the free torsion absorbs in the neighbourhood of 260 cm^{-1}.

The amide V band (670 ± 60 cm^{-1}) is a mixed vibration assigned to the NH_2 wag (ωNH_2 or πNH_2) with a contribution from the C=O out-of-plane deformation. High values are seen in the spectra of $Me_2NC(=O)NH_2$ (725), $(CD_3)_2NC(=O)NH_2$ (720), $MeHNC(=O)C(=O)NH_2$ (724), $MeC(=O)NH_2$ (715), $MeHNC(=O)NH_2$ (712) and $CD_3HNC(=O)NH_2$ (710 cm^{-1}) and low values in those of $N\equiv CC(=O)NH_2$ and $AzC(=O)NH_2$ (620 cm^{-1}). The NH_2 wag, usually absorbing in the region 665 ± 35 cm^{-1} and clearly separated from the twist, is easy to recognize by its broad band structure.

The C=O deformations

Primary amides give the C=O in-plane deformation (amide IV) in the region 610 ± 70 cm^{-1} with a weak to moderate intensity. $MeHNC(=O)C(=O)NH_2$, $R_2NC(=O)C(=O)NH_2$, $EtOC(=O)C(=O)NH_2$, $MeOC(=O)NH_2$ and $Me(CH_2)_9NHC(=O)NH_2$ absorb at about 675 cm^{-1} and $MOC(=O)C(=O)NH_2$

(M = H, K, Na), $H_2NC(=S)NHC(=O)NH_2$ and $N_3C(=O)NH_2$ in the vicinity of 550 cm^{-1}. For the remaining $R-C(=O)NH_2$ compounds the $\delta C{=}O$ is situated at 610 ± 50 cm^{-1}, at the LW side of the amide V absorption.

The weak to medium and sometimes broadish band in the region 560 ± 70 cm^{-1} is assigned to the C=O out-of-plane deformation ($\gamma C{=}O$, $\pi C{=}O$ or amide VI) with a contribution from the NH_2 wag. For $Me_2NC(=O)C(=O)NH_2$ this absorption is observed at 626 cm^{-1}, for $H_2NC(=O)C(=O)NH_2$ at 617 and *526* cm^{-1} (R) and for $NaOC(=O)C(=O)NH_2$ and $KOC(=O)C(=O)NH_2$ at about 490 cm^{-1}. Most of the $\gamma C{=}O$ deformations are active at 550 ± 50 cm^{-1}. The out-of-plane and in-plane C=O deformation can absorb at the same wavenumber, as for is the case in acetamide (584 cm^{-1}). Spectra–structure correlations of aliphatic amides in the 700–250 cm^{-1} region are given by Katon *et al.* [57].

In-plane skeletal deformation

The in-plane skeletal deformation is a mixed vibration described as the external —C—N deformation or ρ—C(=O)—N, comparable with the —C(=O)—O rocking vibration in carboxylic acids or esters. For $XC(=O)C(=O)NH_2$ (X = H_2N, RNH, R_2N, HO, KO, NaO) and $MeHNC(=S)C(=O)NH_2$ this skeletal deformation is observed at 475 ± 25 cm^{-1} and for acetamide, propanamide and butanamide at respectively 466, 479 and 468 cm^{-1}.

Torsion

The $-C(=O)NH_2$ torsion in acetamide occurs at 155 cm^{-1}, and in a few other $R-C(=O)NH_2$ compounds at 145 ± 55 cm^{-1}. In the vapour phase the free NH_2 torsion absorbs at about 260 cm^{-1}.

Table 7.10 Absorption region (cm^{-1}) of the normal vibrations of $-C(=O)NH_2$

Vibration	α—saturated	C=O bound C=S bound	α—unsaturated aromatic	N, O or S bound
$\nu_a NH_2$	3370 ± 40	3380 ± 40	3370 ± 30	3400 ± 50
$\nu_s NH_2$	3205 ± 45	3205 ± 55	3180 ± 30	3210 ± 60
$\nu C{=}O$ (I)	1680 ± 35	1675 ± 25	1660 ± 15	1670 ± 30
δNH_2 (II)	1615 ± 25	1595 ± 15	1610 ± 10	1610 ± 30
$\nu C{-}N$ (III)	1380 ± 80	1395 ± 55	1385 ± 40	1400 ± 70
ρNH_2	1130 ± 40	1100 ± 20	1115 ± 25	1130 ± 40
τNH_2 (VII)	765 ± 45	775 ± 45	780 ± 40	765 ± 55
ωNH_2 (V)	670 ± 50	685 ± 45	660 ± 50	675 ± 55
$\delta C{=}O$ (IV)	600 ± 50	610 ± 70	605 ± 35	610 ± 70
$\gamma C{=}O$	550 ± 50	560 ± 70	535 ± 35	550 ± 40
δ—C—N	430 ± 50	475 ± 25	–	475 ± 55
torsion	130 ± 40	150 ± 50	–	160 ± 40

$R—C(=O)NH_2$ molecules

R = H— [1–10], D— [1], Me— [9–14], CD_3— [11, 13, 14], Et— [15, 16], CH_3CD_2— [15], nPr—, $H_2NC(=O)CH_2CH_2$—, $N{\equiv}CCH_2$— [17, 18], $PhCH_2$— [55], FCH_2— [19–24], $ClCH_2$— [17, 19, 20, 24], ICH_2— [20–23], iPr—, cPr— [25], Br(CN)CH— [18], F_2CH— and Cl_2CH— [24], tBu—, $Cl_2(CN)C$— and $Br_2(CN)C$— [18], F_3C— [21, 23, 24, 26], Cl_3C— [24], 4-ClPhC$(Me)_2$—, $H_2NC(=O)$— [27–29, 40], MeHNC(=O)— and $Me_2NC(=O)$— [27], RNHC(=O)— and $R_2NC(=O)$— [17], HOC(=O)— [17, 28–30], EtOC(=O)— [17], NaOC(=O)— [17, 28, 31], KOC(=O)— [28–31], $H_2NC(=S)$— [32], MeHNC(=S)— and $CD_3HNC(=S)$— [33], $H_2C{=}CH$— [9], Ph— [34–37, 55], 2-FPh—, 2-ClPh—, 2- and 3-Fu— [38], 3-Th— [38], pyrazine— [39], $N{\equiv}C$— [17], H_2N— [41–45, 59], HDN— [41], MeNH— and CD_3NH— [46], $Me(CH_2)_9NH$— [47], MeC(=O)NH— and $CD_3C(=O)NH$— [48], $H_2NC(=O)NH$— [58], $H_2NC(=S)NH$— [49, 58], Me_2N— and $(CD_3)_2N$— [50], Az— [51], N_3— [52], O_2N—, MeO— [53], MeS— and CD_3S— [54].

References

1. I. Suzuki, *Bull. Chem. Soc. Jpn.*, **33**, 1359 (1960).
2. W.J. Orville-Thomas, *J. Mol. Struct.*, **1**, 357 (1967).
3. K. Itoh and T. Shimanouchi, *J. Mol. Spectrosc.*, **42**, 86 (1972).
4. S.T. King, *J. Phys. Chem.*, **75**, 405 (1971).
5. D.J. Gardiner, A.J. Lees and B.P. Straughan, *J. Mol. Struct.*, **53**, 15 (1979).
6. M. Räsänen, *J. Mol. Struct.*, **101**, 275 (1983).
7. M. Räsänen, *J. Mol. Struct.*, **102**, 235 (1983).
8. N. Østergard, P.L. Christiansen and O.F. Nielsen, *J. Mol. Struct.*,**235**, 423 (1991).
9. N. Jonathan, *J. Mol. Spectrosc.*, **6**, 205 (1961).
10. A. Balázs, *Acta Chim. Acad. Sci. Hung.*, **108**, 265 (1981).
11. T. Uno, K. Machida and Y. Saito, *Bull. Chem. Soc. Jpn.*, **42**, 900 (1969).
12. S.T. King, *Spectroschim. Acta, Part A*, **28A**, 165 (1972).
13. T. Uno, K. Machida and Y. Saito, *Spectroschim. Acta, Part A*, **27A**, 833 (1971).
14. I. Suzuki, *Bull. Chem. Soc. Jpn.*, **35**, 1279 (1962).
15. Y. Kuroda, Y. Saito, K. Machida and T. Uno, *Bull. Chem. Soc. Jpn.*, **45**, 2371 (1972).
16. P.U. Bai and K. Venkata Ramiah, *Indian J. Pure Appl. Phys.*, **12**, 143 (1974).
17. H.O. Desseyn, F.K. Vansant and B.J. Van der Veken, *Spectroschim. Acta, Part A*, **31A**, 625 (1975).
18. L. Van Haverbeke and M.A. Herman, *Spectroschim. Acta, Part A*, **31A**, 959 (1975).
19. S. Samdal and R. Seip, *J. Mol. Struct.*, **52**, 195 (1979).
20. F.M. Abid, J.M.L. Katti and M.F. El-Bermani, *J. Mol. Struct.*, **67**, 169 (1980).
21. E.K. Murthy and G.R. Rao, *J. Raman Spectrosc.*, **19**, 359 (1988).
22. E.K. Murthy and G.R. Rao, *J. Raman Spectrosc.*, **19**, 419 (1988).
23. E.K. Murthy and G.R. Rao, *J. Raman Spectrosc.*, **20**, 409 (1989).
24. D. Troitino, E. Sanchez de la Blanca and M.V. Garcia, *Spectroschim. Acta, Part A*, **46A**, 1281 (1990).

25. D.L. Powell and P. Klaboe, *J. Mol. Struct.*, **15**, 217 (1973).
26. E.K. Murthy and G.R. Rao, *J. Raman Spectrosc.*, **19**, 439 (1988).
27. H.O. Desseyn, B.J. Van der Veken and M.A. Herman, *Spectroschim. Acta, Part A*, **33A**, 633 (1977).
28. F. Wallace and E. Wagner, *Spectroschim. Acta*, **34A**, 589 (1978).
29. G.N.R. Tripathi and J.E. Katon, *J. Mol. Struct.*, **54**, 19 (1979).
30. G.N.R. Tripathi and J.E. Katon, *Spectroschim. Acta, Part A*, **35A**, 401 (1979).
31. H.O. Desseyn, B.J. Van der Veken, A.J. Aarts and M.A. Herman, *J. Mol. Struct.*, **63**, 13 (1980).
32. H.O. Desseyn and M.A. Herman, *Spectroschim. Acta, Part A*, **23A**, 2457 (1967).
33. H.O. Desseyn, A.J. Aarts and M.A. Herman, *Spectroschim. Acta, Part A*, **36A**, 59 (1980).
34. S. Weckherlin and W. Lüttke, *Z. Elektrochem.*, **64**, 1228 (1960).
35. R.N. Kniseley, V.A. Fassel, E.L. Farqukar and L.S. Gray, *Spectroschim. Acta*, **18**, 1217 (1962).
36. S. Yoshida, *Chem. Pharm. Bull.*, **11**, 628 (1963).
37. J.H.S. Green and D.J. Harrison, *Spectroschim. Acta, Part A*, **33A**, 583 (1977).
38. G. Alberghina, S. Fisichalla and S. Occhipinti, *Spectroschim. Acta, Part A*, **36A**, 349 (1980).
39. M.J.M. Delgado, F. Marquez, H.I. Suero and J.I. Marcos, *Spectrosc. Lett.*, **21**, 841 (1988).
40. T.A. Scott Jr. and E.L. Wagner, *J. Chem. Phys.*, **30**, 465 (1959).
41. A. Yamaguchi, *J. Chem. Soc. Jpn.*, **78**, 1467 (1957).
42. J.E. Stewart, *J. Chem. Phys.*, **26**, 248 (1957).
43. J. Arenas and R. Parellada, *J. Mol. Struct.*, **10**, 253 (1971).
44. D. Hadzi, J. Kidric, Z.V. Knezevic and B. Barlic, *Spectroschim. Acta, Part A*, **32A**, 693 (1976).
45. W. Kutzelnigg and R. Mecke, *Z. Elektrochem.*, **65**, 109 (1961).
46. Y. Saito, K. Machida and T. Uno, *Spectroschim. Acta, Part A*, **31A**, 1237 (1975).
47. Y. Mido, S. Kimura, Y. Sugano and K. Machida, *Spectroschim. Acta, Part A*, **44A**, 661 (1988).
48. Y. Saito and K. Machida, *Spectroschim. Acta, Part A*, **35A**, 369 (1979).
49. K. Geetharani and D.N. Sathyanarayana, *Spectroschim. Acta, Part A*, **32A**, 227 (1976).
50. Y. Mido, K. Tanase and K. Kido, *Spectroschim. Acta, Part A*, **45A**, 397 (1989).
51. H.L. Spell and J. Laane, *J. Mol. Struct.*, **14**, 39 (1972).
52. W. Buder and A. Schmidt, *Spectroschim. Acta, Part A*, **29A**, 1429 (1973).
53. J.C. Carter and J.E. Devia, *Spectroschim. Acta, Part A*, **29A**, 623 (1973).
54. L. Zhengyan, R. Mattes, H. Schnöckel, M. Thünemann, E. Hunting, U. Hönke and C. Mendel, *J. Mol. Struct.*, **117**, 117 (1984).
55. G. Varsányi, *Assignments for Vibrational Spectra of Seven Hundred Benzene Derivatives*, J.Wiley & Sons, New York (1974).
56. P.G. Puranik and K.W. Ramiah, *Nature*, **191**, 796 (1961).
57. J.E. Katon, W.R. Feairheller Jr. and J.V. Pustinger Jr., *Anal. Chem.*, **36**, 2126 (1964).
58. R.H. Sullivan, J.S. Kwiatkowski and J. Leszczyński, *J. Mol. Struct.*, **295**, 169 (1993).
59. A. Vijav and D.N. Sathyanarayana, *J. Mol. Struct.*, **295**, 245 (1993).

7.2.2 Thiocarbamoyl (aminothiocarbonyl)

Thioformamide provides 12 normal vibrations (9a′ + 3a″). Nine find a place in the table with —C(=S)NH_2 vibrations:

$\nu_a NH_2$	a′	3287	$\nu C{=}S$	a′	843
$\nu_s NH_2$	a′	3165	τNH_2	a″	673
δNH_2	a′	1612	ωNH_2	a″	620
$\nu C{-}N$	a′	1443	$\delta C{=}S$	a′	439
ρNH_2	a′	1125			

and three belong to the —C(=O)H vibrations:

νCH	a′	2905	instead of a torsion
δCH	a′	1325	instead of in-plane deformation
$\omega CH/C{=}S$	a″	985	instead of γ C=S

The C=S vibrations in primary thioamides are mixed in such a way to make them difficult to describe. The literature is not unanimous in assigning these vibrations.

The NH_2 stretching vibrations

The absorption regions of the NH_2 stretching vibrations (ν_a: 3340 ± 60; ν_s: 3160 ± 80 cm^{-1}) are situated ≈50 cm^{-1} lower than those of the corresponding amides. The intensity of these bands is fairly strong. Thioacetamide absorbs at the LW side of these regions: 3295 and 3080 cm^{-1}. An overtone of the NH_2 deformation intensified by Fermni resonance can accompany these normal vibrations. In solution both stretching vibrations move to wavenumbers higher by ≈100 cm^{-1}.

The NH_2 deformation

The NH_2 scissors or thioamide II vibration is observed at 1620 ± 30 cm^{-1} with a moderate to strong intensity. In the spectra of thioacetamide and benzenethioamide, the δNH_2 occurs at respectively 1648 and 1623 cm^{-1}, and in those of R—C(=X)C(=S)NH_2 and R—XC(=S)NH_2 compounds (X = O or S) in the narrow region 1600 ± 10 cm^{-1}.

The C—N stretching vibration

With the exception of the dithiocarbamate ion $H_2NC(=S)S^-$ (1325 cm^{-1}), the νC—N or thioamide III or NCS I band is found in the region 1420 ± 60 cm^{-1} with a moderate intensity. High values (1480 cm^{-1}) are listed for $H_2NHNC(=S)NH_2$ and $MeHNC(=S)NH_2$ and low values for $CD_3C(=S)NH_2$ (1360) and $MeSC(=S)NH_2$ (1365 cm^{-1}). The remaining compounds give this νC—N at 1435 ± 35 cm^{-1}, that is, somewhat higher than for amides.

The NH_2 rocking vibration

The NH_2 rock or NCS II band is variable in the range 1195 ± 110 cm^{-1}. The highest wavenumbers are assigned in the spectra of thioacetamide-d_0 and -d_3 (1303), $MeOC(=S)NH_2$ (1303), $Me_2NC(=S)C(=S)NH_2$ (1286), $H_2NC(=S)NHC(=S)NH_2$ (1280 and *1120*), $H_2NC(=O)NHC(=S)NH_2$ (1285 cm^{-1}) and the lowest in those of thiourea (1084 and 1114) and thioformamide (1125 cm^{-1}). The remaining compounds give the NH_2 rock at 1210 ± 70 cm^{-1}. This region is broader and at higher wavenumbers than that of the corresponding amides (1125 ± 45 cm^{-1}).

The C=S stretching vibration

The C=S stretching vibration (thioamide I or NCS III) is assigned in the broad region 800 ± 130 cm^{-1}, usually with a moderate intensity. Since the S atom is less electronegative than the O atom, the C=S group is not as polar as the C=O group and less prone to form bridges. The νC=S absorption is sharper and not as intense as the νC=O absorption. The broad region results from the fact that this absorption is always mixed, except for thioformamide, for which the band at 843 cm^{-1} is due to the almost pure νC=S. The low wavenumber of the C=S stretching vibration is attributed to the greater contribution of the resonance form (II) [4, 16]:

$$>N-C=S \quad \leftrightarrow \quad >{}^{+}N=C-S^{-}$$

(I) (II)

The C—N bond on the contrary is fortified so that the νC—N and the ρNH_2 that is coupled to it absorb at higher wavenumbers. The C=S stretching vibration is not only responsible for the absorption in the neighbourhood of 780 cm^{-1} but can also contribute to the band at about 980 cm^{-1} (νC—C), 1200 cm^{-1} (ρNH_2) and even to the νC—N around 1420 cm^{-1}, so that all these bands can be associated with the νC=S. In this text the C=S stretching vibration is considered to be located at 780 ± 80 cm^{-1}. The rather strong band at 719 cm^{-1} in the spectrum of thioacetamide is assigned to the νC=S. The high values for the compounds $Me_2NC(=S)C(=S)NH_2$ (929), $MeHNC(=O)C(=S)NH_2$ (924), $KOC(=O)C(=S)NH_2$ (918) and $H_2NC(=O)C(=S)NH_2$ (911 cm^{-1}) [4] are not taken into account, but for the last two compounds the values 802 and 784 cm^{-1} are not excluded.

The NH_2 twisting and wagging vibrations

The NH_2 twist (thioamide VII) is an out-of-plane deformation and gives a band that is weak to moderate but often broad in the region 705 ± 65 cm^{-1}. The weak broad absorption at 760 cm^{-1} in the spectrum of thioacetamide is attributed to this vibration. High wavenumbers originate from the spectra of urea (769 and 729 cm^{-1}) and $CD_3SC(=S)NH_2$ (765 cm^{-1}), and low values from those of $N{\equiv}CC(=S)NH_2$ (648) and thioformamide (673 cm^{-1}). There is a realistic chance that the NH_2 twist occurs at 720 ± 30 cm^{-1}, approximately 50 cm^{-1} lower than the twist in primary amides.

The NH_2 wagging vibration (thioamide V), also an out-of-plane deformation, is weakly to moderately (but broadly) active in the region 645 ± 65 cm^{-1} and penetrates far into the region of the NH_2 twist. Thioacetamide-d_0 and -d_3 (709), $N{\equiv}CCH_2C(=S)NH_2$ (696) and $EtOC(=O)C(=S)NH_2$ (686 cm^{-1}) give the highest values in this region, and $MeSC(=S)NH_2$ (588) and $MeHNC(=S)NH_2$ (600 cm^{-1}, although 637 is considered as well) are responsible for the lowest wavenumbers. The remaining thioamides display the NH_2 wag at 640 ± 30 cm^{-1}, about 25 cm^{-1} lower than the NH_2 wag in primary amides.

The C=S deformations

The weak to moderate absorption in the region 510 ± 90 cm^{-1} is usually assigned to the in-plane C=S deformation (thioamide IV), coupled to the in-plane skeletal deformation and often described as δCS or δNCS. Under the influence of the heavier S atom, the δC=S absorbs at ≈100 cm^{-1} lower wavenumbers than the δC=O. In the spectrum of thioacetamide the C=S in-plane deformation is assigned at 471 cm^{-1}.

Primary thioamides give the C=S out-of-plane deformation in the range 420 ± 100 cm^{-1}, often described as the πNCS. At the HW side of this region thioacetamide absorbs (514 cm^{-1}) and at the LW side $MeSC(=S)NH_2$ (320 cm^{-1}). This region is considerably lower than that of the γC=O of amides.

In-plane skeletal deformation

The in-plane skeletal deformation is weakly to moderately active in the region 325 ± 85 cm^{-1}. This mixed vibration is described as the external —C—N deformation or ρ—C(=S)N and is under the influence of δC=S. For thioacetamide this vibration is assigned at 377 cm^{-1}.

Torsion

The torsion in thioacetamide is located at 162 cm^{-1} and that in $MeSC(=S)NH_2$ at 136 cm^{-1}. A few torsions are observed in the neighbourhood of 120 cm^{-1}.

Table 7.11 Absorption region (cm^{-1}) of the normal vibrations of —C(=S)NH_2

Vibration	Region	Vibration	Region
$\nu_a NH_2$	3340 ± 60	τNH_2 (VII)	705 ± 65
$\nu_s NH_2$	3160 ± 80	ωNH_2 (V)	645 ± 65
δNH_2 (II)	1620 ± 30	δC=S (IV)	510 ± 90
νC—N (III)	1420 ± 60	γC=S (VI)	420 ± 100
ρNH_2	1195 ± 110	δ—C—N	325 ± 85
νC=S (I)	780 ± 80	torsion	≈120

R—C(=S)NH_2 molecules

R = H— [1–4, 40], Me— [3–12], CD_3— [8–10], N≡CCH_2— [3, 13], PhCH_2— [24], CF_3— [14], KOC(=O)— [4, 15], EtOC(=O)— [4], H_2NC(=O)— [4, 16], MeHNC(=O)— and Me_2NC(=O)— [4], H_2NC(=S)— [3, 4, 17–19], MeHNC(=S)— [4, 20], CD_3HNC(=S)— [20], Me_2NC(=S)— [4], N≡C— [3, 4, 21, 22], Ph— [23, 24], H_2N— [3, 25–27], MeHN— [3, 28, 29], EtHN—, Me$(CH_2)_{15}$HN— [29], MeC(=O)HN—, PhHN—, H_2NHN— [3, 30–33], H_2NC(=O)HN— [34, 40], H_2NC(=S)HN— [35, 36, 40], MeO— and CD_3O— [37], MeS— [3, 38], CD_3S— [38], H_2NC(=S)S—, $^-$S— [3, 39].

References

1. M. Davies and W.J. Jones, *J. Chem. Soc.*, 955 (1958).
2. I. Suzuki, *Bull. Chem. Soc. Jpn.*, **35**, 1286 (1962).
3. K.R.G. Devi, D.N. Sathyanarayana and S. Manogaran, *Spectroschim. Acta, Part A*, **37A**, 31 (1981).
4. H.O. Desseyn, B.J. Van der Veken and M.A. Herman, *Appl. Spectrosc.*, **32**, 101 (1978).
5. I. Suzuki, *Bull. Chem. Soc. Jpn.*, **35**, 1449 (1962).
6. W. Walter and H.P. Kubersky, *Liebigs Ann. Chem.*, **694**, 56 (1966).
7. K.A. Jensen and P.H. Nielsen, *Acta Chem. Scand.*, **20**, 597 (1966).
8. A. Ray and D.N. Sathyanarayana, *Bull. Chem. Soc. Jpn.*, **47**, 729 (1974).
9. W. Walter and P. Stäglich, *Spectroschim. Acta, Part A*, **30A**, 1739 (1974).
10. M. Hargittai, S. Samdal and R. Seip, *J. Mol. Struct.*, **71**, 147 (1981).
11. U. Anthoni, P.H. Nielsen and O.F. Nielsen, *J. Mol. Struct.*, **116**, 175 (1984).
12. U. Anthoni, P.H. Nielsen and D.H. Christensen, *Spectroschim. Acta, Part A*, **41A**, 1327 (1985).
13. A. Ray and D.N. Sathyanarayana, *Bull. Chem. Soc. Jpn.*, **46**, 1969 (1973).
14. E. Linder and U. Kunze, *Z. Anorg. Allg. Chem.*, **383**, 255 (1971).
15. H.O. Desseyn, B.J. Van der Veken, A.J. Aarts and M.A. Herman, *J. Mol. Struct.*, **63**, 13 (1980).
16. H.O. Desseyn and M.A. Herman, *Spectroschim. Acta, Part A*, **23A**, 2457 (1967).
17. T.A. Scott Jr. and E.L. Wagner, *J. Chem. Phys.*, **30**, 465 (1959).
18. A. Ray and D.N. Sathyanarayana, *India J. Chem.*, **12**, 1092 (1974).
19. H.O. Desseyn, B.J. Van der Veken and A.J. Aarts, *Can. J. Spectrosc.*, **22**, 84 (1977).
20. H.O. Desseyn, A.J. Aarts and M.A. Herman, *Spectroschim. Acta, Part A*, **36A**, 59 (1980).

21. V.F. Duckworth and R.L. Werner, *Aust. J. Chem.*, **18**, 129 (1965).
22. H.O. Desseyn, F.K. Vansant and B.J. Van der Veken, *Spectroschim. Acta, Part A*, **31A**, 625 (1975).
23. A.J. Aarts, H.O. Desseyn, B.J. Van der Veken and M.A. Herman, *Can. J. Spectrosc.*, **24**, 29 (1979).
24. G. Varsányi, *Assignments for Vibrational Spectra of Seven Hundred Benzene Derivatives*, J.Wiley & Sons, New York (1974).
25. D. Hadzi, J. Kidric, Z.V. Knezevic and B. Barlic, *Spectroschim. Acta, Part A*, **32A**, 693 (1976).
26. J.E. Stewart, *J. Chem. Phys.*, **26**, 248 (1957).
27. A. Yamaguchi, R.B. Penland, S. Mizushima, T.J. Lane, C. Curran and J.V. Quagliano, *J. Amer. Chem. Soc.*, **80**, 527 (1958).
28. K. Dwarakanath and D.N. Sathyanarayana, *Bull. Chem. Soc. Jpn.*, **52**, 2084 (1979).
29. Y. Mido, S. Kimura, Y. Sugano and K. Machida, *Spectroschim. Acta, Part A*, **44A**, 661 (1988).
30. G. Keresztury and M.P. Marzocchi, *Chem. Phys.*, **6**, 117 (1974).
31. G. Keresztury and M.P. Marzocchi, *Spectroschim. Acta, Part A*, **31A**, 275 (1975).
32. D.N. Sathyanarayana, K. Volka and K. Geetharani, *Spectroschim. Acta, Part A*, **33A**, 517 (1977).
33. A. Vijav and D.N. Sathyanarayana, *Spectroschim. Acta, Part A*, **48A**, 1601 (1992).
34. K. Geetharani and D.N. Sathyanarayana, *Spectroschim. Acta, Part A*, **32A**, 227 (1976).
35. W. Malavasi, A. Pignedoli and G. Peyronel, *Spectroschim. Acta, Part A*, **37A**, 663 (1981).
36. G. Peyronel, A. Pignedoli and W. Malavasi, *Spectroschim. Acta, Part A*, **38A**, 971 (1982).
37. L. Zhengyan, R. Mattes, H. Schnöckel, M. Thünemann, E. Hunting, U. Hönke and C. Mendel, *J. Mol. Struct.*, **117**, 117 (1984).
38. R. Mattes, L. Zhengyan, M. Thünemann and H. Schnöckel, *J. Mol. Struct.*, **99**, 119 (1983).
39. J. Knoeck and J. Witt, *Spectroschim. Acta, Part A*, **32A**, 149 (1976).
40. R.H. Sullivan, J.S. Kwiatkowski and J. Leszczyński, *J. Mol. Struct.*, **295**, 169 (1993).

7.3 METHYLAMINO(THIO)CARBONYL COMPOUNDS

N-Monosubstituted amides exist mainly with the NH and C═O bonds in the *trans* configuration. In the C_s structure HC(═X)NHMe (*N*-methylformamide, *N*-methylthioformamide) possesses $3N - 6 = 21$ normal vibrations which are divided into 14a′ and 7a″ species of vibration. Setting aside the external CH vibrations (2a′ + a″), the following 21 (13a′ + 8a″) types of vibrations are considered for the —C(═X)NHC′H_3 structure unit:

a′: νNH, ν_a'Me, ν_sMe, νC═X, δNH, δ_a'Me, δ_sMe, νC—N, νN—C′, ρ'Me, δC═X, δ—C(═X)—N, δC—N—C′;
a″: ν_aMe, δ_aMe, ρMe, γNH, γC═X and three torsions.

The vibrational analysis of MeC(═X)NHMe (*N*-methylacetamide, *N*-methylthioacetamide) reveals 30 normal vibrations (19a′ + 11 a″) of which nine (5a′ + 4a″) are due to the Me group. If one torsion (a″) has been replaced by a C—C stretching (a′) vibration, the remaining 21 vibrations are the same as those mentioned above.

Just as in primary amides, sometimes the typical amide vibrations are treated as amide bands (I–VI) with the skeletal C—N—C′ deformation as amide VII.

7.3.1 Methylcarbamoyl (methylaminocarbonyl)

Table 7.12 Normal vibrations of *trans N*-methylformamide and *N*-methylacetamide

Vibration	HC(=O)NHMe 14 a′	MeC(=O)NHMe 19 a′	Vibration	HC(=O)NHMe 7 a″	MeC(=O)NHMe 11 a″
νNH	3300[a]	3306[a]	ν_aMe		2981
ν'_aMe		2994	ν_aMe	2944[a]	2981[a]
ν'_aMe	2944[a]	2944[a]	δ_aMe	1467[a]	1472[a]
ν_sMe	2880[a]	2915[a]	δ_aMe		1451
ν_sMe		2935	ρ'Me	1040[a]	1114[a]
νCH	2854		ρ'Me		1044
νC=O (I)	1670[a]	1660[a]	ωCH	1015	
δNH (II)	1540[a]	1569[a]	γ NH (V)	710[a]	725[a]
δ'_aMe	1458[a]	1458[a]	γ C=O (VI)	620[a]	600[a]
δ'_aMe		1426	torsion		192
δ_sMe	1410[a]	1414[a]	torsion		143[a,c]
δ_sMe		1374	torsion	103[a,c]	121[a,c]
δCH	1392				
νC—N (III)	1244[a]	1300[a]			
ρMe	1148[a]	1161[a]			
νN—C′	955[a]	1095[a]			
ρMe		1044			
νC—C		980			
δC=O (IV)	763[a]	628[a]			
δ—C—N (VII)		439			
δC—N—C′	368[a]	289[a]			

[a] wavenumbers of the internal vibrations of —C(=O)NHMe
[c] calculated

The NH stretching vibration

The hydrogen bonded NH stretch appears strongly and fairly broad in the region 3315 ± 45 cm^{-1}, with a weaker absorption near 3080 cm^{-1} attributed to an overtone of the amide II band enhanced by Fermi resonance. In solution the band becomes narrower and shifts to higher wavenumbers, for example in R—C(=O)NHMe molecules in which R = Me (3478), F_3C (3470), Cl_3C (3462) and Br_3C (3450 cm^{-1}).

Methyl stretching vibrations

Both methyl antisymmetric stretching vibrations often coincide and absorb between 2995 and 2900 cm^{-1}. The symmetric stretch is found in the region 2870 ± 45 cm^{-1}, clearly separated from the antisymmetric counterpart, with high values for *trans*-MeC(=O)NHMe (2915) and low values for N≡CC(=O)NHMe (2825) and *N*,*N*′-dimethylurea (2840 cm^{-1}). Most R—C(=O)NHMe compounds give a typical sharp peak in the region 2865 ± 25 cm^{-1}.

The C=O stretching vibration

The C=O stretching vibration (amide I) gives rise to a strong band in the range 1680 ± 60 cm^{-1}. The highest values, namely in the field of esters and ketones, are found in the spectra of R—C(=O)NHMe with R = Cl and Br: ≈1736 cm^{-1} (Section 7.1) and the lowest, at about 1625 cm^{-1}, in those of *N*,*N*′-alkylmethylurea. The remaining compounds absorb at 1655 ± 30 cm^{-1}. In dilute solution this band shifts to higher wavenumbers and often gives an overtone in the neighbourhood of 3370 cm^{-1}, which sometimes is taken for the νNH of the *cis* form.

The NH in-plane deformation/C—N stretching vibration

The NH in-plane deformation/C—N stretch (amide II) absorbs only for the *trans* configuration in the region 1550 ± 50 cm^{-1}. In this mixed vibration the δNH is coupled to the νC—N but in most of the secondary amides the δNH makes the larger contribution. On deuteration the band shifts to lower wavenumbers. The often strong pair of bands (I and II) forms a characteristic pattern in the spectra of *trans* secondary (poly)amides and is therefore of diagnostic interest. As a matter of course one has to use great care in view of the NO_2 or CO_2^- absorptions which are also strongly active in the above-mentioned region. The amide II band in *N*-methylacetamide (liquid: 1569; solution: 1525 cm^{-1}) exists to an extent of ≈48% [15, 17] or ≈60% [8] as a NH in-plane deformation. *N*,*N*′-Alkylmethylureas display the amide II band strongly at 1565 ± 35 cm^{-1}. In the less common *cis* configuration these vibrations are less mixed and the δNH is observed at 1465 ± 20 cm^{-1}.

Methyl deformations

The methyl antisymmetric deformations are usually seen as one weak absorption between 1480 and 1410 cm^{-1}. The symmetric deformation appears with a stronger intensity between 1425 and 1375 cm^{-1}, but most R—C(=O)NHMe compounds absorb in the narrower region 1415 ± 10 cm^{-1}.

The C—N stretching vibration/NH in-plane deformation

Just like the amide II band, the νC—N/δNH (amide III) is a mixed vibration which appears moderately to strongly in the region 1270 ± 55 cm^{-1}. For most of the secondary amides it is generally assumed that the νC—N dominates in this absorption. The highest values are found in the spectra of *N*-methylacetamide-d_3 (1325) and *N*-methylacetamide (1300 cm^{-1}) and the lowest in those of BrC(=O)NHMe (1215) and ClC(=O)NHMe (1222 cm^{-1}). The remaining examined compounds give the amide III band at $1265 \pm 35\ cm^{-1}$, a region situated much lower than that of the νC—N in primary amides (1390 ± 40) or secondary thioamides ($1325 \pm 45\ cm^{-1}$). In solution the amide III band shifts to higher wavenumbers.

Methyl rocking vibrations and N—Me stretch

The methyl rocks, coupled to the N—Me stretching vibration, absorb weakly rather than moderately. The ρMe occurs at $1155 \pm 30\ cm^{-1}$ with the highest wavenumber in the spectrum of cHexHNC(=O)NHMe. The region of the ρ'Me is more extensive ($1100 \pm 65\ cm^{-1}$), because of ClC(=O)NHMe (1164), BrC(=O)NHMe (1162) and cHexHNC(=O)NHMe ($1160\ cm^{-1}$) on the one side and sBuHNC(=O)NHMe ($1035\ cm^{-1}$) on the other. The ρ'Me in the remaining compounds tested falls in the region $1085 \pm 50\ cm^{-1}$.

The N—Me stretch, in a sense the symmetric counterpart of the amide III vibration, can be found in the region $1015 \pm 95\ cm^{-1}$. The intensity fluctuates between weak and moderate. Disregarding the high values around $1106\ cm^{-1}$ for $H_2NC(=O)C(=O)NHMe$ and $H_2NC(=O)NHMe$, coincident with the methyl rock, and the low values for cHexHNC(=O)NHMe (920) and BrC(=O)NHMe ($988\ cm^{-1}$), the majority of investigated molecules display the νC—N in the region $1050 \pm 45\ cm^{-1}$.

The NH out-of-plane deformation

The γNH/C=O or ωNH/C=O (amide V) is an out-of-plane skeletal deformation which is moderately but broadly active in the region $735 \pm 60\ cm^{-1}$. With this vibration the O and H atoms move simultaneously out of the plane in the same direction. Whether the γNH or the γC=O contributes in greater degree to this vibration still remains undecided. This deformation may be compared with the ωNH_2/C=O in primary amides. The highest values are furnished by BrC(=O)NHMe and MeHNC(=S)C(=O)NHMe with $793\ cm^{-1}$ followed by ClC(=O)NHMe with $788\ cm^{-1}$, and the lowest by N≡CC(=O)NHMe (675), $H_2NC(=O)C(=O)NHMe$ (687), $HSCH_2C(=O)NHMe$ (693) and finally MeOC(=O)C(=O)NHMe and EtC(=O)NHMe both with $699\ cm^{-1}$. Most R—C(=O)NHMe compounds absorb in the region $745 \pm 35\ cm^{-1}$. *N'*-Alkyl derivatives of *N*-methylurea give this amide V absorption at $765 \pm 15\ cm^{-1}$, and the band is assigned to πC=O/NH [32].

The C=O deformations

The C=O in-plane deformation (amide IV) is observed in the region $695 \pm 75\ cm^{-1}$ with a moderate to strong intensity. The regions of the amide IV and amide V bands overlap. Their wavenumbers are close together and not infrequently both deformations are assigned at the same wavenumber. The highest values (≈770 cm^{-1}) can be found in the spectra of the compounds MeOC(=O)C(=O)NHMe, Me_2NC(=O)C(=O)NHMe and MeHNC(=O)C(=O)NHMe. *N*′-Alkyl-*N*-methylureas R′HNC(=O)NHMe (R′ = Et, nPr, nBu) give this δC=O in the neighbourhood of 760 cm^{-1} [32]. Low values are assigned in the spectra of cHexHNC(=O)NHMe (620), $HSCH_2C$(=O)NHMe (626) and MeC(=O)NHMe (628 cm^{-1}). The remaining compounds show this δC=O at 715 ± 45 cm^{-1}.

The C=O out-of-plane deformation γC=O/NH or πC=O/NH (amide VI) absorbs moderately in the region 600 ± 70 cm^{-1}. With this vibration the O and H atoms move simultaneously out of the plane in the opposite direction. The highest wavenumbers (655 ± 15 cm^{-1}) are assigned in the spectra of R′HNC(=O)NHMe compounds (R′ = Me, Et, nPr, nBu, iPr, iBu, tBu) and the lowest in those of H_2NC(=O)C(=O)NHMe (530) and MeHNC(=O)C(=O)NHMe (532 cm^{-1}). The remaining molecules display the amide VI band in the range 595 ± 45 cm^{-1}.

Skeletal deformations

The external —C(=O)—N skeletal deformation appears weakly to moderately in the wide range 450 ± 100 cm^{-1}. For R′HNC(=O)NHMe the region narrows to 500 ± 50 cm^{-1} with 550 cm^{-1} for R′ = nPr and nBu. For R—C(=O)NHMe compounds (R = OC(=O), NC(=O), Me, Et) the region becomes 420 ± 70 cm^{-1} with the lowest values (350 and 392 cm^{-1}) for MeHNC(=O)C(=O)NHMe. In primary amides the corresponding skeletal deformation is found at 455 ± 75 cm^{-1}.

The lowest a′ vibration is the skeletal C—N—C′H_3 deformation, absorbing in the region 315 ± 55 cm^{-1} with a weak to moderate intensity. The lowest values are assigned in the spectra of MeHNC(=O)C(=O)NHMe (260 and 265 cm^{-1}) and Me_2NC(=O)C(=O)NHMe (280 cm^{-1}).

Table 7.13 Absorption region (cm^{-1}) of the normal vibrations of —C(=O)NHC′H_3

Vibration	Region	Vibration	Region
νNH	3315 ± 45	ρ′Me	1100 ± 65
ν_aMe	2970 ± 30	νN—C′	1015 ± 95
ν'_aMe	2945 ± 45	γNH (V)	735 ± 60
ν_sMe	2870 ± 45	δC=O (IV)	695 ± 75
νC=O (I)	1680 ± 60	γC=O (VI)	600 ± 70
δNH/CN (II)	1550 ± 50	δ—C(=O)—N	450 ± 100
δ_aMe	1450 ± 30	δC—N—C′	315 ± 55
δ'_aMe	1445 ± 35	torsion	230 ± 50
δ_sMe	1400 ± 25	torsion	–
νC—N/δNH (III)	1270 ± 55	torsion	–
ρMe	1155 ± 30		

R—C(=O)NHMe molecules

R = H— [1–6], D— [3], Me— [4–18], CD_3— [11–15], Et— [19], $HSCH_2$— and $HSCH_2CH_2$— [20], X_3C— (X = F, Cl, Br) [21], MeOC(=O)— [22], KOC(=O)— [23], H_2NC(=O)— [24], MeHNC(=O)— [24, 25], Me_2NC(=O)— [24], MeHNC(=S)— [26, 27], N≡C— [28], H_2N— [29], MeHN— [30–32], EtHN—, nPrHN— and nBuHN— [32, 33], iBuHN—, secBuHN— and tBuHN— [33], cHexHN— [34], N_3— [35], Cl— and Br— [36].

References

1. R.L. Jones, *J. Mol. Spectrosc.*, **2**, 581 (1958).
2. H.E. Hallam and Ch.M. Jones, *Trans. Faraday Soc.*, **65**, 2607 (1969).
3. I. Suzuki, *Bull. Chem. Soc. Jpn.*, **35**, 540 (1962).
4. R.L. Jones, *J. Mol. Spectrosc.*, **11**, 411 (1963).
5. S. Ataka, H. Takeuchi and M. Tasumi, *J. Mol. Struct.*, **113**, 147 (1984).
6. A. Balázs, *Acta Chim. Acad. Sci. Hung.*, **108**, 265 (1981).
7. A. Balázs, *J. Mol. Struct.*, **153**, 103 (1987).
8. T. Miyazawa, T. Shimanouchi and S. Mizushima, *J. Chem. Phys.*, **29**, 611 (1958).
9. O.D. Bonner, K.W. Bunzl and G.B. Woolsey, *Spectroschim. Acta*, **22**, 1126 (1966).
10. Y. Grenie, M. Avignon and C. Garrigou-Lagrange, *J. Mol. Struct.*, **24**, 293 (1975).
11. B. Schneider, A. Horeni, H. Pivcová and J. Honzl, *Collect. Czech. Chem. Commun.*, **30**, 2196 (1965).
12. H. Pivcová, B. Schneider and J. Stokr, *Collect. Czech. Chem. Commun.*, **30**, 2215 (1965).
13. A. Warshel, M. Levitt and S. Lifson, *J. Mol. Spectrosc.*,**33**, 84 (1970).
14. J. Jakes and S. Krimm, *Spectroschim. Acta, Part A*, **27A**, 19 (1971).
15. M. Rey-Lafon, M.T. Forel and C. Garrigou-Lagrange, *Spectroschim. Acta, Part A 29A*, 471 (1973).
16. N.G. Mirkin and S. Krimm, *J. Mol. Struct.*, **236**, 97 (1991).
17. N.G. Mirkin and S. Krimm, *J. Am. Chem. Soc.*, **113**, 9742 (1991).
18. T.C. Cheam, *J. Mol. Struct.*, **257**, 57 (1992).
19. J. Jakes and S. Krimm, *Spectroschim. Acta, Part A*, **27A**, 35 (1971).
20. G. Zuppiroli, C. Perchard, M.L. Baron and C. de Lozé, *J. Mol. Struct.*, **69**, 1 (1980).
21. R.A. Nyquist, *Spectroschim. Acta*, **19**, 509 (1963).
22. H.O. Desseyn, B.J. Van der Veken and M.A. Herman, Bull.Soc.Chim.Belg. **84**, 1057 (1975).
23. H.O. Desseyn, B.J. Van der Veken, A.J. Aarts and M.A. Herman, *J. Mol. Struct.*, **63**, 13 (1980).
24. H.O. Desseyn, B.J. Van der Veken and M. Herman, *Spectroschim. Acta*, **33A**, 633 (1977).
25. R.A. Nyquist, R.W. Chrisman, C.L. Putzig, R.W. Woodward and B.R. Loy, *Spectroschim. Acta, Part A*, **35A**, 91 (1979).
26. H.O. Desseyn, J.A. Le Poivre and M.A. Herman, *Spectroschim. Acta, Part A*, **30A**, 503 (1974).
27. H.O. Desseyn, A.J. Aarts and M.A. Herman, *Spectroschim. Acta, Part A*, **36A**, 59 (1980).
28. H.O. Desseyn and B.J. Van der Veken, *Spectroschim. Acta, Part A*, **31A**, 641 (1975).
29. Y. Saito, K. Machida and T. Uno, *Spectroschim. Acta, Part A*, **31A**, 1237 (1975).
30. C.N.R. Rao, G. Chaturvedi and R.K. Gosavi, *J. Mol. Spectrosc.*, **28**, 526 (1968).
31. K.R.G. Devi and D.N. Sathyanarayana, *Bull. Chem. Soc. Jpn.*, **53**, 2993 (1980).

32. Y. Mido, F. Fujita, H. Matsuura and K. Machida, *Spectroschim. Acta, Part A*, **37A**, 103 (1981).
33. Y. Mido, *Spectroschim. Acta, Part A*, **28A**, 1503 (1972).
34. Y. Mido, *Spectroschim. Acta, Part A*, **32A**, 1105 (1976).
35. W. Buder and A. Schmidt, *Spectroschim. Acta, Part A*, **29A**, 1429 (1973).
36. W. Buder and A. Schmidt, *Spectroschim. Acta, Part A*, **29A**, 1419 (1973).

7.3.2 Methylthiocarbamoyl (methylaminothiocarbonyl)

The NH stretching vibration

The NH stretching vibration gives rise to a moderate to strong absorption in the region 3250 ± 70 cm^{-1}. In dilute solutions the band shifts to higher wavenumbers, for example 3425 cm^{-1} for MeC(=S)NHMe and 3450 cm^{-1} for MeOC(=S)NHMe.

Methyl stretching vibrations

Both methyl antisymmetric stretching vibrations absorb between 3000 and 2920 cm^{-1}. The symmetric stretch is active at 2875 ± 45 cm^{-1}, well separated from its antisymmetric counterpart.

The NH in-plane deformation/C—N stretching vibration

The mixed vibration δNH/νC—N (thioamide II) is strongly active in the region 1535 ± 35 cm^{-1}. After deuteration this thioamide II band shifts to lower wavenumbers in such a way as to indicate that the δNH dominates. The vibration bears a good resemblance to the amide II vibration in secondary amides (1550 ± 50 cm^{-1}). *N*-Methylthioformamide {*N*-methylthioacetamide} absorbs at 1550 {1564} cm^{-1}. Extreme values are found in the spectrum of MeHNC(=S)NHMe (1570 and 1506 cm^{-1}).

Methyl deformations

The methyl antisymmetric deformations are observed between 1475 and 1410 cm^{-1} and the symmetric counterpart between 1425 and 1375 cm^{-1}.

The C—N stretching vibration/NH in-plane deformation

The νC—N/δNH (thioamide III) is a mixed vibration occurring in the region 1325 ± 45 cm^{-1} with a moderate to strong intensity. This vibration is assigned as much to the νC—N as to the δNH but the band is less sensible to deuteration than the thioamide II band. *N*-Methylthioformamide {*N*-methylthioacetamide} gives this vibration strongly at 1334 {1356} cm^{-1}.

Methyl rocking vibrations and N—Me stretch

The coupled vibrations ρMe/νC—N and ρ'Me/νC—N are usually seen in the ranges 1145 ± 45 and 1075 ± 40 cm^{-1}, comparable with those of the corresponding secondary amides.

Just as in secondary amides, the N—C′ stretch is in a sense the symmetric counterpart of the thioamide III vibration and is active in the region 1000 ± 50 cm^{-1}. In *N*-methylthioformamide {*N*-methylthioacetamide} this vibration occurs at 987 {950} cm^{-1}. The highest value, by contrast, is found in the spectrum of KOC(=O)C(=S)NHMe (1045 cm^{-1}).

The C=S stretching vibration

The C=S stretching vibration (thioamide I) in secondary thioamides is a mixed vibration (see Section 7.2.2), located in the region 795 ± 110 cm^{-1} with moderate intensity. This region is so extensive because of the extreme values 905 and 688 cm^{-1} in the spectrum of MeOC(=S)NHMe assigned respectively to the *cis* and the *trans* form. High wavenumbers are also found in the spectra of MeHNC(=S)C(=S)NHMe (870) and HC(=S)NHMe (865 cm^{-1}) and low wavenumbers in those of MeC(=S)NHMe with 700 cm^{-1} for the *trans* (95%) and 722 cm^{-1} for the *cis* form (5%). The remaining compounds show the thioamide I band at 780 ± 60 cm^{-1}.

The NH out-of-plane deformation

The γNH/C=S or ωNH/C=S (thioamide V) usually provides a broad band with a moderate to strong intensity in the range 665 ± 55 cm^{-1}. With this vibration the S and the H atoms move simultaneously out of the plane in the same direction. For *N*-methylthioformamide {*N*-methylthioacetamide} the thioamide V absorbs at 700 {690} cm^{-1}. The HW side of the above-mentioned region is limited by 720 cm^{-1} from the spectrum of KOC(=O)C(=S)NHMe and 715 cm^{-1} (coincident with amide V) from H_2NC(=O)C(=S)NHMe. The LW side is covered by MeHNC(=S)NHMe (610 and *642* cm^{-1}). The remaining compounds display the thioamide V band at 670 ± 30 cm^{-1}.

The C=S deformations

The C=S in-plane deformation (thioamide IV) exhibits only weak to medium intensity in the region 585 ± 55 cm^{-1}, that is, 100 cm^{-1} lower than the region of the amide IV absorption. The highest wavenumbers are due to KOC(=O)C(=S)NHMe (640) and MeHNC(=O)C(=S)NHMe (634 cm^{-1}) and the lowest has been traced in the spectrum of H_2NC(=S)C(=S)NHMe: 535 cm^{-1} for both thioamide IV bands. Most RC(=S)NHMe compounds display this thioamide IV band at 580 ± 30 cm^{-1}, with *N*-methylthioformamide (600) and *N*-methylthioacetamide (555 cm^{-1}) as examples.

The γ C=S/NH or πC=S/NH (thioamide VI) is a mixed vibration, weakly to moderately active in the region 470 ± 70 cm^{-1}. With this vibration the S and the H atoms move out of the plane in the opposite direction. By contrast with the NH/C=S wag (thioamide V), the contribution of the C=S in this vibration is greater than the contribution of the NH. A high wavenumber is observed in the spectrum of MeC(=S)NHMe (533) but H_2NC(=S)NHMe (400 cm^{-1}) is responsible for the lowest value. Most RC(=S)NHMe compounds display the thioamide VI band at 470 ± 30 cm^{-1}.

Skeletal deformations

The external —C(=S)—N skeletal deformation is located in the range 395 ± 55 cm^{-1}. The absorption at 370 cm^{-1} in the spectrum of *N*-methylthioacetamide is assigned to this skeletal deformation but described as ρNCS [3], δCS [2] or δCC [1].

The lowest a′ vibration of the —C(=S)NHMe group is the skeletal C—N—C′ deformation at 265 ± 65 cm^{-1} with, as examples HC(=S)NHMe (260) and MeC(=S)NHMe (285 cm^{-1}).

Table 7.14 Absorption regions (cm^{-1}) of the normal vibrations of —C(=S)NHC′H_3

Vibration	Region	Vibration	Region
νNH	3250 ± 70	ρ'Me	1075 ± 40
ν_aMe	2970 ± 30	νN—C′	1000 ± 50
ν'_aMe	2945 ± 25	νC=S (I)	795 ± 110
ν_sMe	2875 ± 45	γ NH (V)	665 ± 55
δNH (II)	1535 ± 35	δC=S (IV)	585 ± 55
δ_aMe	1450 ± 25	γ C=S (VI)	470 ± 70
δ'_aMe	1435 ± 25	δ—C(=S)—N	395 ± 55
δ_sMe	1400 ± 25	δC—N—C′	265 ± 65
νC—N (III)	1325 ± 45	torsion	195 ± 50
ρMe	1145 ± 45	torsion	–
		torsion	–

R—C(=S)NHMe molecules

R = H— [1–3], Me— [1–4], MeHNC(=O)— [3, 5], KOC(=O)—, H_2NC(=O)— and H_2NC(=S)— [3], MeHNC(=S)— [6], Ph— [3], H_2N— [7, 8], MeHN— [9–11], Me$(CH_2)_n$HN— (n= 1–15) [10], MeO— [12], MeS— [13].

References

1. I. Suzuki, *Bull. Chem. Soc. Jpn.*, **35**, 1456 (1962).
2. C.N.R. Rao and G.C. Chaturvedi, *Spectroschim. Acta, Part A*, **27A**, 520 (1971).
3. H.O. Desseyn, A.J. Aarts and M.A. Herman, *Spectroschim. Acta, Part A*, **36A**, 59 (1980).
4. S. Ataka, H. Takeuchi, I. Harada and M. Tasumi, *J. Phys. Chem.*, **88**, 449 (1984).
5. H.O. Desseyn, J.A. Le Poivre and M.A. Herman, *Spectroschim. Acta, Part A*, **30A**, 503 (1974).
6. H.O. Desseyn, A.J. Aarts, E. Esmans and M.A. Herman, *Spectroschim. Acta, Part A*, **35A**, 1203 (1979).
7. K. Dwarakanath and D.N. Sathyanarayana, *Bull. Chem. Soc. Jpn.*, **52**, 2084 (1979).
8. K.R.G. Devi, D.N. Sathyanarayana and S. Manogaran, *Spectroschim. Acta, Part A*, **37A**, 31 (1981).
9. K.R.G. Devi and D.N. Sathyanarayana, *Bull. Chem. Soc. Jpn.*, **53**, 299 (1980).
10. Y. Mido, H. Mizuno and K. Machida, *Spectroschim. Acta, Part A*, **44A**, 445 (1988).
11. R.K. Ritchie, H. Spedding and D. Steele, *Spectroschim. Acta, Part A*, **27A**, 1597 (1971).
12. G.C. Chaturvedi and C.N.R. Rao, *Spectroschim. Acta, Part A*, **27A**, 65 (1971).
13. K.R.G. Devi, D.N. Sathyanarayana and S. Manogaran, *Spectroschim. Acta, Part A*, **37A**, 633 (1981).

7.4 CARBOXYLATE

The —CO_2^- structure unit, occurring in the salts of carboxylic acids or as a dipolar ion in amino acids, has six normal vibrations. According to the position of the plane of symmetry with the O atoms in (i.p.) or perpendicular to (o.p.) this plane, the distribution is as follows:

Point group	$\nu_a CO_2$	$\nu_s CO_2$	δCO_2	ωCO_2	ρCO_2	torsion
C_{2v} (i.p.)	b_2	a_1	a_1	b_1	b_2	a_2
C_s (i.p.)	a'	a'	a'	a''	a'	a''
C_{2v} (o.p.)	b_1	a_1	a_1	b_2	b_1	a_2
C_s (o.p.)	a''	a'	a'	a'	a''	a''

When a carboxylic acid changes into a carboxylate the C=O and C—O are replaced by two equivalent carbon—oxygen bonds which are intermediate in force constant between the C=O and C—O. The νC=O, occurring in a carboxylic acid at 1725 cm^{-1}, shifts to a wavenumber lower by $\approx$130 cm^{-1} and the νC—O, occurring at 1250 cm^{-1}, to $\approx$130 cm^{-1} higher wavenumber to give the CO_2 antisymmetric and symmetric stretching vibration.

The CO_2 stretching vibrations

With the exception of $CF_3CO_2^-$, the CO_2^- antisymmetric stretching vibration (or νCO I) absorbs strongly and broadly in the region 1600 ± 75 cm^{-1}. The highest values originate from the spectra of $X_3CCO_2^-$ (X = F, Cl, Br) with respectively 1680, 1675 and 1664 cm^{-1} and the lowest from those of 2-MePhCO_2^-, 2-Pyr—CO_2^--d$_1$ with 1525 cm^{-1}, 2-Pyr—CO_2^--d$_0$ with 1529 cm^{-1} and sodium benzoate-d$_5$ with 1534 cm^{-1}. The $\nu_a CO_2^-$ in salts of aromatic carboxylic acids is reported in the range 1565 ± 40 cm^{-1}, but the remaining R—CO_2^- compounds absorb at 1600 ± 60 cm^{-1}.

The CO_2^- symmetric stretching vibration (or νCO II) provides a medium to strong band in the region 1385 ± 65 cm^{-1} with the highest wavenumbers from $F_3CCO_2^-$ (1450), $H_2C{=}CHCO_2^-$ (1450) and 2-Pyr—CO_2^- (1445 cm^{-1}). The lowest wavenumbers have been traced in the spectra of sodium oxalate (1321 and 1340) and sodium oxamate $H_2NC({=}O)CO_2^-$ (1356 [36] with νC—N at 1448 or 1323 cm^{-1} [34]). For the remaining compounds, the $\nu_s CO_2^-$ is assigned in the region 1440 ± 40 cm^{-1}. Sodium acetate absorbs at 1422 cm^{-1}, sodium benzoate at 1413 and glycine at 1415 cm^{-1}. A good number of substituted benzoates and amino acids in the zwitterion form show this $\nu_s CO_2^-$ around 1410 cm^{-1}.

Both typical broad absorptions are very useful for identification purposes, although R—NO_2 compounds (Section 9.5) absorb also in these regions but not so broadly.

The CO_2 deformations

The CO_2^- scissors, a mixed vibration, is active in the extensive region 735 ± 125 cm^{-1} with moderate intensity, although a few investigators assign the CO_2^- wag in this region. The HW side is limited by 4-X-substituted sodium benzoates (X = H, D, Me, F, Cl, Br, CO_2^-) (850 ± 10 cm^{-1}) and *cis*-$^-O_2CCH{=}CHCO_2^-$ (855 and *738* cm^{-1}). The lowest values are from MeCO_2^--d$_3$, EtCO_2^--d$_3$ and -d$_5$ and MeHNC(=S)CO_2^--d$_0$, -d$_1$ and -d$_3$ with values in the neighbourhood of 610 cm^{-1}. The remaining molecules display δCO_2^- in the range 725 ± 100 cm^{-1}. Sodium acetate absorbs at 651 cm^{-1}, sodium propanoate at 647 and glycine at 696 cm^{-1}.

The CO_2^- wagging vibration gives rise to a band with variable intensity in the range 575 ± 125 cm^{-1}. A few investigators reserve this region for the CO_2^- scissors. The highest wavenumbers are due to 4-X-substituted sodium benzoates (X = H, D, Me, F, Cl, Br) (690 ± 10 cm^{-1}) and $Cl_3CCO_2^-$ (689 cm^{-1}), and the lowest have been traced in the spectra of a few (thio)oxamates such as $D_2NC({=}O)CO_2^-$ (450), $H_2NC({=}O)CO_2^-$ (485), MeHNC(=S)CO_2^- (465), MeHNC(=O)CO_2^- (490) and in that of $BrCD_2CO_2^-$ (479 cm^{-1}). The remaining compounds absorb in the region 595 ± 90 cm^{-1}, examples being sodium acetate (620), sodium benzoate (683) and glycine (607 cm^{-1}). Although the regions of δCO_2^- and ωCO_2^- overlap, coincident wavenumbers are exceptions rather than the rule.

The CO_2^- rocking vibration is located in the region 470 ± 120 cm^{-1} with a weak to medium intensity, well separated from the wag. Zwitterions (530 ± 60), salts of aromatic carboxylic acids (515 ± 65) and the three isomers of $^-O_2CCH{=}CH{-}CH{=}CH{-}CO_2^-$ (≈540 cm^{-1}) are responsible for the high values. The LW side of the region is limited by $ICH_2CO_2^-$ (359), $N{\equiv}CCH_2CO_2^-$ (371) and $BrCH_2CO_2^-$ (377 cm^{-1}). Most of the rocks have been observed at 460 ± 60 cm^{-1}. Sodium acetate absorbs at 463 cm^{-1}, sodium benzoate at 526 and glycine at 503 cm^{-1}.

Table 7.15 Absorption regions (cm^{-1}) of the normal vibrations of $—CO_2^-$

Vibration	α—saturated	α—halogen	C=O, C=S	α—unsatur.	aromatic	zwitterion
ν_aCO_2	1580 ± 30	1630 ± 45	1640 ± 20	1585 ± 35	1565 ± 40	1585 ± 25
ν_sCO_2	1395 ± 40	1400 ± 50	1385 ± 65	1405 ± 45	1410 ± 35	1410 ± 30
δCO_2	700 ± 90	735 ± 85	700 ± 90	740 ± 115	795 ± 65	739 ± 90
ωCO_2	565 ± 60	600 ± 90	540 ± 90	515 ± 75	670 ± 30	625 ± 55
ρCO_2	430 ± 60	440 ± 80	395 ± 45	480 ± 70	515 ± 65	530 ± 60
torsion	140 ± 60	–	–	–	195 ± 50	–

$R{-}CO_2^-\ M^+$ molecules ($M^+ = Na^+, K^+$)

R = H— [1–4], Me— [4–11], CH_2D—, CHD_2— and CD_3— [5], Et— [9, 12], Et-d_2—, -d_3— and -d_5— [12], $HOCH_2CH_2CH_2$— and $DOCH_2CH_2CH_2$— [13], $HO(O{=})CCH_2CH_2$— [14, 15], $DO(O{=})CCH_2CH_2$— [14], $HO(O{=})CCH_2$— [16], $^-O_2CCH_2$— [17], $MeCHOHCH_2$— and $MeCHODCH_2$— [13], $EtCH{=}CHCH_2$— [18], $N{\equiv}CCH_2$— [9, 19], $HSCH_2$—, EtCHOH— and EtCHOD— [13], cPr— [20], cBu— [21], cPent-d_0— and d_1— [22], tBu— [9], $HO(O{=})CCH_2CH_2C(Me)_2$—, $DO(O{=})CCH_2CH_2C(Me)_2$— and $^-O_2CCH_2CH_2C(Me)_2$— [23], $HO(O{=})CC(Me)_2$— and $^-O_2CC(Me)_2$— [24], FCH_2— [9], $ClCH_2$— [9], $BrCH_2$— [9, 25], $BrCD_2$— [25], ICH_2— [9, 26], Cl_2HC— and Br_2HC— [9], FClHC-d_0— and -d_1— [27], F_3C— [9, 28, 29], Cl_3C— [9, 30], Br_3C— [9], MeC(=O)— [31, 32], $H_2NC({=}O)$— [33–36], $D_2NC({=}O)$— [36], MeHNC(=O)-d_0— and -d_1— [36], $H_2NC({=}S)$-d_0— and -d_2— [36], MeHNC(=S)-d_0—, -d_1— and -d_3— [37], MeOC(=O)— [37], ^-O_2C— [4, 38, 39], $CH_2{=}CH$— [40], nPrCH=CH— [18], HO(O=)CCH=CH— [41, 42], MeO(O=)CCH=CH— [43], $^-O_2CCH{=}CH$— [17, 44], $^-O_2CCH{=}CHCH{=}CH$— [45], HC≡C-d_0— and -d_1— [46], Ph— [47–50], 2-XPh— (X = Me, F, Cl and Br [47], ^-O_2C [51], HO [50]), 3-XPh— (X = Me, F, Cl and Br [47], ^-O_2C [51], H_2N [52]), 4-XPh— (X = D [48], Me, F, Cl and Br [47], ^-O_2C [51, 53, 54]), Ph-d_5— [48], PhF_5— [55], 2-Fu— and 5-X-2-Fu— (X = Cl, Br) [56], 2-Pyr-d_0— and -d_1— [57],

$R{-}CO_2^-$ molecules
R = $^+H_3NCH_2$— and $^+D_3NCH_2$— [58–67], $^+H_2DNCH_2$— and $^+HD_2NCH_2$— [62], $^+H_3NCD_2$— and $^+D_3NCD_2$— [64], $^+H_3NCH_2CH_2$— and $^+D_3NCH_2CH_2$— [68, 69], $^+H_3NCH_2C(=O)NHCH_2$-d_0, -d_2—, -d_4—, -d_6— and -d_8— [70], $^+H_3NCH(Me)$— and $^+D_3NCH(Me)$— [71–75], $^+D_3NCH(CD_3)$— [73], $HOCH_2CH(NH_3^+)$—, $HOCD_2CD(NH_3^+)$— and $DOCH_2CH(ND_3^+)$— [76], $HSCH_2CH(NH_3^+)$— and $DSCH_2CH(ND_3^+)$— [76], $ClCH_2CH(NH_3^+)$— and $ClCH_2CH(ND_3^+)$— [76], $^+H_3NCH_2CH(OH)$— and $D_3NCH_2CH(OD)$— [77], 3-^+H_3NPh— and 3-^+D_3NPh— [52].

References

1. C.J.H. Schutte and K. Buijs, *Spectrochim. Acta*, **20**, 187 (1964).
2. E. Spinner, *Spectrochim. Acta, Part A*, **31A**, 1545 (1975).
3. J.P.M. Maas, *Spectrochim. Acta, Part A*, **34A**, 179 (1978).
4. K. Ito and H.J. Bernstein, *Can. J. Chem.*, **34**, 170 (1956).
5. L.H. Jones and E. McLaren, *J. Chem. Phys.*, **22**, 1796 (1954).
6. J.K. Wilmshurst, *J. Chem. Phys.*, **23**, 2463 (1955).
7. J.K. Wilmshurst, *J. Chem. Phys.*, **25**, 1171 (1956).
8. K. Nakamura, *J. Chem. Soc. Jpn.*, **79**, 1411 (1958).
9. E. Spinner, *J. Chem. Soc.*, 4217 (1964).
10. M. Cadene, *J. Mol. Struct.*, **2**, 193 (1968).
11. A.M. Heyns, *J. Mol. Struct.*, **11**, 93 (1972).
12. E. Spinner, P. Yang, P.T.T. Wong and H.H. Mantsch, *Aust. J. Chem.*, **39**, 475 (1986).
13. M. Morssli, G. Cassanas, L. Bardet, B. Pauvert and A. Terol, *Spectrochim. Acta, Part A*, **47A**, 529 (1991).
14. L. Angeloni, M.P. Marzocchi, D. Hadzi, B. Orel and G. Sbrana, *Spectrochim. Acta, Part A*, **33A**, 735 (1977).
15. J.E. Katon and S.R. Lobo, *J. Mol. Struct.*, **127**, 229 (1985).
16. L. Angeloni, M.P. Marzocchi, S. Detoni, D. Hadzi, B. Orel and G. Sbrana, *Spectrochim. Acta, Part A*, **34A**, 253 (1978).
17. C.B. Baddiel, C.D. Cavendish and W.O. George, *J. Mol. Struct.*, **5**, 263 (1970).
18. K. Tsukamoto, S. Horiuchi, K. Taga, T. Yoshida and H. Okabayashi, *J. Mol. Struct.*, **263**, 75 (1991).
19. D. Sinha and J.E. Katon, *Appl. Spectrosc.*, **26**, 599 (1972).
20. J. Maillols, *J. Mol. Struct.*, **14**, 171 (1972).
21. L. Bardet, J. Maillols, R. Granger and E. Fabregue, *J. Mol. Struct.*, **9**, 433 (1971).
22. L. Bardet, G. Cassanas-Fabre and E. Bourret, *J. Mol. Struct.*, **28**, 45 (1975).
23. S. Yolou, J.L. Delarbre and L. Maury, *J. Raman Spectrosc.*, **23**, 501 (1992).
24. J.L. Delarbre, L. Maury and L. Bardet, *J. Raman Spectrosc.*, **13**, 1 (1982).
25. J.E. Katon and R.L. Kleinlein, *J. Mol. Struct.*, **17**, 239 (1973).
26. J.E. Katon and T.P. Carll, *J. Mol. Struct.*, **7**, 391 (1971).
27. J. Calienni, J.B. Trager, M.A. Davies, U. Gunnia and M. Diem, *J. Phys. Chem.*, **93**, 5049 (1989).
28. W. Klemperer and G.C. Pimentel, *J. Chem. Phys.*, **22**, 1399 (1954).
29. K.O. Christe and D. Naumann, *Spectrochim. Acta, Part A*, **29A**, 2017 (1973).
30. M.S. Soliman, *Spectrochim. Acta, Part A*, **49A**, 183 (1993).
31. J.E. Katon and D.T. Covington, *Spectrosc. Lett.*, **12**, 761 (1979).

32. W.J. Ray and J.E. Katon, *Spectrochim. Acta, Part A*, **36A**, 793 (1980).
33. G.N.R. Tripathi and J.E. Katon, *Spectrochim. Acta, Part A*, **35A**, 401 (1979).
34. F. Wallace and E. Wagner, *Spectrochim. Acta, Part A*, **34A**, 589 (1978).
35. G.N.R. Tripathi and J.E. Katon, *J. Mol. Struct.*, **54**, 19 (1979).
36. H.O. Desseyn, B.J. Van der Veken, A.J. Aarts and M.A. Herman, *J. Mol. Struct.*, **63**, 13 (1980).
37. H.O. Desseyn, A.J. Aarts and M.A. Herman, *Spectrochim. Acta, Part A*, **36A**, 59 (1980).
38. T.A. Shippey, *J. Mol. Struct.*, **65**, 71 (1980).
39. T.A. Shippey, *J. Mol. Struct.*, **67**, 223 (1980).
40. W.R. Feairheller and J.E. Katon, *Spectrochim. Acta, Part A*, **23A**, 2225 (1967).
41. F. Avbelj, B. Orel, M. Klanjsek and D. Hadzi, *Spectrochim. Acta, Part A*, **41A**, 75 (1985).
42. K. Nakamoto, Y.A. Sarma and G.T. Behnke, *J. Chem. Phys.*, **42**, 1662 (1965).
43. J.E. Katon and P.-H.Chu, *J. Mol. Struct.*, **82**, 61 (1982).
44. J. Maillols, L. Bardet and L. Maury *J. Mol. Struct.*, **21**, 185 (1974).
45. P. Sohár and G. Varsányi, *J. Mol. Struct.*, **1**, 437 (1967/68).
46. J.E. Katon and N.T. McDevitt, *Spectrochim. Acta*, **21**, 1717 (1965).
47. J.H.S. Green, *Spectrochim. Acta, Part A*, **33A**, 575 (1977).
48. K. Machida, A. Kuwae, Y. Saito and T. Uno, *Spectrochim. Acta, Part A*, **34A**, 793 (1978).
49. W. Lewandowsky, *J. Mol. Struct.*, **101**, 93 (1983).
50. J.H.S. Green, W. Kynaston and A.S. Lindsey, *Spectrochim. Acta*, **17**, 486 (1961).
51. J.F. Arenas and J.F. Marcos, *Spectrochim. Acta, Part A*, **35A**, 355 (1979).
52. L. Gopal, C.I. Jose and A.B. Biswas, *Spectrochim. Acta, Part A*, **23A**, 513 (1967).
53. F.J. Boerio and P.G. Roth, *Appl. Spectrosc.*, **41**, 463 (1987).
54. G.N.R. Tripathi and S.J. Sheng, *J. Mol. Struct.*, **57**, 21 (1979).
55. J.H.S. Green, D.J. Harrison and C.P. Stockley, *Spectrochim. Acta, Part A*, **33A**, 423 (1977).
56. J.H.S. Green and D.J. Harrison, *Spectrochim. Acta, Part A*, **33A**, 843 (1977).
57. A. Lautié, M.H. Limage and A. Novak, *Spectrochim. Acta, Part A*, **33A**, 121 (1977).
58. M. Tsuboi, T. Onishi, J. Nakagawa, T. Shimanouchi and S.J. Mizushima, *Spectrochim. Acta*, **12**, 253 (1967).
59. D.M. Dodd, *Spectrochim. Acta*, **15**, 1072 (1959).
60. S. Suzuki, T. Shimanouchi and T. Tsuboi, *Spectrochim. Acta*, **19**, 1195 (1965).
61. I. Laulicht, S. Pinchas, D. Samuce and I. Wasserman, *J. Phys. Chem.*, **70**, 2719 (1966).
62. R.K. Khanna, M. Horak and E.R. Lippincott, *Spectrochim. Acta*, **22**, 1759 (1966).
63. U. Stahlberg and E. Steger, *Spectrochim. Acta, Part A*, **23A**, 475 (1967).
64. C. Destrade, C. Garrigou-Lagrange and M.T. Forel, *J. Mol. Struct.*, **10**, 203 (1971).
65. J. Herranz and J.M. Delgado, *Spectrochim. Acta, Part A*, **32A**, 821 (1976).
66. K. Machida, A. Kagayama, Y. Saito, Y. Kuroda and T. Uno, *Spectrochim. Acta, Part A*, **33A**, 569 (1977).
67. S.F.A. Kettle, E. Lugwisha, J. Eckert and N.K. McGuire, *Spectrochim. Acta, Part A*, **45A**, 533 (1989).
68. C. Garrigou-Lagrange, *Can. J. Chem.*, **56**, 663 (1978).
69. R.S. Krishinan and R.S. Katiyar, *Bull. Chem. Soc. Jpn.*, **42**, 2098 (1969).
70. C. Destrade, E. Dupart, M. Joussot-Dubien and C. Garrigou-Lagrange, *Can. J. Chem.*, **52**, 2590 (1974).
71. R.F. Adamowicz and M.L. Sage, *Spectrochim. Acta, Part A*, **30A**, 1007 (1974).
72. K. Machida, A. Kagayama, Y. Saito and T. Uno, *Spectrochim. Acta, Part A*, **34A**, 909 (1978).

73. D.M. Byler and H. Susi, *Spectrochim. Acta, Part A*, **35A**, 1365 (1979).
74. H. Susi and D.M. Byler, *J. Mol. Struct.*, **63**, 1 (1980).
75. K. Fukushima, T. Onishi, T. Shimanouchi and S.-I. Mizushima, *Spectrochim. Acta*, **15**, 236 (1959).
76. H. Susi, D.M. Byler and W.V. Gerasimowicz, *J. Mol. Struct.*, **102**, 63 (1983).
77. K. Machida, M. Izumi and A. Kagayama, *Spectrochim. Acta, Part A*, **35A**, 1333 (1979).

8

Normal Vibrations and Absorption Regions of Alkenes and Alkynes

8.1 ALKENES

Since the normal vibrations of alkenes are derived from those of ethene, it is helpful to have just a look at this spectrum. Ethene belongs to the point group D_{2h}. The twelve normal vibrations are described as in Table 8.1. Only five vibrations are infrared active.

Table 8.1 Normal vibrations of ethene [1–6]

Vibration	D_{2h}	cm^{-1}	Vibration	D_{2h}	cm^{-1}
$\nu_s CH_2$	a_g	3026	ωCH_2	b_{2g}	950
$\nu C{=}C$		1623	$\nu_a CH_2$	b_{2u}	3106[a]
δCH_2		1342	ρCH_2		810[a]
τCH_2	a_u	1027	$\nu_a CH_2$	b_{3g}	3103
$\nu_s CH_2$	b_{1u}	2990 [a]	ρCH_2		1236
δCH_2		1444[a]	ωCH_2	b_{3u}	949[a]

[a] Infrared active

8.1.1 Vinyl (ethenyl)

The simplest compounds with a $—CH{=}CH_2$ structure unit are those of the type $XCH{=}CH_2$ (X = F, Cl, Br and I). In the C_s symmetry the $3N - 6 = 12$ normal vibrations are divided into 9a′ + 3a″ species of vibration. If the νCX (a′) is substituted by a torsion (a″), the twelve normal vibrations are described as follows:

a′: ν_aCH$_2$, νCH, ν_sCH$_2$, νC=C, δCH$_2$, δCH, ρCH$_2$, δ—C=C;
a″: ωCH, ωCH$_2$, τCH$_2$ and torsion.

In some cases the three CH groups are simply conceived as isolated bonds, giving the following CH vibrations: 3νCH, 3δCH and 3ωCH.

The CH stretching vibrations

The CH stretching vibrations occurring above 3000 cm^{-1} normally indicate the presence of sp^2 hybridized CH bonds. Usually ν_aCH$_2$ and νCH absorb above 3000 cm^{-1} and ν_sCH$_2$ around 3000 cm^{-1}. Duncan reported a correlation between the separation of the CH$_2$ stretching frequencies and the HCH angle [147].

The ν_aCH$_2$ is observed in the region 3070 ± 70 cm^{-1}. The highest wavenumbers are those due to PhSCH=CH$_2$ (3140), F$_3$CC(=O)OCH=CH$_2$ (3138), MeOCH=CH$_2$ (3132) and FCH=CH$_2$ (3130 cm^{-1}). The lowest values are from 2-PyCH=CH$_2$ (3005) and 3-XPhCH=CH$_2$ (X = Br and Me) with 3010 cm^{-1}, that is, in the region of the aromatic CH stretching vibrations. Most vinyl compounds absorb at 3085 ± 40 cm^{-1}.

The HW side of the region of νCH (3045 ± 65 cm^{-1}) is limited by values from the spectra of F$_3$CC(=O)OCH=CH$_2$ (3110), FCH=CH$_2$ (3100), ClCH=CH$_2$ (3090), BrCH=CH$_2$ (3087), vinyl formate (3095) and vinyl acetate (3090 cm^{-1}). The LW side is covered by 3-ClPhCH=CH$_2$ (2982), 3-BrPhCH=CH$_2$ (2988), H$_3$SiCH=CH$_2$ (2986) and D$_3$SiCH=CH$_2$ (2983 cm^{-1}). For the remaining compounds the region 3040 ± 40 cm^{-1} applies.

The ν_sCH$_2$ is reported in the range 3000 ± 70 cm^{-1} with, at the HW side, CF$_3$C(=O)OCH=CH$_2$ (3069), FC(=O)OCH=CH$_2$ (3058) and FCH=CH$_2$ (3045 cm^{-1}) and at the LW side 3-MePhCH=CH$_2$, 3-ClPhCH=CH$_2$ and 2- and 4-PyCH=CH$_2$ with ≈2930 cm^{-1}. There is a realistic chance of encountering this vibration at 3000 ± 40 cm^{-1}, sometimes going behind the sp^3 —CH absorptions.

The C=C stretching vibration

The νC=C gives a good group vibration in the region 1625 ± 45 cm^{-1} but the intensity is medium rather than strong. The unsaturated hydrocarbons with terminal vinyl groups usually absorb at 1645 ± 10 cm^{-1}. In conjugated systems, this C=C stretch falls in the range 1610 ± 30 cm^{-1}, that is, in the field of aromatic ring stretching vibrations. 1,3-Butadiene absorbs at 1640 and 1596 cm^{-1}.

The CH in-plane deformations

The δCH$_2$ is located in the region 1400 ± 40 cm^{-1}. The highest values are furnished by F$_3$CCH=CH$_2$ (1440), 1,3-butadiene (1440 and 1385), HOC(=O)CH=CH$_2$ (1430) and H$_2$NC(=O)CH=CH$_2$ (1428 cm^{-1}). The lowest values are from $F_3CC(=O)OCH=CH_2$ (1364), $NaOC(=O)CH=CH_2$ (1370) and $HC(=O)OCH=CH_2$ (1373 cm^{-1}). Most of the vinyl compounds absorb at 1400 $\pm$ 25 cm^{-1} so that this band is usually a good indicator for the presence of the $-CH=CH_2$ fragment.

The δCH appears at 1285 $\pm$ 45 cm^{-1}. Disregarding some extreme values in the spectra of $D_3COCH=CH_2$ (1328) and $MeOCH=CH_2$ (1324 cm^{-1}) on the one side and in those of *trans*-$H_2C=CHCH=CHCH=CH_2$ (1245 and *1280*), *S-trans*-$FCH_2CH=CH_2$ (1243), $BrCH=CH_2$ (1258) and $D_3SiCH=CH_2$ (1260 cm^{-1}) on the other, the δCH is observed at 1295 $\pm$ 25 cm^{-1}.

The vibrational analysis of $R-CH=CH_2$ compounds reveals the second δCH_2, often assigned as the CH_2 rocking vibration, in the region with the greatest spread: 1095 $\pm$ 85 cm^{-1}. This region narrows to 1065 $\pm$ 55 cm^{-1} if some high wavenumbers in the spectra of $H_2C=CHCH=CHCH=CH_2$ (1180 and *1080*), $Me(Cl)CHCH=CH_2$ (1178), $MeCH=CH_2$ (1172), *trans*-$ClCH=CHCH=CH_2$ (1160), *cis*-$ClCH=CHCH=CH_2$ (1130), $(H_2C=CHCH_2)_3P$ (1160), $(H_2C=CHCH_2)_3As$ (1150), $Me_2C=CHCH=CH_2$ (1148), *trans*-$N\equiv CCH=CHCH=CH_2$ (1140) and $X(O=)CCH=CH_2$ (X = H, F, Cl, Br) (1140 $\pm$ 20 cm^{-1}) are not taken into account. The *trans* isomers (conformers) absorb at higher wavenumbers than the *cis* isomers (conformers). The $C-C=$ stretching vibration, occurring in the neighbourhood of 900 cm^{-1}, may also contribute to the absorption in the above-mentioned region, and the ρCH_2 provides a contribution to the absorption at $\approx$900 cm^{-1}, so that some investigators refer this ρCH_2 to the region 900 $\pm$ 50 cm^{-1} [77, 83, 86, 93].

The CH out-of-plane deformations

By far the most useful infrared bands to elucidate the $-CH=CH_2$ structure are the strong absorptions at $\approx$915 and $\approx$990 cm^{-1} with a weak overtone at $\approx$1830 and a very weak one at $\approx$1980 cm^{-1}. Together with the C=C stretching vibration, these out-of-plane deformations are very characteristic.

The CH wagging vibration, occurring at 975 $\pm$ 35 cm^{-1}, is sometimes described as the in-phase *trans* CH=CH wag and is least sensitive to the α-atom. Si-bonded vinyl scores high ($\approx$1000) and halogen-bonded vinyl low ($\approx$950 cm^{-1}), followed by O- and S-bonded vinyl ($\approx$970), $R'C\equiv CCH=CH_2$ and $N\equiv CCH=CH_2$ (970 $\pm$ 10 cm^{-1}). 1,3-Butadiene shows both CH wags at 991 and 967 cm^{-1}. Most of the $R-CH=CH_2$ compounds were found to give this wagging mode in the region 990 $\pm$ 10 cm^{-1}, and alkenes with the formula $Me(CH_2)_nCH=CH_2$ (n = 0–15) even at 990 $\pm$ 5 cm^{-1} [151].

The CH_2 wagging vibration is assigned at 895 $\pm$ 85 cm^{-1} and is more sensitive to the influence of the α-atom. The highest values (960 $\pm$ 20 cm^{-1}) are observed in the spectra of Si-substituted vinyl, followed by carbonyl-substituted vinyl (950 $\pm$ 30) and $F_3CCH=CH_2$ (965 cm^{-1}). The CH_2 wag in halogen-bonded (885 $\pm$ 25) and O- and S-bonded vinyl (855 $\pm$ 45 cm^{-1}) is shifted to lower values. The remaining molecules display this CH_2 wag in the region 915 ± 20 cm^{-1}. For the series $Me(CH_2)_nCH{=}CH_2$ (n = 0–15) this region is reduced to 910 ± 5 cm^{-1} [151].

The ωCH, occurring in the extensive region 565 ± 155 cm^{-1}, is often described as the CH_2 twisting vibration or as the *cis* CH=CH wag and is most under the influence of the α-atom. It is a weak to moderate band of minor importance as a diagnostic tool.

Skeletal —C=C deformation

As a matter of course this external skeletal deformation is very sensitive to the influence of the α-atom, which explains the broad region. The highest values are observed in the spectra of α-saturated vinyl compounds (490 ± 110 cm^{-1}) with 600 cm^{-1} for the *S-cis* conformer of $FCH_2CH{=}CH_2$ (*S-trans*: 430), so that the region 460 ± 80 cm^{-1} also gives satisfaction. Low values are reported in the spectra of Si-bonded vinyl (330 ± 80) and C=O bonded vinyl (345 ± 95 cm^{-1}) examples being $XC(=O)CH{=}CH_2$ (X = F, Cl, Br) compounds (270 ± 20 cm^{-1}).

Table 8.2 Absorption regions (cm^{-1}) of the normal vibrations of $—CH{=}CH_2$

Vibration	α-sat.	C=O bonded	conjugated	aromatic	O and S bonded	Si-bonded	Hal-bonded
ν_aCH_2	3095 ± 15	3105 ± 20	3085 ± 45	3045 ± 45	3115 ± 25	3070 ± 20	3120 ± 10
νCH	3025 ± 25	3040 ± 40	3035 ± 35	3005 ± 25	3070 ± 40	3010 ± 30	3090 ± 10
ν_sCH_2	2990 ± 20	3015 ± 25	2990 ± 40	2965 ± 35	3030 ± 40	2970 ± 20	3035 ± 10
νC=C	1650 ± 20	1630 ± 20	1610 ± 30	1630 ± 10	1620 ± 35	1605 ± 25	1630 ± 25
δCH_2	1420 ± 20	1400 ± 30	1410 ± 30	1410 ± 20	1390 ± 30	1410 ± 20	1375 ± 05
δCH	1275 ± 35	1285 ± 15	1280 ± 40	1305 ± 15	1300 ± 30	1270 ± 20	1280 ± 20
δCH_2	1100 ± 80	1090 ± 70	1095 ± 85	1055 ± 35	1060 ± 50	1020 ± 10	1070 ± 50
ωCH	990 ± 10	985 ± 15	980 ± 20	985 ± 10	970 ± 25	1000 ± 10	950 ± 10
ωCH_2	935 ± 30	950 ± 30	910 ± 40	920 ± 20	855 ± 45	960 ± 20	885 ± 25
τCH_2	600 ± 80	605 ± 60	615 ± 90	600 ± 50	650 ± 70	475 ± 65	645 ± 65
δ—C=C	490 ± 110	345 ± 95	390 ± 100	470 ± 80	480 ± 110	330 ± 80	440 ± 50
torsion	140 ± 60	140 ± 50	150 ± 50	–	140 ± 50	110 ± 40	

$R—CH{=}CH_2$ compounds

R = Me— [6–11], D_3C— [8], Et— [6, 10–13], nBu— [14], $Me(CH_2)_n$— (n = 2–15) [11], $Cl(CH_2)_n$— [15, 16], $Br(CH_2)_n$— [15, 16], $H_2C{=}CHCH_2$— [17, 18], $H_2C{=}CHCH_2CH_2$—, $N{\equiv}CCH_2$— [13, 19–21], $HOCH_2$—, $DOCH_2$— and $HOCD_2$— [22], $MeC(=O)OCH_2$— [23], $ONOCH_2$— and O_2NOCH_2— [24], $OCNCH_2$— and $SCNCH_2$— [146], H_2NCH_2— [25–28], D_2NCH_2— [26, 28], H_2NCD_2— [26], N_3CH_2— [29], $HSCH_2$— [30], $CH_2{=}CHCH_2SSCH_2$— [31], $(H_2C{=}CHCH_2)_2PCH_2$— and $(H_2C{=}CHCH_2)_2AsCH_2$— [32], FCH_2— [33–36], $ClCH_2$— [33, 37–41],

$ClCD_2$— [33, 38], $BrCH_2$— [33, 37, 42], ICH_2— [33, 43], iPr— [11], cHex—, Me(NC)CH— [13], Me(Cl)CH— [44, 45], tBu— [11], F_3C— [9, 46], Cl_3C— [47], H(O=)C— [5, 48–52], D(O=)C— [5, 50, 52], Me(O=)C— [49, 53–56], HO(O=)C— [57-59], MeO(O=)C— [60–65], EtO(O=)C— [62], NaO(O=)C— [57], H_2N(O=)C— [66], Me_2N(O=)C— [67], F(O=)C— [68, 69], Cl(O=)C— [68, 70–72], Br(O=)C— [68], H_2C=CH— [5, 48, 50, 51, 73–81], H_2C=CHCH=C=CH— [82], H_2C=CHCH=CH— [83–87], MeCH=CH— [88], MeCH=CHCH=CH— [84], H_2C=C=CH— [89], ClCH=CH— [90], N≡CCH=CH— [91], H_2C=C(C≡CH)— [92], Me_2C=CH— [93], H_2C=C(Me)— [79, 94], H_2C=C(Cl)— [53, 70, 95, 96], HC≡C— [97–99], DC≡C— [99], NaC≡C— [98], N≡C— [98, 100], ClC≡C— [101], CN— [102], Ph— [11, 103–108], 2-XPh (X = Me [11, 103, 109], Cl and Br [11, 103], F [103, 110]), 3-XPh— (X = Me [103, 109], F, Cl, Br and O_2N [103], Et and HO [11]), 4-XPh— (X = Me [11, 109], Et, iPr, tBu, F, Cl, Br, HO and N≡C [11]), F_5Ph— [111], 2-Py— and 4-Py— [106], MeO— [112–119], CD_3O— [114, 116], EtO— [120–123], F_3CCH_2O— [124], HC(=O)O— [125], FC(=O)O— [126], MeC(=O)O— [61], F_3CC(=O)O— [127], H_2C=CHO— [128], Me_3SiO— and $(CD_3)_3SiO$— [119, 129], HS— [130], MeS— [131, 132], PhS—, H_2C=CHS$(=O)_2$— [148, 149] H_3Si— and D_3Si— [133, 134], MeH_2Si— [135], ClH_2Si— [135, 136], FH_2Si— [135], Me_3Si— [137], $MeCl_2Si$— [138], F_3Si— [46, 139], Cl_3Si— [140], F— [11, 141–143], Cl— [11, 143–145, 150], Br— [11, 143, 145].

References

1. H.C. Allen Jr. and E.K. Plyler, *J. Am. Chem. Soc.*, **80**, 2674 (1958).
2. M.S.J. Dewar and H.S. Rzepa, *J. Mol. Struct.*, **40**, 145 (1977).
3. E. Rytter and D.M. Gruen, *Spectrochim. Acta, Part A*, **35A**, 199 (1979).
4. H. Kollmar and V. Staemmler, *J. Am. Chem. Soc.*, **100**, 4304 (1978).
5. P. Pulay, G. Fogarasi, G. Pongor, J.E. Boggs and A. Vargha, *J. Am. Chem. Soc.*, **105**, 7037 (1983).
6. G. Busca, G. Ramis, V. Lorenzelli, A. Janin and J.-C. Lavalley, *Spectrochim. Acta, Part A*, **43A**, 489 (1987).
7. E.B. Wilson and A.J. Wells, *J. Chem. Phys.*, **9**, 319 (1941).
8. B. Silvi, P. Labarbe and J.P. Perchard, *Spectrochim. Acta, Part A*, **29A**, 263 (1973).
9. I. Tokue, T. Fukuyama and K. Kuchitsi, *J. Mol. Struct.*, **17**, 207 (1973).
10. A.J. Barnes and J.D.R. Howells, *J. Chem. Soc. Faraday Trans. 2*, **69**, 532 (1973).
11. R.A. Nyquist, *Appl. Spectrosc.*, **40**, 196 (1986).
12. J.R. Durig and D.A.C. Compton, *J. Phys. Chem.*, **84**, 773 (1980).
13. D.A.C. Compton and W.F. Murphy, *Spectrochim. Acta, Part A*, **41A**, 1141 (1985).
14. H.W. Schrötter and E.G. Hoffmann, *Liebigs Ann. Chem.*, **672**, 44 (1964).
15. G.A. Crowder and N. Smyrl, *J. Mol. Struct.*, **8**, 255 (1971).
16. G.A. Crowder, *J. Mol. Struct.*, **10**, 290 (1971).
17. E. Gallinella and B. Cadioli, *J. Chem. Soc. Faraday Trans. 2*,, **71**, 781 (1975).

18. F. Inagaki, M. Sakakibara, I. Harara and T. Shimanouchi, *Bull. Chem. Soc. Jpn.*, **48**, 3557 (1975).
19. G.H. Griffith, L.A. Harrah, J.W. Clark and J.R. Durig, *J. Mol. Struct.*, **4**, 255 (1969).
20. A.L. Verma, *J. Mol. Spectrosc.*, **39**, 247 (1971).
21. D.A.C. Compton, S.C. Hsi and H.H. Mantsch, *J. Phys. Chem.*, **85**, 3721 (1981).
22. B. Silvi and J.P. Perchard, *Spectrochim. Acta, Part A*, **32A**, 11 (1976).
23. B. Singh, R. Prasad and R.M.P. Jaiswal, *Proc. Indian Acad. Sci.*, **89**, 201 (1980).
24. R. Maas, Thesis, UIA, Antwerp, 1992.
25. A.L. Verma and P.Venkateswarlu, *J. Mol. Spectrosc.*, **39**, 227 (1971).
26. B. Silvi and J.P. Perchard, *Spectrochim. Acta*, **32A**, 23 (1976).
27. K. Yamanouchi, T. Matsuzawa, K. Kuchitsu, Y. Hamada and M. Stuboi, *J. Mol. Struct.*, **126**, 305 (1985).
28. J.R. Durig, J.F. Sullivan and C.M. Whang, *Spectrochim. Acta, Part A*, **41A**, 129 (1985).
29. P. Klaboe, K. Kosa, C.J. Nielsen, H. Priebe and S.H. Schei, *J. Mol. Struct.*, **176**, 107 (1988).
30. C.S.Hsu, *Spectrosc. Lett.*, **7**, 439 (1974).
31. H. Suzuki, K. Fukushi, S.-I. Ikawa and S. Konaka, *J. Mol. Struct.*,**221**, 141 (1990).
32. G. Davidson and S. Phillips, *Spectrochim. Acta, Part A*, **35A**, 83 (1979).
33. R.D. McLachlan and R.A. Nyquist, *Spectrochim. Acta, Part A*, **24A**, 103 (1968).
34. J.R. Durig, M. Zhen, H.L. Heusel, P.J. Joseph, P. Grower and T.S. Little, *J. Phys. Chem.*, **89**, 2877 (1985).
35. J.R. Durig, T.J. Geyer, T.S. Little and D.T. Durig, *J. Mol. Struct.*, **172**, 165(1988).
36. J. Nieminen, J. Murto and M. Räsänen, *Spectrochim. Acta, Part A*, **47A**, 1495 (1991).
37. C. Sourisseau and P. Pasquier, *J. Mol. Struct.*, **12**, 1 (1972).
38. B. Silvi and C. Sourisseau, *Spectrochim. Acta, Part A*, **31A**, 565 (1975).
39. A.J. Barnes, S. Holroyd, W.O. George, J.E. Goodfield and W.F. Maddams, *Spectrochim. Acta, Part A*, **38A**, 1245 (1982).
40. D.A.C. Compton, S.C. Hsi, H.H. Mantsch and W.F. Murphy, *J. Raman Spectrosc.*, **13**, 30 (1982).
41. J.R. Durig, D.T. Durig, M.R. Jalilian, M. Zhen and T.S. Little, *J. Mol. Struct.*, **194**, 259 (1989).
42. J.R. Durig, Q. Tang and T.S. Little, *J. Mol. Struct.*, **269**, 257 (1992).
43. J.R. Durig, Q. Tang and T.S. Little, *J. Raman Spectrosc.*, **23**, 653 (1992).
44. J.N. Som and G.S. Kastha, *Indian J. Phys.*, **51B**, 77 (1977).
45. S.H. Schei and P. Klaboe, *Acta Chem. Scand., Ser. A*, **37A**, 315 (1983).
46. G.A. Crowder and N. Smyrl, *J. Chem. Phys.*, **53**, 4102 (1970).
47. E.R. Shull, *J. Chem. Phys.*, **27**, 399 (1957).
48. R.K. Harris, *Spectrochim. Acta*, **20**, 1129 (1964).
49. A.J. Bowles, W.O. George and W.F. Maddams, *J. Chem. Soc. B*, 810 (1969).
50. Yu.N. Panchenko, P. Pulay and F. Török, *J. Mol. Struct.*, **34**, 283 (1976).
51. Z. Smith, E.B. Wilson and R.W. Duerst, *Spectrochim. Acta, Part A*, **39A**, 1117 (1983).
52. H.J. Oelichmann, D. Bougeard and B. Schrader, *J. Mol. Struct.*, **77**, 149 (1981).
53. R.K. Harris and R.E. Witkowski, *Spectrochim. Acta*, **20**, 1651 (1964).
54. H.J. Oelichmann, D. Bougeard and B. Schrader, *J. Mol. Struct.*, **77**, 179 (1981).
55. J.R. Durig and T.S. Little, *J. Chem. Phys.*, **75**, 3660 (1981).
56. J. De Smedt, F. Vanhouteghem, C. Van Alsenoy, H.J. Geise, B.J. Van der Veken and P. Coppens, *J. Mol. Struct.*, **195**, 227 (1989).
57. W.R. Feairheller and J.E. Katon, *Spectrochim. Acta, Part A*, **23A**, 2225 (1967).
58. S.W. Charles, F.C. Cullen, N.L. Owen and G.A. Williams, *J. Mol. Struct.*, **157**, 17 (1987).

59. P.F. Krause, J.E. Katon and K.K. Smith, *Spectrochim. Acta, Part A*, **32A**, 960 (1976).
60. W.R. Walton and R.B. Hughes, *J. Am. Chem. Soc.*, **79**, 3985 (1957).
61. W.R. Feairheller and J.E. Katon, *J. Mol. Struct.*, **1**, 239 (1967).
62. A.J. Bowles, W.O. George and D. Cunliffe-Jones, *J. Chem. Soc.*, **B**, 1070 (1970).
63. W.O. George, D.V. Hassed, W.C. Harris and W.F. Maddams, *J. Chem. Soc. Perkin Trans. 2*, 392 (1975).
64. R.M. Moravie and J. Corset, *J. Mol. Struct.*, **30**, 113 (1976).
65. P. Carmona and J. Moreno, *J. Mol. Struct.*, **82**, 177 (1982).
66. N. Jonathan, *J. Mol. Spectrosc.*, **6**, 205 (1961).
67. G.R. Rao and K.V. Ramiah, *Indian J. Pure Appl. Phys.*, **18**, 94 (1980).
68. J.R. Durig, R.J. Berry and P. Groner, *J. Chem. Phys.*, **87**, 6303 (1987).
69. J.R. Durig, J.S. Church and D.A.C. Compton, *J. Chem. Phys.*, **71**, 1175 (1970).
70. D.A.C. Compton, W.O. George, J.E. Goodfield and W.F. Maddams, *Spectrochim. Acta, Part A*, **37A**, 147 (1981).
71. J.E. Katon and W.R. Feairheller Jr., *J. Chem. Phys.*, **47**, 1248 (1967).
72. R.L. Redington and J.R. Kennedy, *Spectrochim. Acta, Part A*, **30A**, 2197 (1974).
73. Yu.N. Panchenko, *Spectrochim. Acta, Part A*, **31A**, 1201 (1975).
74. E. Benedetti, M. Aglietto, S. Pucci, Yu.N. Panchenko, Yu.A. Pentin and O.T. Nikitin, *J. Mol. Struct.*, **49**, 293 (1978).
75. G. Busca, *J. Mol. Struct.*, **117**, 103 (1984).
76. Yu.N. Panchenko and P. Császár, *J. Mol. Struct.*, **130**, 207 (1985).
77. C.W. Bock, Yu.N. Panchenko, S.V. Krasnoshchiokov and V.I. Pupyshev, *J. Mol. Struct.*, **129**, 57 (1985).
78. K.B. Wiberg and R.E. Rosenberg, *J. Am. Chem. Soc.*, **112**, 1509 (1990).
79. D.A.C. Compton, W.O. George and W.F. Maddams, *J. Chem. Soc. Perkin Trans. 2*, 1666 (1976).
80. P. Huber-Wälchli and H.H. Günthard, *Spectrochim. Acta, Part A*, **37A**, 285 (1981).
81. Y. Furukawa, H. Takeuchi, I. Harada and M. Tasumi, *Bull. Chem. Soc. Jpn.*, **56**, 392 (1983).
82. A. Phongsatha, P. Klaboe, H. Hopf, B.N. Cyvin and S.J. Cyvin, *Spectrochim. Acta, Part A*, **34A**, 537 (1978).
83. Yu.N. Panchenko, P. Császár and F. Török, *Acta Chim. Hung.*, **113**, 149 (1983).
84. F.W. Langkilde, R. Wilbrandt, O.F. Nielsen, D.H. Christensen and F.M. Nicolaisen, *Spectrochim. Acta, Part A*, **43A**, 1209 (1987).
85. R. McDiarmid and A. Sabljic, *J. Phys. Chem.*, **91**, 276 (1987).
86. C.W. Bock, Yu.N. Panchenko, S.V. Krasnoshchiokov and V.I. Pupyshev, *J. Mol. Struct.*, **148**, 131 (1986).
87. H. Yoshida, Y. Furukawa and M. Tasumi, *J. Mol. Struct.*, **194**, 279 (1989).
88. D.A.C. Compton, W.O. George and W.F. Maddams, *J. Chem. Soc. Perkin Trans. 2*, 1311 (1977).
89. P. Klaboe, T. Torgrimsen, D.H. Christensen, H. Hopf, A. Erikson, G. Hagen and S.J. Cyvin, *Spectrochim. Acta, Part A*, **30A**, 1527 (1974).
90. A. Borg, Z. Smith, G. Gundersen and P. Klaboe, *Spectrochim. Acta, Part A*, **36A**, 119 (1980).
91. B.H. Thomas, W.J. Orville-Thomas, *J. Mol. Struct.*, **3**, 191 (1969).
92. H. Priebe, C.J. Nielsen, P. Klaboe, H. Hopf and H. Jäger, *J. Mol. Struct.*, **158**, 249 (1987).
93. M.M.A. Aly, M.H. Baron, M.J. Coulange and J. Favrot, *Spectrochim. Acta, Part A*, **42A**, 411 (1986).
94. M. Traetteberg, G. Paulen, S.J. Cyvin, Yu.N. Panchenko and V.I. Mochalov, *J. Mol. Struct.*, **116**, 141 (1984).

95. G.J. Szasz and N. Sheppard, *Trans. Faraday Soc.*, **49**, 358 (1953).
96. Yu.N. Panchenko, O.E. Grikina, V.I. Mochalov, Yu.A. Pentin, N.F. Stepanov, R. Aroca, J. Mink, A.N. Akopyan, A.V. Rodin and V.K. Matveev, *J. Mol. Struct.*, **49**, 17 (1978).
97. N. Sheppard, *J. Chem. Phys.*, **17**, 74 (1949).
98. J. Kanesaka, K. Miyawaki and K. Kawai, *Spectrochim. Acta, Part A*, **32A**, 195 (1976).
99. E. Tørneng, C.J. Nielsen and P. Klaboe, *Spectrochim. Acta, Part A*, **36A**, 975 (1980).
100. F. Halverson, R.F. Stamm and J.J. Whalen, *J. Chem. Phys.*, **16**, 808 (1948).
101. A. Borg and P. Cederbalk, *Acta Chem. Scand., Ser. A*, **40A**, 103 (1986).
102. K. Bolton, N.L. Owen and J. Sheridan, *Spectrochim. Acta, Part A*, **26A**, 909 (1970).
103. W.G. Fately, G.L. Carlson and F.E. Dickson, *Appl. Spectrosc.*, **22**, 651 (1968).
104. W.D. Mross and G. Zundel, *Spectrochim. Acta, Part A*, **26A**, 1109 (1970).
105. D.A. Condirston and J.D. Laposa, *J. Mol. Spectrosc.*, **63**, 466 (1976).
106. J.H.S. Green and D.J. Harrison, *Spectrochim. Acta, Part A*, **33A**, 249 (1977).
107. T.R. Gilson, J.M. Hollas, E. Khalilipour and J.V. Warrington, *J. Mol. Spectrosc.*, **73**, 234 (1978).
108. A. Marchand and J.P. Quintard, *Spectrochim. Acta, Part A*, **36A**, 941 (1980).
109. P.P. Garg and R.M.P. Jaiswal, *Indian J. Pure Appl. Phys.*, **27**, 75 (1989).
110. J.M. Hollas and M.Z. Bin Hussein, *J. Mol. Spectrosc.*, **136**, 31 (1989).
111. J.H.S. Green, D.J. Harrison and C.P. Stockley, *Spectrochim. Acta, Part A*, **33A**, 423 (1977).
112. I.S. Ignatyev, A.N. Lazarev, M.B. Smirnov, M.L. Alpert and B.A. Trofimov, *J. Mol. Struct.*, **72**, 25 (1981).
113. B. Cadioli, E. Gallinella and U. Pincelli, *J. Mol. Struct.*, **78**, 215 (1982).
114. W. Pyckhout, P. Van Nuffel, C. Van Alsenoy, L. Van Den Enden and H.J. Geise, *J. Mol. Struct.*, **102**, 333 (1983).
115. T. Beech, R. Gunde, P. Felder and H.H. Günthard, *Spectrochim. Acta, Part A*, **41A**, 319 (1985).
116. J.F. Sullivan, T.J. Dickson and J.R. Durig, *Spectrochim. Acta, Part A*, **42A**, 113 (1986).
117. N.L. Owen and N. Sheppard, *Trans. Faraday Soc.* **60**, 634 (1964).
118. P. Cahill, L.P. Gold and N.L. Owen, *J. Chem. Phys.*, **48**, 1620 (1968).
119. A.N. Lazarev, I.S. Ignat'ev, L.L. Schukovskaya and R.I. Pal'chik, *Spectrochim. Acta, Part A*, **27A**, 2291 (1971).
120. N.L. Owen and N. Sheppard, *Spectrochim. Acta*, **22**, 1101 (1966).
121. M. Sakakibara, F. Inagaki, I. Harada and T. Shimanouchi, *Bull. Chem. Soc. Jpn.*, **49**, 46 (1976).
122. J.R. Durig and D.J. Gerson, *J. Mol. Struct.*, **71**, 131 (1981).
123. N.L. Owen and G.O. Sørensen, *J. Phys. Chem.*, **83**, 1483 (1979).
124. S.W. Charles, F.C. Cullen and N.L. Owen, *J. Chem. Soc. Faraday Trans. 2*, **70**, 483 (1974).
125. W. Pyckhout, C. Van Alsenoy, H.J. Geise, B. Van der Veken, P. Coppens and M. Traetteberg, *J. Mol. Struct.*, **147**, 85 (1986).
126. J.R. Durig, J. Lin and B.J. Van der Veken, *J. Raman Spectrosc.*, **23**, 287 (1992).
127. G.A. Crowder, *Spectrochim. Acta, Part A*, **28A**, 1625 (1972).
128. J.M. Comerford, P.C. Anderson, W.H. Snyder and H.S. Kimmel, *Spectrochim. Acta, Part A*, **33A**, 651 (1977).
129. J. Dedier and A. Marchand, *Spectrochim. Acta, Part A*, **38A**, 339 (1982).
130. V. Almond, S.W. Charles, J.N. Macdonald and N.L. Owen, *J. Mol. Struct.*, **100**, 223 (1983).
131. J. Fabian, H. Krober and R. Mayer, *Spectrochim. Acta, Part A*, **24A**, 727 (1968).
132. S. Samdal, H.M. Seip and T. Torgrimsen, *J. Mol. Struct.*, **57**, 105 (1979).
133. S.G. Frankiss, *Spectrochim. Acta*, **22**, 295 (1966).

134. V.F. Kalasinsky, S.E. Rodgers and J.A.S. Smith, *Spectrochim. Acta, Part A*, **41A**, 155 (1985).
135. J.R. Durig, J.F. Sullivan and M.A. Qtaitat, *J. Mol. Struct.*, **243**, 239 (1991).
136. J.R. Durig, J.F. Sullivan, G.A. Guirgis and M.A. Qtaitat, *J. Phys. Chem.*, **95**, 1563 (1991).
137. J.R. Durig, W.J. Natter and M. Johnson-Streusand, *Appl. Spectrosc.*, **34**, 60 (1980).
138. K. Taga, T. Yoshida, H. Okabayashi, K. Ohno and H. Matsuura, *J. Mol. Struct.*, **192**, 63 (1989).
139. J.R. Durig, and K.L. Hellams, *J. Mol. Struct.*, **6**, 315 (1970).
140. E.R. Shull, R.A. Thursack and C.M. Birdsall, *J. Chem. Phys.*, **24**, 147 (1956).
141. B. Bak and D.H. Christensen, *Spectrochim. Acta*, **12**, 355 (1958).
142. F.A. Andersen and K.A. Jensen, *J. Mol. Struct.*, **60**, 165 (1980).
143. D.C. McKean, *Spectrochim. Acta, Part A*, **31A**, 1167 (1975).
144. E. Enomoto and M. Asahina, *J. Mol. Spectrosc.*, **19**, 117 (1966).
145. C.W. Gullikson and J.R. Nielsen, *J. Mol. Spectrosc.*, **1**, 158 (1957).
146. T. Torgrimsen, P. Klaboe and F. Nicolaisen, *J. Mol. Struct.*, **20**, 213 (1974).
147. J.L. Duncan, *Spectrochim. Acta, Part A*, **26A**, 430 (1970).
148. I. Hargittai, B. Rozsondai, B. Nagel, P. Bulcke, G. Robinet and J.-F. Labarre, *J. Chem. Soc. Dalton Trans.*, 861 (1978).
149. N.L. Allinger and Y. Fan, *J. Comput. Chem.*, **14**, 655 (1993).
150. E. Diana, O. Gambino, R. Rossetti and P.L. Stanghellini, *Spectrochim. Acta, Part A*, **49A**, 1247 (1993).
151. C.J. Pouchert, *The Aldrich Library of FT-IR Spectra*, Aldrich Chemical Company, 1st edn. (1985).

8.1.2 Vinylidene (ethenylidene)

The compound $Cl_2C{=}CH_2$ belongs to the point group C_{2v} and the twelve normal vibrations are divided among $5a_1 + a_2 + 2b_1 + 4b_2$ types of vibration. The a_2 vibration is forbidden.

Table 8.3 Normal vibrations of 1,1-dichloroethene

Vibration	C_{2v}	cm^{-1}	Vibration	C_{2v}	cm^{-1}
ν_sCH_2	a_1	3035	ωCH_2	b_1	872
$\nu C{=}C$		1614	ωCCl_2		435
δCH_2		1391	ν_aCH_2	b_2	3130
ν_sCCl_2		601	ρCH_2		1076
δCCl_2		298	ν_aCCl_2		793
τCH_2	a_2	–	ρCCl_2		370

If the two halogen atoms are different, the symmetry is lowered from C_{2v} to C_s and the $a_1 + b_2$ vibrations become a' and the $a_2 + b_1$ vibrations become a'' vibrations. In this work only one free bond in the $CH_2{=}C{<}$ group is taken into consideration, so that for this fragment $3N - 6 = 9$ vibrations are taken into account:

$\nu_a CH_2$, $\nu_s CH_2$, $\nu C{=}C$, δCH_2, ρCH_2, ωCH_2, τCH_2, γ—C=C and δ—C=C

or simply: νCH (2), νC=C, δCH (2), ωCH (2) and skeletal deformations (2).

The twelve vibrations of $CH_2{=}C(X)$— are found by adding one C—X stretching vibration, one C=C—X deformation and one torsion to the nine vibrations of $CH_2{=}C{<}$. The absorption frequency of the C=C—X deformation is comparable with that of the similar external skeletal deformation and is added to the table as 'alternative skeletal deformation'.

The CH stretching vibrations

The CH stretching vibrations are observed between 3150 and 2990 cm^{-1} with a weak to moderate intensity. Although the absorption regions overlap each other, the two CH stretching vibrations are rarely assigned at the same wavenumber.

The highest CH stretching vibration, often called $\nu_a CH_2$ (3105 ± 45 cm^{-1}), is assigned at 3144 cm^{-1} in the spectrum of $FC({=}O)(F)C{=}CH_2$, followed by the values in the spectra of $Me(F)C{=}CH_2$, $Cl(Br)C{=}CH_2$ and $H_2C{=}CCl(Br)C{=}CH_2$ with 3140 cm^{-1} and $(NC)_2C{=}CH_2$ and $Cl(NC)C{=}CH_2$ with 3135 cm^{-1}. The lowest values appear in the spectra of $H_2C{=}CF(F)C{=}CH_2$ (3062), $Me(NCCH_2)C{=}CH_2$ (3066), $H_2C{=}CCl(F)C{=}CH_2$ and $F(Cl)C{=}CH_2$ with 3069 cm^{-1} and $H_2CH(Cl)C{=}CH_2$ with 3070 cm^{-1}. The remaining compounds absorb at 3100 ± 30 cm^{-1}.

The CH stretching vibration absorbing in the region 3030 ± 40 cm^{-1}, is considered as the $\nu_s CH_2$. The HW side of this region is limited by $FC({=}O)(F)C{=}CH_2$, $Et(Me)C{=}CH_2$, $iPr(Me)C{=}CH_2$, $HOCH_2(Me)C{=}CH_2$ and $BrCH_2(Me)C{=}CH_2$ with values in the neighbourhood of 3070 cm^{-1}. Low wavenumbers ($\approx$2990 cm^{-1}) are observed in the spectra of $Me_2C{=}CH_2$, $(BrCH_2)_2C{=}CH_2$, $H_2C{=}CH(Me)C{=}CH_2$, $MeCH{=}CH(Me)C{=}CH_2$, $HC{\equiv}C(Me)C{=}CH_2$ and $HOC({=}O)(Me)C{=}CH_2$.

The C=C stretching vibration

With the exception of high values for $F_2C{=}CH_2$ (1730) and $F(Me)C{=}CH_2$ (1687 cm^{-1}), the C=C stretching vibration occurs at 1625 ± 50 cm^{-1} with a moderate intensity. In consequence, misinterpretation of a fluorine-substituted double bond as a carbonyl group is not excluded. High values originate also from the spectra of $H_2C{=}CF(F)C{=}CH_2$ (1675) and $MeO(Me)C{=}CH_2$ (1670 cm^{-1}). The lowest wavenumbers have been traced in the spectra of conjugated alkenes such as $H_2C{=}CH(HC{\equiv}C)C{=}CH_2$ (1576), $H_2C{=}CBr(Br)C{=}CH_2$ (1577 and *1603*), $H_2C{=}CCl(Cl)C{=}CH_2$ (1581 and *1607*) and $H_2C{=}CH(Cl)C{=}CH_2$ (1585 and *1635*) and in the spectrum of $Br_2C{=}CH_2$ (1593 cm^{-1}). The remaining molecules display this νC=C at 1630 ± 30 cm^{-1}.

The CH in-plane deformations

The δCH_2 is reported in the range 1395 $\pm$ 45 cm^{-1}. Ph(Me)C=CH_2 and 4-FPh(Me)C=CH_2 show this CH in-plane deformation at 1440 cm^{-1} and HOC(=O)(Me)C=CH_2 at 1432 cm^{-1}. The compounds H_2C=CX(X)C=CH_2 with X = Br (1352 and *1392*), Cl (1357 and *1386*) and F (1365 and *1385*) and H_2C=CH(Cl)C=CH_2 (1365 or *1420* cm^{-1}) are responsible for the low values. Usually this δCH_2 is assigned in the region 1400 $\pm$ 30 cm^{-1}. As the δCH_2 in R—CH=CH_2 compounds also absorbs in this region, the CH in-plane deformation in 2-substituted 1,3-butadienes is assigned without specification of the group.

Generally the CH_2 rocking vibration is assigned in the region 1000 $\pm$ 90 cm^{-1}. A few investigators situate this ρCH_2 at higher wavenumbers: Et(Cl)C=CH_2 (1171) and Et(Br)C=CH_2 (1161) [63], N≡C(Cl)C=CH_2 (1165) [48], $ClCH_2$(Cl)C=CH_2 (1132) [35, 37], $ClCH_2$(Br)C=CH_2 (1127) [38] and $(cPr)_2$C=CH_2 (1172 cm^{-1}) [39]. For the above-mentioned compounds the values of, respectively, 1007, 996, 930, 938 [34], 941 [36] and 958 cm^{-1} may also be considered for this ρCH_2. The CH_2 rock as well as the C—C stretching vibration contribute to both absorptions. High values for this ρCH_2 are found in the spectra of Cl_2C=CH_2 (1088) and $BrCH_2$(Br)C=CH_2 (1090 cm^{-1}) and low wavenumbers in those of $BrCH_2$(Me)C=CH_2 (910) and Me(I)C=CH_2 (927 cm^{-1}). Such compounds as H_2C=Cl(Cl)C=CH_2 are active at both sides of this region (1080 and 915 cm^{-1}). RR′C=CH_2 molecules in which R and R′ equals CH_2 or CH_3 show this ρCH_2 in a narrower region (975 $\pm$ 15), for instance: $(Me)_2$C=CH_2 (974), Et(Me)C=CH_2 (989), nPr(Me)C=CH_2 (990), $HOCH_2$(Me)C=CH_2 (964), N≡CCH_2(Me)C=CH_2 (964), $(ClCH_2)_2$C=CH_2 (985) and $(BrCH_2)_2$C=CH_2 (985 cm^{-1}).

The CH out-of-plane deformations

The CH_2 wagging vibration gives rise to a moderate to strong band in the region 875 $\pm$ 75 cm^{-1} and for —C(—C)C=CH_2 compounds in the smaller region 895 $\pm$ 50 cm^{-1}, often around 890 cm^{-1} with a weak overtone at ≈1780 cm^{-1}. The region for 1,1-halogen-substituted derivatives is situated at lower wavenumbers (845 $\pm$ 45 cm^{-1}). The highest wavenumbers are observed in the spectra of C=O-substituted vinylidene compounds such as HC(=O)(Me)C=CH_2 (948), HOC(=O)(Me)C=CH_2 (948) and MeOC(=O)(Me)C=CH_2 (945 cm^{-1}). The lowest values are from 1,1-difluoroethene (803), F(Cl)C=CH_2 (836), H_2C=CCl(F)C=CH_2 (836) and MeO(Me)C=CH_2 (830 cm^{-1}). The remaining compounds absorb in the region 890 $\pm$ 45 cm^{-1}. This ωCH_2 is, together with the νC=C, the most useful infrared band for recognizing the vinylidene group.

The CH_2 twist, absorbing in the region 705 $\pm$ 105 cm^{-1}, is of minor importance as a diagnostic tool. With the exception of the highest and lowest values summarized in Table 8.4, the τCH_2 is assigned at 705 $\pm$ 55 cm^{-1}.

Table 8.4 CH_2 twisting vibration in $RR'C{=}CH_2$ compounds

R	R′	cm^{-1}	R	R′	cm^{-1}
Me	FC(=O)	808	F	Cl	607
FCH_2	Cl	804	F	$H_2C{=}CCl$	607
cPr	cPr	771	F	Me	629
Me	ClC(=O)	769	Cl	$ClCH_2$	633
Me	HC≡C	765	Br	$BrCH_2$	637
$ClCH_2$	$ClCH_2$	765	Ph	Ph	638

Skeletal deformations

The external out-of-plane skeletal deformation is assigned in a region (475 ± 85 cm^{-1}) limited by the compounds summarized in Table 8.5.

Table 8.5 γ—C=C in $RR'C{=}CH_2$ compounds

R	R′	cm^{-1}	R	R′	cm^{-1}
Br	$H_2C{=}CBr$	559	Br	FCH_2	390 (*s—cis*)
Me	$H_2C{=}CMe$	558	Br	$BrCH_2$	397 (*s—tr*)
Me	$HOCH_2$	557	Br	$ClCH_2$	398 (*s—tr*)
N≡C	Cl	557	I	Me	400
Me	HO(O=)C	555	Br	Br	404
Cl	$H_2C{=}CCl$	554	Br	Me	409

The external in-plane skeletal deformations can be found in the region 325 ± 145 cm^{-1}: generally a δ—C=C at 395 ± 75 and a second 'alternative' δ—C=C at 290 ± 110 cm^{-1}. The lowest values are observed in the spectra of halogen-substituted vinylidene compounds.

Table 8.6 Absorption regions (cm^{-1}) of the normal vibrations of $RR'C{=}CH_2$

R R′	saturated saturated	saturated conjugated	conjugated conjugated	saturated halogen	conjugated halogen	halogen halogen
$\nu_a CH_2$	3095 ± 35	3105 ± 20	3005 ± 30	3120 ± 20	3105 ± 45	3100 ± 40
$\nu_s CH_2$	3030 ± 40	3030 ± 40	3025 ± 35	3035 ± 35	3030 ± 40	3035 ± 25
νC=C	1650 ± 25	1630 ± 25	1595 ± 20	1640 ± 20	1625 ± 50	1625 ± 35
δCH_2	1410 ± 20	1420 ± 20	1400 ± 15	1405 ± 25	1390 ± 35	1395 ± 20
ρCH_2	960 ± 50	990 ± 50	970 ± 40	1005 ± 85	1000 ± 90	1015 ± 75
ωCH_2	890 ± 45	905 ± 45	915 ± 25	890 ± 45	875 ± 40	845 ± 45
τCH_2	720 ± 60	715 ± 95	690 ± 60	720 ± 90	690 ± 85	660 ± 55
γ—C=C	490 ± 70	485 ± 75	510 ± 50	475 ± 85	485 ± 75	475 ± 75
δ—C=C	420 ± 50	385 ± 65	405 ± 65	380 ± 60	370 ± 50	380 ± 60
alternative δ—C=C	350 ± 50	310 ± 50	340 ± 50	320 ± 60	270 ± 80	275 ± 95

$RR'C{=}CH_2$ compounds

R	R′
Me—	Me— [1, 2, 4], Et— [3], nPr— [3], $N{\equiv}CCH_2$— [4–8], $HOCH_2$—, $ClCH_2$— [8], $BrCH_2$— [7, 9], iPr, HC(=O)— [10, 11], HOC(=O)—, MeOC(=O)—, nBuOC(=O)— and nOctOC(=O)— [12], nDodecylOC(=O)— [13], FC(=O)— [14, 15], ClC(=O)— [16], BrC(=O)— [17], $H_2C{=}CH$— [18, 19], $H_2C{=}C(Me)$— [20], MeCH=CH— [21], Ph— [22], 4-FPh—, $HC{\equiv}C$—, $N{\equiv}C$— [23], F— [24], Cl— [25, 26], Br— [27], I— [28], MeO— [29, 30];
FCH_2—	Cl— [31], Br— [31];
$ClCH_2$—	$ClCH_2$— [32, 33], Cl— [34–37], Br— [36, 38];
$BrCH_2$—	$BrCH_2$— [32], Br— [35, 36];
cPr—	cPr— [39];
$H_2C{=}CH$—	$HC{\equiv}C$— [40], Cl— [18, 41–43];
$H_2C{=}C(Cl)$—	F—, Cl— [42, 44], Br—;
Ph—	Ph— [45];
$N{\equiv}C$—	$N{\equiv}C$— [46], EtOC(=O)— [47], Cl— [48];
MeS—	MeS—[49];
F—	$H_2C{=}CF$— [44], FC(=O)— [50], F— [51–56], Cl— [56];
Cl—	CD_3— [26], Et— [63], $Cl_2C{=}C(Cl)$— [42], Cl— [54–62], Br— [62];
Br—	Et— [63], $H_2C{=}C(Br)$— [44], Br— [53, 54, 56, 62].

References

1. C.M. Pathak and W.H. Fletcher, *J. Mol. Spectrosc.*, **31**, 32 (1969).
2. A.J. Barnes and J.D.R. Howells, *J. Chem. Soc. Faraday Trans. 2*, **69**, 532 (1973).
3. T. Shimanouchi, Y. Abe and M. Mikami, *Spectrochim. Acta, Part A*, **24A**, 1037 (1968).
4. D.A.C. Compton, S.C. Hsi and H.H. Mantsch, *J. Phys. Chem.*, **85**, 3721 (1981).
5. A.O. Diallo, *Spectrochim. Acta, Part A*, **35A**, 1189 (1979).
6. D.A.C. Compton, W.F. Murphy and H.H. Mantsch, *Spectrochim. Acta, Part A*, **37A**, 453 (1981).
7. S.H. Schei, *Spectrochim. Acta, Part A*, **39A**, 327 (1983).
8. D.A.C. Compton, S.C. Hsi, H.H. Mantsch and W.F. Murphy, *J. Raman Spectrosc.*, **13**, 30 (1982).
9. A.O. Diallo, *Spectrochim. Acta, Part A*, **36A**, 799 (1980).
10. H.J. Oelichamn, D. Bougeard and B. Schrader, *J. Mol. Struct.*, **77**, 179 (1981).
11. J.R. Durig, J. Qiu, B. Dehoff and T.S. Little, *Spectrochim. Acta, Part A*, **42A**, 89 (1986).
12. T.R. Manley and C.G. Martin, *Spectrochim. Acta, Part A*, **32A**, 357 (1976).
13. E. Butchert and T.R. Manley, *Spectrochim. Acta, Part A*, **34A**, 781 (1978).
14. J.R. Durig, P.A. Brletic and J.S. Church, *J. Chem. Phys.*, **76**, 1723 (1982).
15. B.C. Laskowski, R.L. Jaffe and A. Komornicki, *J. Chem. Phys.*, **82**, 5089 (1985).
16. J.R. Durig, P.A. Brletic, Y.S. Li, A.-Y. Wang and T.S. Little, *J. Mol. Struct.*, **223**, 291 (1990).

17. J.R. Durig, W. Zhao, R.J. Berry and T.S. Little, *J. Mol. Struct.*, **212**, 169 (1989).
18. R.K. Harris and R.E. Witkowsky, *Spectrochim. Acta*, **20**, 1651 (1964).
19. M. Traetteberg, G. Paulen, S.J. Cyvin, Yu.N. Panchenko and V.I. Mochalov, *J. Mol. Struct.*, **116**, 141 (1984).
20. C.W. Bock and Yu.N. Panchenko, *J. Mol. Struct.*, **221**, 159 (1990).
21. M.M.A. Aly, M.H. Baron, M.J. Coulange and J. Favrot, *Spectrochim. Acta, Part A*, **42A**, 411 (1986).
22. R.M.P. Jaiswal and P.P. Garg, *India J. Phys.*, **58B**, 307 (1984).
23. J. Bragin, K.L. Kizer and J.R. Durig, *J. Mol. Spectrosc.*, **38**, 289 (1971).
24. G.A. Crowder and N. Smyrl, *J. Mol. Spectrosc.*, **40**, 117 (1971).
25. H. Hunziker and H.H. Günthard, *Spectrochim. Acta*, **21**, 51 (1965).
26. R. Meyer, H. Hunziker and H.H. Günthard, *Spectrochim. Acta, Part A*, **23A**, 1775 (1967).
27. R. Meyer and H.H. Günthard, *Spectrochim. Acta, Part A*, **23A**, 2341 (1967).
28. R. Meyer, H. Hunziker and H.H. Günthard, *Spectrochim. Acta, Part A*, **25A**, 295 (1969).
29. A.O. Diallo, *Spectrochim. Acta, Part A*, **37A**, 529 (1981).
30. E. Gallinella, U. Pincelli and B. Cadioli, *J. Mol. Struct.*,**99**, 31 (1983).
31. P. Klaboe, T. Torgrimsen and D.H. Christensen, *J. Mol. Struct.*, **23**, 15 (1974).
32. R. Gaufrés and C. Roulph, *J. Mol. Struct.*, **9**, 107 (1971).
33. G.A. Crowder, *J. Mol. Struct.*, **10**, 294 (1971).
34. G.A. Crowder, *J. Mol. Spectrosc.*, **20**, 430 (1966).
35. T. Torgrimsen and P. Klaboe, *J. Mol. Struct.*, **20**, 229 (1974).
36. S.H. Schei, *Spectrochim. Acta, Part A*, **39A**, 1043 (1983).
37. D.T. Durig, G.A. Guirgis and J.R. Durig, *J. Raman Spectrosc.*, **23**, 37 (1992).
38. S.H. Schei and P. Klaboe, *J. Mol. Struct.*, **96**, 9 (1982).
39. A.B. Nease and C.J. Wurrey, *J. Phys. Chem.*, **83**, 2135 (1979).
40. H. Priebe, C.J. Nielsen, P. Klaboe, H. Hopf and H. Jäger, *J. Mol. Struct.*, **158**, 249 (1987).
41. G.J. Szasz and N. Sheppard, *Trans. Faraday Soc.*, **49**, 358 (1953).
42. Yu.N. Panchenko, O.E. Grikina, V.I. Mochalov, Yu.A. Pentin, N.F. Stepanov, R. Aroca, J. Mink, A.N. Akopyan, A.V. Rodin and V.K. Matveev, *J. Mol. Struct.*, **49**, 17 (1978).
43. D.A.C. Compton, W.O. George, J.E. Goodfield and W.F. Maddams, *Spectrochim. Acta, Part A*, **37A**, 147 (1981).
44. J.P. Toth and D.F. Koster, *Spectrochim. Acta, Part A*, **31A**, 1891 (1975).
45. A. Bree and R. Zwarich, *J. Mol. Struct.*, **75**, 213 (1981).
46. A. Rosenberg and J.P. Devlin, *Spectrochim. Acta*, **21**, 1613 (1965).
47. S. Reynolds, D.P. Oxley and R.G. Pritchard, *Spectrochim. Acta, Part A*, **38A**, 103 (1982).
48. S.B. Lie and P. Klaboe, *Spectrochim. Acta, Part A*, **26A**, 1191 (1970).
49. P. Jandal, H.M. Seip and T. Torgrimsen, *J. Mol. Struct.*, **32**, 369 (1976).
50. J.R. Durig, A.-I. Wang, T.S. Little, P.A. Brletic and J.R. Bucenell, *J. Chem. Phys.*, **91**, 7361 (1989).
51. D.C. Smith, J.R. Nielsen and H.H. Claassen, *J. Chem. Phys.*, **18**, 326 (1950).
52. W.F. Edgell and C.J. Ultee, *J. Chem. Phys.*, **22**, 1983 (1954).
53. J.R. Scherer and J. Overend, *J. Chem. Phys.*, **32**, 1720 (1960).
54. J.M. Freeman and T. Henshall, *Can. J. Chem.*, **47**, 935 (1969).
55. S. Jeyapandian and G.A.S. Raj, *J. Mol. Struct.*, **8**, 97 (1971).
56. D.E. Mann, N. Acquista and E.K. Plyler, *J. Chem. Phys.*, **23**, 2122 (1955).
57. P. Joyner and G. Glockler, *J. Chem. Phys.*, **20**, 302 (1952).
58. S. Enomoto and S. Echinohe, *Nippon Kagaku Zasshi*, 1343 (1958).
59. F. Winther and D.O. Hummel, *Spectrochim. Acta, Part A*, **23A**, 1839 (1967).
60. F. Winther, *Z. Naturforsch., Teil A*, **25A**, 1912 (1970).

61. E. D'Alessio, E. Silberman and E.A. Jones, *J. Mol. Struct.*, **9**, 393 (1971).
62. R.A. Nyquist and J.W. Thompson, *Spectrochim. Acta, Part A*, **33A**, 63 (1977).
63. G.A. Crowder and N. Smyrl, *J. Mol. Struct.*, **10**, 373 (1971).

8.1.3 Vinylene (ethenylene)

The X—CH=CH—X molecule in the *cis* {*trans*} configuration belongs to the point group C_{2v} {C_{2h}}. The twelve normal vibrations are being divided into $5a_1 + 2a_2 + b_1 + 4b_2$ {$5a_g + 2a_u + b_g + 4b_u$} species of vibration. The a_2 {a_g, b_g} vibrations are not active in the infrared.

Table 8.7 Normal vibrations of *cis* and *trans* 1,2-dichloroethene

Vibration	C_{2v}	C_{2h}	*cis*	*trans*	Vibration	C_{2v}	C_{2h}	*cis*	*trans*
νCH	a_1	a_g	3080	3073R	ωCCl	a_2	a_u	406R	225
νC=C			1590	1580R	ωCH	b_1	b_g	701	760R
δCH			1180	1272R	νCH	b_2	b_u	3080	3090
νCCl			711	844R	δCH			1296	1200
δCCl			173	349R	νCCl			845	820
ωCH	a_2	a_u	876R	896	δCCl			568	260

The molecule X—CH=CH—Y belongs to the point group C_s. Considering one free bond, the twelve normal vibrations (9a′ + 3a″) are reduced to the following $3N - 6 = 9$ vibrations for the —CH=CH— structure unit:

νCH (2), νC=C, δCH (2), ωCH (2), δ—C=C and torsion.

CH stretching vibrations

With the exception of the high values observed in the spectra of FCH=CHF (*cis:* 3136; *trans*: 3114), F_5PhCH=CHPhF_5 (3105) and *trans*-F_3CCH=CHCF_3 (3104 cm^{-1}), the RCH=CHR′ compounds display the highest νCH in the region 3055 $\pm$ 45 cm^{-1}.

The ν'CH (*cis*: 3020 $\pm$ 40; *trans*: 3025 $\pm$ 35 cm^{-1}) penetrates in the field of the sp^3-CH stretching vibrations. The high wavenumbers from FCH=CHF (*cis*: 3122; *trans*: 3111), *trans*-F_5PhCH=CHPhF_5 (3085), *trans*-F_3CCH=CHCF_3 (3084), *trans*-BrCH=CHBr (3081) and ClCH=CHCl (*cis*: 3080; *trans*: 3073 cm^{-1}) have not been taken into account. The lowest wavenumber (2980 cm^{-1}) is noted in the spectrum of *cis*-MeCH=CHMe.

The C=C stretching vibration

In the infrared the *cis* isomers absorb moderately but more strongly than the *trans* isomers. In delimiting the absorption region (1620 $\pm$ 60 cm^{-1}), the very high values from FCH=CHF (*cis*: 1715; *trans*: 1694) and $F_3CCH=CHCF_3$ (*cis*: 1696; *trans*: 1715R) and the extremely low values from ICH=CHI (*cis*: 1545; *trans*: 1537 cm^{-1}) are not taken into account. For the *cis* molecules high wavenumbers are observed in the spectra of MeOCH=CHOMe (1680), MeCH=CHOEt (1669) and MeCH=CHOMe (1668 cm^{-1}) and low values in those of XCH=CHC(=O)Y compounds (X = Cl or Br; Y = Cl or Br) (1575 $\pm$ 15), BrCH=CHBr (1585) and ClCH=CHCl (1590 cm^{-1}). Usually the *cis* isomers reveal this C=C stretching vibration at 1630 $\pm$ 35 cm^{-1}.

In *trans* isomers the intensity of the νC=C decreases strongly in proportion as the symmetry of the molecule increases. In symmetrical RCH=CHR compounds this vibration is not infrared active but is strongly Raman active. For the *trans* molecules the region (1625 $\pm$ 55 cm^{-1}) is limited by MeCH=CHMe (1676) and MeOCH=CHOMe (1670 cm^{-1}) at the HW side and by XCH=CHC(=O)Y compounds (X = Cl or Br; Y = Cl or Br) (1585 $\pm$ 15), ClCH=CHCl (1580) and BrCH=CHBr (1581 cm^{-1}) at the LW side. The remaining molecules display the νC=C at 1635 $\pm$ 35 cm^{-1}. Often the *trans* isomers absorb at higher wavenumbers than the *cis* isomers, and the difference in wavenumber becomes smaller in conjugated dienes, for which this vibration splits into an in-phase and an out-of-phase component: MeCH=CHMe (*cis*: 1662; *trans*: 1676R), MeCH=CHEt (*cis:* 1658; *trans*: 1673R), HO(O=)CCH=CHC(=O)OH (*cis*: 1648; *trans*: 1664R), HO(O=)CCH=CHCH=CHC(=O)OH (*cis–cis*: ν_s 1640, ν_a 1595; *trans–trans*: ν_s 1640; ν_a 1615), ClCH=CHCH=CHCl (*cis–cis*: ν_s 1625; ν_a 1574; *trans–trans*: ν_s 1625; ν_a 1572 cm^{-1}). In the compounds XCH=CHX (X = F, Cl, Br, I), MeCH=CHOMe (*cis*: 1668; *trans*: 1659), MeCH=CHOEt (*cis*: 1669; *trans*: 1659) and MeOCH=CHOMe (*cis*: 1680; *trans*: 1670 cm^{-1}), however, the *cis* isomer absorbs at higher wavenumbers than the *trans* isomer.

The CH in-plane deformations

The absorption region of δCH for *cis* isomers is 1345 $\pm$ 80 or generally 1345 $\pm$ 55 cm^{-1} if the extreme values from $F_3CCH=CHCF_3$ (1425), MeCH=CHMe (1425), $HOCH_2CH=CHCH_2OH$ (1417), DO(O=)CCH=CHC(=O)OD (1416), *cis,cis*-ClCH=CHCH=CHCl (1415 and *1305*) and *cis,cis*-MeCH=CHCH=CHMe (1265 and *1344* cm^{-1}) are not taken into account. The very low values 1219 and 1254 cm^{-1} for ICH=CHI and BrCH=CHBr fall outside the above-mentioned region.

The absorption region of δCH for *trans* isomers is 1300 $\pm$ 40 or generally 1300 $\pm$ 30 cm^{-1}. The highest wavenumbers (1337 cm^{-1}) are due to MeCH=CHOMe and MeCH=CHOEt and the lowest values ($\approx$1266 cm^{-1}) are from MeO(O=)CCH=CHC(=O)X compounds (X = Cl, ONa, OH and OD). The low values 1225 for ICH=CHI and 1251 cm^{-1} for BrCH=CHBr fall outside the region.

The absorption region of δ'CH for *cis* isomers is 1240 $\pm$ 55 or generally 1240 $\pm$

40 cm^{-1}. The region is limited by MeOCH=CHOMe and MeCH=CHC(=O)Me with1292 cm^{-1} and by *cis,cis*-ClCH=CHCH=CHCl with *1230* and 1185 cm^{-1}, if the low values for ICH=CHI (1120), BrCH=CHBr (1150) and ClCH=CHCl (1180 cm^{-1}) are not taken into account.

The absorption region of δ'CH for *trans* isomers is 1260 ± 45 or generally 1260 ± 30 cm^{-1}. The upper limit is given by MeCH=CHMe (1305), F_5PhCH=CHPhF_5 (1305) and Ph-d_5CH=CHPh-d_5 (1302 cm^{-1}) and the lower limit by HO(O=)CCH=CHC(=O)OK (1218) and ClCH=CHCH=CH_2 (1227 cm^{-1}). The compounds ICH=CHI (1128), BrCH=CHBr (1160) and ClCH=CHCl (1200 cm^{-1}) absorb outside this region.

The CH out-of-plane deformations

For *cis*-1,2-dichloroethene, the out-of-plane out-of-phase CH deformation is forbidden in the infrared but active in the Raman at 876 cm^{-1}. In the spectra of *cis* isomers this ωCH (925 ± 75 cm^{-1}) has only a weak intensity and is of minor importance as a diagnostic tool. Disregarding the extreme values (≈1000 cm^{-1}) from HO(O=)CCH=CHC(=O)OH and DO(O=)CCH=CHC(=O)OD and from MeCH=CHMe (852 cm^{-1}), the region is reduced to 930 ± 50 cm^{-1}.

The *trans* isomers of FCH=CHF (875), ClCH=CHCl (896), BrCH=CHBr (899) and ICH=CHI (907 cm^{-1}) absorb strongly in the infrared but these low values fall outside the region 955 ± 45 cm^{-1}. High wavenumbers in the vicinity of 1000 cm^{-1} are found in the spectra of HO(O=)CCH=CHC(=O)OK and MeO(O=)CCH=CHC(=O)OMe. Colthup [62] reported that fatty acids and derivatives with *trans* polyene groups absorb strongly in the neighbourhood of 1000 cm^{-1}, *trans,trans,trans*-trienes near 994 and *trans,trans*-dienes near 986 cm^{-1}. In the spectrum of *trans,trans*-ClCH=CHCH=CHCl these CH out-of-plane deformations are assigned at 955 and 913 cm^{-1}. Most of the *trans* RCH=CHR′ compounds show the in-phase CH wagging vibration moderately to strongly in the region 950 ± 30 cm^{-1}. This CH out-of-plane deformation provides the most useful infrared band to elucidate the *trans*-CH=CH— structure, especially if the intensity of the C=C stretching vibration fails. The strong band at 961 cm^{-1} in the spectrum of *trans*-PhCH=CHMe is assigned to this ωCH. Colthup situates this ωCH in the spectra of conjugated polyenes with *cis* and *trans* fragments in the region 950 ± 50 cm^{-1} [62].

Cis-1,2-Dichloroethene absorbs strongly at 701 cm^{-1} and *cis*-1,2-difluoroethene at 756 cm^{-1}. Together with the C=C stretching vibration this in-phase CH wag (735 ± 35 cm^{-1}) is the most characteristic vibration of the *cis*-CH=CH— group, but the intensity of this band is not always sufficiently strong. Therefore this vibration is qualitatively inferior compared with the *trans* 955 cm^{-1} absorption. The low values from ICH=CHI (645) and BrCH=CHBr (670 cm^{-1}) fall outside this region.

The ωCH in *trans*-1,2-dichloroethene is forbidden in the infrared and appears at 760 cm^{-1} in the Raman. This vibration is weakly active or inactive in the infrared and is of minor importance as a diagnostic tool. The upper limit of the absorption region (825 ± 75 cm^{-1}) is formed by MeO(O=)CCH=CHC(=O)X compounds (X = HO, NaO, Cl) with values in the neighbourhood of 898 cm^{-1} and by X(O=)CCH=CHC(=O)X compounds (X = HO, DO, MeO) with wavenumbers near 895 cm^{-1} in the Raman. Low values are found in the Raman spectra of MeCH=CHMe (755), ClCH=CHCl (760), MeCH=CHCH=CHMe (750) and ClCH=CHCH=CHCl (760 cm^{-1}). Generally, the region 820 ± 50 cm^{-1} is useful for this out-of-phase wag. The very low values from BrCH=CHBr (736R) and ICH=CHI (592R) are not taken into account.

Skeletal deformations

The vinylene group provides two external skeletal deformations: an in-plane deformation and an out-of-plane deformation or torsion. The other side of the —CH=CH— unit gives an 'alternative' skeletal deformation. Without further information it is not possible to assign these deformations unambiguously. Table 8.8 gives these extensive absorption regions with a few examples.

Table 8.8 Vinylene skeletal deformations

Vibration	Absorption regions (cm^{-1})		MeCH=CHMe [3, 5]		MeCH=CHC≡N [19]	
	cis-molecules	*trans*-molecules	*cis*	*trans*	*cis*	*trans*
δ—C=C	555 ± 120	525 ± 95	566	501	655	555
torsion —C=C	420 ± 100	360 ± 110	396	260	517	461
alternative δC=C—	280 ± 115	300 ± 100	291	294	394	398

Aly *et al.* have studied the influence of the configuration of —CH=CH— molecules on the skeletal deformations [15].

R—CH=CH—R′ compounds

R	R′
Me—	Me— [1–6], H(O=)C— [7, 8, 9], Cl(O=)C— [10], Me(O=)C— [7], MeO(O=)C— [11, 12], EtO(O=)C— [11], H_2C=CH— [13], H_2C=C(Me)— [14], MeHC=CH— [15, 16], H_2C=CHCH=CH— [17], H(O=)CCH=CH—, H(O=)CCH=CHCH=CH— and H(O=)CCH=C(Me)CH=CH— [18], N≡C— [19, 20], Ph—, 4-MeOPh—, MeO— [21], EtO— [21, 22], MeCH=CHO— [23], Cl— [24];

Table 8.9 Absorption regions (cm^{-1}) of the normal vibration of *cis*-R—CH=CH—R′

R R′	saturated saturated	saturated conjugated (C=C; C=O)	conjugated conjugated (C=C; C=O)	conjugated halogen
νCH	3050 ± 40	3045 ± 25	3055 ± 25	3080 ± 20
ν'CH	3010 ± 30	3025 ± 25	3005 ± 25	3040 ± 20
νC=C	1660 ± 20	1630 ± 30	1625 ± 25	1595 ± 35
δCH	1390 ± 35	1330 ± 65	1350 ± 60	1360 ± 55
δ'CH	1260 ± 35	1255 ± 35	1245 ± 45	1235 ± 50
ωCH	915 ± 65	930 ± 50	960 ± 40	940 ± 40
ω'CH	735 ± 55	735 ± 55	750 ± 40	740 ± 40
δ—C=C	515 ± 75	580 ± 80	555 ± 120	575 ± 95
torsion —C=C	405 ± 85	435 ± 85	365 ± 45	425 ± 75
alternative δC=C—	265 ± 45	345 ± 50	230 ± 65	245 ± 70

Absorption regions (cm^{-1}) of the normal vibrations of *trans*-R—CH=CH—R′

R R′	saturated saturated	saturated conjugated (C=C; C=O)	conjugated conjugated (C=C; C=O)	conjugated halogen
νCH	3040 ± 25	3040 ± 25	3055 ± 40	3075 ± 25
ν'CH	3025 ± 25	3015 ± 20	3025 ± 35	3035 ± 15
νC=C	1665 ± 15	1640 ± 30	1640 ± 30	1600 ± 30
δCH	1320 ± 20	1310 ± 20	1295 ± 35	1305 ± 35
δ'CH	1285 ± 20	1260 ± 30	1260 ± 45	1260 ± 30
ωCH	955 ± 20	955 ± 30	970 ± 30	945 ± 35
ω'CH	800 ± 50	800 ± 50	830 ± 70	810 ± 50
δ—C=C	530 ± 90	540 ± 90	490 ± 60	525 ± 75
torsion —C=C	330 ± 80	365 ± 106	350 ± 100	350 ± 100
alternative δC=C—	270 ± 40	320 ± 80	270 ± 70	300 ± 100

Et—	Et— [25], $^{-}O_2CCH_2$— [26], N≡C— [20];
nPr—	$^{-}O_2C$— [26];
$HOCH_2$—	$HOCH_2$—;
$ClCH_2$—	$ClCH_2$— [27], Cl— [28];
F_3C—	F_3C— [29];
H(O=)C—	NH_2— [30];
Cl(O=)C—	Cl(O=)C— [31], Cl—, Br— and I— [32];
Br(O=)C—	Cl— and Br— [32];

HO(O=)C—	HO(O=)C— [33], KO(O=)C— [34], HO(O=)CCH=CH— [35];
DO(O=)C—	DO(O=)C— [33];
MeO(O=)C—	Cl(O=)C— [36, 37], HO(O=)C—, DO(O=)C— and NaO(O=)C— [37], MeO(O=)C— [36, 38, 39], Ph— [40];
EtO(O=)C—	EtO(O=)C— [38];
nBuO(O=)C—	nBuO(O=)C— [38];
KO(O=)C—	KO(O=)C— [41];
H_2C=CH—	H_2C=CH— [17, 42–44], N≡C— [45], Cl— [46];
O=C=N—	O=C=N— [47];
N≡C—	N≡C— [48, 49];
Ph—	Ph— [50–53];
Ph-d_5—	Ph-d_5— [50];
2-, 3- and 4-dPh—	2-, 3- and4-dPh— [50];
4-FPh—	4-FPh— [52];
F_5Ph—	F_5Ph— [52];
MeO—	MeO— [54, 55];
F—	F— [3, 56, 57];
Cl—	ClCH=CH— [58, 59], Cl— [60, 61];
Br—	Br— [60];
I—	I— [60].

References

1. A.J. Barnes and J.D.R. Howells, *J. Chem. Soc. Faraday Trans. 2*, **69**, 532 (1973).
2. I.W. Levin and R.A.R. Pearce, *J. Mol. Spectrosc.*, **49**, 91 (1974).
3. I.W. Levin, R.A.R. Pearce and W.C. Harris, *J. Chem. Phys.*, **59**, 3048 (1973).
4. I.W. Levin and R.A.R. Pearce, *Vibrational Spectra and Structure*, Vol. 4 (J.R. Durig, Ed.), Elsevier Scientific Publishing Company, Amsterdam, Oxford, New York (1975), pp.162–169.
5. D.C. McKean, M.W. Mackenzie, A.R. Morrison, J.C. Lavalley, A. Janin, V. Fawcett and H.G.M. Edwards, *Spectrochim. Acta, Part A*, **41A**, 435 (1985).
6. G. Busca, G. Ramis, V. Lorenzelli, A. Janin and J.-C. Lavalley, *Spectrochim. Acta, Part A*, **43A**, 489 (1987).
7. A.J. Bowles, W.O. George and W.F. Maddams, *J. Chem. Soc.*, B, 810 (1969).
8. J.R. Durig, S.C. Brown, V.F. Kalasinsky and W.O. George, *Spectrochim. Acta, Part A*, **32A**, 807 (1976).
9. H.J. Oelichmann, D. Bougeard and B. Schrader, *J. Mol. Struct.*, **77**, 179 (1981).
10. R.K. Gupta, R. Prasad and H.L. Bhatnagar, *Spectrochim. Acta, Part A*, **45A**, 595 (1989).
11. A.J. Bowles, W.O. George and D.B. Cunliffe-Jones, *J. Chem. Soc.*, B, 1070 (1970).
12. P. Carmona and J. Moreno, *J. Mol. Struct.*, **82**, 177 (1982).
13. D.A.C. Compton, W.O. George and W.F. Maddams, *J. Chem. Soc. Perkin Trans. 2*, 1311 (1977).
14. M.M.A. Aly, M.H. Baron, M.J. Coulange and J. Favrot, *Spectrochim. Acta, Part A*, **42A**, 411 (1986).
15. M.M.A. Aly, M.H. Baron, J. Favrot, F. Romain and M. Revault, *Spectrochim. Acta, Part A*, **40A**, 1037 (1984).

16. M.M.A. Aly, M.H. Baron, M.J. Coulange and J. Favrot, *J. Mol. Struct.*, **142**, 407 (1986).
17. F.W. Langkilde, R. Wilbrandt, O.F. Nielsen, D.H. Christensen and F.M. Nicolaisen, *Spectrochim. Acta, Part A*, **43A**, 1209 (1987).
18. M.M.A. Aly, M.H. Baron, J. Favrot, J. Belloc and M. Revault, *Can. J. Chem.*, **63**, 1587 (1985).
19. J.R. Durig, C.K. Tong, C.W. Hawley and J. Bragin, *J. Phys. Chem.*, **75**, 44 (1971).
20. D.A.C. Compton and W.F. Murphy, *J. Phys. Chem.*, **85**, 482 (1981).
21. S.W. Charles, F.C. Cullen and N.L. Owen, *J. Mol. Struct.*, **18**, 183 (1973).
22. F. Marsault-Herail, G.S. Chiglien, J.P. Dorie and M.L. Martin, *Spectrochim. Acta, Part A*, **29A**, 151 (1973).
23. H.C. Hollein and W.H. Snyder, *J. Mol. Struct.*, **82**, 187 (1982).
24. J.R. Durig and G.A. Guirgis, *J. Raman Spectrosc.*, **13**, 160 (1982).
25. H.W. Schrötter and E.G. Hoffmann, *Liebigs Ann. Chem.*, **672**, 44 (1964).
26. K. Tsukamoto, S. Horiuchi, K. Taga, T. Yoshida and H. Okabayashi, *J. Mol. Struct.*, **263**, 75 (1991).
27. P. Piaggio, G. Viviano and G. Dellepiane, *J. Mol. Struct.*, **20**, 243 (1974).
28. J.R. Durig, T.G. Costner, T.S. Little and D.T. Durig, *J. Phys. Chem.*, **96**, 7194 (1992).
29. H. Bürger, G. Pawelke and H. Oberhammer, *J. Mol. Struct.*, **84**, 49 (1982).
30. J. Terpiński and J. Dabrowski, *J. Mol. Struct.*, **4**, 285 (1969).
31. J.M. Landry and J.E. Katon, *Spectrochim. Acta, Part A*, **40A**, 871 (1984).
32. K. Kamieńska-Trela, H. Barańska and A. Labudzińska, *J. Mol. Struct.*, **54**, 59 (1979).
33. J. Maillols, L. Bardet and L. Maury, *J. Mol. Struct.*, **30**, 57 (1976).
34. F. Avbelj, B. Orel, M. Klanjsek and D. Hadzi, *Spectrochim. Acta, Part A*, **41A**, 75 (1985).
35. P. Sohár and G. Varsányi, *J. Mol. Struct.*, **1**, 437 (1967–68).
36. J.E. Katon and P.H. Chu, *J. Mol. Struct.*, **78**, 141 (1982).
37. J.E. Katon and P.H. Chu, *J. Mol. Struct.*, **82**, 61 (1982).
38. D.A.C. Compton, W.O. George and A.J. Porter, *J. Chem. Soc. Perkin Trans. 2*, 400 (1975).
39. C. Téllez, R. Knudsen and O. Sala, *J. Mol. Struct.*, **67**, 189 (1980).
40. M.D.G. Faria and J.J.C. Teixeira-Dias, *J. Raman Spectrosc.*, **22**, 519 (1991).
41. J. Maillols, L. Bardet and L. Maury, *J. Mol. Struct.*, **21**, 185 (1974).
42. E.R. Lippincott and T.E. Kenney, *J. Am. Chem. Soc.*, **84**, 3641 (1962).
43. R. McDiarmid and S. Sabljic, *J. Phys. Chem.*, **91**, 276 (1987).
44. H. Yoshida, Y. Furukawa and M. Tasumi, *J. Mol. Struct.*, **194**, 279 (1989).
45. B.H. Thomas and W.J. Orville-Thomas, *J. Mol. Struct.*, **3**, 191 (1969).
46. A. Borg, Z. Smith, G. Gundersen and P. Klaboe, *Spectrochim. Acta, Part A*, **36A**, 119 (1980).
47. G. L.Carlson, *Spectrochim. Acta*, **20**, 1781 (1964).
48. F.A. Miller, O. Sala, P. Devlin, J. Overend, E. Lippert, W. Luder, H. Moser and J. Varchmin, *Spectrochim. Acta, Part A*, **20**, 1233 (1964).
49. A. Rosenberg and J.P. Devlin, *Spectrochim. Acta*, **21**, 1613 (1965).
50. Z. Meić and H. Güsten, *Spectrochim. Acta, Part A*, **34A**, 101 (1978).
51. A. Bree and R. Zwarich, *J. Mol. Struct.*, **75**, 213 (1981).
52. Z. Meić and H. Güsten, *Spectrochim. Acta, Part A*, **36A**, 1021 (1980).
53. K. Palmö, *Spectrochim. Acta, Part A*, **44A**, 341 (1988).
54. H.S. Kimmel, J.T. Waldron and W.H. Snyder, *J. Mol. Struct.*, **21**, 445 (1974).
55. J.M. Comerford, P.G. Anderson, W.H. Snyder and H.S. Kimmel, *Spectrochim. Acta, Part A*, **33A**, 651 (1977).
56. N.C. Craig and E.A. Entemann, *J. Chem. Phys.*, **36**, 243 (1962).
57. R.A.R. Pearce and I.W. Levin, *J. Chem. Phys.*, **59**, 2698 (1973).

58. E. Benedetti, M. Aglietto, P. Vergamini, R. Aroca, A.V. Rodin, Y.N. Panchenko and Y.A. Pentin, *J. Mol. Struct.*, **34**, 21 (1976).
59. G. Gundersen, P. Klaboe, A. Borg and Z. Smith, *Spectrochim. Acta, Part A*, **36A**, 843 (1980).
60. J.Lecomte, *Encyclopedia of Physics, Light and Matter II*, Springer-Verlag, Heidelberg (1958).
61. K. Tanabe and S. Saëki, *Bull. Chem. Soc. Jpn.*, **47**, 2545 (1974).
62. N.B. Colthup, *Appl. Spectrosc.*, **25**, 368 (1971).

8.2 ALKYNES

The 15 normal vibrations of $XCH_2C{\equiv}CH$ (X = F, Cl, Br, I) (Section 3.5.6) are divided, according to C_s symmetry, into 10a′ + 5a″ vibrations. Nine vibrations belong to the XCH_2— group and six to the —C≡CH fragment:

a′: νCH, νC≡C, δCH, δ—C≡C;
a″: γCH, γ—C≡C;

and the six vibrations of the —C≡CX fragment are:

a′: νCX, νC≡C, δCX, δ—C≡C;
a″: γCX, γ—C≡C.

8.2.1 Ethynyl

CH stretching vibration

With the exception of FC≡CH (3355 cm^{-1}), the νCH gives rise to a sharp and strong absorption band in the region 3300 ± 40 cm^{-1}. The intensity of this band is so strong that it is clearly observable superimposed on the broad OH. . .O stretching band in alcohols or acids. The influence of the halogen is illustrated by the high values observed in the spectra of ClC≡CH (3340), $FCH_2C{\equiv}CH$ (3338) and F(O=)CC≡CH (3337 cm^{-1}). This characteristic absorption is shifted to lower wavenumbers in conjugated RC≡CH compounds such as Me(O=)CC≡CH (3262), MeO(O=)CC≡CH (3271), HC≡CC(=O)C≡CH (3275) and NaO(O=)CC≡CH (3278 cm^{-1}). The sp-CH stretch in the remaining compounds absorbs at 3310 ± 25 cm^{-1} and takes a unique place as a group vibration.

The C≡C stretching vibration

Because of the compounds $F_3CC{\equiv}CH$ (2165) and NaO(O=)CC≡CH (2095), the absorption region of the C≡C stretch (2130 ± 35) is somewhat larger than that of propynyl (2125 ± 25 cm^{-1}), although this latter region is also usable for most of the RC≡CH compounds. This νC≡C is a good group vibration but the intensity is moderate to weak and decreases progressively when the triple bond moves away from the terminal position. In symmetrical disubstituted ethynes the νC≡C is infrared-forbidden, but, very prominent in Raman spectra. The above-mentioned region does not included: FC≡CH (2255), Br≡CH (2085) and IC≡CH (2060 cm^{-1}) (see Sections 8.2.3 and 8.2.4) and the butadiynes with, as examples, ClC≡CC≡CH (2071), BrC≡CC≡CH (2095) and IC≡CC≡CH (2060 cm^{-1}). For compounds containing the —C≡CC≡C— fragment the two νC≡C bands are described as symmetric and antisymmetric —C≡CC≡C— stretchings, for example in the spectra of HC≡CC≡CH (ν_s: 2172; ν_a: 2005) and MeC≡CC≡CH (ν_s: 2239; ν_a: 2071 cm^{-1}).

The CH deformations

The in-plane and out-of-plane CH deformations are active in the regions 675 ± 55 and 655 ± 45 cm^{-1}, with an intensity that varies from moderate to strong. Usually, but not always the in-plane deformation is assigned at a higher wavenumber than the out-of-plane deformation. The splitting between the two bendings is greater in unsaturated and carbonyl-bonded ethynyl than in RC≡CH compounds in which R is a saturated fragment, for which the two deformations often give rise to one broad absorption (Section 3.5.6). In the spectra of molecules with axial symmetry about the C≡CH group, these bending modes are degenerate and only one band will be observed: MeC≡CH (642), FC≡CH (578), ClC≡CH (604), BrC≡CH (618) and IC≡CH (630 cm^{-1}). The highest wavenumbers are observed in the spectra of HC≡CC(=O)C≡CH (730 and 712), F(O=)CC≡CH (701 and *695*), Cl(O=)CC≡CH (703 and *696*), NaO(O=)CC≡CH (704 and *652*) and Me(O=)CC≡CH (700 and *649* cm^{-1}) and the lowest in those of cPrC≡CH (610 and *648*), MeC≡CC≡CH (614 and *696*), H_2C=CHC≡CH (615 and *629*), PhC≡CH (613 and *653*), 4-HC≡CPhC≡CH (618, 615, *646* and *632*) and 2-HC≡CPhC≡CH (615, *630* and *640* cm^{-1}). There is a large chance of encountering both deformations at 660 ± 40 cm^{-1}.

Skeletal deformations

In cases of no axial symmetry about the C≡C bond, the two skeletal —C≡C deformations are active in the regions 295 ± 75 and 215 ± 75 cm^{-1}. Usually the higher wavenumber is assigned to the out-of-plane deformation and the lower wavenumber to the in-plane deformation. The lowest wavenumbers are found in the spectra of carbonyl-bonded ethynyl compounds such as Cl(O=)CC≡CH (224 and 157)), F(O=)CC≡CH (229 and 189) and Me(O=)CC≡CH (228 and 183 cm^{-1}). In the spectra of compounds with axial symmetry about the C≡CH group, only one band is observed: MeC≡CH (336), FC≡CH (367), ClC≡CH (326), BrC≡CH

(295) and IC≡CH (262 cm^{-1}). The two skeletal deformations of the diacetylenes are described as antisymmetric and symmetric deformations, but the wavenumber of the antisymmetric deformation falls outside the higher mentioned region: HC≡CC≡CH (484 and *230*), MeC≡CC≡CH (484 and *324*), ClC≡CC≡CH (463 and *335*), BrC≡CC≡CH (470 and *355*), IC≡CC≡CH (473 and *357*) and 4-HC≡CPhC≡CH (477 and *325* cm^{-1}).

Table 8.10 Absorption regions (cm^{-1}) of the normal vibrations of —C≡CH

Vibration	propynyl	α-saturated	C=O bonded	α-unsaturated	aromatic
νCH	3315 ± 25	3310 ± 30	3300 ± 40	3315 ± 25	3305 ± 25
νC≡C	2125 ± 25	2130 ± 35	2120 ± 25	2110 ± 15	2110 ± 10
δCH	660 ± 30	660 ± 30	705 ± 25	660 ± 40	645 ± 15
γCH	635 ± 15	630 ± 20	670 ± 30	620 ± 10	620 ± 10
γ—C≡C	330 ± 35	295 ± 75	280 ± 60	290 ± 50	345 ± 25
δ—C≡C	195 ± 45	195 ± 45	215 ± 75	195 ± 45	180 ± 30

R—C≡CH compounds

R = R′CH_2— (see Section 3.5.6).

R = Me— [1–11, 68], CD_3— [9], MeCD_2— and CD_3CD_2— [12], cPr— [13], cBu— [14], cHex—, HOCHD— and HOCD_2— [15], tBu—, F_3C— [11, 16–18], HC(=O)— [19–22], DC(=O)— [19, 20], FC(=O)— [22–24], ClC(=O)— [24–26], MeC(=O)— [27], HC≡CC(=O)— [28], HOC(=O)— [29], MeOC(=O)— [30, 31], EtOC(=O)— [32], NaOC(=O)— [29], H_2C=CH— [33–35], H_2C=C(Me)— [36], MeHC=CH— [36], H_2C=CHC(=CH_2)— [37], N=N=CH— and N=N=CD— [38], HC≡C— [39, 40], MeC≡C— [41, 42], ClC≡C—, BrC≡C— and IC≡C— [43–45], Ph— [1, 46, 47], Ph-d_5— [47], 2-, 3- and 4-HC≡CPh— [48, 49], 4-MePh—, 4-ClPh—, MeS— [50–52], F— [53–55], Cl— and Br— [53, 55], I— [53].

8.2.2 Chloroethynyl

The C≡C stretching vibration

For the small collection of R—C≡CCl compounds the νC≡C is found in the region 2230 ± 40 cm^{-1}, that is, 100 cm^{-1} higher than for R—C≡CH compounds. The low value (2110 cm^{-1}) in the spectrum of chloroethyne falls outside this region (Section 8.2.1). Disregarding the low wavenumbers from IC≡CCl (2191) and N≡CC≡CCl (2194 cm^{-1}), the region is reduced to 2245 ± 25 cm^{-1}.

The C—Cl stretching vibration

The C—Cl stretch is active in the extensive region 595 ± 165 cm^{-1}, for which the high values in the spectra of dihaloethyne compounds are not taken into account: ClC≡CCl (988 and *477*), BrC≡CCl (923) and IC≡CCl (886 cm^{-1}). High wavenumbers are also assigned in the spectra of HC≡CCl (756) and F_3CC≡CCl (723 cm^{-1}) and low wavenumbers in those of MeC≡CC≡CCl (437) and H(O=)CC≡CCl (473 cm^{-1}). The narrower region 590 ± 70 cm^{-1} makes possible the use of this νC—Cl.

Skeletal deformations

According to the scarce data, the two external skeletal —C≡C deformations appear in the ranges 395 ± 75 and 310 ± 50 cm^{-1}. Usually the higher {lower} wavenumber is assigned to the out-of-plane {in-plane} deformation. In the case of axial symmetry only one band is observed. Self-evidently, the H—C≡C deformation (604 cm^{-1}) in the spectrum of HC≡CCl is outside this region (Section 8.2.1).

The ≡C—Cl deformations

The two ≡C—Cl deformations are assigned in the regions 230 ± 105 and 140 ± 50 cm^{-1}, but it is difficult to determine which wavenumber is respectively responsible for the in-plane or out-of-plane deformation. The compound H(O=)CC≡CH shows these two deformations at 152 and 114 cm^{-1} and the higher wavenumber is assigned to the out-of-plane ≡C—Cl deformation. In the spectra of molecules with axial symmetry about the C≡C—Cl group, the bending modes are degenerate and make their appearance at the same wavenumber: MeC≡CCl (184), ClC≡CCl (172 and 333) and N≡CC≡CCl (129 cm^{-1}).

R—C≡C—Cl compounds
R = H— [53], Me— [2, 3, 8, 9, 56–58], CD_3— [2, 9], $ClCH_2$— [3, 59], F_3C— [60, 61], H(O=)C— [21, 62], H_2C=CH— [63], HC≡C— [43–45], MeC≡C— [41], N≡C— [44], Ph— [64], Cl—, Br— and I— [65, 66].

8.2.3 Bromoethynyl

The C≡C stretching vibration

The C≡C stretching vibration in R—C≡C—Br compounds appears in the region 2200 ± 50 cm^{-1}. Bromoethyne (2085 cm^{-1}) absorbs more in the direction of the R—C≡CH compounds (Section 8.2.1). Ignoring the low wavenumbers in the spectra of N≡CC≡CBr (2150) and IC≡CBr (2166 cm^{-1}) this region narrows to

$2215 \pm 35\ cm^{-1}$, and occupies the second place in the series C≡CCl > C≡CBr > C≡CI > C≡CH.

The C—Br stretching vibration

The C—Br stretching vibration is located in the region $520 \pm 170\ cm^{-1}$, if the extreme values in the spectra of BrC≡CBr (832 and 267) and IC≡CBr ($782\ cm^{-1}$) are not taken into account. Ignoring the high wavenumbers from F_3CC≡CBr (686) and BrC≡CCH_2CH_2C≡CBr (630 and *525*) and the low values from MeC≡CC≡CBr ($356\ cm^{-1}$), the region is reduced to $520 \pm 95\ cm^{-1}$.

Skeletal deformations

The out-of-plane and in-plane external skeletal deformations are assigned in the same regions as for R—C≡CCl compounds: 395 ± 75 and $310 \pm 50\ cm^{-1}$. The H—C≡C deformations in HC≡CBr are found in Table 8.10.

The ≡C—Br deformations

Vibrational analysis reveals that the two ≡C—Br deformations are active in the regions 225 ± 100 and $120 \pm 50\ cm^{-1}$. In the spectrum of H(O═)CC≡CBr, the two deformations are situated at 143 and $105\ cm^{-1}$. The compounds with axial symmetry give rise to only one deformation, for example MeC≡CBr (171) and BrC≡CBr (137 and $311\ cm^{-1}$).

R—C≡C—Br compounds
R = H— [53], Me— [2, 3, 8, 9, 56–58], CD_3— [9], BrC≡CCH_2CH_2— [67], $BrCH_2$— [3], F_3C— [60, 61], H(O═)C— [21, 62], HC≡C— [43–45], MeC≡C— [41], N≡C— [44], Ph— [64], Cl— and Br— [65, 66], I— [65].

8.2.4 Iodoethynyl

The C≡C stretching vibration

Except for the low value of $2060\ cm^{-1}$ from the spectrum of iodoethyne (Section 8.2.1), this C≡C stretching vibration is found in the region $2170 \pm 50\ cm^{-1}$, and occupies the third place in the series C≡CCl > C≡CBr > C≡CI > C≡CH.

The C—I stretching vibration

With the exception of the extreme values from di-iodoethyne (720 and $190\ cm^{-1}$), the νCI falls in the region $485 \pm 175\ cm^{-1}$. Disregarding the high wavenumber of $660\ cm^{-1}$ from F_3CC≡CI and the low value of $310\ cm^{-1}$ from MeC≡CC≡CI, this region is narrowed to $425 \pm 65\ cm^{-1}$.

Skeletal deformations

The external skeletal deformations in R—C≡CI compounds are located in the same regions as for R—C≡CCl and R—C≡CBr compounds: 395 ± 75 and 310 ± 50 cm^{-1}.

The ≡C—I deformations

When the molecule has no axial symmetry the two ≡C—I deformations may be observed in the regions 200 ± 100 and 120 ± 50 cm^{-1}.

R—C≡C—I compounds
R = H— [53], Me— [2, 8, 9, 57], CD_3— [9], F_3C— [60, 61], H(O=)C— [21, 62], HC≡C— [43–45], MeC≡C— [41], N≡C— [44], Cl— and I— [65, 66], Br— [65].

Table 8.11 Absorption regions (cm^{-1}) of the normal vibrations of —C≡C—X

Vibration	—C≡C—Cl	—C≡C—Br	—C≡C—I
νC≡C	2230 ± 40	2200 ± 50	2170 ± 50
νC—X	595 ± 165	520 ± 170	485 ± 175
γ—C≡C	395 ± 75	395 ± 75	395 ± 75
δ—C≡C	310 ± 50	310 ± 50	310 ± 50
δ ≡C—X	230 ± 105	225 ± 100	200 ± 100
γ ≡C—X	140 ± 50	120 ± 50	120 ± 50

References

1. R.A. Nyquist and W.J. Potts, *Spectrochim. Acta*, **16**, 419 (1960).
2. P.N. Daykin, S. Sundaram and F.F. Cleveland, *J. Chem. Phys.*, **37**, 1087 (1963).
3. R.A. Nyquist, A.L. Johnson and Y.S. Lo, *Spectrochim. Acta*, **21**, 77 (1965).
4. D.R.J. Boyd and H.W. Thompson, *Trans. Faraday Soc.*, **48**, 493 (1952).
5. D.R.J. Boyd and H.W. Thompson, *Trans. Faraday Soc.*, **49**, 141 (1952).
6. R.J. Grisenthwaite and H.W. Thompson, *Trans. Faraday Soc.*, **50**, 212 (1954).
7. M.T. Christensen and H.W. Thompson, *Trans. Faraday Soc.*, **52**, 1439 (1956).
8. J.L. Duncan, *Spectrochim. Acta*, **20**, 1197 (1964).
9. J.C. Whitmer, *J. Mol. Struct.*, **21**, 173 (1974).
10. A. Natarajan and J.S.P. Ebenezer, *Can. J. Spectrosc.*, **31**, 158 (1986).
11. C.V. Berney, L.R. Cousins and F.A. Miller, *Spectrochim. Acta*, **19**, 2019 (1963).
12. J. Saussey, J. Lamotte and J.C. Lavalley, *Spectrochim. Acta, Part A*, **32A**, 763 (1976).
13. G. Schrumpf and A.W. Klein, *Spectrochim. Acta, Part A*, **41A**, 1251 (1985).
14. J.R. Durig, M.J. Lee, T.S. Little, M. Dakkouri and A. Grünvogel-Hurst, *Spectrochim. Acta, Part A*, **48A**, 691 (1992).
15. J. Travert, J.C. Lavalley and D. Chenery, *Spectrochim. Acta, Part A*, **35A**, 291 (1979).
16. V. Galasso and A. Bigotto, *Spectrochim. Acta*, **21**, 2085 (1965).

17. R.H. Sanborn, *Spectrochim. Acta, Part A*, **23A**, 1999 (1967).
18. Y.S. Park and H.S. Shurvell, *Can. J. Spectrosc.*, **35**, 60 (1990).
19. G.W. King and D. Moule, *Spectrochim. Acta*, **17**, 286 (1961).
20. J.C.D. Brand and D.G. Williamson, *Discuss. Faraday Soc.*, **35**, 184 (1963).
21. P. Klaboe and G. Kremer, *Spectrochim. Acta, Part A*, **33A**, 947 (1977).
22. W.J. Balfour, S.G. Fougere and D. Klapstein, *Spectrochim. Acta, Part A*, **47A**, 1127 (1991).
23. W.J. Balfour, D. Klapstein and S. Visaisouk, *Spectrochim. Acta, Part A*, **31A**, 1085 (1975).
24. W.J. Balfour and M.K. Phibbs, *Spectrochim. Acta, Part A*, **35A**, 385 (1979).
25. E. Augdahl, E. Kloster-Jensen and A. Rogstad, *Spectrochim. Acta, Part A*, **30A**, 399 (1974).
26. W.J. Balfour, R.H. Mitchell and S. Visaisouk, *Spectrochim. Acta, Part A*, **31A**, 967 (1975).
27. G.A. Crowder, *Spectrochim. Acta, Part A*, **29A**, 1885 (1973).
28. F.A. Miller, B.M. Harney and J. Tyrrell, *Spectrochim. Acta, Part A*, **27A**, 1003 (1971).
29. J.E. Katon and N.T. McDevitt, *Spectrochim. Acta*, **21**, 1717 (1965).
30. J.E. Katon and T.B. BenKinney, *Spectrochim. Acta, Part A*, **39A**, 877 (1983).
31. G. Williams and N.L. Owen, *Trans. Faraday Soc.*, **67**, 950 (1971).
32. S.W. Charles, G.I.L. Jones, N.L. Owen and L.A. West, *J. Mol. Struct.*, **26**, 249 (1975).
33. N. Sheppard, *J. Chem. Phys.*, **17**, 74 (1949).
34. J. Kanesaka, K. Miyawaki and K. Kawai, *Spectrochim. Acta, Part A*, **32A**, 195 (1976).
35. E. Tørneng, C.J. Nielsen and P. Klaboe, *Spectrochim. Acta, Part A*, **36A**, 975 (1980).
36. T.G.V. Yakovleva and A.A. Petrov, *Opt. Spectrosc.*, **11**, 320 (1961).
37. H. Priebe, C.J. Nielsen, P. Klaboe, H. Hopf and H. Jäger, *J. Mol. Struct.*, **158**, 249 (1987).
38. F.K. Chi and G.E. Leroi, *Spectrochim. Acta, Part A*, **31A**, 1759 (1975).
39. J.L. Hardwick, D.A. Ramsay, J.M. Garneau, J. Lavogne and A. Cabane, *J. Mol. Spectrosc.*, **76**, 492 (1979).
40. I. Freund and R.S. Halford, *J. Chem. Phys.*, **42**, 4131 (1965).
41. L. Benestad, E. Augdahl and E. Kloster-Jensen, *Spectrochim. Acta, Part A*, **31A**, 1329 (1975).
42. B.M. Nikolova, *J. Mol. Struct.*, **273**, 291 (1992).
43. P. Klaboe, E. Kloster-Jensen and S.J. Cyvin, *Spectrochim. Acta, Part A*, **23A**, 2733 (1967).
44. D.H. Christensen, I. Johnson, P. Klaboe and E. Kloster-Jensen, *Spectrochim. Acta, Part A*, **25A**, 1569 (1969).
45. M.K. Phibbs, *Spectrochim. Acta, Part A*, **29A**, 599 (1973).
46. J.C. Evans and R.A. Nyquist, *Spectrochim. Acta*, **16**, 918 (1960).
47. G.W. King and S.P. So, *J. Mol. Spectrosc.*, **36**, 468 (1970).
48. G.W. King and A.A.G. Van Putten, *J. Mol. Spectrosc.*, **70**, 53 (1978).
49. J.F. Arenas, J.I. Marcos and F.J. Ramirez, *Spectrochim. Acta, Part A*, **45A**, 781 (1989).
50. H.J. Boonstra and L.C. Rinzema, *Recl. Trav. Chim. Pays-Bas*, **79**, 962 (1960).
51. A.G. Moritz, *Spectrochim. Acta, Part A*, **23A**, 167 (1967).
52. D.H. Christensen and D. den Engelsen, *Spectrochim. Acta, Part A*, **26A**, 1747 (1970).
53. A. Rogstad and S.J. Cyvin, *J. Mol. Struct.*, **20**, 373 (1974).
54. H. Bürger, W. Schneider, S. Sommer and W. Thiel, *J. Chem. Phys.*, **95**, 5660 (1991).
55. G.R. Hunt and M.K. Wilson, *J. Chem. Phys.*, **34**, 1301 (1961).
56. F.F. Cleveland and H.J. McMurry, *J. Chem. Phys.*, **11**, 450 (1943).
57. A.G. Meister, *J. Chem. Phys.*, **16**, 950 (1948).
58. R.A. Nyquist, *Spectrochim. Acta*, **21**, 1245 (1965).

59. D. Christen, F. Gleisberg, G. Kremer and W. Zeil, *J. Mol. Spectrosc.*, **70**, 179 (1978).
60. E. Augdahl, E. Kloster-Jensen, V. Devarajan and S.J. Cyvin, *Spectrochim. Acta, Part A*, **29A**, 1329 (1973).
61. H.B. Friedrich, D.J. Burton and P.A. Schemmer, *Spectrochim. Acta, Part A*, **45A**, 181 (1989).
62. E. Lagset, P. Klaboe, E. Kloster-Jensen, S.J. Cyvin and F.M. Nicolaisen, *Spectrochim. Acta, Part A*, **29A**, 17 (1973).
63. A. Borg and P. Cederbalk, *Acta Chem. Scand., Ser. A*, **40A**, 103 (1986).
64. R.D. McLachlan, *Spectrochim. Acta, Part A*, **26A**, 919 (1970).
65. D.H. Christensen, T. Stroyer-Hansen, P. Klaboe, E. Kloster-Jensen and E. Tucker, *Spectrochim. Acta, Part A*, **28A**, 944 (1972).
66. P. Klaboe, E. Kloster-Jensen, D.H. Christensen and I. Johnson, *Spectrochim. Acta, Part A*, **26A**, 1567 (1970).
67. D.L. Powell, P. Klaboe, B.N. Cyvin and H. Hopf, *J. Mol. Struct.*,**41**, 215 (1977).
68. P.L. Stanghellini and R. Rossetti, *Inorg. Chem.*, **29**, 2047 (1990).

9

Normal Vibrations and Absorption Regions of Nitrogen Compounds

9.1 AMINO

Methanamine in the pyramidal sp^3 structure belongs to the point group C_s. The 15 normal vibrations differentiate between 9a′ and 6a″ vibrational modes. After subtraction of eight methyl vibrations (5a′ + 3a″) and the C—N stretching mode (a′), 3a′ + 3a″ normal vibrations for NH_2 remain. Hamburg *et al.* [24] report that in benzenamine the amine group and the plane of the benzene ring form an angle of 38° and prefer the C_s structure to the plane C_{2v} structure.

Table 9.1 Amine vibrations of methanamine and benzenamine

	$\nu_a NH_2$	$\nu_s NH_2$	δNH_2	$\rho/\tau NH_2$	ωNH_2	torsion
C_{2v}	b_1	a_1	a_1	b_1	b_2	a_2
C_s	a″	a′	a′	a″	a′	a″
Methanamine vapor(C_s)	3427	3361	1623	1195	780	264
Benzenamine(C_s, C_{2v})	3480	3395	1619	1050	670	245

Calculated according to the equation of Götze and Garbe [79] $>HNH = 0.241\ \Delta\nu + 91.5$, in which $\Delta\nu$ represents the difference between the two NH stretching vibrations, the HNH angle in methanamine {benzenamine} is 107° {112°}.

Amine stretching vibrations

In α-saturated liquid amines, the NH_2 stretching vibrations give rise to a nearly symmetrical doublet with a weak to moderate intensity. A shoulder (3180 $\pm$ 20 cm^{-1}) on the LW side of the $\nu_s NH_2$ is attributed to the overtone of the NH_2 scissoring vibration, enhanced by Fermi resonance [80–83]. In the vapour state and in solution the absorptions move to higher wavenumbers, become sharper and are well separated. In primary amines containing two equivalent NH bonds, the change of force constant has the same influence on the symmetric and antisymmetric stretching vibration. Bellamy and Williams [84] found a correlation between the two vibrations: $\nu_s = 0.876\ \nu_a + 345.5$, and since the value 345.5 approximates to $\nu_a/10$, this equation has been reduced to $\nu_s = 0.98\ \nu_a$ [85, 86]. For molecules in which a neighbouring group interacts with one hydrogen of the amine group, this relationship no longer holds.

In liquid benzenamines the symmetric stretching vibration (LW band) absorbs more strongely than the antisymmetric counterpart (HW band). The overtone band of the NH_2 scissors (3210 $\pm$ 50 cm^{-1}) is clearly separated from the stretchings. Electron attracting {releasing} groups on the 2- and 4-positions increase {lower} the wavenumber of both stretchings [86, 91, 94, 95, 97]. This effect is less pronounced for substituents on the 3 position. In solution the wavenumbers {intensities} of the stretchings increase {weaken} in order of the following solvents: acetonitrile, nitromethane, benzene, carbon disulfide and tetrachloromethane [95]. An empirical correlation between the two stretching vibrations was given by Krueger [87] for benzenamines in the free state: $\nu_s = 0.682\ \nu_a + 1023$, with the exception of some 2-substituted benzenamines in which the substituent interacts with a hydrogen of the amine group.

In associated aliphatic and alicyclic primary amines, the NH_2 antisymmetric stretching vibration occurs at 3365 $\pm$ 25 cm^{-1} [88]. 1-Butanamine absorbs at 3370 cm^{-1}, $F_3CCH_2NH_2$ takes the HW side with 3385 cm^{-1}, and in amines with branching on the α-carbon atom this band tends to shift to lower wavenumbers: $tBuNH_2$ (3348), $iPrNH_2$ (3356) and $cPentNH_2$ (3357 cm^{-1}). In dilute solution or in the vapour state, the $\nu_a NH_2$ appears at 3420 $\pm$ 40 cm^{-1}. In benzenamines the $\nu_a NH_2$ is located at higher wavenumbers: 3410 $\pm$ 70 cm^{-1}, but more usually at 3440 $\pm$ 40 cm^{-1} except for 2-, 3- and 4-aminophenol, absorbing respectively at 3390, 3364 and 3345 cm^{-1}. In dilute solutions this region becomes 3470 $\pm$ 50 cm^{-1}. For a series of 3-X- and 4-X-substituted benzenamines, Nyquist [97] found 3480 $\pm$ 30 cm^{-1} in solution and 3500 $\pm$ 20 cm^{-1} for a gas, with the highest values for CF_3-, NO_2- and N$\equiv$C-substituted benzenamines and the lowest for H_2N-, Me_2N- and MeO-substituted benzenamines. In N- or S-bonded NH_2 this band is observed at 3350 $\pm$ 40 cm^{-1} as in hydrazines [71, 73, 89], thiosemicarbazides [77, 78], *N*-aminoheterocyclic aromatic compounds [83] and sulfonamides [25, 74–76].

Aliphatic and alicyclic primary amines in the associated state display the $\nu_s NH_2$ in the region 3290 $\pm$ 30 cm^{-1}, and at 3350 $\pm$ 40 cm^{-1} as a gas or in dilute solution. For the associated benzenamines the region becomes 3320 $\pm$ 70 cm^{-1} or 3345 $\pm$ 45 cm^{-1} if the low values in the spectra of 4-HOPhNH_2 (3285) and 3-H(O=)CPhNH_2 (3254 cm^{-1}) are not taken into account. In dilute solutions the benzenamines absorb at 3370 $\pm$ 50 cm^{-1}. For a series of 3-X- and 4-X-substituted benzenamines Nyquist [97] found 3395 $\pm$ 20 cm^{-1} in solution and 3415 $\pm$ 15 cm^{-1} as a gas, with the highest {lowest} values for CF_3-, O_2N- and N≡C- {H_2N-, Me_2N- and MeO-} substituted benzenamines.

Amine deformation

The NH_2 scissoring vibration gives rise to a broad strong band in the region 1600 $\pm$ 50 cm^{-1}. Disregarding the high value found in the gas phase spectrum of $H_2N(F_3C)_2CNH_2$ (1640 and *1614* cm^{-1}), the associated aliphatic and alicyclic amines absorb at 1615 $\pm$ 20 cm^{-1}. In benzenamines, electron withdrawing {releasing} substituents cause a slight shift to higher {lower} wavenumbers [90, 91]. The highest {lowest} wavenumbers for the NH_2 scissors are assigned in the spectra of thioamides (1620 $\pm$ 30) {sulfonamides (1565 $\pm$ 15 cm^{-1})}.

Amine rocking/twisting vibration

The literature does not agree in assigning the amine rock (C_{2v}) or twist (C_s) coupled to another b_1 (C_{2v}) or a$''$ (C_s) vibration. In methanamine this NH_2 twist (1195 cm^{-1}) is coupled to the methyl rock. In aliphatic amines this vibration is mixed with the methylene twist and also with the C—N stretching vibration. In this work we accept that the $\rho/\tau NH_2$ provides the greatest contribution to the absorption at 1160 $\pm$ 140 cm^{-1}. Thioamides absorb at high wavenumbers (1195 $\pm$ 110 cm^{-1}) and benzenamines in the region 1070 $\pm$ 50 cm^{-1}, or 1060 $\pm$ 40 cm^{-1} if the value of 1050 cm^{-1} for benzenamine [21, 23–25] is preferred to 1115 cm^{-1} [7].

Amine wagging vibration

Associated α-saturated primary amines show a characteristic very broad diffuse band between 1000 and 700 cm^{-1}, with maximum absorption at 840 $\pm$ 55 cm^{-1}. For amines in the unbonded state this wag appears sharply at lower wavenumbers (770 $\pm$ 30 cm^{-1}). In the spectra of aromatic amines and sulfonamides this band is weak, diffuse and affected by other vibrations in such a way to make it difficult to determine the exact wavenumber, which explains the extensive region (620 $\pm$ 100 cm^{-1}). For benzenamine, various wavenumbers are proposed: 700 [7], 670 [23, 25, 96], 605 [24] and 570 cm^{-1} [21].

Amine torsion

According to the available data, the NH_2 torsion is assigned in the region 290 $\pm$ 130 cm^{-1}. In α-saturated amines the region is reduced to 280 $\pm$ 70 cm^{-1} [92, 93].

Table 9.2 Absorption regions (cm^{-1}) of the normal vibrations of NH_2

Vibration	α-saturated	aromatic	amides	thioamides	N-bonded	sulfonamides
$\nu_a NH_2$	3365 ± 25	3410 ± 70	3390 ± 60	3340 ± 60	3350 ± 40	3355 ± 35
$\nu_s NH_2$	3290 ± 30	3320 ± 70	3210 ± 60	3160 ± 80	3250 ± 70	3250 ± 20
δNH_2	1615 ± 20	1620 ± 20	1610 ± 30	1620 ± 30	1620 ± 20	1565 ± 15
$\rho/\tau NH_2$	1195 ± 90	1070 ± 50	1125 ± 45	1195 ± 110	1195 ± 90	1160 ± 30
ωNH_2	840 ± 55	620 ± 100	670 ± 60	645 ± 65	830 ± 50	690 ± 40
torsion NH_2	280 ± 70	230 ± 70			280 ± 70	355 ± 65

R—NH_2 compounds

R = R′CH_2— (see 3.4.1.)

R = Me— [1–9], CD_3— [1–3], iPr— [10, 19], cPr— [11, 12], cBu— [13], cPent— [14], $H_2N(F_3C)_2C$— [15], $H_2NCD_2CD_2$— [16, 17], H_2C=$CHCD_2$— and H_2C=$CDCD_2$— [18], H_2C=CH— [69, 70], Ph— [7, 21–25, 96], XPh— (X = 2-, 3- and 4-Me [25–27], 2-Et [25], 3- and 4-F_3C [28, 29], 2-, 3- and 4-H(O=)C [30], 3- and 4-Me(O=)C [25], 2-, 3- and 4-HO(O=)C [25], 4-N≡C [25], 2-, 3- and 4-HO [25, 31], 2-, 3- and 4-MeO [25], 2- and 4-HS [25], 2-, 3- and 4-O_2N [22, 25, 32, 33], 2- and 4-H_2N [25, 34], 2-, 3- and 4-F [25, 35–39], 2-, 3- and 4-Cl [20, 25, 40, 41], 2- and 4-Br [25, 42]), X—2-MePh— (X = 3-Me [43, 44], 4-Me [43, 45], 5- and 6-Me [25], 3-Cl [46], 3-O_2N [47], 4-Br [48], 4-HO [49], 4, 6-Me_2 [50], 6-Cl [51]), X—2-EtPh— (X = 6-Et [25]), X—2-HOPh— (X = 4-Me [52], 5-Me [53], 5-O_2N [49], 5-Cl [53]), X—2-MeOPh— (X = 5-MeO [25], 5-Cl [46]), X—2-O_2NPh— (X = 4-Me, 4-MeO and 6-O_2N [25], 4-O_2N [22, 25], 4-Cl [54, 55], 4, 6-$(O_2N)_2$ [22]), X—2-FPh— (X = 4-F [56], 5-Me [57], 5-F [25, 37]), X—2-ClPh— (X = 3-Cl [46, 58], 4-Cl [25], 4-O_2N [54, 55], 5-Cl [25, 37, 46, 58], 6-Cl [25, 37, 58–60], 4, 6-Cl_2 and 4, 6-Br_2 [61]), X—2-IPh— (X = 4-I [25]), X—3-MePh— (X = 4-Me [43], 4-Br [48]), X—3-FPh— (X = 4-Me [57], 4-Cl [62], 4-F [63]), X—3-Cl Ph— (X = 4-Cl [59], 4-MeO [46], 5-Cl [46, 58]), F_5Ph— [64], Cl_5Ph— [65], X—Pym— (X = 2-H_2N [66], 4, 5-$(H_2N)_2$ [67]), X—Pyrazine (X = 2-H_2N [68]),

R = R′C(=O)— (see Section 7.2.1).

R = R′C(=S)— (see Section 7.2.2).

R = H_2N— [71], H_2NC(=S)NH— [77, 78], H(O=)CNHC(=S)NH— [72], 2,4-$(O_2N)_2$PhNH— [25, 73], $MeSO_2$— and CD_3SO_2— [74], H_2NSO_2— [76], Me_2NSO_2— [75], 4-$MePhSO_2$— [25].

References

1. P. Pulay and F. Török, *J. Mol. Struct.*, **29**, 239 (1975).
2. E.L. Wu, G. Zerbi, S. Califano and B. Crawford Jr., *J. Chem. Phys.*, **35**, 2060 (1961).

3. A.P. Gray and R.C. Lord, *J. Chem. Phys.*, **26**, 690 (1957).
4. H. Wolff and D. Staschewski, *Ber. Bunsenges. Phys. Chem.*, **68**, 135 (1964).
5. H. Wolff and H. Ludwig, *Ber. Bunsenges. Phys. Chem.*, **68**, 143 (1964).
6. H. Wolff and H. Ludwig, *Ber. Bunsenges. Phys. Chem.*, **70**, 474 (1966).
7. M. Tsuboi, *Spectrochim. Acta*, **16**, 505 (1960).
8. H. Wolff and H. Ludwig, *J. Chem. Phys.*, **56**, 5278 (1972).
9. G. Dellepiane and G. Zerbi, *J. Chem. Phys.*, **48**, 3573 (1968).
10. A. Piart-Goypiron, M.H. Baron, H. Zine, J. Belloc and M.J. Coulange, *Spectrochim. Acta, Part A*, **49A**, 103 (1993).
11. V.F. Kalasinsky, D.E. Powers and W.C. Harris, *J. Phys. Chem.*, **83**, 509 (1979).
12. A.O. Diallo, Nguyen-Van-Thanh and I. Rossi, *Spectrochim. Acta, Part A*, **43A**, 415 (1987).
13. V.F. Kalasinsky, G.A. Guirgis and J.R. Durig, *J. Mol. Struct.*, **39**, 51 (1977).
14. V.F. Kalasinsky and T.S. Little, *J. Raman Spectrosc.*, **9**, 224 (1980).
15. K.E. Blick, F.C. Nahm and K. Niedenzu, *Spectrochim. Acta, Part A*, **27A**, 777 (1971).
16. A.-L. Borring and K. Rasmussen, *Spectrochim. Acta, Part A*, **31A**, 889 (1975).
17. M.G. Giorgini, M.R. Pellett, G. Paliani and R.S. Cataliotti, *J. Raman Spectrosc.*, **14**, 16 (1983).
18. B. Silvi and J.P. Perchard, *Spectrochim. Acta, Part A*, **32A**, 23 (1976).
19. J.R. Durig, G.A. Guirgis and D.A.C. Compton, *J. Phys. Chem.*, **83**, 1313 (1979).
20. M.M. Szostak, *Croat. Chem. Acta*, **61**, 633 (1988).
21. Y. Tanaka and K. Machida, *J. Mol. Spectrosc.*, **51**, 508 (1974).
22. E. Schmelz, B. Dolabdjian and H.L. Schmidt, *Spectrochim. Acta, Part A*, **34A**, 221 (1978).
23. J.C. Evans, *Spectrochim. Acta*, **16**, 428 (1966).
24. E. Hamburg, R. Grecu and M. Fernea, *Rev. Roum. Chim.*, **17**, 1845 (1972).
25. G.Varsányi, *Assignments for Vibrational Spectra of Seven Hundred Benzene Derivatives*, J.Wiley and Sons (1974).
26. A.K. Ansari and P.K. Verma, *Indian J. Pure Appl. Phys.*, **16**, 454 (1978).
27. N. Abasbegovic, L. Colombo and P. Bleckmann, *J. Raman Spectrosc.*, **6**, 92 (1977).
28. R.A. Yadav and I.S. Singh, *Spectrochim. Acta, Part A*, **41A**, 191 (1985).
29. R.A. Amma, K.P.R. Nair and M.P. Srivastava, *Indian J. Pure Appl. Phys.*, **10**, 58 (1972).
30. M.P. Srivastava, B.B. Lal and I.S. Singh, *Indian J. Pure Appl. Phys.*, **10**, 50 (1972).
31. V.N. Verma and D.K. Rai, *Appl. Spectrosc.*, **24**, 447 (1970).
32. M. Harrand, *J. Raman Spectrosc.*, **4**, 53 (1975).
33. M.M. Szostak, *J. Raman Spectrosc.*, **8**, 43 (1979).
34. E.E. Ernstbrunner, R.B. Girling, W.E.L. Grossman, E. Mayer, K.P.J. Williams and R.E. Hester, *J. Raman Spectrosc.*, **10**, 161 (1981).
35. M.A. Shashidhar, K.S. Rao and E.S. Jayadevappa, *Indian J. Pure Appl. Phys.*, **4**, 170 (1966).
36. M.A. Shashidhar, K.S. Rao and E.S. Jayadevappa, *Spectrochim. Acta, Part A*, **26A**, 2373 (1970).
37. S.N. Sinh and N.L. Singh, *Indian J. Pure Appl. Phys.*, **7**, 250 (1969).
38. P.K. Verma, *Indian J. Phys.*, **51B**, 58 (1977).
39. P.K. Verma, *Indian J. Pure Appl. Phys.*, **6**, 144 (1968).
40. V.B. Singh, R.N. Singh and I.S. Singh, *Spectrochim. Acta*, **22**, 927 (1966).
41. G.N.R. Tripathi and J.E. Katon, *J. Chem. Phys.*, **70**, 1383 (1979).
42. R.M.P. Jaiswal, J.E. Katon and G.N.R. Tripathi, *Spectrochim. Acta, Part A*, **39A**, 275 (1983).
43. M. Prasad, *Indian J. Pure Appl. Phys.*, **13**, 718 (1975).

44. A.R. Shukla, C.M. Pathak, N.G. Dongre, B.P. Asthana and J. Shamir, *J. Raman Spectrosc.*, **17**, 299 (1986).
45. A.R. Shukla, C.M. Pathak, N.G. Dongre, B.P. Asthana and J. Shamir, *Proc. Indian Acad. Sci. (Chem. Sci.)*, **97**, 97 (1986).
46. P. Venkatacharyulu, V.L.N. Prasad, Nallgonda and D. Premaswarup, *Indian J. Pure Appl. Phys.*, **19**, 1178 (1981).
47. N.S. Sundar, *Spectrochim. Acta, Part A*, **41A**, 905 (1985).
48. M. Rangacharyulu and D. Premaswarup, *Indian J. Phys.*, **54B**, 567 (1980).
49. N.S. Sundar, *Spectrochim. Acta, Part A*, **41A**, 1449 (1985).
50. J.A. Faniran and H.F. Shurvell, *Spectrochim. Acta, Part A*, **38A**, 1155 (1982).
51. A.R. Shukla, C.M. Pathak, N.G. Dongre, B.P. Asthana and J. Shamir, *Proc. Indian Acad. Sci. (Chem. Sci.)*, **97**, 593 (1986).
52. R.K. Goel, K.P. Kansal and S.N. Sharma, *Indian J. Pure Appl. Phys.*, **17**, 778 (1979).
53. R.K. Goel, S. Sharma, K.P. Kansal and S.N. Sharma, *Indian J. Pure Appl. Phys.*, **18**, 281 (1980).
54. V.N. Verma and K.P.R. Nair, *Indian J. Pure Appl. Phys.*, **8**, 682 (1970).
55. V.N. Verma, *Spectrosc. Lett.*, **6**, 23 (1973).
56. R.B. Singh, N.P. Singh and D.K. Rai, *Indian J. Pure Appl. Phys.*, **19**, 740 (1981).
57. S.N. Sharma and C.P.D. Dwivedi, *Indian J. Pure Appl. Phys.*, **13**, 570 (1975).
58. R.K. Goel, S.K. Gupta, R.M.P. Jaisawal and P.P. Garg, *Indian J. Pure Appl. Phys.*, **18**, 223 (1980).
59. M.V.F. Dotes, C. Siguënza and P.F. González-Díaz, *Spectrochim. Acta, Part A*, **42A**, 1029 (1986).
60. P.K. Bishui, *Indian J. Pure Appl. Phys.*, **10**, 637 (1972).
61. J.A. Faniran, H.F. Shurvell, D.A. Raeside, B.U. Petelenz and J. Korppi-Tomola, *Can. J. Spectrosc.*, **24**, 148 (1979).
62. N.K. Sanyal, S.L. Srivastava and R.K. Goel, *Indian J. Pure Appl. Phys.*, **16**, 719 (1978).
63. R. Rao, M.K. Aralakkanavar, K.S. Rao and M.A. Shashidhar, *Spectrochim. Acta, Part A*, **45A**, 103 (1989).
64. J.A. Faniran and H.F. Shurvell, *Spectrochim. Acta, Part A*, **31A**, 1127 (1975).
65. J.A. Faniran, I. Iweibo and R.A. Oderinde, *J. Raman Spectrosc.*, **11**, 477 (1981).
66. M. Maehara, S. Nakama, Y. Nibu, H. Shimada and R. Shimada, *Bull. Chem. Soc. Jpn.*, **60**, 2769 (1987).
67. S.L. Srivastava and Rohitashava, *Indian J. Phys.*, **55B**, 455 (1981).
68. A.K. Kalkar and C.C. Ars, *Bull. Chem. Soc. Jpn.*, **59**, 3223 (1986).
69. Y. Hamada, K. Hashiguchi, M. Tsuboi, Y. Koga and S. Kondo, *J. Mol. Spectrosc.*, **105**, 93 (1984).
70. Y. Hamada, N. Sato and M. Tsuboi, *J. Mol. Spectrosc.*, **124**, 172 (1987).
71. D.N. Sathyanarayana and D. Nicholls, *Spectrochim. Acta, Part A*, **34A**, 263 (1978).
72. S.K. Sinha, S. Ram and D.P. Lamba, *Spectrochim. Acta, Part A*, **44A**, 713 (1988).
73. J. Shukla and K.N. Upadhya, *Indian J. Pure Appl. Phys.*, **11**, 787 (1973).
74. K. Hanai, T. Okuda, T. Uno and K. Machida, *Spectrochim. Acta, Part A*, **31A**, 1217 (1975).
75. Y. Tanaka, Y. Tanaka, Y. Saito and K. Machida, *Spectrochim. Acta, Part A*, **39A**, 159 (1983).
76. T. Uno, K. Machida and K. Hanai, *Spectrochim. Acta*, **22**, 2065 (1966).
77. G. Keresztury and M.P. Marzocchi, *Spectrochim. Acta, Part A*, **31A**, 275 (1975).
78. G. Keresztury and M.P. Marzocchi, *Chem. Phys.*, **6**, 117 (1974).
79. H.J. Götze and W. Garbe, *Spectrochim. Acta, Part A*, **35A**, 461 (1979).

80. H. Wolff, U. Schmidt and E. Wolff, *Spectrochim. Acta, Part A*, **36A**, 899 (1970).
81. L.K. Dyall, *Spectrochim. Acta, Part A*, **25A**, 1423 (1969).
82. L.K. Dyall, *Spectrochim. Acta, Part A*, **25A**, 1717 (1969).
83. L.K. Dyall, *Spectrochim. Acta, Part A*, **44A**, 283 (1988).
84. L.J. Bellamy and R.L. Williams, *Spectrochim. Acta*, **9**, 341 (1957).
85. J.E. Stewart, *J. Chem. Phys.*, **30**, 1259 (1959).
86. E. Sacher, *Spectrochim. Acta, Part A*, **43A**, 747 (1987).
87. P.J. Krueger, *Nature*, **194**, 1077 (1962).
88. L. Segal and F.V. Eggerton, *Appl. Spectrosc.*, **15**, 112 (1961).
89. D. Hadzi, J. Jan and A. Ocvirk, *Spectrochim. Acta, Part A*, **25A**, 97 (1969).
90. S.Ca lifano and R. Moccia, *Gazz. Chim. Ital.*, **87**, 805 (1957).
91. A.R. Katritzky and R.A. Jones, *J. Chem. Soc.*, 3674 (1959).
92. S.M. Craven and F.F. Bentley, *Appl. Spectrosc.*, **26**, 449 (1972).
93. S.M. Craven, F.F. Bentley and D.F. Pensenstadler, *Appl. Spectrosc.*, **26**, 647 (1972).
94. C. Laurence and B. Wojtkowiak, *Bull. Soc. Chim. Fr.*, 3124 (1971).
95. G. Varsányi, *Vibrational Spectra of Benzene Derivatives*, Academic Press (1969). pp. 376–377.
96. D.A. Thornton, *J. Coord. Chem.*, **24B**, 261 (1991).
97. R.A. Nyquist, *Appl. Spectrosc.*, **47**, 411 (1993).

9.2 METHYLAMINO

According to the simplest approximation, the CNHCH fragment in EtNHMe is in a plane and this compound belongs to the point group C_s. The 15 vibrations in —$NHCH_3$ differentiate between 9a′ + 6a″ vibrational modes:

a′: νNH, ν_a'Me, ν_sMe, δ_a'Me, δ_sMe, δNH, νN—C, ρ'Me, δ—N—C;
a″: ν_aMe, δ_aMe, ρMe, γNH and two torsions.

The NH stretching vibration

The associated NH stretching vibration in α-saturated, N- or S-bonded —NHMe compounds gives rise to a weak to moderate but broadish band in the region 3265 $\pm$ 50 cm^{-1}. In dilute solutions or in the vapour state this band narrows and shifts to higher wavenumbers: 3395 $\pm$ 45 cm^{-1}. In α-unsaturated and aromatic compounds the associated νNH appears moderately to strongly in the region 3400 $\pm$ 40 cm^{-1} and at 3450 $\pm$ 30 cm^{-1} in the unbonded state, with the highest {lowest} wavenumbers in the spectra of *N*-methylbenzenamines with an electron attracting {releasing} group on the 4-position, for example 4-XPhNHMe in which X = O_2N (3445), HO(O═)C (3444), H (3434), Me (3429) and MeO (3423 cm^{-1}) [8, 14, 15]. In 2-substituted compounds or in the associated state, hydrogen bridges disturb this rule.

Methyl stretching vibrations

Except for the high wavenumbers in the spectrum of O_2NNHMe (3016 and *2953* cm^{-1}), the antisymmetric stretching vibrations absorb in the regions 2970 ± 30 cm^{-1} and 2945 ± 45 cm^{-1}. The band intensity varies from weak in *N*-methyl-substituted sulfonamides to moderate in *N*-methyl-substituted amines. The methyl symmetric stretch absorbs, sharply and clearly separated from the antisymmetric counterparts, in the region 2855 ± 70 cm^{-1} with a moderate to strong intensity in amines and with a weak intensity in sulfonamides. These low wavenumbers, attributed to a reduced force constant of the CH bond under the influence of the free electron pair of the nitrogen atom, is typical for *N*-Me compounds [8, 16–20] but also for methyl ethers. *N*-methylmethanamine with 2785 cm^{-1} in the solid state and 2791 cm^{-1} as a gas is responsible for the lowest value and O_2NNHMe exhibits the highest wavenumber: 2923 cm^{-1}.

The NH in-plane deformation

In *N*-methylamines the δNH should be expected in the region 1530 ± 50 cm^{-1}, but the very weak intensity of the band hinders the assignment. α-Saturated secondary amines show an unspectacular small shoulder (≈1500 cm^{-1}) on the high-frequency wing of the methyl and methylene deformations. In *N*-substituted benzenamines this δNH almost disappears in the stretching vibrations of the ring 19a (19b) (≈1510 cm^{-1}) [21] or 8a (8b) (1640 ± 10 cm^{-1}) [13]. The NH deformation in *N*-methyl-substituted sulfonamides is expected at lower wavenumbers, so that the region 1395 ± 25 cm^{-1} [11, 13, 22] is preferred to 1635 ± 15 cm^{-1} [23]. In *N*-methyl-substituted (thio)amides this δNH (amide II), coupled with the νC—N, exhibits a stronger absorption (Section 7.3).

Methyl deformations

Both methyl antisymmetric deformations absorb weakly to moderately and often coincide. The region 1410 ± 35 cm^{-1} of the methyl symmetric deformation is typical for *N*-methyl-substituted amines and (thio)amides alike [24] and approaches that of R—C(=O)OMe (1435 ± 15) and ROMe (1445 ± 15 cm^{-1}) compounds. The intensity of the symmetric deformation slightly exceeds that of the antisymmetric modes.

Methyl rocks and N—C stretching vibration

The three weak to moderate absorptions in the regions 1145 ± 45, 1095 ± 75 and 1015 ± 95 cm^{-1} are due to two methyl rocking vibrations and a N—C stretching vibration strongly coupled among themselves. The bands in the above-mentioned regions are assigned to the N—C stretch as well as to a methyl rock [21, 25]. The absorption near 1135 {1070} cm^{-1} in the spectra of *N*-methylsulfonamides is attributed to the methyl rock {N—C stretch} [11, 12].

The NH wagging vibration

Associated *N*-methyl-substituted aliphatic amines exhibit a very broad diffuse band between 950 and 650 cm^{-1} with a maximum absorption at 725 ± 20 cm^{-1}, attributed to the γ (or ω)NH. In the spectra of *N*-methyl-substituted benzenamines and sulfonamides this NH wag is observed at lower wavenumbers: 635 ± 35 cm^{-1}. The broad band at 670 cm^{-1} in the spectrum of 4-$MePhSO_2NHMe$ is assigned to this wag [13], although Goldstein *et al.* [12] preferred the absorption at 722 cm^{-1}.

Skeletal deformation and torsions

In *N*-methyl-substituted aliphatic amines and sulfonamides the δC—N—C is found in the region 360 ± 50 cm^{-1}. In *N*-methylbenzenamines this vibration is coupled to the substituent-sensitive Ph—N deformation. The methyl torsion is often assigned in the region 230 ± 30 cm^{-1} and the NHMe torsion may be expected in the neighbourhood of 100 cm^{-1} [26, 27].

Table 9.3 Absorption regions (cm^{-1}) of the normal vibrations of —NHMe

Vibration	α-saturated N-bonded	aromatic α-unsaturated	—SO_2NHMe	—C(=O)NHMe	—C(=S)NHMe
νNH	3265 ± 50	3400 ± 40	3265 ± 50	3315 ± 45	3250 ± 70
ν_aMe	2965 ± 25	2965 ± 25	2965 ± 25	2970 ± 30	2970 ± 30
ν_a'Me	2950 ± 25	2950 ± 25	2950 ± 25	2945 ± 45	2945 ± 25
ν_sMe	2855 ± 70	2855 ± 70	2855 ± 70	2870 ± 45	2875 ± 45
δNH	1530 ± 50	1530 ± 50	1395 ± 25	1550 ± 50	1535 ± 35
δ_aMe	1470 ± 15	1470 ± 15	1470 ± 15	1450 ± 30	1450 ± 25
δ_a'Me	1460 ± 15	1460 ± 15	1460 ± 15	1445 ± 35	1435 ± 25
δ_sMe	1410 ± 35	1410 ± 35	1410 ± 35	1400 ± 25	1400 ± 25
ρMe/νN—C	1150 ± 30	1140 ± 15	1135 ± 15	1155 ± 30	1145 ± 45
ρ'Me/νN—C	1085 ± 65	1055 ± 25	1070 ± 15	1100 ± 65	1075 ± 40
νN—C/ρMe	995 ± 75	985 ± 65	1020 ± 50	1015 ± 95	1000 ± 50
$\gamma(\omega)$NH	725 ± 20	635 ± 35	635 ± 35	735 ± 60	665 ± 55
δC—N—C	360 ± 50	–	360 ± 50	315 ± 55	265 ± 65
torsion Me	230 ± 30	230 ± 30	230 ± 30	230 ± 50	195 ± 50
torsion NHMe	100 ± 30	<100	–	–	–

R—NHMe compounds

R = R′C(=O)NHMe (see Section 7.3.1).
R′C(=S)NHMe (see Section 7.3.2).

R = Me— [1–4], $H_2C{=}CH$— [5], Ph— [6–8, 13], 2-HO(O=)CPh— [9], 4-MeOPh— [13], 2- and 4-FPh— [13], O_2N— [10], PhNH— [13], $MeSO_2$— and CD_3SO_2— [11], $PhSO_2$— [12], 4-$MePhSO_2$— [12, 13].

References

1. G. Gamer and H. Wolff, *Spectrochim. Acta, Part A*, **29A**, 129 (1973).
2. M.J. Buttler and D.C. McKean, *Spectrochim. Acta*, **21**, 465 (1965).
3. W.G. Fateley and F.A. Miller, *Spectrochim. Acta*, **18**, 980 (1962).
4. A.A. Chalmers and D.C. McKean, *Spectrochim. Acta*, **21**, 1387 (1965).
5. Y. Amatatsu, Y. Hamada, M. Tsuboi and M. Sugie, *J. Mol. Spectrosc.*, **111**, 29 (1985).
6. A. Perrier-Datin and J.M. Lebas, *J. Chim. Phys.*, **69**, 591 (1972).
7. A.K. Ansari and P.K. Verma, *Indian J. Pure Appl. Phys.*, **16**, 454 (1978).
8. A.R. Katritzky and R.A. Jones, *J. Chem. Soc.*, 3674 (1959).
9. A. Tramer, *J. Mol. Struct.*, **4**, 313 (1969).
10. M.I. Dakhis, V.G. Dashevsky and V.G. Avakyan, *J. Mol. Struct.*, **13**, 339 (1972).
11. A. Noguchi, K. Hanai and T. Okuda, *Spectrochim. Acta, part A*, **36A**, 829 (1980).
12. M. Goldstein, M.A. Russell and H.A. Willis, *Spectrochim. Acta, Part A*, **25A**, 1275 (1969).
13. G. Varsányi, *Assignments for Vibrational Spectra of Seven Hundred Benzene Derivatives*, J.Wiley & Sons, New York (1974).
14. C. Laurence and M. Berthelot, *Spectrochim. Acta, Part A*, **34A**, 1127 (1978).
15. L.K. Dyall and J.E. Kemp, *Spectrochim. Acta*, **22**, 467 (1966).
16. J.P. Perchard, M.-T. Forel and M.-L. Josien, *J. Chim. Phys.*, **61**, 660 (1964).
17. D.C. McKean and I.A. Ellis, *J. Mol. Struct.*, **29**, 81 (1975).
18. I.A. Degen, *Appl. Spectrosc.*, **23**, 239 (1969).
19. R.D. Hill and G.D. Meakins, *J. Chem. Soc.*, 760 (1958).
20. J.T. Braunholtz, E.A.V. Ebsworth, F.G. Mann and N. Sheppard, *J. Chem. Soc.*, 2780 (1958).
21. D. Hadzi and M. Skrbljak, *J. Chem. Soc.*, 843 (1957).
22. D. Hadzi, *J. Chem. Soc.*, 847 (1957).
23. T. Momose, Y. Ueda and T. Shoji, *Chem. Pharm. Bull.*, **7**, 734 (1959).
24. M. Beer, H.B. Kesseler and G.B.B.M. Sutherland, *J. Chem. Phys.*, **29**, 1097 (1958).
25. P.N. Gates, D. Steele and R.A.R. Pearce, *J. Chem. Soc. Perkin Trans. 2*, 1607 (1972).
26. R.A. Kydd and A.R.C. Dunham, *J. Mol. Struct.*, **98**, 39 (1983).
27. R. Cervellati, G. Corbelli, A. Dal Borgo and D.G. Lister, *J. Mol. Struct.*, **73**, 31 (1981).

9.3 ACETYLAMINO

In the C_s symmetry the $3N - 6 = 21$ normal vibrations of —NHC(=O)Me are divided into 13a′ + 8a″ species of vibration:

a′: νNH, ν'_aMe, ν_sMe, νC=O, δNH, δ'_aMe, δ_sMe, νC—N, ρ'Me, νC—C, δC=O, δN—C—C, δ—N—C;
a″: ν_aMe, δ_aMe, ρMe, ωNH, γC=O and three torsions.

The NH stretching vibration

The NH stretching vibration in N-substituted acetamides appears strongly and broadly in the region 3280 ± 60 cm^{-1} and shifts to higher wavenumbers (3410 ± 70 cm^{-1}) in dilute solutions.

Methyl stretching vibrations

The methyl antisymmetric stretching vibrations exhibit a weak band in the regions 2990 ± 20 and 2965 ± 35 cm^{-1}. The methyl symmetric stretch appears weakly in the range 2900 ± 45 cm^{-1}.

The C=O stretching vibration

The C=O stretching vibration (amide I) absorbs strongly in the region 1690 ± 45 cm^{-1}, with the highest values from the spectra of imides: MeC(=O)NHC(=O)Me (1735 and *1701*) and MeC(=O)NHC(=O)Et (1734 cm^{-1}). The remaining compounds show this νC=O in a region typical for amides: 1675 ± 30 cm^{-1}.

The NH in-plane deformation

The NH in-plane deformation (amide II), coupled with the νC—N, gives a strong band in the region 1540 ± 60 cm^{-1} for the *trans* configuration only. Together with the amide I absorption, this δNH shows a characteristic pair of bands. The highest wavenumber is assigned in the spectrum of *N*-chloroacetamide (1597 cm^{-1}). The lowest values are found in the spectra of imides (1500 ± 20 cm^{-1}): H_2NC(=O)NHC(=O)Me (1482) and MeC(=O)NHC(=O)Me (1505 cm^{-1}). Most of the N—substituted acetamides display this δNH at 1545 ± 25 cm^{-1}.

Methyl deformations

The methyl antisymmetric deformations provide a weak to moderate band in the regions: 1450 ± 30 and 1420 ± 20 cm^{-1}, but they usually coincide. As contrasted with the very weak stretchings, the methyl symmetric deformation appears more strongly in the region 1365 ± 10 cm^{-1}.

The C—N stretching vibration

The C—N stretching vibration (amide III), coupled with the δNH, is moderately to strongly active in the region 1275 ± 55 cm^{-1}. The imides are responsible for the extensive region: MeC(=O)NHC(=O)Me gives rise to an antisymmetric NC_2

at 1310 and a symmetric NC_2 stretch at 1222 cm^{-1}. The *N*-phenyl-substituted acetamides (acetanilides) absorb at the HW side of the region (1310 ± 20 cm^{-1}) and *N*-alkyl-substituted acetamides at 1290 ± 35 cm^{-1}.

Methyl rocking vibrations and C—C stretching vibration

The methyl rocks are observed as weak to medium bands in the regions 1090 ± 40 and 1015 ± 35 cm^{-1}. The νC—C absorbs weakly to moderately in the region 915 ± 65 cm^{-1}. *N*-Phenyl-substituted acetamides display this C—C stretching vibration near 965 cm^{-1}.

The NH out-of-plane deformation

The γNH/C=O or ωNH/C=O (amide V) is moderately active with a broad band in the region 790 ± 70 cm^{-1}. The *N*-phenyl-substituted acetamides and the *trans,cis* imides absorb at the HW side of this region (840 ± 20 cm^{-1}) and the *N*-alkyl-substituted amides and the *trans,trans* imides at the LW side (760 ± 40 cm^{-1}).

The C=O deformations

The C=O in-plane deformation (amide IV) is located at 625 ± 70 cm^{-1} as a moderate band. The highest wavenumber is observed in the spectrum of MeC(=O)NHCl (693) and the lowest in that of MeC(=O)NHC(=O)Me (560 and *648* cm^{-1}). *N*-Methylacetamide absorbs at 628 cm^{-1}.

Usually the C=O out-of-plane deformation γC=O/NH (amide VI) is assigned in the region 540 ± 80 cm^{-1}. High values originate from the spectra of EtNHC(=O)Me and nBuNHC(=O)Me (620 cm^{-1}). A few 4-XPhNHC(=O)Me compounds (X = H, Br, HO, EtO) absorb at 485 ± 25 cm^{-1}.

Skeletal deformations and torsions

Acetylamino compounds display the in-plane skeletal N—C—C deformation in the region 420 ± 55 cm^{-1}, for example RNHC(=O)Me compounds in which R = Me (429), Et (427) and nBu (450cm^{-1}). *N*-Phenyl-substituted acetamides absorb in the neighbourhood of 385 cm^{-1}.

The external —N—C deformation is found in the range 310 ± 65 cm^{-1}. The lowest external —N—C deformation (a$''$) or —NHC(=O)Me torsion depends largely upon the R substituent, but for a few molecules the torsion is assigned at 225 ± 65 cm^{-1}. Usually the methyl torsion absorbs at 200 ± 65 cm^{-1} and the C(=O)Me torsion at lower wavenumbers, probably at 100 ± 40 cm^{-1}.

Table 9.4 Absorption regions (cm^{-1}) of the normal vibrations of —NHC(=O)Me

Vibration	Region	Vibration	Region
νNH	3280 ± 60	ρMe	1090 ± 40
ν_aMe	2990 ± 20	ρ'Me	1015 ± 35
ν'_aMe	2965 ± 35	νC—C	915 ± 65
ν_sMe	2900 ± 45	$\gamma(\omega)$NH (V)	790 ± 70
νC=O (I)	1690 ± 45	δC=O (IV)	625 ± 70
δNH (II)	1540 ± 60	γC=O (VI)	540 ± 80
δ_aMe	1450 ± 30	δN—C—C	420 ± 55
δ'_aMe	1420 ± 20	δ—N—C	310 ± 65
δ_sMe	1365 ± 10	torsion NHC(=O)Me	225 ± 65
νC—N (III)	1275 ± 55	torsion Me	200 ± 65
		torsion C(=O)Me	100 ± 40

R—NHC(=O)Me molecules

R = Me— [1–15], CD_3— [8–12], Et— [1, 16], nPr— and nBu— [16], $HOCH_2CH_2$—, $HSCH_2CH_2$— [17], MeC(=O)— [18–20, 27], EtC(=O)— [21], H_2NC(=O)— [22], Ph— [24, 26], 4-MePh— [26], 4-H_2NPh— [24], 2-MeC(=O)NHPh— [26], 2-, 3- and 4-O_2NPh— [26], 4-HOPh—, 4-EtOPh—, 2-ClPh— [26], 4-ClPh— [23], 4-BrPh— [24], 2,6-Cl_2Ph— [26], Cl— [25].

References

1. T. Miyazawa, T. Shimanouchi and S.-I. Mizushima, *J. Chem. Phys.*, **24**, 408 (1956).
2. T. Miyazawa, T. Shimanouchi and S.-I.Mizushima, *J. Chem. Phys.*, **29**, 611 (1958).
3. R.L. Jones, *J. Mol. Spectrosc.*, **11**, 411 (1963).
4. O.D. Bonner, K.W. Bunzl and G.B. Woolsey, *Spectrochim. Acta*, **22**, 1125 (1966).
5. S.E. Krikorian and M. Mahpour, *Spectrochim. Acta., Part A*, **29A**, 1233 (1973).
6. S. Ataka, H. Takeuchi and M. Tasumi, *J. Mol. Struct.*, **113**, 147 (1984).
7. Y. Grenie, M. Avignon and C. Garrigou-Lagrange, *J. Mol. Struct.*, **24**, 293 (1975).
8. M. Rey-Lafon, M.-T. Forel and C. Garrigou-Lagrange, *Spectrochim. Acta, Part A*, **29A**, 471 (1973).
9. A. Warshel, M. Levitt and S. Lifson, *J. Mol. Spectrosc.*, **33**, 84 (1970).
10. B. Schneider, A. Horeni, H. Pivcová and J. Honzl, *Collect. Czech. Chem. Commun.*, **30**, 2196 (1965).
11. H. Pivcová, B. Schneider and J. Stokr, *Collect. Czech. Chem. Commun.*, **30**, 2215 (1965).
12. J. Jakes and S. Krimm, *Spectrochim. Acta, Part A*, **27A**, 19 (1971).
13. A. Balázs, *Acta Chim. Acad. Sci. Hung.*, **108**, 265 (1981).
14. A. Balázs, *J. Mol. Struct.*, **153**, 103 (1987).
15. N.G. Mirkin and S. Krimm, *J. Mol. Struct.*, **236**, 97 (1991).
16. J. Jakes and S. Krimm, *Spectrochim. Acta, Part A*, **27A**, 35 (1971).
17. G. Zuppiroli, C. Perchard, M.L. Baron and C. de Lozé, *J. Mol. Struct.*, **69**, 1 (1980).
18. Y. Kuroda, Y. Saito, K. Machida and T. Uno, *Spectrochim. Acta, Part A*, **27A**, 1481 (1971).

19. Y. Kuroda, Y. Saito, K. Machida and T. Uno, *Spectrochim. Acta, Part A*, **29A**, 411 (1973).
20. T. Uno and K. Machida, *Bull. Chem. Soc. Jpn.*, **34**, 545 (1961).
21. T. Uno and K. Machida, *Bull. Chem. Soc. Jpn.*, **34**, 551 (1961).
22. Y. Saito and K. Machida, *Spectrochim. Acta, Part A*, **35A**, 369 (1979).
23. P. Venkatacharyulu, I.V.S.R.M. Sarma and D. Premaswarup, *Indian J. Pure Appl. Phys.*, **20**, 670 (1982).
24. S. Tariq, N. Ali and P.K. Verma, *Indian J. Pure Appl. Phys.*, **22**, 265 (1984).
25. J.E. Devia and J.C. Carter, *Spectrochim. Acta, Part A*, **29A**, 613 (1973).
26. G. Varsányi, *Assignments for Vibrational Spectra of Seven Hundred Benzene Derivatives*, J.Wiley & Sons, New York (1974).
27. F. Ramondo, S. Nunziante and L. Bencivenni, *J. Mol. Struct.*, **291**, 219 (1993).

9.4 DIMETHYLAMINO

With the H and N atom of *N*-methylmethanamine in the symmetry plane and both methyl groups situated symmetrically with respect to this plane, the CNC angle is bisected. According to C_s, the 24 normal vibrations differentiate between 13a′ + 11a″ vibrational modes. Substitution of the NH stretching vibration (a′) by a torsion (a″) results in 12a′ + 12a″ normal vibrations for the —NMe_2 fragment:

a′: ν_aMe (2), ν_sMe, δ_aMe (2), δ_sMe, ρMe (2), $\nu_s NC_2$, δNC_2, ωNC_2, torsion Me;
a″: ν_aMe (2), ν_sMe, δ_aMe (2), δ_sMe, ρMe (2), $\nu_a NC_2$, ρNC_2, torsion Me and torsion NMe_2.

The molecular fragments —$PhNMe_2$ [9, 10] and —SO_2NMe_2 [48, 51] are studied in this way. If the molecular skeleton and both methyl groups form part of the symmetry plane, the normal vibrations are divided into 14a′ + 10a″ species of vibration. In that case the dimethylamino fragment possesses the plane structure, as in HC(=O)NMe_2 [19, 20, 22] and MeC(=O)NMe_2 [25]:

a′: ν_aMe (2), ν_sMe (2), δ_aMe (2), δ_sMe (2), $\nu_a NC_2$, ρMe (2), $\nu_s NC_2$, δNC_2, ρNC_2;
a″: ν_aMe (2), δ_aMe (2), ρMe (2), ωNC_2, torsion Me (2), torsion NMe_2.

Methyl stretching vibrations

Although the NMe_2 group provides four methyl antisymmetric stretching vibrations, usually only two are observed: 2990 ± 30 and 2950 ± 25 cm^{-1}. The in-phase(a′) and the out-of-phase (a″) vibrations rarely give rise to separate bands. In delimiting these absorption regions, the high wavenumbers in the spectrum of O_2NNMe_2 (3033 and 2993 cm^{-1}) are not taken into account. Not infrequently tertiary (sulfon)amides overstep the limit of 3000 cm^{-1}, for instance: $MeSO_2NMe_2$ and $PhSO_2NMe_2$

(≈3020 cm^{-1}), N≡$CNMe_2$ (3017 cm^{-1}) and CD_3C(=O)NMe_2 (3016 cm^{-1}). Tertiary amines of the type R—NMe_2 (R = alkyl or aryl) are active at 2980 ± 15 cm^{-1}.

The free pair of electrons of the N atom weakens the CH bond parallel to the axis of this orbital, producing the typical lowering of the 'symmetric' stretching vibrations, which approach near the region in which the overtones and combination bands of the methyl deformations are active [59–63]. The literature does not agree in assigning these symmetrical stretching vibrations, which explains the extensive regions (2855 ± 65 and 2835 ± 65 cm^{-1}). At the HW side, near 2920 cm^{-1}, one finds a few N, N-dimethylbenzenamines such as 4-$XPhNMe_2$ (X = H(O=)C, HO(O=)C, O_2N), Me_2NC(=X)NMe_2 (X = S and Se) and O=$NNMe_2$. Low wavenumbers are observed in the spectra of $MeNMe_2$ and HC(=O)NMe_2 (≈2775 cm^{-1}), $HNMe_2$, $EtNMe_2$ and HC(=S)NMe_2, which absorb near 2790 cm^{-1}. The highest value (2948 cm^{-1}), assigned in the spectrum of O_2NNMe_2, is not taken into account.

Methyl deformations

Occasionally the in-plane (a′) and out-of-plane (a″) vibrations are observed separately so that only two bands instead of four emerge (1465 ± 25 and 1445 ± 25 cm^{-1}). The high wavenumber (1495 cm^{-1}) in the spectrum of N_3C(=O)NMe_2 falls outside this region. The HW side is covered by $RNMe_2$ (R = N≡C, 4-N≡CPh and N≡CC(=O)) with the value 1488 cm^{-1}. Low values in the neighbourhood of 1425 cm^{-1} have been traced in the spectra of Me_2NC(=Se)NMe_2 and N≡$CNMe_2$.

In the spectrum of N≡$CNMe_2$ the methyl symmetric deformations are assigned at 1342 and 1333 cm^{-1} and in that of F_2PNMe_2 at *1443* and 1313 cm^{-1}. Disregarding these low values, the methyl symmetric deformations are observed separately in the regions 1410 ± 35 and 1385 ± 30 cm^{-1}. High values (≈1442 cm^{-1}) originate from the spectra of $HNMe_2$, $HONMe_2$ and $H_2NSO_2NMe_2$-d_0 and -d_2. The LW side is covered by MeC(=S)NMe_2, with 1360 cm^{-1}.

Skeletal stretching vibrations and methyl rocks

Between 1300 and 700 cm^{-1} four methyl rocks and two skeletal NC_2 stretching vibrations are observed separately in the regions:

$\rho Me/\nu_a NC_2$	$\nu_a NC_2/\rho Me$	ρMe	ρMe	ρMe	$\nu_s NC_2$
1250 ± 50	1165 ± 35	1115 ± 65	1060 ± 40	1005 ± 65	850 ± 150 cm^{-1}

The highest methyl rock is coupled with the NC_2 stretching vibration. The *N,N*-dimethylamides are responsible for the extensive region in which the $\nu_s NC_2$ is assigned because the region 750 ± 30 cm^{-1} is often attributed to this vibration [7, 30–32, 38]. Mielke and Barnes [25] and Garrigou-Lagrange and Forel [28] attribute the wavenumber 960 {740} cm^{-1} in the spectrum of *N,N*-dimethylacetamide not only to the $\nu_s NC_2$ {νC—C}, but also for a part to the νC—C {$\nu_s NC_2$}. Most investigators, however, hold the same view that the νC—C makes the greatest contribution to the absorption at 750 ± 30 cm^{-1}, so that the more convenient region 900 ± 80 cm^{-1} is reserved for the NC_2 symmetric stretching vibration. In the spectra of *N,N*-dimethylbenzenamines the $\nu_s NC_2$ is found at 950 ± 20 cm^{-1}.

Skeletal deformations

The NC_2 in-plane deformation can be found in the region 460 ± 65 cm^{-1}. The extreme values are from HC(=S)NMe_2 with ≈520 cm^{-1} and HC(=O)NMe_2 with ≈400 cm^{-1}. The vibrational analysis of R—NMe_2 compounds reveals another two partly external skeletal NC_2 deformations: the NC_2 wagging vibration (360 ± 50 cm^{-1}) and the NC_2 rocking vibration (300 ± 75 cm^{-1}).

Table 9.5 Absorption regions (cm^{-1}) of the normal vibrations of —NMe_2

Vibration	Region	Vibration	Region
ν_aMe	2990 ± 30	ρMe/$\nu_a NC_2$	1250 ± 50
ν_aMe	2990 ± 30	$\nu_a NC_2$/ρMe	1165 ± 35
ν_aMe	2950 ± 25	ρMe	1115 ± 65
ν_aMe	2950 ± 25	ρMe	1060 ± 40
ν_sMe	2855 ± 65	ρMe	1005 ± 65
ν_sMe	2835 ± 65	$\nu_s NC_2$	900 ± 80
δ_aMe	1465 ± 25	δNC_2	460 ± 65
δ_aMe	1465 ± 25	ωNC_2	360 ± 50
δ_aMe	1445 ± 25	ρNC_2	300 ± 75
δ_aMe	1445 ± 25	torsion Me	245 ± 50
δ_sMe	1410 ± 35	torsion Me	185 ± 55
δ_sMe	1385 ± 30	torsion NC_2	120 ± 50

R—NMe_2 molecules

R = H— [1–4], D— [1, 2], Me— [3, 5, 6], Et— [7], Ph— [8–11, 18], 4-H(O=)CPh— [12, 18], 2-HO(O=)CPh— [13, 14], 4-HO(O=)CPh— and 4-EtO(O=)CPh— [15], 4-N≡CPh— [10], 4-O_2NPh— and 4-(O=)NPh— [18], 4-Me_2NPh— and 4-(4-Me_2NPh)—Ph— [11], 2- and 4-Py— [16], N≡C— [17], HC(=O)— [19–24], MeC(=O)— [24–29], CD_3C(=O)— [27–29], EtC(=O)— [30, 31], nPrC(=O)— [32], N≡CC(=O)— [33], Me_2NC(=O)— [34, 35],

$N_3C(=)$— [36], $H_2NC(=O)C(=O)$— and $MeHNC(=O)C(=O)$— [37], $H_2C{=}CHC(=O)$— [38], $H_2NC(=O)$— [39], $HC(=S)$— [21, 40], $MeC(=S)$— [40, 41], $N{\equiv}CC(=S)$— [42], $Me_2NC(=S)$— [43, 44], $(CD_3)_2NC(=S)$— [44], $HC(=Se)$— [45], $Me_2NC(=Se)$— and $(CD_3)_2NC(=Se)$— [46], $FS(=O)_2$— and $BrS(=O)_2$— [47], $ClS(=O)_2$— [47, 48], $MeS(=O)_2$— and $PhS(=O)_2$— [50], $H_2NS(=O)_2$— and $D_2NS(=O)_2$— [51], $H_2C{=}N$— [52], O_2N— [53], $C{=}N$— [54], HO and DO— [55, 56], Cl— [57], F_2P— [58], $(Me_2N)_2P$— and $(Me_2N)_2As$— [49].

References

1. G. Gamer and H. Wolff, *Spectrochim. Acta, Part A*, **29A**, 129 (1973).
2. M.J. Buttler and D.C. McKean, *Spectrochim. Acta*, **21**, 465 (1965).
3. W.G. Fateley and F.A. Miller, *Spectrochim. Acta*, **18**, 980 (1962).
4. A.A. Chalmers and D.C. McKean, *Spectrochim. Acta*, **21**, 1387 (1965).
5. G. Dellepiane and G. Zerbi, *J. Chem. Phys.*, **48**, 3573 (1968).
6. J.R. Barcelo and J. Bellanato, *Spectrochim. Acta*, **8**, 27 (1956).
7. J.R. Durig and F.O. Fox, *J. Mol. Struct.*, **95**, 85 (1982).
8. A.K. Ansari and P.K. Verma, *Indian J. Pure Appl. Phys.*, **16**, 454 (1978).
9. A. Perrier-Datin and J.-M. Lebas, *J. Chim. Phys.*, **69**, 591 (1972).
10. P.N. Gates, D. Steele and R.A.R. Pearce, *J. Chem. Soc. Perkin Trans. 2*, 1607 (1972).
11. V. Guichard, A. Bourkba, M.-F. Lautie and O. Poizat, *Spectrochim. Acta, Part A*, **45A**, 187 (1989).
12. J.G. Rosencrance and P.N. Jagodzinski, *Spectrochim. Acta, Part A*, **42A**, 869 (1986).
13. K.R.K. Rao and C.I. Jose, *J. Mol. Struct.*, **18**, 447 (1973).
14. A. Tramer, *J. Mol. Struct.*, **4**, 313 (1969).
15. M. Forster and R.E. Hester, *J. Chem. Soc. Faraday Trans. 2*, **77**, 1535 (1981).
16. S.P. Gupta, A. Salik and R.K. Goel, *Indian J. Phys.*, **61B**, 427 (1987).
17. F.B. Brown and W.H. Fletcher, *Spectrochim. Acta*, **19**, 915 (1963).
18. G. Varsányi, *Assignments for Vibrational Spectra of Seven Hundred Benzene Derivatives*, J.Wiley & Sons, New York (1974).
19. Z. Mielke, H. Ratajczak, M. Wiewiorowski, A.J. Barnes and S.J. Mitson, *Spectrochim. Acta, Part A*, **42A**, 63 (1986).
20. T.C. Jao, I. Scott and D. Steele, *J. Mol. Spectrosc.*, **92**, 1 (1982).
21. G. Durgaprasad, D.N. Sathyanarayana and C.C. Patel, *Bull. Chem. Soc. Jpn.*, **44**, 316 (1971).
22. G. Kaufmann and M.J.F. Leroy, *Bull. Soc. Chim. Fr.*, **II**, 402 (1967).
23. D. Steele and A. Quatermain, *Spectrochim. Acta, Part A*, **43A**, 781 (1987).
24. R.L. Jones, *J. Mol. Spectrosc.*, **11**, 411 (1963).
25. Z. Mielke and A.J. Barnes, *J. Chem. Soc. Faraday Trans. 2*, **82**, 437 (1986).
26. G. Durgaprasad, D.N. Sathyanarayana, C.C. Patel and H.S. Randhawa, *Spectrochim. Acta, Part A*, **28A**, 2311 (1972).
27. C. Garrigou-Lagrange, C. De Lozé, P. Bacelon, P. Combelas and J. Dagaut, *J. Chim. Phys.*, **67**, 1936 (1970).
28. C. Garrigou-Lagrange and M.-T. Forel, *J. Chim. Phys.*, **68**, 1329 (1971).
29. A.M. Dwivedi, S. Krimm and S. Mierson, *Spectrochim. Acta, Part A*, **45A**, 271 (1989).
30. K.V. Ramiah, V.V. Chalapathi and C.A.I. Chary, *Curr. Sci.*, **35**, 350 (1966).

31. V.V. Chalapathi and K.V. Ramiah, *Curr. Sci.*, **37**, 453 (1968).
32. K.V. Ramiah and S.K. Sayee, *Curr. Sci.*, **38**, 457 (1969).
33. H.O. Desseyn and J.A. Le Poivre, *Spectrochim. Acta, Part A*, **31A**, 635 (1975).
34. H.L. Spell and J. Laane, *Spectrochim. Acta, Part A*, **28A**, 295 (1972).
35. K. Ravindranath and K.V. Ramiah, *Indian J. Pure Appl. Phys.*, **15**, 182 (1977).
36. W. Buder and A. Schmidt, *Spectrochim. Acta, Part A*, **29A**, 1429 (1973).
37. H.O. Desseyn, B.J. Van der Veken and M.A. Herman, *Spectrochim. Acta, Part A*, **33A**, 633 (1977).
38. G.R. Rao and K.V. Ramiah, *Indian J. Pure Appl. Phys.*, **18**, 94 (1980).
39. Y. Mido, K. Tanase and K. Kido, *Spectrochim. Acta, Part A*, **45A**, 397 (1989).
40. C.A.I. Chary and K.V. Ramiah, *Proc. Indian Acad. Sci.*, **69A**, 18 (1969).
41. A. Ray and D.N. Sathyanarayana, *Bull. Chem. Soc. Jpn.*, **45**, 2712 (1972).
42. H.O. Desseyn and J.A. Le Poivre, *Spectrochim. Acta, Part A*, **31A**, 647 (1975).
43. R.K. Gosavi, U. Agarwala and C.N.R. Rao, *J. Am. Chem. Soc.*, **89**, 235 (1967).
44. U. Anthoni, P.H. Nielsen, G. Borch, J. Gustavsen and P. Klaboe, *Spectrochim. Acta, Part A*, **33A**, 403 (1977).
45. U. Anthoni, L. Hendriksen, P.H. Nielsen, G. Borch and P. Klaboe, *Spectrochim. Acta, Part A*, **30A**, 1351 (1974).
46. U. Anthoni, P.H. Nielsen, G. Borch and P. Klaboe, *Spectrochim. Acta, Part A*, **34A**, 955 (1978).
47. H. Bürger, K. Burczyk, A. Blaschette and H. Safari, *Spectrochim. Acta, Part A*, **27A**, 1073 (1971).
48. Y. Tanaka, Y. Tanaka, Y. Saito and K. Machida, *Bull. Chem. Soc. Jpn.*, **51**, 1324 (1978).
49. G. Davidson and S. Philips, *Spectrochim. Acta, Part A*, **35A**, 141 (1979).
50. M. Goldstein, M.A. Russell and H.A. Willis, *Spectrochim. Acta, Part A*, **25A**, 1275 (1969).
51. Y. Tanaka, Y. Tanaka, Y. Saito and K. Machida, *Spectrochim. Acta, Part A*, **39A**, 159 (1983).
52. W.C. Harris, F.L. Glenn and L.B. Knight, *Spectrochim. Acta, Part A*, **31A**, 11 (1975).
53. C. Trinquecoste, M. Rey-Lafon and M.-T. Forel, *Spectrochim. Acta, Part A*, **30A**, 813 (1974).
54. P. Rademacher and W. Lüttke, *Spectrochim. Acta, Part A*, **27A**, 715 (1971).
55. H. Böhlig, W. Muller-Sachs, J. Fruwert and G. Geiseler, *Z. Phys. Chem.*, **266**, 415 (1985).
56. H. Böhlig, S. Franke and J. Fruwert, *Z. Phys. Chem.*, **268**, 355 (1987).
57. J.R. Durig, N.E. Lindsay and T.J. Hizer, *J. Phys. Chem.*, **91**, 5027 (1987).
58. M.A. Fleming, R.J. Wyma and R.C. Taylor, *Spectrochim. Acta*, **21**, 1189 (1965).
59. R.D. Hill and G.D. Meakins, *J. Chem. Soc.*, 760 (1958).
60. A.R. Katritzky and R.A. Jones, *J. Chem. Soc.*, 3674 (1959).
61. J.P. Perchard, M.-T. Forel and M.-L. Josien, *J. Chim. Phys.*, **61**, 660 (1964).
62. I.A. Degen, *Appl. Spectrosc.*, **23**, 239 (1969).
63. D.C. McKean and I.A. Ellis, *J. Mol. Struct.*, **29**, 81 (1975).

9.5 NITRO

According to whether the NO_2 plane is in the symmetry plane (i.p) or perpendicular to this plane (o.p), the distribution of the normal vibrations is as follows:

Point group	$\nu_a NO_2$	$\nu_s NO_2$	δNO_2	ωNO_2	ρNO_2	torsion
C_{2v} (i.p)	b_2	a_1	a_1	b_1	b_2	a_2
C_s (i.p)	a′	a′	a′	a″	a′	a″
C_{2v} (o.p)	b_1	a_1	a_1	b_2	b_1	a_2
C_s (o.p)	a″	a′	a′	a′	a″	a″

The most characteristic bands in the spectra of nitro compounds are due to the NO_2 stretching vibrations, which are the two most useful group frequencies, not only because of their spectral position but also for their strong intensity. It is very interesting to compare the absorption regions of NO_2 with those of CO_2^-.

Antisymmetric stretching vibration

In saturated nitro compounds the antisymmetric stretching vibration is located in the region 1580 ± 80 cm^{-1}. Electron withdrawing {releasing} groups on the α-carbon, such as NO_2 or halogen, result in a high {low} frequency shift [16, 38, 93], for example: F_3CNO_2 (1620), Br_3CNO_2 (1606), $ClCH_2NO_2$ (1577), $EtNO_2$ (1560), $iPrNO_2$ (1535) and $tBuNO_2$ (1530 cm^{-1}). Evidently the wavenumber decreases with increased branching on the α-atom. The highest values (1630 ± 30 cm^{-1}) are furnished by tri- and poly-nitro alkanes and the lowest (1560 ± 60 cm^{-1}) by nitroamines. For molecules without NO_2 or halogen in the vicinity, the region becomes more user friendly: 1570 ± 20 cm^{-1}. Nitromethane gives the $\nu_a NO_2$ near 1562 cm^{-1} and at $\approx$1586 cm^{-1} in the vapour state. Molecules with heavy N or O isotopes absorb at lower wavenumbers, for example: $Me^{15}NO_2$ (1524) [1, 11] and $MeN^{18}O_2$ (1522 cm^{-1}) [11].

With the exception of the compounds substituted by a strong electron releasing group on the 2- or 4-position, the nitrobenzene derivatives display the $\nu_a NO_2$ in the region 1535 ± 30 cm^{-1} and the 3-nitropyridines at 1530 ± 20 cm^{-1} [49]. The strong band is easy to recognize among the aromatic ring stretching vibrations. A strong electron donating group on the 2- or 4-position lowers this region to 1500 ± 20 cm^{-1}, such as in nitrobenzenamines, nitrophenols, nitrophenol ethers and R′HN substituted 3-nitropyridines, for example: 4-H_2NPhNO_2 (1481), 4-Me_2NPhNO_2 (1484), 2-Me(O=)CHN—3-O_2N—Py (1486), 2-H_2NPhNO_2 (1510), 4-$EtOPhNO_2$ (1515) and 4-$HOPhNO_2$ (1516 cm^{-1}). Nitrobenzene absorbs at 1523 cm^{-1} and 3-nitropyridine at 1530 cm^{-1}. The conjugated nitroalkenes also show the $\nu_a NO_2$ at 1535 ± 30 cm^{-1} [38] with $H_2C{=}CHNO_2$ to the HW side (1565 cm^{-1}).

Symmetric stretching vibration

Saturated nitro compounds without NO_2 or halogen on the α-carbon display the NO_2 symmetric stretching vibration at 1380 ± 20 cm^{-1}. A shift to lower wavenumbers occurs when the α-carbon is substituted by NO_2 or halogen, for example:

$CH_2(NO_2)_2$ (1330 and 1292), F_3CNO_2 (1310) and Br_3CNO_2 (1311 cm^{-1}). The lowest wavenumbers are observed in the spectra of tri- and poly-nitro compounds: 1295 ± 80 cm^{-1}. This $\nu_s NO_2$ is, more than the $\nu_a NO_2$, coupled with other vibrations of the molecule, particularly with the C—N stretching vibration. The intensity of the $\nu_s NO_2$ is somewhat weaker than that of the $\nu_a NO_2$. Nitromethane absorbs in the vicinity of 1400 cm^{-1}.

In substituted nitrobenzenes the $\nu_s NO_2$ appears strongly at 1345 ± 30 cm^{-1}, in 3-nitropyridines at 1350 ± 20 cm^{-1} [49] and in conjugated nitroalkenes at 1345 ± 15 cm^{-1} [38]. With the exception of 2-H_2N—5-$ClPhNO_2$ (*1350* [75] or 1371 [76]), 2-HO(O=)$CPhNO_2$ (1371), 3,4-Cl_2PhNO_2 (1370), 3-Cl—4-H_2NPhNO_2 (1367) and 2-Me—3-H_2NPhNO_2 (1367 cm^{-1}), this vibration in nitrobenzenes is found at 1345 ± 20 cm^{-1}. Nitrobenzene itself absorbs at 1348 cm^{-1}. Nitrobenzenes with a strong electron donating substituent (NH_2, NMe_2, OH, OMe ...) give rise to a second band at 1305 ± 20 cm^{-1}, next to the absorption at 1345 ± 20 cm^{-1}. This phenomenon, occurring with 1, 4- and to a lesser extent also with 1,2-substitution, may be due to self-association [57], Fermi resonance [58] or a mixed vibration with νC—N [91]. The contribution of the $\nu_s NO_2$ to this second band explains the extension of the region to lower wavenumbers: 1325 ± 40 cm^{-1}. The extremely low values from the spectra of 2-H_2N-1,3-$(O_2N)_2Ph$ (1268 and *1364*) and 1,3,5-$(O_2N)_3$-2,4,6-$(H_2N)_3Ph$ (1230, *1303 and 1322* cm^{-1}) are not taken into account.

Deformation

In the spectra of non-conjugated nitro compounds the NO_2 in-plane deformation is assigned in the region 690 ± 85 cm^{-1}, usually as a weak to moderate band. For nitroamines this region becomes 765 ± 10 cm^{-1} [10, 11]. High wavenumbers are also found in the spectra of $cPrNO_2$ (770), F_3CNO_2 (750), $tBuNO_2$ (731) and $iPrNO_2$ (724 cm^{-1}). For most of the saturated, halogenated or NO_2-substituted nitro compounds this region is reduced to 650 ± 45, or even 655 ± 25 cm^{-1}, if the varying values from tetranitromethane (690, *668* and 606 cm^{-1}) are not taken into account. The region 755 ± 10 cm^{-1} in the spectra of the esters of nitric acid is assigned to ωNO_2 [38].

The NO_2 scissors occurs at higher frequencies (850 ± 60 cm^{-1}) when conjugated to C=C or aromatic molecules, according to some investigators with a contribution of the νC—N which is expected near 1120 cm^{-1} [3, 48, 92]. For nitrobenzene this δNO_2 is reported at 852 cm^{-1}, for H_2C=$CHNO_2$ at 890 cm^{-1}. Most of the nitrobenzenes absorb in the region 855 ± 40 cm^{-1}, but the following compounds fall outside this region: 1,3-dinitrobenzene (904 and *834*), 3-N≡$CPhNO_2$ and 3-H(O=)$CPhNO_2$ with ≈900 cm^{-1} at the HW side and 2,4,6-trinitromethylbenzene (814 [72], *874* and 805 [73]), 2-$ThNO_2$ (813), F_5PhNO_2 (812), 1,3-$(O_2N)_2$-5-HO(O=)CPh (811), 2-Me-1,5-$(O_2N)_2Ph$ (*836* and 792) and 2-Me-1,3-$(O_2N)_2Ph$ (*841* and 791 cm^{-1}) at the LW side.

Wagging vibration

The ωNO_2 in aliphatic nitro compounds is weakly to moderately active in the range 620 $\pm$ 110 cm^{-1}, disregarding the values around 400 cm^{-1} in the spectra of X_3CNO_2 (X = F, Cl, Br) assigned as much to the wag as to the rock [3, 22, 23]. Except for $PhC(=O)CH_2NO_2$-d_7 (510 cm^{-1}) the $R'CH_2NO_2$ compounds display the ωNO_2 at 570 $\pm$ 45 cm^{-1}. Secondary and tertiary compounds give this wag at 610 $\pm$ 40 cm^{-1} and nitroamines in the region 660 $\pm$ 70 cm^{-1}. Nitromethane absorbs near 605 cm^{-1}. The region 700 $\pm$ 20 cm^{-1} in the spectra of nitrates is assigned to δNO_2 [38].

In α-unsaturated and aromatic compounds the ωNO_2 is assigned at 740 $\pm$ 50 cm^{-1} with a moderate to strong intensity, a region in which also the aromatic γCH is active. In the spectrum of nitrobenzene the absorption at 794 {703} cm^{-1} is assigned to the γCH {ωNO_2} although the two vibrations are coupled [5, 14, 40, 43]. $H_2C{=}CHNO_2$ absorbs at 714 cm^{-1}. The highest values have been observed in the spectra of 2-Me-5-$ClPhNO_2$ (790), 4-H_2NPhNO_2 (785 [57] or *750* [45, 48]), 2-$MePhNO_2$ (784 [43] or *729* [45]) and 2-$HOPhNO_2$ (781 [53] or *747* [45] cm^{-1}). The lowest values are from nitrobenzene-d_4 (690), 1,2-dinitrobenzene (701 and *728*), 1,4-dinitrobenzene (710 and *772*), 2-HO-6-$BrCH_2C(=O)PhNO_2$ and nitrobenzene with 703 cm^{-1}. Many ωNO_2 vibrations are assigned in the region 745 $\pm$ 35 cm^{-1}.

Rocking vibration

In aliphatic nitro compounds the ρNO_2 falls somewhere in the large region 495 $\pm$ 125 cm^{-1}. For molecules without NO_2 or halogen on the α-carbon the region becomes 480 $\pm$ 50 cm^{-1}, a region that splits up into 500 $\pm$ 30 cm^{-1} for secondary and tertiary compounds and 465 $\pm$ 35 cm^{-1} for primary compounds. For example, $cPrNO_2$ absorbs at 528 cm^{-1} and nitromethane at 480 cm^{-1}. The highest wavenumbers come from nitroamines: 590 $\pm$ 30 cm^{-1}.

In aromatic nitro compounds the ρNO_2 is active in the region 545 $\pm$ 45 cm^{-1}, mostly even at 540 $\pm$ 30 cm^{-1}. Nitrobenzene show this rock at 531 cm^{-1} but the ρNO_2 in $H_2C{=}CHNO_2$ falls outside this region, at 652 cm^{-1}.

Torsion

In aliphatic compounds the NO_2 torsion absorbs probably in the neighbourhood of 60 cm^{-1} [30]. For aromatic compounds Varsányi *et al.* [63] found 70 $\pm$ 20 cm^{-1} and Suryanarayana *et al.* [85] 65 $\pm$ 10 cm^{-1}. The value 139 cm^{-1} in the spectrum of nitrobenzene [41] is considered to be an overtone [90].

Table 9.6 Absorption regions (cm^{-1}) of the normal vibrations of —NO_2

Vibration	α-saturated	α-halogen	dinitro	tri- and poly-nitro
$\nu_a NO_2$	1570 ± 20	1590 ± 35	1575 ± 35	1630 ± 30
$\nu_s NO_2$	1380 ± 20	1340 ± 35	1345 ± 60	1295 ± 80
δNO_2	700 ± 70	695 ± 55	660 ± 30	650 ± 45
ωNO_2	580 ± 70	590 ± 70	580 ± 70	580 ± 70
ρNO_2	480 ± 50	450 ± 80	440 ± 70	440 ± 70
torsion NO_2	≈60			

Vibration	nitroamines	nitrates	α-unsaturated aromatic	aromatic +I
$\nu_a NO_2$	1560 ± 60	1635 ± 20	1535 ± 30	1500 ± 20
$\nu_s NO_2$	1325 ± 25	1285 ± 15	1345 ± 30	1325 ± 40
δNO_2	765 ± 10	755 ± 10 (ω)	850 ± 60	855 ± 40
ωNO_2	660 ± 70	700 ± 20 (δ)	740 ± 50	760 ± 30
ρNO_2	590 ± 30	535 ± 35	545 ± 45	540 ± 30
torsion NO_2		≈80	70 ± 20	

+I = strong electron releasing group on the 2- or 4- position.

R—NO_2 compounds

R = Me— [1–12], CD_3— [8–12], Et— [2, 4, 7, 13, 14], $MeCD_2$—, MeCHD—, CD_3CH_2— and CD_3CD_2— [13], nPr— [2, 4], nBu—, nPent—, nHex—, nOct—, nNon— and nDec— [4], $HOCH_2CH_2$— [15], $O_2NCH_2CH_2$— [7], $PhCH_2$— [16], PhC(=O)CH_2-d_0—, -d_2—, -d_5— and -d_7— [17], iPr— [14, 18], cPr— [14, 19], tBu— [14], $ClCH_2$— $ClCD_2$— $BrCH_2$— and $BrCD_2$— [20], EtCHClCHCl—, EtCHBrCHBr—, $MeCCl_2$— and $EtCHClCCl_2$— [16], $HOCH_2CHCl$— and $HOCH_2CCl_2$— [15], O_2NCCl_2— [25], F_3C— [3, 21–23], Cl_3C— [22, 23, 24], Br_3C— [22, 23], O_2NCH_2— [7, 26], O_2NCMe_2— [27, 28], $O_2NCMe_2CMe_2$— [27, 29], $O_2NCH(Me)$— and $O_2NCH(Et)$— [16], $(O_2N)_2CH$— [7], $(O_2N)_3C$— [7, 30], $(O_2N)_2C(Me)C(NO_2)Me$— [27], $(O_2N)_3CC(NO_2)_2$— [7], MeHN—, MeDN—, CD_3HN— and CD_3DN— [10, 31], Me_2N— and $(CD_3)_2N$— [11, 31], MeO— and CD_3O— [32–34], EtO— [35, 36], nPentO—, $PhCH_2O$— and $EtC(Me)_2O$— [38], CH_2=$CHCH_2O$— [37], iPrO— [37], H_2C=CH— [5, 14], MeCH=CH— [39], EtCH=CH— [38], Ph— [3, 5, 14, 40–43], Ph-d_5— [40, 44], XPh— (X = 2-, 3- and 4-Me [43, 45–50], 4-HO(O=)CCH_2 [48], 4-N≡CCH_2 [45], 3- and 4-$HOCH_2$ [48], 2-, 3- and 4-$ClCH_2$ [48], 3- and 4-$BrCH_2$ [45, 48], 4-F_3C [45], 3- and 4-N≡C [45, 47, 48, 50], 2-, 3- and 4-H(O=)C [45, 47, 48, 51], 3- and 4-Me(O=)C [45, 48, 50], 3-nPr(O=)C, 2-, 3- and 4-HO(O=)C [45, 48, 52], 3- and 4-MeO(O=)C [45, 47, 48, 50], 3- and 4-EtO(O=)C [45, 47, 50], 2-, 3- and 4-Cl(O=)C and 2-, 3- and 4-$BrCH_2$(O=)C [45, 48], 2-, 3- and 4-H_2N [45, 47–49, 56–58], 2-, 3- and 4-Me(O=)CNH [45], 4-Me_2N, 4-Et_2N and 4-H_2NNH [45],

2-, 3- and 4-O_2N [45, 47, 48, 50, 59–64, 85], 2-, 3- and 4-HO [3, 45, 47, 48, 53, 54], 3- and 4-MeO [45, 47, 48, 50], 2- and 4-EtO [55], 2- and 4-Me(O=)CO [45], 2-MeOS [45], 2-, 3- and 4-F [43, 45, 48, 50, 65], 2-, 3- and 4-Cl [43, 45, 47, 48, 50, 64, 66], Br [43, 45, 47, 48, 50, 66], I [43, 45, 47, 48]), X-2-MePh— (X = 3-Cl [67], 5-Cl[68], 3-O_2N [63], 5-O_2N [63, 69], 3-H_2N [70], 4-HO [71], 5-Me(O=)C [45], 3, 5-$(O_2N)_2$ [72, 73, 74]), X-2-HOPh— (X = 3-O_2N [45, 63], 4-H(O=)C, 4-, 5- and 6-$BrCH_2$(O=)C [45], 5-Me(O=)C, 5-Cl), X-2-H_2NPh— (X = 3-O_2N [45, 63], 5-Cl [75, 76], 5-Me, 5-MeO [45], 3,5-$(NO_2)_2$—4,6-$(NH_2)_2$ [77, 78]), X-2-MeOPh— (X = 5-Me(O=)C [45]), X-2-ClPh— (X = 3-, 4- and 5-Cl [85]), X-2-BrPh— (X = 5-Br [79]), X-3-HOPh— (X = 4-$BrCH_2$(O=)C [45]), X-3-MeOPh— (X = 4-Br [80]), X-3-H_2NPh— (X = 4-HO [71]), X-3-O_2NPh— (X = 4-HO(O=)CCH_2, 4-MeO(O=)CCH_2 [63], 5-HO(O=)C [45], 5-Cl(O=)C [45, 63], 4-H_2N, 4-H_2NNH, 5-O_2N [45, 63, 81, 82], 4-HO, 4-$BrCH_2CH_2O$, 4-F [45, 63, 83, 84, 85], 4-Cl [45, 66, 84, 85], 4-Br [45, 66, 85]), X-3-ClPh— (X = 4- and 5-Cl [79, 85], 4-H_2N [75, 76]), X-3-BrPh— (X = 4-MeO [80]), X-3-$^-O_3$SPh— (X = 4-Cl [45]), X-4-HOPh— (X = 3-$BrCH_2$(O=)C and 3-Me(O=)C [45]), 3,5-$(O_2N)_3$Ph-d_3— [45, 81, 82], 2-$(O_2N)_2$-Me_4Ph— [86], F_5Ph— [87], 4-Py— [88], 4-Py-*N*-oxide-d_0— and -d_4— [89], 2-Th—.

References

1. G. Malewsky, M. Pfeiffer and P. Reich, *J. Mol. Struct.*, **3**, 419 (1969).
2. D.C. Smith, C.J. Pan and J.R. Nielsen, *J. Chem. Phys.*, **18**, 706 (1950).
3. J.H.S. Green, W. Kynaston and A.S. Lindsey, *Spectrochim. Acta*, **17**, 486 (1961).
4. G. Geiseler and H. Kessler, *Ber. Bunsenges. Phys. Chem.*, **68**, 571 (1964).
5. G. Varsányi, S. Holly and L. Imre, *Spectrochim. Acta, Part A*, **23A**, 1205 (1967).
6. D.C. McKean and R.A. Watt, *J. Mol. Spectrosc.*, **61**, 184 (1976).
7. A. Loewenschuss, N. Yellin and A. Gabai, *Spectrochim. Acta, Part A*, **30A**, 371 (1974).
8. F.D. Verderame, J.A. Lannon, L.E. Harris, W.G. Thomas and E.A. Lucia, *J. Chem. Phys.*, **56**, 2638 (1972).
9. D. Papousek, K. Sarka, V. Spirko and B. Jordanov, *Collect. Czech. Chem. Commun.*, **36**, 890 (1971).
10. M.I. Dakhis, V.G. Dashevsky and V.G. Avakyan, *J. Mol. Struct.*, **13**, 339 (1972).
11. C. Trinquecoste, M. Rey-Lafon and M.-T. Forel, *Spectrochim. Acta, Part A*, **30A**, 813 (1974).
12. J.R. Hill, D.S. Moore, S.C. Schmidt and C.B. Storm, *J. Phys. Chem.*, **95**, 3037 (1991).
13. P. Groner, R. Meyer and H.H. Günthard, *Chem. Phys.*, **11**, 63 (1975).
14. J.R. Durig, F. Sun and Y.S. Li, *J. Mol. Struct.*, **101**, 79 (1983).
15. Z. Eckstein, P. Gluzińsky, W. Sobótka and T. Urbański, *J. Chem. Soc.*, 1370 (1961).
16. W.H. Lunn, *Spectrochim. Acta*, **16**, 1088 (1960).
17. M. Maltese and C. Ercolani, *J. Chem. Soc.*, **B**, 1147 (1970).
18. J.R. Durig, J.A.S. Smith, Y.S. Li and F.M. Wasacz, *J. Mol. Struct.*, **99**, 45 (1983).
19. J.R. Holtzclaw, W.C. Harris and S.F. Bush, *J. Raman Spectrosc.*, **9**, 257 (1980).
20. P. Gluzińsky and Z. Eckstein, *Spectrochim. Acta, Part A*, **24A**, 1777 (1968).
21. J. Mason and J. Dunderdale, *J. Chem. Soc.*, 759 (1956).
22. A. Castelli, A. Palm and C. Alexander, *J. Chem. Phys.*, **44**, 1577 (1966).

23. B. Vizi, B.N. Cyvin and S.J. Cyvin, *Acta Chim. Acad. Sci. Hung.*, **83**, 303 (1974).
24. M.S. Soliman, *Spectrochim. Acta, Part A*, **49A**, 183 (1993).
25. A.O. Diallo, *Spectrochim. Acta, Part A*, **27A**, 239 (1971).
26. K. Singh, *Spectrochim. Acta, Part A*, **23A**, 1089 (1967).
27. A.O. Diallo, *Spectrochim. Acta, Part A*, **30A**, 1505 (1974).
28. Z. Buczkowski and T. Urbański, *Spectrochim. Acta*, **22**, 227 (1966).
29. B.G. Tan, L.H.L. Chia, H.H. Huang, M.-H. Kuok and S.-H. Tang, *J. Chem. Soc. Perkin Trans. 2*, 1407 (1984).
30. P.H. Lindenmeyer and P.M. Harris, *J. Chem. Phys.*, **21**, 408 (1953).
31. C. Trinquecoste, Thesis, Bordeaux, 1973.
32. J.A. Lannon, L.E. Harris, F.D. Verderame, W.G. Thomas, E.A. Lucia and S. Kowiers, *J. Mol. Spectrosc.*, **50**, 68 (1974).
33. J.R. Durig, N.E. Lindsay and B.J. Van der Veken, *Indian J. Pure Appl. Phys.*, **26**, 223 (1988).
34. R.Odeurs, Thesis, UIA, Antwerp, 1984.
35. J.R. Durig and T.G. Sheehan, *J. Raman Spectrosc.*, **21**, 635 (1990).
36. J.R. Durig and N.E. Lindsay, *Spectrochim. Acta, Pazrt A*, **46A**, 1125 (1990).
37. R. Maas, Thesis UIA, Antwerp, 1992.
38. J.F. Brown Jr., *J. Am. Chem. Soc.*, **77**, 6341 (1955).
39. A.G. Turner and W.R. Carper, *Spectrochim. Acta, Part A*, **43A**, 975 (1987).
40. J.D. Laposa, *Spectrochim. Acta, Part A*, **35A**, 65 (1979).
41. C.V. Stephenson, W.C. Coburn Jr. and W.S. Wilcow, *Spectrochim. Acta*, **17**, 933 (1961).
42. S. Pinchas, D. Samuel and B.L. Silver, *Spectrochim. Acta*, **20**, 179 (1964).
43. J.H.S. Green and D.J. Harrison, *Spectrochim. Acta, Part A*, **26A**, 1925 (1970).
44. A. Kuwae and K. Machida, *Spectrochim. Acta, Part A*, **35A**, 27 (1979).
45. G. Varsányi, *Assignments for Vibrational Spectra of Seven Hundred Benzene Derivatives*, J.Wiley & Sons, New York (1974).
46. N. Abasbegović, L. Colombo and P. Bleckmann, *J. Raman Spectrosc.*, **6**, 92 (1977).
47. C. Garrigou-Lagrange, M. Chehata and J. Lascombe, *J. Chim. Phys.*, **63**, 552 (1966).
48. O. Exner, S. Kovác and E. Solcániová, *Collect. Czech. Chem. Commun.*, **37**, 2156 (1972).
49. A. Perjéssy, D. Rasala, P. Tomasik and R. Gawinecki, *Collect. Czech. Chem. Commun.*, **50**, 2443 (1985).
50. C. Garrigou-Lagrange, M. Chehata and G. Sourisseau, *J. Chim. Phys.*, **62**, 261 (1965).
51. C.J.W. Brooks and J.F. Morman, *J. Chem. Soc.*, 3372 (1961).
52. R.C. Gupta and M. Saxena, *Indian J. Phys.*, **46**, 76 (1972).
53. Y. Kishore, S.N. Sharma and C.P.D. Dwivedi, *Indian J. Phys.*, **48**, 412 (1974).
54. M. Horák, J. Smolíkova and J. Pitha, *Collect. Czech. Chem. Commun.*, **26**, 2891 (1961).
55. K.M. Mathur, D.P. Juyal and R.N. Singh, *Indian J. Pure Appl. Phys.*, **9**, 756 (1971).
56. M.M. Szostak, *J. Raman Spectrosc.*, **8**, 43 (1979).
57. M. Harrand, *J. Raman Spectrosc.*, **4**, 53 (1975).
58. J.F. Bertran, M. Hernández and B. La Serna, *Spectrochim. Acta, Part A*, **38A**, 149 (1982).
59. J.V. Shukla and K.N. Upadhya, *Indian J. Pure Appl. Phys.*, **7**, 830 (1969).
60. J.V. Shukla, V.B. Singh and K.N. Upadhya, *Indian J. Phys.*, **42**, 511 (1968).
61. G.N. Andreen, B. Jordanov, I.N. Juchnovski and B. Schrader, *J. Mol. Struct.*, **115**, 375 (1984).
62. J.E. Katon, K. Hanai and G.N.R. Tripathi, *J. Chem. Phys.*, **73**, 697 (1980).
63. G. Varsányi, E. Molnar-Paal, K. Kosa and G. Keresztury, *Acta Chim. Acad. Sci. Hung.*, **100**, 481 (1979).
64. J.H.S. Green and H.A. Lauwers, *Spectrochim. Acta, Part A*, **27A**, 817 (1971).
65. K.C. Medhi, *Spectrochim. Acta*, **20**, 675 (1964).
66. E.F. Mooney, *Spectrochim. Acta*, **20**, 1021 (1964).

67. R.B. Singh and D.K. Rai, *Indian J. Phys.*, **53B**, 144 (1979).
68. R.B. Singh, N.P. Singh and D.K. Rai, *Indian J. Phys.*, **56B**, 62 (1982).
69. J. Shukla and K.N. Upadhya, *Indian J. Pure Appl. Phys.*, **11**, 787 (1973).
70. N.S. Sundar, *Spectrochim. Acta, Part A*, **41A**, 905 (1985).
71. N.S. Sundar, *Spectrochim. Acta, Part A*, **41A**, 1449 (1985).
72. J.J.P. Stewart, S.R. Bosco and W.R. Carper, *Spectrochim. Acta, Part A*, **42A**, 13 (1986).
73. W.R. Carper, S.R. Bosco and J.J.P. Stewart, *Spectrochim. Acta, Part A*, **42A**, 461 (1986).
74. W.R. Carper and J.J.P. Stewart, *Spectrochim. Acta, Part A*, **43A**, 1249 (1987).
75. V.N. Verma and K.P.R. Nair, *Indian J. Pure Appl. Phys.*, **8**, 682 (1970).
76. V.N. Verma, *Spectrosc. Lett.*, **6**, 23 (1973).
77. T.G. Towns, *Spectrochim. Acta, Part A*, **39A**, 801 (1983).
78. B.L. Deopura and U.D. Gupta, *J. Chem. Phys.*, **54**, 4013 (1971).
79. A.P. Upadhya and K.N. Upadhya, *Indian J. Phys.*, **55B**, 213 (1981).
80. M. Rangacharyulu and D. Premaswarup, *Indian J. Pure Appl. Phys.*, **19**, 166 (1981).
81. H.F. Shurvell, J.A. Faniran and E.A. Symons, *Can. J. Chem.*, **45**, 117 (1967).
82. H.F. Shurvell, A.R. Norris and D.E. Irish, *Can. J. Chem.*, **47**, 2515 (1969).
83. A.K. Ansari and P.K. Verma, *Spectrochim. Acta, Part A*, **35A**, 35 (1979).
84. K.C. Medhi, *Indian J. Phys.*, **35**, 583 (1961).
85. V. Suryanarayana, A.P. Kumar, G.R. Rao and G.C. Pandey, *Spectrochim. Acta, Part A*, **48A**, 1481 (1992).
86. P. Sgarabotto, M. Braghetti, R.S. Cataliotti, G. Paliani, S. Sorriso, U. Caia and S. Santini, *Can. J. Chem.*, **65**, 2122 (1987).
87. J.H.S. Green, D.J. Harrison and C.P. Stockley, *Spectrochim. Acta, Part A*, **33A**, 423 (1977).
88. M. Joyeux and N.Q. Dao, *J. Raman Spectrosc.*, **19**, 441 (1988).
89. M. Joyeux and N.Q. Dao, *Spectrochim. Acta, Part A*, **44A**, 1447 (1988).
90. G. Varsányi, *Vibrational Spectra of Benzene Derivatives*, Academic Press, New York (1969).
91. L.M. Epstein, E.S. Shubina, L.D. Ashkinadze and L.A. Kazitsyna, *Spectrochim. Acta, Part A*, **38A**, 317 (1982).
92. D.S. Ranga Rao and G. Thyagarajan, *Indian J. Pure Appl. Phys.*, **16**, 941 (1978).
93. R.A. Nyquist, *Appl. Spectrosc.*, **42**, 624 (1988).

10

Normal Vibrations and Absorption Regions of Oxy Compounds

10.1 R′-OXY COMPOUNDS

10.1.1 Hydroxy

The OH group provides three normal vibrations: νOH, δOH and γ OH, of which not only the stretching vibration but also the out-of-plane deformation are good group vibrations.

OH stretching vibration

The competition between steric hindrance and inter- and intra-molecular hydrogen bridges makes it possible that, in the spectra of R—OH compounds in the liquid or solid state, the OH stretching vibration appears in the extensive region 3275 $\pm$ 370 cm^{-1}. This region is reduced to 3380 $\pm$ 200 cm^{-1} if some high values in the spectra of phenols and low values in the spectra of carboxylic acids are not taken into account.

Sterically unhindered alcohols, including most of the primary alcohols but not aromatic alcohols and polyols, show this νOH. . .O at 3370 $\pm$ 50 cm^{-1}, usually even at 3345 $\pm$ 15 cm^{-1}, as a strong broad band. The ring substituted aromatic primary alcohols (benzyl alcohols) cover a wider region (3300 $\pm$ 120 cm^{-1}), limited by 3,5-$(MeO)_2PhCH_2OH$ (3420) on the one side and 2-$IPhCH_2OH$ and 1,3-$(HOCH_2)_2Ph$ (3180 cm^{-1}) on the other. The little shoulder (3550 cm^{-1}) on the broad band

(3380 cm^{-1}) in the spectrum of benzenemethanol gives away the presence of a monomer.

Alicyclic OH compounds are active at 3330 ± 50 cm^{-1} and secondary alcohols, being sterically hindered to a certain degree, usually above 3350 cm^{-1} (3370 ± 30 cm^{-1}). Cyclohexanol and cyclopentanol absorb at ≈3333 cm^{-1}, cyclobutanol at 3304 cm^{-1} and tBuCH(OH))X (X = Me and Et) near the upper limit: ≈3391 cm^{-1}.

The OH stretching vibration of the more strongly sterically hindered tertiary alcohols appears at higher wavenumbers: 3470 ± 110 cm^{-1}. In the following examples the OH stretching vibration shows more and more the character of a free νOH, the band becomes sharper and the wavenumber increases: Me_3COH-d_0 (3365) and -d_9 (3380), Et_3COH (3405), Oct_3COH (3410), Ph_3COH and $cHex_3COH$ (3475), Ph(Me)C(OH)C(OH)(Me)Ph (3495) and $(Ph)_2C(OH)C(OH)(Ph)_2$ (3577 cm^{-1}).

Polyols, if they are not sterically hindered, show low absorption bands (3280 ± 100 cm^{-1}), particularly compounds with an intense formation of hydrogen bridges such as $EtC(CH_2OH)_3$ (3231), $HOCH_2CH(OH)CH(OH)CH_2OH$ (3215) and 3-$HOCH_2PhCH_2OH$ (3180 cm^{-1}). Likewise, at the LW side one finds the νOH...O in the spectra of hydroxylamines (R_1R_2N—OH) and oximes (R=N—OH) in the region 3255 ± 75 cm^{-1} and in those of silanols ($R_1R_2R_3Si$—OH) ath 3260 ± 40 cm^{-1}.

Phenols cover the widest range: 3390 ± 255 cm^{-1}. The highest OH stretching vibrations are found in the spectra of 2,6-disubstituted phenols, with 2,6-tBu_2PhOH (3645 cm^{-1}) as an example. The lowest values come from C=O substituted phenols (3190 ± 55 cm^{-1}) [58], 2-nitrophenols (3230 ± 40 cm^{-1}) [41] and 2,4,6-trinitrophenol with 3150 cm^{-1} [41]. Most of the other phenols were found to give this OH stretching vibration in the region 3390 ± 70 cm^{-1}, in which the 2-substituted {4-substituted} phenols are responsible for the highest {lowest} values:

	2-X-PhOH	3-X-PhOH [62]	4-X-PhOH [62]	Naphthols
νOH	3475 ± 90	3360 ± 40	3310 ± 90	3310 ± 90

Examples are: 2-Ph-PhOH (3565), 2-MeOPhOH (3512), 2-MePhOH (3448), 3-MePhOH (3335), 4-iBuPhOH (3221), Naph-1-OH (3290), 2-, 3- and 4-tBuPhOH (respectively 3535, 3360 and 3240 cm^{-1}) and 2-Me-Naph-1-OH (3390 cm^{-1}). Compounds falling outside the above-mentioned regions are: 2-HOPhPh-2-OH (3155), 3-HOPhOH (3184) and 4-cPentPhOH (3184 cm^{-1}).

The associated OH...O stretching vibration in organic acids gives rise to a very broad band with maximum absorption at 3050 ± 150 cm^{-1} (Section 7.1.7).

In the absence of intramolecular hydrogen bridges, the broad band of the OH stretching vibration disappears in dilute solution or in the vapour state and returns as a sharp peak at 3610 $\pm$ 30 cm^{-1}. In acids the broad band does not disappear completely and the absorption in the neighbourhood of 3500 cm^{-1} shows part of the free OH.

The OH in-plane deformation

Primary alcohols in the associated state exhibit a weak to moderate, often diffuse band in the region 1400 $\pm$ 40 cm^{-1}, attributed to the OH. . .O in-plane deformation. Sometimes this region is extended to lower wavenumbers (1310 $\pm$ 130 cm^{-1}) because of the coupling of this vibration with the methylene wag and twist. In this case the OH in-plane deformation is conceived as a mixed vibration: $\delta(OH + CH_2)$ [59–61]. In secondary alcohols also this δOH (1380 $\pm$ 50 cm^{-1}) is coupled to the δCH. Tertiary alcohols absorb in the region 1370 $\pm$ 40 cm^{-1}. In dilute solutions the OH in-plane deformation shifts to lower wavenumbers and is difficult to detect among the CH deformations.

In the spectra of silanols the δOH appears as a weak to moderate band at 1065 $\pm$ 35 cm^{-1}.

The moderate to strong absorption at 1350 $\pm$ 40 cm^{-1} in the spectra of phenols is assigned to the δOH. . .O. The band with the strongest intensity in the region 1220 $\pm$ 40 cm^{-1} is due to the νC—O with a contribution of this OH in-plane deformation [34, 55, 57]. Phenols give the free OH in-plane deformation as a sharp peak with moderate intensity in the region 1170 $\pm$ 30 cm^{-1} [55].

The region 1395 $\pm$ 55 cm^{-1} in the spectra of carboxylic acids is also assigned as well to this OH in-plane deformation [56].

The OH out-of-plane deformation

With the exception of acids, the OH out-of-plane deformation in the associated state exhibits a broad, diffuse band in the extensive region 685 $\pm$ 115 cm^{-1}. In the spectra of higher alcohols (octadecanol upwards) and poly-substituted phenols, in which the OH group contributes only in a small amount, the weak band, on which other absorptions are superimposed, is difficult to observe. Primary alcohols absorb at 640 $\pm$ 70, secondary at 630 $\pm$ 30 and tertiary alcohols at 620 $\pm$ 30 cm^{-1}.

The OH out-of-plane deformation in the spectra of phenols takes up the whole region: 685 $\pm$ 115 cm^{-1}. Carbonyl substituted phenols give high values: 750 $\pm$ 50 cm^{-1}, for example 2-MeC(=O)PhOH (800), 4-MeC(=O)PhOH (780) and 3-HC(=O)PhOH (710 cm^{-1}). The remaining phenols restrict themselves to the region 650 $\pm$ 80 cm^{-1} with a weak, broad band disturbed by ring vibrations.

Carboxylic acids display this γ OH. . .O in the range 905 $\pm$ 65 cm^{-1} as a weak to moderate band with the appearance of a V.

The free OH out-of-plane deformation is described as a torsion and absorbs at low wavenumbers: 300 $\pm$ 80 cm^{-1} in the spectra of alcohols [4–11] and 330 $\pm$ 80 cm^{-1} in the spectra of phenols [29, 30, 32].

Table 10.1 Absorption regions (cm^{-1}) of the normal vibrations of the associated —OH

Vibration	Non aromatic primary alcohols	Aromatic primary alcohols	Secondary and alicyclic	Tertiary alcohols	Polyols
νOH	3370 ± 50	3300 ± 120	3340 ± 60	3470 ± 110	3280 ± 100
δOH	1400 ± 40	1400 ± 40	1380 ± 50	1370 ± 40	1400 ± 35
γ OH	640 ± 70	660 ± 20	630 ± 30	620 ± 30	650 ± 70
	Hydroxylamines and oximes	Silanols	Phenols	Carboxylic acids	
νOH	3255 ± 75	3260 ± 40	3390 ± 255	3050 ± 150	
δOH	1400 ± 30	1065 ± 35	1350 ± 40	1395 ± 55	
γ OH	740 ± 60	670 ± 30	685 ± 115	905 ± 65	

R—CH_2OH compounds (see Section 3.2.1).
R—CH_2CH_2OH compounds (see Section 3.5.4).
R—CHOH—R′ compounds (see Section 5.3).
R—OH compounds
R = cBu— [1], cPent— [2], cHex—, cHept—, cOct—, tBu— [3–5], $(CD_3)_3$C— [4, 5], Et_3C—, Oct_3C—, $cHex_3$C—, $F_3C(Me)_2$C— [6], $(F_3C)_2$MeC— [7], $(F_3C)_3$C— [8], $Cl_3C(F_3C)_2$C— [9], $F_2C{=}C(CF_3)$— [10], 4-HO$(Me)_2$C—Ph—C$(Me)_2$— [11], F_3C— [12], $(Ph)_3$C— [13, 14], Ph(Me)C(OH)—C(Me)Ph—, $(Ph)_2$C(OH)—C$(Ph)_2$—, Me_2N— [15, 16], MeCH=N— [17], Me_2C=N—, Me(EtO)C=N—, MeC(=O)NH—, $(Me)_3$Si— [18], Ph_2(HO)Si—, Ph_3Si—, 4-HO$(Me)_2$Si—Ph—Si$(Me)_2$— [19, 20], Ph— [21–29], 2-, 3- and 4-XPh— (X = Me [29–34], Et and iPr [34], tBu [26], HC(=O) [34–36], MeC(=O) [34, 36], PhC(=O), HO [25, 34, 37, 38], H_2N [34, 36, 39], O_2N [34, 40, 41], F, Cl, Br, I [30, 31, 34, 36]), 4-XPh— (X = iBu, cPent, $BrCH_2$C(=O) [34], N≡C [34, 36], HO [28]), 2- and 4-XPh— (X = EtC(=O), nPrC(=O) [34]), 3- and 4-XPh— (X = MeOC(=O) [34], HOC(=O) [34, 36]), 2-XPh— (X = Ph, 2-HOPh, HONH(O=)C [54]), Me_2Ph— [34, 42], Me_3Ph— [43], 2,6-tBu_2Ph—, Cl_2Ph— [34, 42, 44], Cl_3Ph— [44], Cl_4Ph— [44, 45], Cl_5Ph— [44, 46], 2,4,6-Br_3Ph— [47], Br_5Ph— [46], 2,4,6-I_3Ph— [48], 2,3-$(HO)_2$Ph— [49], $(O_2N)_2$Ph— [34], $(O_2N)_3$Ph— [41], 2-H_2N-5-MePh— [50], 2-H_2N-4-MePh— and 2-H_2N-4-ClPh— [51], 2-H_2N-4-O_2NPh— [52], Naph-1-, Naph-2- and 6-HO—Naph-2— [53].

References

1. J.R. Durig and W.H. Green, *Spectrochim. Acta, Part A*, **25A**, 849 (1969).
2. J.R. Durig, J.M. Karriker and W.C. Harris, *Spectrochim. Acta, Part A*, **27A**, 1955 (1971).

3. J.G. Pritchard and H.M. Nelson, *J. Phys. Chem.*, **64**, 795 (1960).
4. J. Korppi-Tommola, *J. Mol. Struct.*, **40**, 13 (1977).
5. J. Korppi-Tommola, *Spectrochim. Acta, Part A*, **34A**, 1077 (1978).
6. J. Korppi-Tommola, *Acta Chem. Scand., Ser. A*, **31A**, 563 (1977).
7. J. Korppi-Tommola, *Acta Chem. Scand., Ser. A*, **31A**, 568 (1977).
8. J. Murto, A. Kivinen, J. Korppi-Tommola, R. Viitala and J. Hyömäki, *Acta Chem. Scand.*, **27**, 107 (1973).
9. J. Murto, A. Kivinen, K. Kajander, J. Hyömäki and J. Korppi-Tomola, *Acta Chem. Scand.*, **27**, 96 (1973).
10. J. Murto, A. Kivinen, R. Henriksson, A. Aspiala and J. Partanen, *Spectrochim. Acta, Part A*, **36A**, 607 (1980).
11. B. Zelei and S. Dobos, *Spectrochim. Acta, Part A*, **35A**, 915 (1979).
12. J.S. Francisco, *Spectrochim. Acta, Part A*, **40A**, 923 (1984).
13. R.E. Weston Jr., A. Tsukamoto and N.N. Lichtin, *Spectrochim. Acta*, **22**, 433 (1966).
14. W. Saffioti and N. Le Calvé, *Spectrochim. Acta, Part A*, **28A**, 1435 (1972).
15. H. Böhlig, S. Franke and J. Fruwert, *Z. Phys. Chem.*, **268**, 355 (1987).
16. H. Böhlig, W. Müller-Sachs, J. Fruwert and G. Geiseler, *Z. Phys. Chem.*, **266**, 415 (1985).
17. G. Geiseler, H. Böhlig and J. Fruwert, *J. Mol. Struct.*, **18**, 43 (1973).
18. J. Rouviere, V. Tabacik and G. Fleury, *Spectrochim. Acta, Part A*, **29A**, 229 (1973).
19. B. Zelei, S. Dobos and R. Righini, *Spectrochim. Acta, Part A*, **34A**, 343 (1978).
20. G. Sbrana, N. Neto, M. Muniz-Miranda and M. Nocentini, *Spectrochim. Acta*, **39**, 295 (1983).
21. J.H.S. Green, *J. Chem. Soc.*, 2236 (1961).
22. H.D. Bist, J.C.D. Brand and D.R. Williams, *J. Mol. Spectrosc.*, **21**, 76 (1966).
23. H.D. Bist, J.C.D. Brand and D.R. Williams, *J. Mol. Spectrosc.*, **24**, 402 (1967).
24. J.C. Evans, *Spectrochim. Acta*, **16**, 1382 (1960).
25. A. Hidalgo and C. Otero, *Spectrochim. Acta*, **16**, 528 (1960).
26. R. Soda, *Bull. Chem. Soc. Jpn.*, **34**, 1482 (1961).
27. H.W. Wilson, R.W. MacNamee and J.R. Durig, *J. Raman Spectrosc.*, **11**, 252 (1981).
28. M. Kubinyi, F. Billes, A. Grofcsik and G. Keresztury, *J. Mol. Struct.*, **266**, 339 (1992).
29. R.J. Jakobsen and J.W. Brasch, *Spectrochim. Acta*, **21**, 1753 (1965).
30. J.H.S. Green, D.J. Harrison and W. Kynaston, *Spectrochim. Acta, Part A*, **27A**, 2199 (1971).
31. C. Garrigou-Lagrange, M. Chehata and J. Lascombe, *J. Chim. Phys.*, **63**, 552 (1966).
32. R.J. Jakobsen, *Spectrochim. Acta*, **21**, 433 (1965).
33. H. Takeuchi, N. Watanabe and I. Harada, *Spectrochim. Acta, Part A*, **44A**, 749 (1988).
34. G. Varsányi, *Assignments for Vibrational Spectra of Seven Hundred Benzene Drivatives*, J.Wiley & Sons, New York (1974).
35. A.P. Upadhyay and K.N. Upadhyay, *Indian J. Phys.*, **55B,** 232 (1981).
36. R.J. Jakobsen and E.J.Brewer, Appl.Spectrosc. **16**, 32 (1962).
37. H.W. Wilson, *Spectrochim. Acta*, **30A**, 2141 (1974).
38. E.Steger, U.Stahlberg and N.T.Q.Dieu, *Spectrochim. Acta*, **24A**, 1023 (1968).
39. V.N.Verma and D.K.Rai, Appl.Spectrosc. **24**, 445 (1970).
40. Y.Kishore, S.N.Sharma and C.P.D.Dwivedi, *Indian J. Phys.*, **48**, 412 (1974).
41. M.Horák, J. Smoliková and J. Pitha, *Collect. Czech. Chem. Commun.*, **26**, 2891 (1961).
42. J.H.S. Green, D.J. Harrison and W. Kynaston, *Spectrochim. Acta, Part A*, **28A**, 33 (1972).
43. J.A. Faniran and H.F. Shurvell, *Spectrochim. Acta, Part A*, **38A**, 1155 (1982).

44. J.-P. Bayle, J. Jullien, H. Stahl-Larivière, N. Le Calvé and B. Pasquier, *Spectrochim. Acta, Part A*, **39A**, 677 (1983).
45. N.S. Sundar and C. Santhamma, *Spectrochim. Acta, Part A*, **44A**, 69 (1988).
46. J.A. Faniran, *Spectrochim. Acta, Part A*, **35A**, 1257 (1979).
47. J.A. Faniran and H.F. Shurvell, *J. Raman Spectrosc.*, **9**, 73 (1980).
48. I. Iweibo and J.A. Faniran, *Spectrochim. Acta, Part A*, **37A**, 375 (1981).
49. A.K. Ansari and P.K. Verma, *Indian J. Phys.*, **53B**, 136 (1979).
50. R.K. Goel, K.P. Kansal and S.N. Sharma, *Indian J. Pure Appl. Phys.*, **17**, 778 (1979).
51. R.K. Goel, S. Sharma, K.P. Kansal and S.N. Sharma, *Indian J. Pure Appl. Phys.*, **18**, 281 (1980).
52. N.S. Sundar, *Spectrochim. Acta, Part A*, **41A**, 1449 (1985).
53. O.P. Sharma and R.D. Singh, *Indian J. Phys.*, **52B**, 93 (1977).
54. P.V. Khadikar, B. Pol and S.M. Ali, *Spectrochim. Acta, Part A*, **42A**, 755 (1986).
55. G. Varsányi, *Vibrational Spectra of Benzene Derivatives*, Academic Press, New York (1969).
56. G. Varsányi and P. Sohár, *Acta Chim. Acad. Sci. Hung.*, **74**, 315 (1972).
57. G. Varsányi and P. Sohár, *Acta Chim. Acad. Sci. Hung.*, **76**, 243 (1973).
58. G. Varsányi, G. Horváth, L. Imre, J. Schawartz, P. Sohár and F. Sóti, *Acta Chim. Acad. Sci. Hung.*, **93**, 315 (1977).
59. A.V. Stuart and G.B.B.M. Sutherland, *J. Chem. Phys.*, **24**, 559 (1956).
60. S. Krimm, C.Y. Liang and G.B.B.M. Sutherland, *J. Chem. Phys.*, **25**, 778 (1956).
61. P. Tarte and R. Deponthière, *J. Chem. Phys.*, **26**, 962 (1957).
62. C. Laurence and B. Wojtkowiak, *Bull. Soc. Chim. Fr.*, 3124 (1971).

10.1.2 Methoxy

The CH_3O group yields $3N-6=12$ normal vibrations, of which nine belong to the methyl group. The remaining three may be described as: C—O stretching vibration, C—O— deformation and MeO— torsion. Often the MeO group, which is capable of taking up different positions with respect to the rest of the molecule, gives rise to extra bands, mainly in the ν_sMe region. Moreover, overtones and combination bands of the deformations may occur in the stretching mode area, which obstructs an unambiguous assignment. Because of the non-equivalence of the CH bonds, the methyl vibrations can also be described as independent CH vibrations.

Methyl stretching vibrations

To all appearances the MeO molecules display the ν_aMe at a wavenumber lower than 3050 cm^{-1} (3000 $\pm$ 50 cm^{-1}). At the HW side of this region absorption is shown by MeOC(=O)OC(=O)OMe (3049) and also by esters in which the halogen clearly asserts its influence: FC(=O)OMe (3047 cm^{-1}) and ClC(=O)OMe, $ClCH_2C$(=O)OMe, Cl_2CHC(=O)OMe, Cl_3CC(=O)OMe and ClC(=O)C(=O)OMe, which all absorb in the neighbourhood of 3040 cm^{-1}. At the LW side one finds MeOCH=CHOMe (2955) and H_2C=CHOMe (2959 cm^{-1}). There is a realistic chance of finding the ν_aMe with weak to moderate intensity in the region 2995 $\pm$ 35 cm^{-1}.

The ν_a'Me is located in the region 2975 ± 55 cm^{-1} so that both stretchings often coincide. Compounds such as MeOC(=O)OC(=O)OMe and FC(=O)OMe reveal this ν_a'Me near 3025 cm^{-1} and the methyl esters of the chlorinated acetic acids in the neighbourhood of 3020 cm^{-1}. Self—evidently one has to proceed with caution in assigning these stretchings in unsaturated or aromatic methyl ethers and esters because of the =CH stretching vibrations being expected also in the above mentioned regions. In the LW area methoxymethane-d_3 and -d_1 (2917 and 2922), MeS(=O)OMe (2925) and H_2C=CHOMe (2927 cm^{-1}) are active. The remaining studied MeO compounds absorb usually in the region 2975 ± 40 cm^{-1} with a weak to moderate intensity.

The methyl symmetric stretching vibration in methyl esters is reported in the extensive range 2920 ± 80 cm^{-1}, but 2900 ± 60 cm^{-1} is also useful if the high values in the spectra of KOC(=O)OMe (2996) and FC(=O)OMe (2974 cm^{-1}) are disregarded. In the region 2850 ± 30 cm^{-1} one finds the ν_sMe of saturated, unsaturated and aromatic methyl ethers. The lowest values (≈2820 cm^{-1}) have been observed in the spectra of EtOMe-d_2, MeCH=CHOMe, $MeOCH_2CH_2OMe$, cBuOMe and in those of a few aromatic methyl ethers such as 2-ClPhOMe (2830), PhOMe (2834), 2-MeOPy (2850) and the isomers of $(MeO)_2$PhC(=O)H and $(MeO)_3$PhC(=O)H with 2850 ± 10 cm^{-1}. The compounds in which the MeO group is joined to a halogenated C atom display this ν_sMe at the HW side of the above-mentioned region.

Methyl deformations

The regions of the methyl deformations overlap each other, so that these normal vibrations are not always observed separately.

The δ_aMe is reported in the region 1460 ± 25 cm^{-1} with, at the HW side, MeOMe and MeS(=O)OMe with 1485 cm^{-1}, H_2C=CHOMe with 1482 cm^{-1} and methyl acetate with 1480 cm^{-1}. The lowest wavenumbers are due to DO(O=)CCH=CHC(=O)OMe (1438) and HO(O=)CCH=CHC(=O)OMe (1440 cm^{-1}). Disregarding these extreme values, most of the MeO compounds show this ν_aMe at 1460 ± 20 cm^{-1}.

The δ_a'Me is active in the region 1455 ± 20 cm^{-1}. The highest wavenumbers come from H_2C=CHOMe (1471) and Me_2POMe (1470 cm^{-1}) and the lowest from DO(O=)CCH=CHC(=O)OMe (1438) and HO(O=)CCH=CHC(=O)OMe (1439 cm^{-1}). Methoxymethane absorbs at 1459 cm^{-1}.

In some methoxy compounds the methyl symmetric deformation (1435 ± 35 cm^{-1}) is scarcely or not at all separated from its antisymmetric counterpart, such as in the spectra of Me_2P(=O)OMe (1470), 4-N≡CPhOMe (1467), 3-N≡CPhOMe (1462), *cis*-MeOCH=CHOMe (1460) and FCH_2OMe (1453 cm^{-1}). Low values originate from the spectra of H_2NC(=S)OMe-d_0 and -d_2 (1400 and 1410), Me_2P(=S)OMe (1417), Cl_2P(=S)OMe (1422) and MeOOMe (1424 cm^{-1}). The majority of the investigated molecules were found to give this symmetric deformation in a region (1445 ± 15 cm^{-1}) relatively high compared with that for the saturated hydrocarbons (1375 ± 15 cm^{-1}). Nevertheless, some investigators assign the absorption near 1390 cm^{-1} in di- and tri-methoxybenzaldehydes to the δ_sMe instead of the δCH of the aldehyde function [52, 57].

Methyl rocking vibrations and C—O stretch

Many of the R—OMe compounds shows the ρMe in the region 1195 ± 45 cm^{-1} with variable intensity, except MeOMe (1250 and *1181*) and $Cl_2NC(=O)OMe$ (1071 cm^{-1}). High values originate from the spectra of $MeOCH_2OMe$ (1232), cBuOMe (1228), $H_2C=CHOMe$ (1219) and XCH_2CH_2OMe (X = Cl, Br, I) with values near 1210 cm^{-1}. In the spectra of some RC(=O)OMe compounds the ρMe is assigned at or near the same wavenumber as that of the C(=O)—O stretching vibration, which appears as a strong band in the region 1255 ± 60 cm^{-1}. In the LW area one finds MeHNC(=S)OMe (1150), 2-ThOMe (1151), MeOOMe (1156) and Cl_2HCCF_2OMe (1159 cm^{-1}). Most of the R—OMe molecules give this ρMe, which is least coupled to the C—O stretching vibration, in the region 1180 ± 20 cm^{-1}.

The ρ'Me/νC—O is more coupled to the C—O stretching vibration and absorbs in the region 1155 ± 35 cm^{-1}. A few compounds show both methyl rocking vibrations at the same wavenumber, such as $Me_2P(=O)OMe$ (1187), ClS(=O)OMe (1186), Me_2POMe-d_0 and -d_6 (1183), ICH_2OMe (1180), $Cl_2P(=S)OMe$ (1175) and MeS(=O)OMe (1173 cm^{-1}). In addition, the HW side of the above-mentioned region is limited by $BrCH_2OMe$ (1190), H_3SiOMe (1185) and $F_3CC(=O)OMe$ (1178 cm^{-1}). The lowest values come from $H_2NC(=O)OMe$ (1120), MeCH=CHOMe (1121), 2-BrPhOMe (1121), 4-N≡CPhOMe (1122) and 2-ClPhOMe (1129 cm^{-1}). Most of the R—OMe compounds show this normal vibration at 1150 ± 25 cm^{-1}.

The vibrational analysis of R—OMe molecules reveals the νC—O/ρ'Me in the region 975 ± 125 cm^{-1}, usually with moderate intensity. The majority of the investigated molecules were found to give this stretching mode in the narrower region 980 ± 80 cm^{-1}, exceptions being $H_2NC(=S)OMe$ (1100), $MeSiH_2OMe$ (1092), Me_2POMe (1068) and CD_3OMe (853), $H_2NC(=O)OMe$ (880) and $H_2C=CHOMe$ (896 cm^{-1}). The ν_sCOC in the spectrum of methoxymethane is assigned at 919 cm^{-1} and methoxyethane shows this νC—O at 1019 cm^{-1}.

Skeletal deformation

The skeletal C—O— deformation can be found in the extensive region 415 ± 165 cm^{-1}. This band rarely possesses moderate intensity, and is usually very weak to weak. The highest wavenumbers (460 ± 120 cm^{-1}) are assigned in the spectra of saturated methyl ethers and the lowest in those of esters (320 ± 70), aromatic methyl ethers (320 ± 50) and methoxysilanes (320 ± 25 cm^{-1}).

Table 10.2 Absorption regions (cm^{-1}) of the normal vibrations of —OMe

Vibration	Saturated	Unsaturated	Aromatic	(Thio)esters	S(=O)—	P—bonded	Si—bonded
ν_qMe	3000 ± 25	2985 ± 35	2985 ± 20	3020 ± 30	3015 ± 25	3020 ± 30	2975 ± 15
ν_a'Me	2955 ± 35	2945 ± 20	2955 ± 20	2990 ± 40	3000 ± 25	2985 ± 35	2940 ± 15
ν_sMe	2845 ± 25	2850 ± 30	2845 ± 15	2920 ± 80	2940 ± 25	2930 ± 30	2835 ± 15
δ_aMe	1465 ± 20	1470 ± 15	1465 ± 10	1460 ± 25	1465 ± 20	1460 ± 15	1465 ± 10
δ_a'Me	1455 ± 20	1460 ± 15	1460 ± 15	1450 ± 15	1455 ± 10	1455 ± 15	1460 ± 10
δ_sMe	1445 ± 15	1450 ± 10	1450 ± 20	1425 ± 25	1445 ± 15	1445 ± 25	1450 ± 15
ρMe	1195 ± 40	1190 ± 30	1190 ± 45	1185 ± 35	1195 ± 25	1185 ± 15	1185 ± 15
ρ'Me/νC—O	1160 ± 30	1145 ± 25	1150 ± 30	1155 ± 35	1165 ± 25	1165 ± 25	1160 ± 25
νC—O/ρ'Me	940 ± 85	945 ± 50	1025 ± 30	975 ± 125	990 ± 20	1055 ± 30	1070 ± 25
δC—O—	460 ± 120	430 ± 100	320 ± 50	320 ± 70	475 ± 60	425 ± 75	320 ± 25
Me torsion	225 ± 40	235 ± 15	–	225 ± 65	–	220 ± 50	190 ± 40
MeO torsion	160 ± 50	185 ± 50	–	–	–	185 ± 15	–

R—OMe compounds

R = R′CH_2— (see Section 3.2.2).

R = Me— [1–12], CD_3— [1, 11, 13], DCH_2— [1], $MeCD_2$— [14], $ClCD_2$— [10, 15, 16], CD_3CD_2— [14], $BrCH_2CH$(MeO)— [17], iPr— [4, 18], Me(CF_3)CH— [19], $(CF_3)_2CH$— [20], cBu— [21], tBu— [22], Me_2(MeO)C— [23], Me$(MeO)_2$C— [24], Cl_2HCCF_2— [25], H_2C=CH— [26–31], CH_2=C=CH— [32], H_2C=C(Me)— [33, 34], MeCH=CH— [35], MeOCH=CH— [36, 37], Ph— [38–41], Ph-d_5— [39], XPh— (X = 2-, 3- and 4-Me [50], 4-$HOCH_2CH_2$, 3-$ClCH_2$, 4-$HOCH_2$ [42], 3-F_3C, 2-, 3- and 4-H(O=)C [43, 44, 50], 2-, 3- and 4-Me(O=)C [50], 3- and 4-Cl(O=)C, 4-MeCH=CH, 2-, 3- and 4-N≡C [45, 50], 2-, 3- and 4-H_2N [46, 50], 2- and 3-O_2N [50], 2-, 3- and 4-HO [50], 2-, 3- and 4-MeO [50], 2-, 3- and 4-HS, 2-, 3- and 4-F [38, 47, 48, 50], 2-, 3- and 4-Cl [38, 47, 50], 2-, 3- and 4-Br [38, 47, 50], 2- and 3-I), X-2-HOPh— (X = 3-, 4- and 5-H(O=)C [49, 50]), X-2-MeOPh— (X = 3-Me [51], 3- and 4-H(O=)C [52], 3- and 4-MeO [50, 53]), X-2-H_2NPh— (X = 4-Cl [54]), X-3-MeOPh— (X = 4-H(O=)C [52], 5-MeO [50, 53]), X-2-ClPh— (X = 4-H_2N [54], 3-Cl [55], 5-Cl [47, 48], 6-Cl [50]), X-3-FPh— (X = 4-F [56]), 2,3-$(MeO)_2$—4-H(O=)CPh—, 2,4-$(MeO)_2$—5-H(O=)CPh— and 3,5-$(MeO)_2$—2-H(O=)CPh— [57], 2-Py— [58], 2-Fu— [59], 2-Th— [60].

R = R′C(=O)— (see Section 7.1.8).

R = H_2NC(=S)—, HDNC(=S)— and D_2NC(=S)— [68], MeHNC(=S)— [61], KOC(=S)— [62], MeS(=O)— [63], MeOS(=O)— [64–67], ClS(=O)— [64], MeS$(=O)_2$— [69], MeOS$(=O)_2$— [67, 70], FS$(=O)_2$—, ClS$(=O)_2$— [69, 71], 4-MePhS$(=O)_2$—, Me_2P— [72, 73], $(CD_3)_2$P— [72], F_2P— [74], Me_2P(=O)— [75, 76], $(CD_3)_2$P(=O)— [76], Me(MeO)P(=O)— [75, 77, 78], $(MeO)_2$P(=O)— [75, 79], F_2P(=O)— [80], Cl_2P(=O)— [81–83], Me_2P(=S)— [75], Me(MeO)P(=S)— [75],

$(MeO)_2P(=S)$— [75, 84, 85], $Cl(MeO)P(=S)$— [84], $Cl_2P(=S)$— [86, 87], H_3Si— and D_3Si— [88, 89], $MeSiH_2$— and $MeSiD_2$— [90], Me_3Si— and $(CD_3)_3Si$— [91], MeO— [92, 93], O=N— [94].

References

1. A. Allan, D.C. McKean, J.-P. Perchard and M.-L. Josien, *Spectrochim. Acta, Part A*, **27A**, 1409 (1971).
2. A.A. Chalmers and D.C. McKean, *Spectrochim. Acta*, **21**, 1387 (1965).
3. J.-P. Perchard, M.-T. Forel and M.-L. Josien, *J. Chim. Phys.*, **61**, 632 (1964).
4. R.G. Snyder and G. Zerbi, *Spectrochim. Acta, Part A*, **23A**, 391 (1967).
5. P. Labarbe, M.-T. Forel and G. Bessis, *Spectrochim. Acta, Part A*, **24A**, 2165 (1968).
6. J.M. Freeman and T. Henshall, *J. Mol. Struct.*, **1**, 31 (1967).
7. J. Deroualt, M. Fouassier and M.-T. Forel, *J. Mol. Struct.*, **11**, 423 (1972).
8. A. Loutellier, L. Schriver, A. Burneau and J.-P. Perchard, *J. Mol. Struct.*, **82**, 165 (1982).
9. H.F. Hameka, *J. Mol. Struct.*, **226**, 241 (1991).
10. H.R. Linton and E.R. Nixon, *Spectrochim. Acta*, **15**, 146 (1959).
11. C.E. Blom, C. Altona and A. Oskam, *Mol. Phys.*, **34**, 557 (1977).
12. N.L. Allinger, M. Rahman and J.H. Lii, *J. Am. Chem. Soc.*, **112**, 8293 (1990).
13. J. Derouault, J. Le Calve and M.-T. Forel, *Spectrochim. Acta, Part A*, **28A**, 359 (1972).
14. J.-P. Perchard, *Spectrochim. Acta, Part A*, **26A**, 707 (1970).
15. D.C. McKean, I. Torto and A.R. Morrisson, *J. Mol. Struct.*, **99**, 101 (1983).
16. R.G. Jones and W.J. Orville-Thomas, *J. Chem. Soc.*, 692 (1974).
17. J.E. Katon and P.D. Miller, *Appl. Spectrosc.*, **29**, 501 (1975).
18. A.D.H. Clague and A. Danti, *Spectrochim. Acta, Part A*, **24A**, 439 (1968).
19. J.R. Durig, R.A. Larsen, R. Kelley, F.-Yi Sun and Y.S. Li, *J. Raman Spectrosc.*, **21**, 109 (1990).
20. Y.S. Li, R.A. Larsen, F.O. Cox and J.R. Durig, *J. Raman Spectrosc.*, **20**, 1 (1989).
21. J.R. Durig, G.A. Guirgis and V.F. Kalasinsky, *J. Mol. Struct.*, **52**, 27 (1979).
22. A. Sawa, H. Otha and S. Konaka, *J. Mol. Struct.*, **172**, 275 (1988).
23. K. Kumar and A.L. Verma, *J. Mol. Struct.*, **22**, 173 (1974).
24. K. Kumar, *J. Mol. Struct.*, **12**, 19 (1972).
25. Y.S. Li and J.R. Durig, *J. Mol. Struct.*, **81**, 181 (1982).
26. W. Pyckhout, P. Van Nuffel, C. Van Alsenoy, L. Van den Enden and H.J. Geise, *J. Mol. Struct.*, **102**, 333 (1983).
27. I.S. Ignatyev, A.N. Lazarev, M.B. Smirnov, M.L. Alpert and B.A. Trofimov, *J. Mol. Struct.*, **72**, 25 (1981).
28. B. Cadiolo, E. Gallinella and U. Pincelli, *J. Mol. Struct.*, **78**, 215 (1982).
29. T. Beech, R. Gunde, P. Felder and H.H. Günthard, *Spectrochim. Acta, Part A*, **41A**, 319 (1985).
30. N.L. Owen and N. Sheppard, *Trans. Faraday Soc.*, **60**, 634 (1964).
31. A.N. Lazarev, I.S. Ignatyev, L.L. Schukovskaya and R.I. Palčhik, *Spectrochim. Acta, Part A*, **27A**, 2291 (1971).
32. S.V. Eroshchenko, L.M. Sinegovskaya, O.A. Tarasova, Yu.L. Frolov, B.A. Tropimov and I.S. Ignatyev, *Spectrochim. Acta, Part A*, **46A**, 1505 (1990).
33. A.O. Diallo, *Spectrochim. Acta, Part A*, **37A**, 529 (1981).
34. E. Gallinella, U. Pincelli and B. Cadioli, *J. Mol. Struct.*, **99**, 31 (1983).
35. S.W. Charles, F.C. Cullen and N.L. Owen, *J. Mol. Struct.*, **18**, 183 (1973).

36. H.S. Kimmel, J.T. Waldron and W.H. Snyder, *J. Mol. Struct.*, **21**, 445 (1974).
37. J.M. Comerford, P.G. Anderson, W.H. Snyder and H.S. Kimmel, *Spectrochim. Acta, Part A*, **33A**, 651 (1977).
38. N.L. Owen and R.E. Hester, *Spectrochim. Acta, Part A*, **25A**, 343 (1969).
39. W.J. Balfour, *Spectrochim. Acta, Part A*, **39A**, 795 (1983).
40. J.H.S. Green, *Spectrochim. Acta*, **18**, 48 (1962).
41. H. Tylli and H. Konschin, *J. Mol. Struct.*, **42**, 7 (1977).
42. S. Chakravorti, A.K. Sarkar, K. Mallick and S.B. Banerjee, *Indian J. Phys.*, **56B**, 96 (1982).
43. M.P. Srivastava, O.N. Singh and I.S. Singh, *Curr. Sci.*, **37**, 100 (1968).
44. C.P.D. Dwivedi, *Indian J. Pure Appl. Phys.*, **6**, 440 (1968).
45. R.K. Goel and M.L. Agarwal, *Spectrochim. Acta, Part A*, **38A**, 583 (1982).
46. V.B. Singh and A.K. Sinha, *Indian J. Phys.*, **61B**, 344 (1987).
47. B. Laksmaiah and G.R. Rao, *J. Raman Spectrosc.*, **20**, 449 (1989).
48. B. Laksmaiah and G.R. Rao, *J. Raman Spectrosc.*, **20**, 439 (1989).
49. S.P. Gupta, C. Gupta, S. Sharma and R.K. Goel, *Indian J. Pure Appl. Phys.*, **24**, 111 (1986).
50. G. Varsányi, *Assignments for Vibrational Spectra of Seven Hundred Benzene Drivatives*, J.Wiley & Sons, New York (1974).
51. O.P. Singh, R.P. Singh and R.N. Singh, *Spectrochim. Acta, Part A*, **49A**, 517 (1993).
52. S.J. Singh and R. Singh, *Indian J. Pure Appl. Phys.*, **16**, 939 (1978).
53. A.K. Sarkar, S. Chakravorti and S.B. Banerjee, *Indian J. Phys.*, **51B**, 71 (1977).
54. P. Venkatacharyulu, V.L.N. Prasad, Nallgonda and D. Premaswarup, *Indian J. Pure Appl. Phys.*, **19**, 1178 (1981).
55. R.K. Goel and S.K. Mathur, *Proc. Natl. Acad. Sci. India*, **51A**, 190 (1981).
56. R. Rao, M.K. Aralakkanavar, K.S. Rao and M.A. Shashidhar, *Spectrochim. Acta, Part A*, **45A**, 103 (1989).
57. P. Venkoji, *Acta Chim. Acad. Sci. Hung.*, **117**, 163 (1984).
58. K.C. Medhi, *Bull. Chem. Soc. Jpn.*, **57**, 261 (1984).
59. M. Sénechal and P. Saumagne, *J. Chim. Phys.*, **69**, 1246 (1972).
60. J.J. Peron, P. Saumagne and J.M. Lebas, *Spectrochim. Acta, Part A*, **26A**, 1651 (1970).
61. G.C. Chaturvedi and C.N.R. Rao, *Spectrochim. Acta, Part A*, **27A**, 65 (1971).
62. R. Mattes and K. Scholten, *Spectrochim. Acta, Part A*, **31A**, 1307 (1975).
63. G.E. Binder and A. Schmidt, *Spectrochim. Acta, Part A*, **33A**, 816 (1977).
64. P.V. Huong and E. Raducanu, *J. Mol. Struct.*, **23**, 81 (1974).
65. A.B. Remizov, A.I. Fishman and I.S. Pominov, *Spectrochim. Acta, Part A*, **35A**, 901 (1979).
66. A.J. Barnes and B.J. Van der Veken, *J. Mol. Struct.*, **157**, 119 (1987).
67. A.B. Remizov, A.I. Fishman and I.S. Pominov, *Spectrochim. Acta, Part A*, **35A**, 909 (1979).
68. L. Zhengyan, R. Mattes, H. Schnöckel, M. Thünemann, E. Hunting, U. Höhnke and C. Mendel, *J. Mol. Struct.*, **117**, 117 (1984).
69. A. Simon, H. Kriegsmann and H. Dutz, *Chem. Ber.*, **89**, 2378 (1956).
70. K.O. Christe and E.C. Curtis, *Spectrochim. Acta, Part A*, **28A**, 1889 (1972).
71. B. Nagel, J. Stark, J. Fruwert and G. Geiseler, *Spectrochim. Acta, Part A*, **32A**, 1297 (1976).
72. B.J. Van der Veken, T.S. Little, Y.S. Li and M.E. Harris, *Spectrochim. Acta,, Part A*, **42A**, 123 (1986).
73. J.R. Durig and F.F.D. Daeyaert, *J. Mol. Struct.*, **261**, 133 (1992).
74. J.R. Durig and B.J. Streusand, *Appl. Spectrosc.*, **34**, 65 (1980).
75. W.D. Von Burkhardt, E.G. Hohn and J. Goubeau, *Z. Anorg. Allg. Chem.*, **442**, 19 (1978).

76. B.J. Van der Veken, R.L. Odeurs, M.A. Herman and J.R. Durig, *Spectrochim. Acta, Part A*, **40A**, 565 (1984).
77. R.M. Moravie, F. Froment and J. Corset, *Spectrochim. Acta, Part A*, **45A**, 1015 (1989).
78. B.J. Van der Veken, Thesis UIA, Antwerp, 1979.
79. V. von Hornung, O. Aboulwafa, A. Lentz and J. Goubeau, *Z. Anorg. Allg. Chem.*, **380**, 137 (1971).
80. G.H. Pieters, B.J. Van der Veken, A.J. Barnes, T.S. Little, W.Y. Zhao and J.R. Durig, *Spectrochim. Acta, Part A*, **43A**, 657 (1987).
81. G.H. Pieters, B.J. Van der Veken and M.A. Herman, *J. Mol. Struct.*, **102**, 27 (1983).
82. G.H. Pieters, B.J. Van der Veken and M.A. Herman, *J. Mol. Struct.*, **102**, 221 (1983).
83. G.H. Pieters, B.J. Van der Veken, A.J. Barnes, T.S. Little and J.R. Durig, *J. Mol. Struct.*, **125**, 243 (1984).
84. J.R. Durig and J.S. Diyorio, *J. Mol. Struct.*, **3**, 179 (1969).
85. O. Aboulwafa, A. Lentz and J. Goubeau, *Z. Anorg. Allg. Chem.*, **380**, 128 (1971).
86. R.A. Nyquist, *Spectrochim. Acta, Part A*, **28A**, 285 (1972).
87. R.A. Nyquist and W.W. Muelder, *Spectrochim. Acta*, **22**, 1563 (1966).
88. I.S. Ignatyev, *J. Mol. Struct.*, **172**, 139 (1988).
89. G.S. Weiss and E.R. Nixon, *Spectrochim. Acta*, **21**, 903 (1965).
90. K. Ohno, K. Taga and H. Murata, *J. Mol. Struct.*, **55**, 7 (1979).
91. T.F. Tenisheva, A.N. Lazarev and R.I. Uspenskaya, *J. Mol. Struct.*, **37**, 173 (1977).
92. K.O. Christe, *Spectrochim. Acta, Part A*, **27A**, 463 (1971).
93. M.E.B. Bell and J. Laane, *Spectrochim. Acta, Part A*, **28A**, 2239 (1972).
94. H.D. Stidham, G.A. Guirgis, B.J. Van der Veken, T.G. Sheehan and J.R. Durig, *J. Raman Spectrosc.*, **21**, 615 (1990).

10.1.3 Ethoxy

The EtO group provides 21 normal vibrations. Eighteen are attributed to ethyl (Section 3.5.1). The remaining three normal vibrations are: a C—O stretching vibration, a C—O— deformation and an EtO— torsion.

Methyl and methylene stretching vibrations

The five ethyl CH stretching vibrations absorb between 2995 and 2855 cm^{-1} with a moderate to strong intensity in ethyl ethers and a weak to moderate intensity in ethyl esters. These normal vibrations are usually arranged in order of descending wavenumber:

$$\nu_a\text{Me} \geq \nu_a'\text{Me} \geq \nu_a\text{CH}_2 > \nu_s\text{Me} \geq \nu_s\text{CH}_2.$$

Methyl and methylene deformations

The methyl symmetric deformation gives rise to a moderate to strong band in the region 1385 ± 15 cm^{-1}. The other three deformations absorb weakly to moderately and often coincide. In the HW region one finds the δCH_2 in the spectrum of ethoxy ethane at 1494 cm^{-1} [7] or 1490 cm^{-1} [10]. In unsaturated and aromatic ethers the methylene scissors stays at 1480 ± 10 cm^{-1}. The methyl antisymmetric deformations are active between 1480 and 1425 cm^{-1}. A low wavenumber (≈1425

cm^{-1}) is assigned in the spectrum of ethoxyethane, but most EtO compounds show the methyl antisymmetric deformations at 1455 ± 25 cm^{-1}.

Methylene wagging and twisting vibrations

With a range of 1350 ± 40 cm^{-1} and a moderate to strong intensity, the methylene wag in R—OEt compounds comes in the region of the methyl symmetric deformation. In this text we accept that the δ_sMe {ωCH_2} possesses the highest {lowest} wavenumber, although the literature does not agree concerning this assignment.

The methylene twist is located in the region 1285 ± 45 cm^{-1} with a weak to moderate intensity. Wavenumbers in the neighbourhood of 1325 cm^{-1} are observed in the spectra of ethyl benzoate and the ethyl esters of pyridinecarboxylic acids. Low values have been traced in the spectra of EtO(O=)CCH=CHC(=O)OEt (1244 cm^{-1}) and in those of a few ethyl ethers such as $MeSCH_2OEt$, 4-H(O=)CPhOEt and 2-FPhOEt with wavenumbers near 1260 cm^{-1}.

Methyl rocking vibrations and CC/CO stretching vibrations

These four vibrations are coupled in such a way as to make it difficult to determine which vibration contributes in the highest degree to a distinct absorption. The intensities of these vibrations are mostly weak or moderate, rarely strong.

The highest absorption region (1165 ± 30 cm^{-1}) is assigned to the ρMe. Wavenumbers in the neighbourhood of 1195 cm^{-1} are found in the spectra of EtOEt and *trans*-MeCH=CHOEt. In ethyl esters this vibration can hide in the C(=O)O stretching vibration which appears in the region 1245 ± 65 cm^{-1}.

The ρ'Me is located in the region 1120 ± 40 cm^{-1}. Ethoxyethane shows these rocks at 1155 and 1120 cm^{-1} and ethyl acetate at 1098 cm^{-1}. In this region also the antisymmetric COC stretching vibration in ethers is active, so that one has to proceed with caution in assigning these vibrations.

The region 1060 ± 40 cm^{-1} is attributed to a skeletal stretching vibration with a contribution of the C—C and the C—O bond. For most of the ethyl esters the C—O stretching vibration is assigned in this region. In ethyl ethers, however, the C—C stretching vibration dominates in this absorption.

In the spectra of ethyl esters, the stretching vibration in the region 875 ± 65 cm^{-1} is attributed mainly to the C—C bond. In the spectra of ethyl ethers this vibration is often called the COC symmetric stretch, with a contribution of the C—C bond. EtOEt gives this ν_s COC at 850 cm^{-1}.

Methylene rocking vibration

The methylene rock is active with a weak to moderate intensity in the region 790 ± 50 cm^{-1}, which narrows to 800 ± 25 cm^{-1} if the values from the following compounds are not taken into account: 4-H(O=)CPhOEt (838), 4-BrPhOEt (830), 1, 4-$(EtO)_2$Ph (748) and 2-ThC(=O)OEt (750 cm^{-1}).

Skeletal deformations

The skeletal O—C—C deformation occurs at 390 ± 85 cm^{-1} with a weak to moderate intensity. Ethoxyethane absorbs at 450 cm^{-1}, methoxyethane at 470 cm^{-1} and ethylacetate at 378 cm^{-1}. The very extensive region (390 ± 140 cm^{-1}) for the external C—O—C deformation is due to the fact that ethers (450 ± 80) absorb at higher wavenumbers than esters (310 ± 60 cm^{-1}).

Table 10.3 Absorption regions (cm^{-1}) of the normal vibrations of —OEt

Vibration	Saturated	Unsaturated	Aromatic	Esters
ν_aMe	2985 ± 10	2985 ± 10	2985 ± 10	2985 ± 10
ν'_aMe	2965 ± 25	2975 ± 10	2975 ± 10	2975 ± 15
$\nu_a CH_2$	2940 ± 10	2935 ± 15	2935 ± 10	2945 ± 15
ν_sMe	2910 ± 30	2905 ± 15	2895 ± 15	2910 ± 20
$\nu_s CH_2$	2875 ± 20	2875 ± 15	2875 ± 15	2885 ± 25
δCH_2	1475 ± 20	1480 ± 10	1480 ± 10	1475 ± 15
δ_aMe	1465 ± 15	1460 ± 10	1470 ± 10	1460 ± 15
δ'_aMe	1445 ± 20	1440 ± 10	1445 ± 15	1450 ± 15
δ_sMe	1385 ± 15	1385 ± 10	1385 ± 10	1385 ± 15
ωCH_2	1350 ± 30	1335 ± 25	1345 ± 25	1360 ± 25
τCH_2	1285 ± 25	1290 ± 20	1280 ± 20	1285 ± 45
ρMe	1165 ± 30	1170 ± 25	1160 ± 15	1165 ± 30
ρ'Me	1120 ± 40	1120 ± 10	1110 ± 20	1115 ± 35
νC—O/C—C	1065 ± 35	1080 ± 20	1070 ± 30	1060 ± 40
νC—C/C—O	875 ± 65	870 ± 30	885 ± 50	890 ± 50
ρCH_2	805 ± 20	800 ± 35	790 ± 50	800 ± 25
δC—O—C	470 ± 60	420 ± 50	420 ± 50	310 ± 60
δO—C—C	395 ± 75	390 ± 50	390 ± 50	350 ± 45
torsion Me	230 ± 30	235 ± 15	235 ± 25	245 ± 35
torsion Et	150 ± 50	150 ± 50	–	160 ± 40
torsion OEt	–	–	–	–

R—OEt molecules

R = H— and D— [1–3], Me— [3–8], CD_3— [4–6], Et— [7–10], iPr— [7], $ClCH_2CH_2$—, HC≡CCH_2— [11], $MeSCH_2$— [12], H_2C=CH— [13–16], MeCH=CH— [17, 18], Ph— [19, 25], 2- and 4-XPh— (X = F, Cl and Br [20–22, 25], O_2N and H(O=)C [23, 24]), 4-MeC(=O)NHPh—, 3- and 4-EtOPh— [25], 1,$3Cl_2$Ph— [25].

References

1. Y. Mikawa, J.W. Brasch and R.J. Jakobsen, *Spectrochim. Acta, Part A*, **27A**, 529 (1971).
2. J.R. Durig, W.E. Bucy, C.J. Wurrey and L.A. Carreira, *J. Phys. Chem.*, **79**, 988 (1975).
3. N.L. Allinger, M. Rahman and J.-H. Lii, *J. Am. Chem. Soc.*, **112**, 8293 (1990).
4. T. Kitagawa, K. Ohno, H. Sugata and T. Miyazawa, *Bull. Chem. Soc. Jpn.*, **45**, 969 (1972).
5. J.P. Perchard, *Spectrochim. Acta, Part A*, **26A**, 707 (1970).
6. J.R. Durig and D.A.C. Compton, *J. Chem. Phys.*, **69**, 4713 (1978).
7. R.G. Snyder and G. Zerbi, *Spectrochim. Acta, Part A*, **23A**, 391 (1967).
8. A.D.H. Clague and A. Danti, *Spectrochim. Acta, Part A*, **24A**, 439 (1968).
9. H. Wieser, W.G. Laidlaw, P.J. Krueger and H. Fuhrer, *Spectrochim. Acta, Part A*, **24A**, 1055 (12968).
10. H. Wieser and P.J. Krueger, *Spectrochim. Acta, Part A*, **26A**, 1349 (1970).
11. S.W. Charles, F.C. Cullen and N.L. Owen, *J. Chem. Soc.*,Faraday Trans II **72**, 351 (1976).
12. H. Matsuura, H. Murata and M. Sakakibara, *J. Mol. Struct.*, **96**, 267 (1983).
13. J.R. Durig and D.J. Gerson, *J. Mol. Struct.*, **71**, 131 (1981).
14. M. Sakakibara, F. Inagaki, I. Harada and T. Shimanouchi, *Bull. Chem. Soc. Jpn.*, **49**, 46 (1976).
15. N.L. Owen and G.O. Sørensen, *J. Phys. Chem.*, **83**, 1483 (1979).
16. N.L. Owen and N. Sheppard, *Spectrochim. Acta*, **22**, 1101 (1966).
17. S.W. Charles, F.C. Cullen and N.L. Owen, *J. Mol. Struct.*, **18**, 183 (1973).
18. F. Marsault-Herail, G.S. Chiglien, J.P. Dorie and M.L. Martin, *Spectrochim. Acta, Part A*, **29A**, 151 (1973).
19. J.H.S. Green, *Spectrochim. Acta*, **18**, 39 (1962).
20. E.F. Mooney, *Spectrochim. Acta*, **19**, 877 (1963).
21. S.P. Sinha and C.L. Chatterjee, *Spectrosc. Lett.*, **9**, 455 (1976).
22. R.C. Maheshwari and M.M. Shukla, *Indian J. Pure Appl. Phys.*, **13**, 135 (1975).
23. K.M. Mathur, D.P. Juyal and R.N. Singh, *Indian J. Pure Appl. Phys.*, **9**, 756 (1971).
24. P. Venkoji, *Spectrochim. Acta, Part A*, **42A**, 1301 (1986).
25. G. Varsányi, *Assignments for Vibrational Spectra of Seven Hundred Benzene Derivatives*, J.Wiley & Sons, New York (1974).

10.2 R′-YLOXY COMPOUNDS

The R′C(=O)O— group in which R′ = H (formates), Cl (chloroformates) or Me (acetates) provides for R′ = H and Cl {Me} nine {eighteen} normal vibrations of which six {fifteen} are studied with the —C(=O)H group (see Section 7.1.1), the —C(=O)Cl group (see Section 7.1.3) {or the —C(=O)Me group (see Section 7.1.5)}.

10.2.1 Formylox

The CH stretching vibration of formates of the type HC(=O)O—R usually absorbs at 2935 ± 35 cm^{-1} with a weak to moderate intensity. The most characteristic band of the formates arises from the C=O stretching vibration which absorbs strongly in the region 1730 ± 40 cm^{-1}. The HW side of this region is limited by $HC(=O)OCH_2Cl$ (vapour 1770 [15], liquid 1740 cm^{-1} [9, 13]) and the LW side by HC(=O)OMe (solid 1692 [8], liquid 1754 cm^{-1} [2, 4, 6, 7, 10]). For the remaining formates this region narrows to 1740 ± 20 cm^{-1}. The CH in-plane deformation is a weak to moderate absorption in the range 1365 ± 15 cm^{-1} with the exception of $HC(=O)OSiD_3$ with 1331 cm^{-1} [19]. The C(=O)—O stretching vibration appears strongly at 1165 ± 45 cm^{-1} with extreme values of 1207 and 1210 cm^{-1} in the spectra of HC(=O)OMe-d_0 and -d_3 and 1120 cm^{-1} in the spectrum of $HC(=O)OCH_2Cl$. Most of the HC(=O)O—R compounds show the CH out-of-plane deformation at 1040 ± 30 cm^{-1}, except $HC(=O)OSiH_3$-d_0 and -d_3 with respectively 1104 and 1095 cm^{-1}. The band with moderate intensity in the region 720 ± 50 cm^{-1} is due to the C=O in-plane deformation, which is to some degree coupled to the δCH. The external skeletal C—O—R in-plane deformation is assigned in the region 320 ± 90 cm^{-1}, with 325 cm^{-1} for HC(=O)OMe. Some vibrational analysis of formates reveals the out-of-plane skeletal deformation or torsion at 290 ± 70 cm^{-1} and the —OC(=O)H torsion at lower wavenumbers: 105 ± 40 cm^{-1}.

Table 10.4 Absorption regions (cm^{-1}) of the normal vibrations of —OC(=O)H

Vibration	Region	Vibration	Region
νCH	2935 ± 35	δC=O	720 ± 50
νC=O	1730 ± 40	δC—O—	320 ± 90
δCH	1365 ± 15	torsion	290 ± 70
νC(=O)—O	1165 ± 45	torsion	105 ± 40
γ CH	1040 ± 30		

R—OC(=O)H compounds
R = Me— [1–10], CD_3— [7–10], Et— [9, 11], HC≡CH_2— [9, 12], ClH_2C— [9, 13–16], ClD_2C— [9, 13, 15], tBu— [9, 17], H_2C=CH— [18], H_3Si— and D_3Si— [19].

10.2.2 Chloroformyloxy (chlorocarbonyloxy)

Six normal vibrations of chloroformyloxy are treated with the chloroformyl group (see Section 7.1.3). The C(=O)—O stretching vibration gives rise to a strong band in the region 1160 ± 45 cm^{-1}. Durig and Griffin [22] assigned the band at 1202 cm^{-1} in the spectrum of MeOC(=O)Cl to this stretching vibration and that at 1159 cm^{-1} to a methyl rock, whereas Nyquist [21] preferred the contrary. The lowest value comes from $ClCH_2OC(=O)Cl$ (1117 cm^{-1}). The external skeletal C—O—R in-plane deformation absorbs weakly at 275 ± 25 cm^{-1} and the —OC(=O)Cl torsion near 70 cm^{-1}.

Table 10.5 Absorption regions (cm^{-1}) of the normal vibrations of —OC(=O)Cl

Vibration	Region	Vibration	Region
νC=O	1780 ± 20	δC—Cl	365 ± 50
νC(=O)—O	1160 ± 45	δC—O—	275 ± 25
νC—Cl	790 ± 60	torsion	160 ± 20
γC=O	675 ± 15	torsion	≈70

R—OC(=O)Cl compounds
R = Me— [20–22], CD_3— [22], Et— [11], nPr—, iPr—, nBu— and iBu— [38], H_2C=$CHCH_2$—, HC≡CCH_2— [21], $PhCH_2$—, $ClCH_2$—, ClCHD— and $ClCD_2$— [14, 23], Ph—.

10.2.3 Acetyloxy (acetoxy)

Methyl vibrations

In acetates the methyl stretching vibrations give rise to very weak absorptions. The antisymmetric deformations, however, absorb with moderate intensity and the methyl symmetric deformation with moderate to strong intensity. The Me—C stretching vibration is coupled to the methyl rock.

The C=O stretching vibration

With the exception of MeC(=O)OCl (1818 cm^{-1}) the C=O stretching vibration in R—OC(=O)Me compounds occurs at 1750 ± 20 cm^{-1}. As examples, MeC(=O)OCH_2Cl absorbs at 1770 cm^{-1}, MeC(=O)OPh at 1765 cm^{-1} (the formula isomer PhC(=O)OMe at 1724 cm^{-1}) and MeC(=O)OCH_2CH=CH_2 at 1743 cm^{-1}.

The C(=O)—O— vibrations

The C(=O)—O stretching vibration absorbs strongly in the region 1235 ± 30 cm^{-1} with high wavenumbers for MeC(=O)OCD_3 (1265) and MeC(=O)$OSiH_3$ (1257) and low values for MeC(=O)OPh (1205) and MeC(=O)OCH=CH_2 (1217 cm^{-1}). The skeletal C—O—R in-plane deformation is active in the region 275 ± 45 cm^{-1}.

Table 10.6 Absorption regions (cm^{-1}) of the normal vibrations of —OC(=O)Me

Vibration	Region	Vibration	Region
ν_aMe	3010 ± 30	ρ'Me	975 ± 45
ν'_aMe	2970 ± 30	νC—C	860 ± 50
ν_sMe	2910 ± 40	δC=O	620 ± 30
νC=O	1750 ± 20	γC=O	600 ± 20
δ_aMe	1445 ± 20	δC—C—O	415 ± 50
δ'_aMe	1435 ± 15	δC—O—	275 ± 45
δ_sMe	1370 ± 10	torsion	160 ± 50
νC(=O)—O	1235 ± 30	torsion	–
ρMe	1050 ± 30	torsion	–

R—OC(=O)Me compounds

R = Me— [2, 4, 6, 24–27], CD_3— [2, 5, 24, 28], Et— [29–31], H_2C=$CHCH_2$— [32], $ClCH_2$— [14, 33, 34], $ClCD_2$— [14, 34], H_2C=CH— [35], Ph— [36, 39], Ph-d_5— [36], 4-FPh—, 2- and 4-MeOPh— [39], 2- and 4-O_2NPh— [39], Cl— [37], H_3Si— and D_3Si— [19].

References

1. A. Hadni, J. Deschamps and M.-L. Josien, *C.R. Acad. Sci.*, **242**, 1014 (1956).
2. J.K. Wilmshurst, *J. Mol. Spectrosc.*, **1**, 201 (1957).
3. H. Susi and J.R. Scherer, *Spectrochim. Acta, Part A*, **25A**, 1243 (1969).
4. P. Matzke, O. Chacon and C. Andrade, *J. Mol. Struct.*, **9**, 255 (1971).
5. J. Derouault, J. Le Calve and M.-T. Forel, *Spectrochim. Acta, Part A*, **28A**, 359 (1972).
6. R.M. Moravie and J. Corset, *J. Mol. Struct.*, **30**, 113 (1976).
7. H. Susi and T. Zell, *Spectrochim. Acta*, **19**, 1933 (1963).
8. W.C. Harris, D.A. Coe and W.O. George, *Spectrochim. Acta, Part A*, **32A**, 1 (1976).
9. M.G. Dahlqvist and K. Euranto, *Spectrochim. Acta, Part A*, **34A**, 863 (1978).
10. E.B. Marmar, C. Pouchan, A. Dargelos and M. Chaillet, *J. Mol. Struct.*, **57**, 189 (1979).
11. S.W. Charles, G.I.L. Jones, N.L. Owen, S.J. Cyvin and B.N. Cyvin, *J. Mol. Struct.*, **16**, 225 (1973).
12. G.I.L. Jones, D.G. Lister and N.L. Owen, *Trans. Faraday Soc.*, **71**, 1330 (1975).
13. M.G. Dahlqvist, *Spectrochim. Acta, Part A*, **36A**, 37 (1980).
14. F. Daeyaert, Thesis, UIA, Antwerp, 1988.
15. F. Daeyaert and B.J. Van der Veken, *J. Mol. Struct.*, **213**, 97 (1989).
16. M. Räsänen, H. Kunttu, J. Murto and M. Dahlqvist, *J. Mol. Struct.*, **159**, 65 (1987).
17. Y. Umemura, J. Corset and R.M. Moravie, *J. Mol. Struct.*,**52**, 175 (1979).
18. W. Pyckhout, C. Alsenoy, H.J. Heise, B.J. Van der Veken, P. Coppens and M. Traetteberg, *J. Mol. Struct.*, **147**, 85 (1986).
19. A.G. Robiette and J.C. Thompson, *Spectrochim. Acta*, **21**, 2023 (1965).
20. J.C. Evans and J. Overend, *Spectrochim. Acta*, **19**, 701 (1963).
21. R.A. Nyquist, *Spectrochim. Acta, Part A*, **28A**, 285 (1972).
22. J.R. Durig and M.G. Griffin, *J. Mol. Spectrosc.*, **64**, 252 (1977).
23. F. Daeyaert and B.J. Van der Veken, *J. Mol. Struct.*, **198**, 239 (1989).

24. W.O. George, T.E. Houston and W.C. Harris, *Spectrochim. Acta, Part A*, **30A**, 1035 (1974).
25. H. Hollenstein and H.H. Günthard, *J. Mol. Spectrosc.*, **84**, 457 (1980).
26. R. Fausto and J.J.C. Teixeira-Dias, *J. Mol. Struct.*, **144**, 215 (1986).
27. D. Steele and A. Muller, *J. Phys. Chem.*, **95**, 6163 (1991).
28. B. Nolin and R.N. Jones, *Can. J. Chem.*, **34**, 1382 (1956).
29. M.A. Raso, M.V. Garcia and J. Morcillo, *J. Mol. Struct.*, **115**, 449 (1984).
30. Y. Mido, H. Shiomi, H. Matsuura, M.A. Raso, M.V. Garcia and J. Morcillo, *J. Mol. Struct.*, **176**, 253 (1988).
31. T.-K. Ha, C. Pal and P.N. Ghosh, *Spectrochim. Acta, Part A*, **48A**, 1083 (1992).
32. B. Singh, R. Prasad and R.M.P. Jaiswal, *Proc. Indoan Acad. Sci.* (*Chem. Sci.*), **89**, 201 (1980).
33. S.W. Charles, G.I.L. Jones, N.L. Owen and L.A. West, *J. Mol. Struct.*, **32**, 111 (1976).
34. F. Daeyaert, H.O. Desseyn and B.J. Van der Veken, *Spectrochim. Acta, Part A*, **44A**, 1165 (1988).
35. W.R. Fairheller and J.E. Katon, *J. Mol. Struct.*, **1**, 239 (1967).
36. Y. Kim, H. Noma and K. Machida, *Spectrochim. Acta, Part A*, **42A**, 891 (1986).
37. J.C. Evans, G.Y.S. Lo and Y.L. Liang, *Spectrochim. Acta*, **21**, 973 (1965).
38. H.A. Ory, *Spectrochim. Acta*, **16**, 1488 (1960).
39. G. Varsányi, *Assignments for Vibrational Spectra of Seven Hundred Benzene Derivatives*, J.Wiley & Sons, New York (1974).

11

Normal Vibrations and Absorption Regions of Sulfur Compounds

11.1 THIO COMPOUNDS

11.1.1 Methylthio

The MeS group yields twelve normal vibrations, of which nine belong to the methyl group. The remaining three may be described as: C—S stretching vibration, C—S— deformation and MeS— torsion. Since the CH bonds of the methyl group take up different positions with respect to the C—S bond and the methyl group strictly does not possess C_{3v} symmetry, the methyl vibrations have to be described in terms of individual CH vibrations. For convenience, however, the usual terms antisymmetric and symmetric are used.

Methyl stretching vibrations

The ν_aMe appears weakly to moderately in the region 3005 ± 25 cm^{-1} with the highest value (3030 cm^{-1}) for N≡CSMe and EtSSMe, followed by ClC(=O)SMe with 3025 and HC≡CSMe with 3017 cm^{-1}. In the LW range MeSMe (2982) and MeHgSMe (2984 cm^{-1}) show absorption.

The region of the ν_a'Me (2980 ± 45 cm^{-1}) partly overlaps that of the ν_aMe and often both methyl stretchings occur at the same wavenumber. The highest values are those due to ClC(=O)SMe (3025), N≡CSMe (3022), O=NSMe (3012) and HC≡CSMe (3012 cm^{-1}). Low wavenumbers are found in the spectra of

$H_2NC({=}S)SMe$ (2940) and MeSMe (2960 cm^{-1}) but most R—SMe compounds show this ν_a'Me at 2980 ± 20 cm^{-1}.

The ν_sMe absorbs moderately to strongly in the region 2925 ± 20 cm^{-1}. A few examples are: HC≡CSMe-d_0 and -d_1 (2944), ClC(=O)SMe (2941), HSMe (2930), $H_2NC({=}S)SMe$ (2910), MeHNC(=S)SMe (2910) and MeSMe (2917 and 2904 cm^{-1} [13]).

Methyl deformations

The antisymmetric deformations δ_aMe and δ_a'Me are assigned in the respective regions 1445 ± 25 and 1430 ± 30 cm^{-1}, so that there is a realistic chance of finding both deformations at the same wavenumber. The δ_sMe in MeS compounds occurs with moderate intensity at ≈120 cm^{-1} lower wavenumbers (1320 ± 20 cm^{-1}) than the δ_sMe in MeO compounds (Section 10.1.2).

Methyl rocking vibrations

In delimiting the absorption region of the ρMe (995 ± 40 cm^{-1}) the high value 1065 cm^{-1} for MeSH has not been taken into account. In the HW region, absorption is shown by such compounds as $MeSCH_2CH_2SMe$ (1035), MeSMe (1032) and $MeOCH_2SMe$ (1008 cm^{-1}). The LW side of the above-mentioned region is limited by 955 cm^{-1} from the spectra of MeSSMe, EtSSMe, tBuSSMe and MeHgSMe.

The ρ'Me is assigned at 940 ± 40 cm^{-1} with 976 cm^{-1} (coincident with ρMe) for ClC(=O)SMe and 950 and 906 cm^{-1} for MeSMe. Both rocking vibrations absorb at ≈200 cm^{-1} lower wavenumbers than the corresponding rocks in MeO compounds.

The C—S stretching vibration

The C—S stretching vibration in R—SMe compounds gives rise to a weak to moderate, rarely strong band in the region 725 ± 50 cm^{-1}. The highest values have been observed in the spectra of $MeOCH_2CH_2SMe$ (773), $MeS(CH_2)_nSMe$ (n = 3–5) (≈770) and MeSMe (742) as ν_aCSC. The *gauche* conformer of EtSMe absorbs at 676 cm^{-1}. The remaining observed C—S stretching modes absorb at 720 ± 30 cm^{-1}. The compounds tBuSSMe, MeSSMe, MeHgSMe and $N{\equiv}CCH_2SMe$ are active in the neighbourhood of 693 cm^{-1}. Methylthiomethane shows the ν_sCSC at 694 cm^{-1}.

Skeletal deformation

Vibrational studies suggest that the skeletal C—S— deformation occurs at 270 ± 70 cm^{-1} as a weak band. In the spectra of MeC(=S)SMe this band is found at 338 cm^{-1}, in that of MeSMe at 284, MeSSMe at 274 and Me_2PSMe at 207 cm^{-1}.

Table 11.1 Absorption regions (cm^{-1}) of the normal vibrations of —SMe

Vibration	Saturated Cl bonded	Unsaturated aromatic	Esters	Disulfides	N and P bonded	Si and Hg bonded
ν_aMe	2990 ± 10	3005 ± 15	3005 ± 25	3010 ± 20	3003 ± 10	2990 ± 10
ν'_aMe	2980 ± 20	2985 ± 30	2980 ± 45	2985 ± 10	2995 ± 15	2982 ± 10
ν_sMe	2920 ± 15	2930 ± 15	2925 ± 20	2915 ± 10	2920 ± 15	2920 ± 10
δ_aMe	1440 ± 15	1450 ± 20	1430 ± 10	1432 ± 10	1435 ± 10	1435 ± 10
δ'_aMe	1420 ± 20	1440 ± 20	1425 ± 10	1425 ± 15	1423 ± 10	1430 ± 10
δ_sMe	1320 ± 20	1320 ± 10	1310 ± 10	1310 ± 10	1315 ± 15	1315 ± 10
ρMe	1000 ± 35	995 ± 30	967 ± 10	970 ± 15	977 ± 10	975 ± 20
ρ'Me	940 ± 35	960 ± 10	945 ± 35	950 ± 10	960 ± 20	965 ± 10
νC—S	725 ± 50	715 ± 25	715 ± 20	715 ± 25	720 ± 20	700 ± 10
δC—S—	250 ± 40	290 ± 35	270 ± 70	280 ± 50	240 ± 40	–
torsion Me	195 ± 35	–	150 ± 25	155 ± 20	–	–
torsion SMe	145 ± 35	–	–	100 ± 30	–	–

R—SMe molecules

R = R′CH_2— (see Section 3.3.2).

R = H— [1–5], Me— [6–13], Cl— [53], CH_2=CH— [14, 15], CH_2=C(SMe)— [16], HC≡C— and DC≡C— [17–19], N≡C— [20], Ph— [21–26], Ph-d_5— [23, 24], 4-XPh— (X = H(O=)C, HO(O=)C, MeS [26], Cl and Br [26, 27]), 2-Pym— [28], H(O=)C— [29], Cl(O=)C— [30–32], Me(O=)C— [33], H_2N(O=)C— [34], Me(S=)C— [38], MeS(S=)C— [38], H_2N(S=)C— and D_2N(S=)C— [34–36], MeHN(S=)C— [37], MeS— [39–45], EtS— [44, 46], tBuS— [43, 47], F(O=)CS— [48], O=N— [49, 50], Me_2P— [51], Cl_2P(=O)— [52], MeH_2Si— [54], Me_2HSi— [55], MeHg— [56].

References

1. J. Wagner, *Z. Phys. Chem., Abt. B*, **40B**, 36 (1938).
2. H.W. Thompson and N.P. Skerrett, *Trans. Faraday Soc.*, **36**, 812 (1940).
3. H. Siebert, *Z. Anorg. Allg. Chem.*, **271**, 65 (1952).
4. I.W. May and E.L. Pace, *Spectrochim. Acta, Part A*, **24A**, 1605 (1968).
5. A.J. Barnes, H.E. Hallam and J.D.R. Howells, *J. Chem. Soc Faraday Trans. 2*, **68**, 737 (1972).
6. J.P. McCullough, W.N. Hubbard, F.A. Frow, I.A. Hossenlopp and G. Waddington, *J. Am. Chem. Soc.*, **79**, 561 (1957).
7. J.R. Allkins and P. Hendra, *Spectrochim. Acta*, **22**, 2075 (1966).
8. J.M. Freeman and T. Henshall, *J. Mol. Struct.*, **1**, 31 (1967).
9. S.G. Frankiss, *J. Mol. Struct.*, **3**, 89 (1969).
10. N.L. Owen and R.E. Hester, *Spectrochim. Acta, Part A*, **25A**, 345 (1969).
11. G. Geiseler and G. Hanschmann, *J. Mol. Struct.*, **8**, 293 (1971).
12. M. Tranquille, P. Labarbe, M. Fouassier and M.-T. Forel, *J. Mol. Struct.*, **8**, 273 (1971).
13. I.W. Levin, R.A.R. Pearce and R.C. Spiker, *Spectrochim. Acta, Part A*, **31A**, 41 (1975).
14. J. Fabian, H. Krober and R. Mayer, *Spectrochim. Acta, Part A*, **24A**, 727 (1968).
15. S. Samdal, H.M. Seip and T. Torgrimsen, *J. Mol. Struct.*, **57**, 105 (1979).

16. P. Jandal, H.M. Seip and T. Torgrimsen, *J. Mol. Struct.*, **32**, 369 (1976).
17. H.J. Boonstra and L.C. Rinzema, *Recl. Trav. Chim. Pays-Bas*, **79**, 962 (1960).
18. A.G. Moritz, *Spectrochim. Acta, Part A*, **23A**, 167 (1967).
19. D.H. Christensen and D. den Engelsen, *Spectrochim. Acta, Part A*, **26A**, 1747 (1970).
20. N.S. Ham and J.B. Willis, *Spectrochim. Acta*, **16**, 279 (1960).
21. J.H.S. Green, *Spectrochim. Acta*, **18**, 39 (1962).
22. J.H.S. Green, *Spectrochim. Acta, Part A*, **24A**, 1627 (1968).
23. M. Bouquet, G. Chassaing, J. Corset, J. Favrot and J. Limouzi, *Spectrochim. Acta, Part A*, **37A**, 727 (1981).
24. W.J. Balfour, K.S. Chandrasekhar and S.P. Kyca, *Spectrochim. Acta, Part A*, **42A**, 39 (1986).
25. G. Paliani and S. Santini, *J. Raman Spectrosc.*, **19**, 161 (1988).
26. G. Varasńyi, *Assignments for Vibrational Spectra of Seven Hundred Benzene Derivatives*, J.Wiley & Sons, New York (1974).
27. J.H.S. Green, D.J. Harrison, W. Kynaston and D.W. Scott, *Spectrochim. Acta, Part A*, **26A**, 1515 (1970).
28. G. Mille, M. Guiliano, J. Kister, J. Chouteau and J. Metzger, *Spectrochim. Acta, Part A*, **36A**, 713 (1980).
29. G.I.L. Jones, D.G. Lister, N.L. Owen, M.C.L. Gerry and P. Palmieri, *J. Mol. Spectrosc.*, **60**, 348 (1976).
30. T. Miyazawa and K.S. Pitzer, *J. Chem. Phys.*, **30**, 1076 (1959).
31. J.C. Evans and J. Overend, *Spectrochim. Acta*, **19**, 701 (1963).
32. R.A. Nyquist, *J. Mol. Struct.*, **1**, 1 (1967).
33. A. Smolders, G. Maes and T. Zeegers-Huyskens, *J. Mol. Struct.*,**172**, 23 (1988).
34. L. Zhengyan, R. Mattes, H. Schnöckel, M. Thünemann, E. Hunting, U. Hönke and C. Mendel, *J. Mol. Struct.*, **117**, 117 (1984).
35. K.R.G. Devi, D.N. Sathyanarayana and S. Manogaran, *Spectrochim. Acta, Part A*, **37A**, 31 (1981).
36. R. Mattes, L. Zhengyan, M. Thünemann and H. Schnöckel, *J. Mol. Struct.*, **99**, 119 (1983).
37. K.R.G. Devi, D.N. Sathyanarayana and S. Manogaran, *Spectrochim. Acta, Part A*, **37A**, 633 (1981).
38. K. Herzog, E. Steger, P. Rosmus, S. Scheithauer and R. Mayer, *J. Mol. Struct.*, **3**, 339 (1969).
39. H. Gerding and R. Westrik, *Recl. Trav. Chim. Pays-Bas*, **61**, 412 (1942).
40. I.F. Trotter and H.W. Thompson, *J. Chem. Soc.*, 481 (1946).
41. D.W. Scott, H.L. Finke, M.E. Gross, G.B. Guthrie and H.M. Huffman, *J. Am. Chem. Soc.*, **72**, 2424 (1950).
42. S.G. Frankiss, *J. Mol. Struct.*, **3**, 89 (1969).
43. H. Sugeta, *Spectrochim. Acta, Part A*, **31A**, 1729 (1975).
44. W. Zhao and S. Krimm, *J. Mol. Struct.*, **224**, 7 (1990).
45. M. Meyer, *J. Mol. Struct.*, **273**, 99 (1992).
46. K.G. Allum, J.A. Creighton, J.H.S. Green, G.J. Minkoff and L.J.S. Price, *Spectrochim. Acta, Part A*, **24A**, 927 (1968).
47. H. Sugeta, A. Go and T. Miyazawa, *Bull. Chem. Soc. Jpn.*, **46**, 3407 (1973).
48. C.O. Della Védova, *Spectrochim. Acta, Part A*, **47A**, 1619 (1991).
49. D.H. Christensen, N. Rastrup-Andersen, D. Jones, P. Klaboe and E.R. Lippincott, *Spectrochim. Acta, Part A*, **24A**, 1581 (1968).
50. D.M. Byler and H. Susi, *J. Mol. Struct.*, **77**, 25 (1981).
51. J.R. Durig, D.F. Smith, D.A. Barron, R.J. Harlan and H.V. Phan, *J. Raman Spectrosc.*, **23**, 107 (1992).

52. R.A. Nyquist, *Spectrochim. Acta, Part A*, **27A**, 697 (1971).
53. F. Winther, A. Guarnieri and O.F. Nielsen, *Spectrochim. Acta, Part A*, **31A**, 689 (1975).
54. K. Taga, K. Ohno and H. Murata, *J. Mol. Struct.*,**67**, 199 (1980).
55. K. Taga, *J. Mol. Struct.*, **82**, 1 (1982).
56. R.A. Nyquist and J.R. Mann, *Spectrochim. Acta, Part A*, **28A**, 511 (1972).

11.1.2 Ethylthio

Just like OEt, the SEt fragment displays 21 fundamental vibrations. Eighteen vibrations belong to the ethyl group (Section 3.5.1). To these vibrations are added a C—S stretching vibration, a C—S— deformation and an EtS torsion.

Methyl and methylene stretching vibrations

Usually the five CH stretching vibrations are assigned between 2995 and 2850 cm^{-1} with a moderate intensity. The absorption at 2875 cm^{-1} in the spectrum of EtSCN is assigned to the symmetric stretching vibration [27–29] and also to an overtone [30, 31] (see Section 10.1.2).

Methyl and methylene deformations

The antisymmetric methyl deformations give rise to a medium band between 1480 and 1440 cm^{-1} and the methylene scissors absorbs between 1445 and 1415 cm^{-1}, as for example in the spectrum of ethylthioethane (δ_aMe 1480 and 1460 δ_a'Me, 1455 and 1445 and δCH_2 1441 and 1425 cm^{-1}). The lowest wavenumber for the methylene scissors is found in the spectrum of MeSSEt and EtSSEt: 1418 cm^{-1}. The region of the methyl symmetric deformation (1380 $\pm$ 10 cm^{-1}) agrees with that of the $Me(CH_2)_n$— fragments (Sections 3.5.1 and 3.5.5).

Methylene wagging and twisting vibrations

The methylene wagging vibration exhibits a moderate to strong band in the region 1280 $\pm$ 30 cm^{-1}, usually even at 1275 $\pm$ 10 cm^{-1}, that is $\approx$80 cm^{-1} lower than in OEt compounds. The methylene twist absorbs scarcely 25 cm^{-1} lower than the wag. Ethylthioethane shows the wag at 1282 and 1271 cm^{-1}, EtSCN at 1273 cm^{-1} and MeSEt at 1264 cm^{-1}, and the twist at respectively 1261 and 1246, 1244 and 1249 cm^{-1}.

Methyl rocks and C—C stretching vibration

The methyl rocks are reported between 1100 and 1010 cm^{-1} with a weak to moderate intensity. Ethanethiol absorbs near 1100 and 1048 cm^{-1} and CD_3SEt near 1040 (*trans*) or 1014 cm^{-1} (*gauche*). The C—C stretching vibration is active in the range 975 $\pm$ 25 cm^{-1} with an intensity varying between weak and strong. As contrasted with the OEt compounds, in which the methyl rocks are strongly coupled to the C—C and the C—O stretching vibration, these rocks are not coupled to the νC—S and hardly to the νC—C.

Methylene rocking vibration

The methylene rock in R—SEt compounds is assigned at relatively low wavenumbers (765 ± 35 cm^{-1}) with a weak to moderate intensity. The highest value comes from EtSEt (798 and *762*) and the lowest from HSEt (*tr*: 782; *g*: 735 cm^{-1}). The remaining compounds show this ρCH_2 at 775 ± 25 cm^{-1}. This vibration is somewhat coupled to the νC—S and sensitive to conformation.

The C—S stretching vibration

The C—S stretching vibration occurs at 670 ± 35 cm^{-1} with a weak to moderate intensity, that is 50 cm^{-1} lower than the Me—S stretching vibration, but agrees with the ν_sC—S—C (680 ± 45 cm^{-1}) in R—CH_2SMe compounds (Section 3.3.2) [39]. Ethylthioethane shows both C—S stretching vibrations at 696 and 638 cm^{-1}, EtSSEt at 670 and 640 cm^{-1} and EtSCN at 686 cm^{-1}. This vibration is sensitive to the conformational state of the molecule. Compounds such as iPrSEt and tBuSEt absorb at 695 cm^{-1} in the *trans* conformation and at 670 cm^{-1} in the *gauche* conformation.

Skeletal deformations

The highest deformation is considered as a S—C—C deformation occurring at 350 ± 40 cm^{-1}. In ethylthioethane the two S—C—C deformations absorb at 395 and 336 cm^{-1}. The lowest deformation is the external R—S—C skeletal deformation (235 ± 70 cm^{-1}). High wavenumbers are due to EtSEt (305 and 294) and low wavenumbers to EtSCN (165) and MeSEt (*tr*: 215; *g*: 270 cm^{-1}). In many R—SEt compounds this external skeletal deformation is located at 265 ± 35 cm^{-1}.

Table 11.2 Absorption regions (cm^{-1}) of the normal vibrations of —SEt

Vibration	Region	Vibration	Region
ν_aMe	2980 ± 15	ρMe	1075 ± 30
ν'_aMe	2965 ± 10	ρ'Me	1035 ± 25
$\nu_a CH_2$	2940 ± 20	νC—C	975 ± 25
ν_sMe	2920 ± 25	ρCH_2	765 ± 35
$\nu_s CH_2$	2880 ± 30	νC—S	670 ± 35
δ_aMe	1465 ± 15	δS—C—C	350 ± 40
δ'_aMe	1450 ± 10	δ—S—C	235 ± 70
δCH_2	1430 ± 15	torsion Me	245 ± 35
δ_sMe	1380 ± 10	torsion Et	185 ± 30
ωCH_2	1280 ± 30	torsion SEt	75 ± 30
τCH_2	1250 ± 20		

R—SEt molecules
R = H— [1–10], D— [5], Me— [6–17], CD_3— [11, 12], Et— [7–12, 18–22], nPr— [21], $ClCH_2CH_2$— [22], $PhCH_2$— [23], $MeOCH_2$— [24], iPr— [10, 25, 26], tBu— [10, 26], N≡C— [27–31], Ph— [32, 33], MeS— [34–37], EtS— [10, 35–38].

References

1. N. Sheppard, *J. Chem. Phys.*, **17**, 79 (1949).
2. A.J. Barnes, H.E. Hallam and J.D.R. Howells, *J. Chem. Soc.*,Faraday II **68**, 737 (1972).
3. D. Smith, J.P. Devlin and D.W. Scott, *J. Mol. Spectrosc.*, **25**, 174 (1968).
4. J.R. Durig, W.E. Bucy, C.J. Wurrey and L.A. Carreira, *J. Phys. Chem.*, **79**, 988 (1975).
5. H. Wolff and J. Szydlowsky, *Can. J. Chem.*, **63**, 1708 (1985).
6. D.W. Scott, H.L. Finke, J.P. McCullough, M.E. Gross, K.D. Williamson, G. Waddington and H.M. Huffman, *J. Am. Chem. Soc.*, **73**, 261 (1951).
7. D.W. Scott and M.Z. El-Sabban, *J. Mol. Spectrosc.*, **30**, 317 (1969).
8. R. Fausto, J.J.C. Teixeira-Dias and P.R. Carey, *J. Mol. Struct.*, **159**, 137 (1987).
9. W.O. George, J.H.S. Green and D.J. Harrison, *Spectrochim. Acta, Part A*, **24A**, 367 (1968).
10. D.W. Scott and J.P. McCullough, *J. Am. Chem. Soc.*, **80**, 3554 (1958).
11. M. Ohsaku, Y. Shiro and H. Murata, *Bull. Chem. Soc. Jpn.*, **46**, 1399 (1973).
12. N. Nogami, H. Sugeta and T. Miyazawa, *Bull. Chem. Soc. Jpn.*, **48**, 3573 (1975).
13. M. Hayashi, T. Shimanouchi and S. Mizushima, *J. Chem. Phys.*, **26**, 608 (1957).
14. M. Ohsaku, Y. Shiro and H. Murata, *Bull. Chem. Soc. Jpn.*, **45**, 954 (1972).
15. J.R. Durig, D.A.C. Compton and M.R. Jalilian, *J. Phys. Chem.*, **83**, 511 (1979).
16. M. Sakakibara, H. Matsuura, I. Harada and T. Shimanouchi, *Bull. Chem. Soc. Jpn.*, **50**, 111 (1977).
17. J.R. Durig, M.S. Rollins and H.V. Phan, *J. Mol. Struct.*, **263**, 95 (1991).
18. D.W. Scott, H.L. Finke, W.N. Hubbard, J.P. McCullough, G.D. Olivier, M.E. Gross, C. Katz, K.D. Williamson, G. Waddington and H.M. Huffman, *J. Am. Chem. Soc.*, **74**, 4656 (1952).
19. M. Ohsaku, Y. Shiro and H. Murata, *Bull. Chem. Soc. Jpn.*, **45**, 956 (1972).
20. R.S. Cataliotti, G. Paliani and S. Santini, *Can. J. Phys.*, **64**, 100 (1986).
21. M. Otha, Y. Ogawa, H. Matsuura, I. Harada and T. Shimanouchi, *Bull. Chem. Soc. Jpn.*, **50**, 380 (1977).
22. S.D. Christesen, *J. Raman Spectrosc.*, **22**, 459 (1991).
23. K. Doerffel and B. Adler, *Wiss. Z. Tech. Hochsch. Chem. Leuna-Merseburg*, **10**, 7 (1968).
24. H. Matsuura, H. Murata and M. Sakakibara, *J. Mol. Struct.*, **96**, 267 (1983).
25. M. Ohsaku, H. Murata and Y. Shiro, *Spectrochim. Acta, Part A*, **33A**, 467 (1977).
26. M. Sakakibara, I. Harada, H. Matsuura and T. Shimanouchi, *J. Mol. Struct.*, **49**, 29 (1978).
27. R.P. Hirschmann, R.N. Kniseley and V.A. Fassel, *Spectrochim. Acta*, **20**, 809 (1964).
28. G.A. Crowder, *J. Mol. Struct.*, **7**, 147 (1971).
29. O.H. Ellestad and T. Torgrimsen, *J. Mol. Struct.*, **12**, 79 (1972).
30. J.R. Durig, J.F. Sullivan and H.L. Heusel, *J. Phys. Chem.*, **88**, 374 (1984).
31. G.O. Braathen and A. Gatial, *Spectrochim. Acta, Part A*, **42A**, 615 (1986).
32. J.H.S. Green, *Spectrochim. Acta*, **18**, 39 (1962).
33. G. Varsányi, *Assignments for Vibrational Spectra of Seven Hundred Benzene Derivatives*, J.Wiley & Sons, New York (1974).

34. K.G. Allum, J.A. Creighton, J.H.S. Green, G.J. Minkoff and L.J.S. Prince, *Spectrochim. Acta, Part A*, **24A**, 927 (1968).
35. H. Sugeta, A. Go and T. Miyazawa, *Bull. Chem. Soc. Jpn.*, **46**, 3407 (1973).
36. H. Sugeta, *Spectrochim. Acta, Part A*, **31A**, 1729 (1975).
37. W. Zhao and S. Krimm, *J. Mol. Struct.*, **224**, 7 (1990).
38. D.W. Scott, H.L. Finke, J.P. McCullough, M.E. Gross, R.E. Pennington and G. Waddington, *J. Am. Chem. Soc.*, **74**, 2478 (1952).
39. M. Ohsaku, *Bull. Chem. Soc. Jpn.*, **48**, 707 (1975).

11.2 METHYLSULFINYL

The —S(=O)Me fragment, just like the —C(=O)Me group, accounts for fifteen normal vibrations, of which nine belong to the Me group. The remaining six are described as follows: νS=O, δS=O, γS=O, νC—S, δC—S— and MeS(=O)— torsion. In the C_s configuration these normal vibrations are divided in 9a′ and 6a″ species of vibration:

a′: ν'_aMe, ν_sMe, δ'_aMe, δ_sMe, νS=O, ρ'Me, νC—S, δS=O, δC—S—;
a″: ν_aMe, δ_aMe, ρMe, γS=O and two torsions.

In the spectrum of MeS(=O)Me the 24 normal vibrations are divided among 13a′ and 11a″ types of vibration (Table 11.3).

Table 11.3 Normal vibrations and assignments of MeS(=O)Me in the liquid state [2]

Vibration (a′)	cm^{-1}	Vibration (a″)	cm^{-1}
ν_aMe	2998	νS=O	1058
ν_aMe	2998	ρMe	1022
ν'_aMe	2998	ρMe	954
ν'_aMe	2998	ρ'Me	932
ν_sMe	2914	ρ'Me	897
ν_sMe	2914	ν_aCSC	699
δ_aMe	1438	ν_sCSC	669
δ_aMe	1420	δS=O	382
δ'_aMe	1420	γS=O	332
δ'_aMe	1407	δC—S—C	309
δ_sMe	1310	torsion Me	248
δ_sMe	1295	torsion Me	214

In delimiting the absorption regions of the —S(=O)Me fragment, only a few interpretations of R—S(=O)Me spectra are available, so that the limits are only indicative and far from representative.

Methyl stretching vibrations and deformations

The stretching vibrations and the deformations give rise to weak or moderate absorptions. Often both antisymmetric modes coincide. The methyl symmetric deformation is shifted to lower values (1305 $\pm$ 15 cm^{-1}) as compared with those for saturated hydrocarbons, for example in the spectra of PhS(=O)Me (1295), ClS(=O)Me (1297), 4-MePhS(=O)Me (1300) and MeOS(=O)Me (1302 cm^{-1}).

The S=O stretching vibration

The νS=O absorbs strongly in the region 1095 $\pm$ 50 cm^{-1}. The highest wavenumbers are found in the spectra of ClS(=O)Me (1145) and MeOS(=O)Me (1130 cm^{-1}). The narrower region 1060 $\pm$ 15 cm^{-1} is useful for many simple sulfoxides [7]. The S=O stretching vibration is sensitive to dissolution [9, 11] and more than one band in the above-mentioned region can be attributed to different conformers [7]. The wavenumber of the νS=O is easily derived from the environment. If for instance MeOS(=O)OMe absorbs at 1208 and MeS(=O)Me at 1058 cm^{-1}, a compound such as MeOS(=O)Me is expected to absorb at $\sqrt{1208} \times \sqrt{1058} = 1133$ cm^{-1}. In a similar way the νS=O of ClS(=O)Me (1145 cm^{-1}) is calculated from those of MeS(=O)Me (1058) and ClS(=O)Cl (1240 cm^{-1}).

Methyl rocking vibrations

The methyl rocks give rise to weak or moderate bands between 1025 and 895 cm^{-1}. Both limits come from the spectrum of sulfinylbismethane (Table 11.3). ClS(=O)Me shows these rocks at 948 and 932 cm^{-1}.

The C—S stretching vibration

In the spectra of the tested compounds the νC—S, which in nearly symmetrical compounds is assigned as ν_aCSC and ν_sCSC, occurs at 680 $\pm$ 20 cm^{-1} with a weak to moderate intensity.

The S=O deformations

Kresze *et al.* [9] report that the S=O in-plane deformation of 4-XPhS(=O)Me compounds (X = Me, MeO, Cl, O_2N) occurs at 530 $\pm$ 10 cm^{-1}. As the δS=O in PhS(=O)Me absorbs at 497 cm^{-1} and in MeS(=O)Me at 382 cm^{-1}, the absorption region of the S=O in-plane deformation is 460 $\pm$ 80 cm^{-1}. The S=O out-of-plane deformation is located in the region 355 $\pm$ 25 cm^{-1}, with 360 $\pm$ 10 cm^{-1} for 4-XPhS(=O)Me compounds [9].

Skeletal C—S— deformation

The C—S— skeletal deformation exhibits a very weak band at 300 $\pm$ 20 cm^{-1} in infrared spectra, sometimes visible only in the Raman spectra, such as in those of PhS(=O)Me (320R), 4-MePhS(=O)Me (282R) and 4-MeOPhS(=O)Me (285R).

Table 11.4 Absorption regions (cm^{-1}) of the normal vibrations of —S(=O)Me

Vibration	Region	Vibration	Region
ν_aMe	3005 ± 20	ρ'Me	930 ± 35
ν'_aMe	2995 ± 15	νC—S	680 ± 20
ν_sMe	2925 ± 25	δS=O	460 ± 80
δ_aMe	1425 ± 15	γS=O	355 ± 25
δ'_aMe	1415 ± 15	δC—S—	300 ± 20
δ_sMe	1305 ± 15	torsion	190 ± 80
νS=O	1095 ± 50	torsion	100 ± 60
ρMe	985 ± 40		

R—S(=O)Me molecules

R = Me— [1–6], Et—, iPr— and tBu— [7], $MeSCH_2$—, Ph— [8, 9], Ph-d_5— [8], 4-MePh— [7, 9], 4-XPh— (X = MeO, Cl, O_2N) [9], MeO— [10], Cl— [10].

References

1. W.D. Horrocks Jr. and F.A. Cotton, *Spectrochim. Acta*, **17**, 134 (1961).
2. M.-T. Forel and M. Tranquille, *Spectrochim. Acta, Part A*, **26A**, 1023 (1970).
3. G. Geiseler and G. Hanschmann, *J. Mol. Struct.*, **8**, 293 (1971).
4. G. Geiseler and G. Hanschmann, *J. Mol. Struct.*, **11**, 283 (1972).
5. M. Tranquille, P. Labarbe, M. Fouassier and M.-T. Forel, *J. Mol. Struct.*, **8**, 273 (1971).
6. S. Bianco, R.S. Cataliotti, S.Chieli, F. Guerrini, C. Gaburri, G. Paliani, A. Peraio, M. Scamosci and C. Taratza, *Spectrochim. Acta, Part A*, **42A**, 855 (1986).
7. M. Oki, I. Oka and K. Sakaguchi, *Bull. Chem. Soc. Jpn.*, **42**, 2944 (1969).
8. M. Bouquet, G. Chassaing, J. Corset, J. Favrot and J. Limouzi, *Spectrochim. Acta, Part A*, **37A**, 727 (1981).
9. G. Kresze, E. Ropte and B. Schrader, *Spectrochim. Acta*, **21**, 1633 (1965).
10. G.E. Binder and A. Schmidt, *Spectrochim. Acta, Part A*, **33A**, 815 (1977).
11. T. Cairns, G. Eglinton and D.T. Gibson, *Spectrochim. Acta*, **20**, 31 (1964).

11.3 SULFONYL COMPOUNDS

11.3.1 Methylsulfonyl

The —S(=O)$_2$Me fragment possesses $3N-6 = 18$ normal vibrations of which nine are methyl vibrations and the remaining ones belong to the —S(=O)$_2$C skeleton. In the C_s configuration the distribution is as follows:

a′: ν_a'Me, ν_sMe, δ_a'Me, δ_sMe, $\nu_s SO_2$, ρ'Me, νC—S, δSO_2, ωSO_2, δC—S—;
a″: ν_aMe, δ_aMe, $\nu_a SO_2$, ρMe, τSO_2, ρSO_2 and two torsions.

Table 11.5 Normal vibrations and assignments of MeS(=O)$_2$Me

Vibration	cm^{-1}	Vibration	cm^{-1}
ν_aMe	3027	ρMe	1011
ν_aMe	3024	ρMe	987
ν_a'Me	3015	ρ'Me	947
ν_a'Me	3015	ρ'Me	934
ν_sMe	2935	ν_aC—S—C	763
ν_sMe	2935	ν_sC—S—C	700
δ_aMe	1439	δSO_2	501
δ_aMe	1427	ωSO_2	457
δ_a'Me	1427	τSO_2	383
δ_a'Me	1408	ρSO_2	326
δ_sMe	1335	δC—S—C	300
δ_sMe	1314	torsion Me	271
$\nu_a SO_2$	1298	torsion Me	–
$\nu_s SO_2$	1135		

Methyl stretching vibrations and deformations

The methyl stretching vibrations are observed between 3050 and 2920 cm^{-1} and the deformations between 1460 and 1300 cm^{-1}. Both antisymmetric modes often coincide. The methyl symmetric deformation, in most cases occurring at 1325 $\pm$ 15 cm^{-1}, is often obscured by the strong band of the SO_2 stretching vibration.

The SO_2 stretching vibrations

The most characteristic absorptions of the —SO_2Me group are the SO_2 stretching vibrations. With the exception of $MeSO_2ONa$ (1247 cm^{-1}), the antisymmetric mode appears strongly in the region 1330 $\pm$ 60 cm^{-1}. The highest value (1390 cm^{-1}) is furnished by $MeSO_2F$ and $MeSO_2Br$, followed by 1366 cm^{-1} from $MeSO_2Cl$. Low values come from 4-$H_2NNHPhSO_2Me$ (1270), $EtSO_2Me$ (1274) and $PhSO_2Me$ (1285 cm^{-1}). The remaining molecules show this strong band at 1330 $\pm$ 30 cm^{-1}. Sulfur dioxide absorbs near 1361 cm^{-1}.

The symmetric counterpart absorbs also strongly in the region 1180 $\pm$ 45 cm^{-1}, with high wavenumbers in the spectra of $MeSO_2F$ (1223), $MeSO_2ONa$ (1190) and HC≡CCH(Me)SO_2Me (1181 cm^{-1}) and low wavenumbers in those of $EtSO_2Me$ (1132), $MeSSO_2Me$ (1134), H_2C=CHSO_2Me (1135) and $MeSO_2Me$ (1135 cm^{-1}).

Most of the R—S(=O)$_2$Me compounds give the $\nu_s SO_2$ in the narrow region 1160 $\pm$ 20 cm^{-1}. Sulfur dioxide stretches symmetrically at 1151 cm^{-1}.

Methyl rocking vibrations

For 24 aromatic methylsulfones Momose *et al.* [25] assign the methyl rocks in the regions 965 $\pm$ 15 and 955 $\pm$ 10 cm^{-1} and the C—S stretching vibration at 770 $\pm$ 20 cm^{-1}. Merian [24] supposes that the region 970 $\pm$ 5 cm^{-1} may be reserved for a methyl rocking vibration. Often both rocks are assigned at the same wavenumber. They are active with a weak to moderate intensity in the regions 985 $\pm$ 35 and 940 $\pm$ 40 cm^{-1}. The former region narrows to 970 $\pm$ 20 cm^{-1} if the values of 1011 cm^{-1} in the spectrum of sulfonylbismethane and 1017 cm^{-1} in that of HC≡CH(Me)SO$_2$Me are disregarded.

The C—S stretching vibration

The C—S stretching vibration appears moderately to strongly in the region 745 $\pm$ 45 cm^{-1}. The HW side is limited by CD$_3$NHSO$_2$Me and CD$_3$NDSO$_2$Me with 790 cm^{-1} and by MeSO$_2$ONa and H$_2$C=CHSO$_2$Me with 788 cm^{-1}. The lowest values are observed in the spectra of MeSO$_2$Me (ν_sCSC 700), MeOSO$_2$Me (721), nPrOSO$_2$Me (724), ClCH$_2$CH$_2$OSO$_2$Me (729), FSO$_2$Me (730), EtOSO$_2$Me (733) and BrSO$_2$Me (733 cm^{-1}). The majority of the investigated molecules were found to give this stretching mode at 760 $\pm$ 25 cm^{-1}.

The SO$_2$ deformations

The SO$_2$ scissors can readily be detected in the region 535 $\pm$ 40 cm^{-1}, mainly by the moderate to strong intensity. Sulfur dioxide ‘snips’ at 519 cm^{-1}.

The SO$_2$ wagging vibration appears weakly to moderately at 485 $\pm$ 50 cm^{-1} but the vibration in this region is sometimes reported as a rock [10].

If for all investigated molecules it is accepted that the rock absorbs at a lower wavenumber than the twist, the region 405 $\pm$ 65 cm^{-1} is considered for the twisting vibration and the region 320 $\pm$ 40 cm^{-1} for the rocking vibration. In the spectrum of ClCH$_2$SO$_2$Me the absorptions at 507, 452, 374 and 309 cm^{-1} are assigned respectively to the SO$_2$ deformation, wag, twist and rock [12].

Skeletal C—S— deformation

The skeletal C—S— deformation can be found as a weak to moderate absorption in the region 295 $\pm$ 40 cm^{-1}. The C—S—N bend in MeNHSO$_2$Me is assigned at 331 cm^{-1} [16] and the C—S—Cl bend in ClSO$_2$Me at 256 cm^{-1} [23].

Table 11.6 Absorption regions (cm^{-1}) of the normal vibrations of $—SO_2Me$

Vibration	Region	Vibration	Region
$\nu_a Me$	3030 ± 20	$\rho' Me$	940 ± 40
$\nu'_a Me$	3025 ± 25	$\nu C—S$	745 ± 45
$\nu_s Me$	2940 ± 20	δSO_2	535 ± 40
$\delta_a Me$	1430 ± 30	ωSO_2	485 ± 50
$\delta'_a Me$	1415 ± 15	τSO_2	405 ± 65
$\delta_s Me$	1325 ± 25	ρSO_2	320 ± 40
$\nu_a SO_2$	1330 ± 60	$\delta C—S—$	295 ± 40
$\nu_s SO_2$	1180 ± 45	torsion	–
ρMe	985 ± 35	torsion	–

$R—S(=O)_2Me$ molecules

R = Me— [1–10], CD_3— [9], Et— [11, 70], $ClCH_2$— and $BrCH_2$— [12], HC≡CCH(Me)—, H_2C=CH— [2], Ph— [13, 69, 70], Ph-d_5— [13], 4-XPh— (X = Me, MeO, Cl and O_2N [14], HO(O=)C, H_2NNH), H_2N— and D_2N— [15], MeHN— [16, 17], MeHN-d_1—, -d_3— and -d_4— [16], iPrHN—, tBuHN—, Me_2N—, Et_2N— and iPr_2N— [17], PhNH-d_0, -d_1—, -d_5— and -d_6— [18], 3-XPhNH— (X = Me, Cl, O_2N) [19], 4-XPhNH— (X = Me, MeC(=O), N≡C, O_2N, MeO, Cl and $MeSO_2$) [19], MeO— and EtO— [20], nPrO—, $ClCH_2CH_2O$—, NaO— [50], MeS—, F— [10, 21, 22], Cl— [10, 21, 22, 23], Br— [21].

11.3.2 Fluorosulfonyl

In the C_s symmetry the nine normal vibrations of the $—SO_2F$ fragment are divided into 5a′ + 4a″ types of vibration:

a′: $\nu_s SO_2$, $\nu S—F$, δSO_2, ωSO_2, $\delta F—S—$;
a″: $\nu_a SO_2$, ρSO_2, τSO_2 and torsion.

The literature does not always agree in assigning $\delta SO_2 > \omega SO_2 > \tau SO_2 > \rho SO_2$ with decreasing wavenumbers.

The SO_2 stretching vibrations

The SO_2 antisymmetric stretching vibration in $R—S(=O)_2F$ compounds absorbs strongly in the region 1445 ± 60 cm^{-1} with high values for R = F (1502), FO (1501), F_2NO (1492) and ClO (1481 cm^{-1}) and a low value (1385 cm^{-1}) for R = 3-H_2NPh and 4-H_2NPh. The remaining molecules restrict themselves to 1440 ± 40 cm^{-1}, with 1415 ± 10 cm^{-1} for the aromatic sulfonyl fluorides.

The SO_2 symmetric stretching vibration appears strongly in the region 1230 ± 40 cm^{-1}. If the highest values for R = F (1269), F_2NO (1254), FO (1248), ClO

(1248) and HO (1243 cm^{-1}) are not taken into account the region narrows to 1215 $\pm$ 25 cm^{-1}.

The F—S stretching vibration

The νS—F possesses a moderate intensity and occurs in the region 825 $\pm$ 75 cm^{-1}, of which the upper limit is taken by $HOSO_2F$ (896), FSO_2F (887), $ClCH_2SO_2F$ (882) and $BrCH_2SO_2F$ (853 cm^{-1}).

The SO_2 deformations

The SO_2 scissors in R—SO_2F compounds is located in the region 555 $\pm$ 70 cm^{-1}, with high values for R = F (625), 4-$ClSO_2Ph$ (620), 4-Cl(O=)CPh (617) and 4-HO(O=)CPh (608 cm^{-1}) and low values for R = Br (489), Cl (505), Me (528) and Me_2N (531 cm^{-1}), so that the remaining compounds absorb at 570 $\pm$ 35 cm^{-1}.

The SO_2 wagging vibration makes its appearance, separated from the scissors, in the region 510 $\pm$ 60 cm^{-1}, except for F_2NOSO_2F, in which the scissors vibration probably coincides with the wag. The spectrum of 3-$OCNPhSO_2F$ gives the highest value (567 cm^{-1}). The lowest wavenumbers are found in the spectra of $BrSO_2F$ (458), Me_2NSO_2F (461), $ClSO_2F$ (480) and $ClCH_2SO_2F$ (486 cm^{-1}). The remaining observed wags fit into the region 525 $\pm$ 35 cm^{-1}.

With the exception of the low wavenumber (310 cm^{-1}) in the spectrum of $BrSO_2F$, the SO_2 twisting vibration is located in the region 470 $\pm$ 70 cm^{-1}. For R = Me_2N, OH or Cl the twist is assigned at respectively 406, 409 or 430 cm^{-1} and for R = F the twist appears at 539 cm^{-1}. Most of the sulfonyl fluorides give this twist at 480 $\pm$ 35 cm^{-1}.

The SO_2 rocking vibration has been observed in the region 375 $\pm$ 85 cm^{-1} with low values for R = Br (290), Cl (300) and $BrCH_2$ (304 cm^{-1}). The remaining R—SO_2F compounds absorb at 390 $\pm$ 70 cm^{-1}.

Skeletal F—S— deformation

Only a few values for this deformation are known, so that the δF—S— probably absorbs near 300 $\pm$ 30 cm^{-1}.

Table 11.7 Absorption regions (cm^{-1}) of the normal vibrations of —SO_2F

Vibration	Region	Vibration	Region
$\nu_a SO_2$	1445 $\pm$ 60	τSO_2	470 $\pm$ 70
$\nu_s SO_2$	1230 $\pm$ 40	ρSO_2	375 $\pm$ 85
νF—S	825 $\pm$ 75	δF—S—	300 $\pm$ 30
δSO_2	555 $\pm$ 70	torsion	–
ωSO_2	510 $\pm$ 60		

R—S(=O)$_2$F molecules
R = Me— [10, 21, 22], ClCH$_2$— and BrCH$_2$— [26], PhCH$_2$—, 2-XPh— (X = O$_2$N, ClSO$_2$), 3-XPh— (X = Me(O=)C, HO(O=)C, Cl[O=)C, H$_2$N, O=C=N, ClSO$_2$), 4-XPh— (X = Me, HO(O=)C, Cl(O=)C, H$_2$N, ClSO$_2$), Me$_2$N— [27], HO— [29–32], MeO— [31], F$_2$NO— [28, 31], FO— [31, 37], ClO— [31], F— [32–36], Cl— [32, 35, 38], Br— [39, 40].

11.3.3 Chlorosulfonyl

The SO$_2$ stretching vibrations

The $\nu_a SO_2$ gives rise to a strong band in the region 1385 ± 35 cm^{-1}. The R′OSO$_2$Cl compounds (R′ = H, alkyl, aryl) take the upper limit with 1410 ± 10 cm^{-1}. Benzenesulfonyl chlorides absorb at 1390 ± 15 cm^{-1} [43, 44].

With the exception of F$_3$CSO$_2$Cl (1237 cm^{-1}) the $\nu_s SO_2$ appears strongly in the region 1180 ± 30 cm^{-1}, usually even at 1175 ± 25 cm^{-1} except for HOSO$_2$Cl (1209), (CD$_3$)$_2$NSO$_2$Cl (1207) and 4-MePhOSO$_2$Cl (1205 cm^{-1}). For benzenesulfonyl chlorides the region narrows to 1185 ± 15 cm^{-1} [43, 44].

The SO$_2$ deformations

The region 570 ± 60 cm^{-1} is attributed to the SO$_2$ scissors but also to the Cl—S stretching vibration, which is strongly coupled to this SO$_2$ in-plane deformation. The upper limit is taken by FSO$_2$Cl with 625 cm^{-1}; ClSO$_2$CH$_2$SO$_2$Cl shows both deformations at 610 and 528 cm^{-1} and ClSO$_2$CD$_2$SO$_2$Cl at 609 and 510 cm^{-1}. The remaining molecules display this SO$_2$ scissors at 560 ± 30 cm^{-1}.

The SO$_2$ wagging vibration is active in the region 510 ± 40 cm^{-1}, formed by methanedisulfonylchloride (546 and 494) and methanedisulfonylchloride-d$_2$ (533 and 473 cm^{-1}).

The SO$_2$ twisting vibration absorbs in the region 410 ± 80 cm^{-1}, with 490 cm^{-1} for MeSO$_2$Cl and 357 and 333 cm^{-1} for ClSO$_2$CD$_2$SO$_2$Cl.

The SO$_2$ rocking vibration occurs at 300 ± 30 cm^{-1}, with 290 cm^{-1} for methanesulfonylchloride.

The Cl—S stretching vibration and deformation

In R—SO$_2$Cl compounds the Cl—S stretching vibration is coupled to the SO$_2$ deformation. Birchall and Gillespie [32] assign the absorption near 625 cm^{-1} in the spectrum of FSO$_2$Cl to the νCl—S and that at 427 cm^{-1} to the δSO_2 with a contribution of νCl—S. According to Pfeiffer [35] or Craig and Futamura [38], the band at 427 cm^{-1} is due to the νCl—S and that at 625 cm^{-1} to the δSO_2 and the νCl—S. Hanai *et al.* [23] assign the band at 373 cm^{-1} in the spectrum of MeSO$_2$Cl to the Cl—S stretch and that at 533 cm^{-1} to the SO$_2$ deformation as well as to the Cl—S stretching vibration. The region 395 ± 35 cm^{-1} may be reserved for the Cl—S stretch, although this vibration seems to be active also in the region of the δSO_2 (570 ± 60 cm^{-1}).

The vibrational analysis in the literature reveals only a few wavenumbers for the Cl—S— skeletal deformation.

Table 11.8 Absorption regions (cm^{-1}) of the normal vibrations of $—SO_2Cl$

Vibration	Region	Vibration	Region
$\nu_a SO_2$	1385 ± 35	νCl—S	395 ± 35
$\nu_s SO_2$	1180 ± 30	ρSO_2	300 ± 30
δSO_2	570 ± 60	δCl—S—	235 ± 45
ωSO_2	510 ± 40	torsion	
τSO_2	410 ± 80		

$R—S(=O)_2Cl$ molecules

R = Me— [10, 21–23], CD_3— [23], Et— [48], iPr—, $ClCH_2$— and $BrCH_2$— [26], nBu—, $Cl(CH_2)_3$—, $ClSO_2CH_2$— and $ClSO_2CD_2$— [41], $PhCH_2$— [44], 4-O_2NPhCH_2—, F_3C—, Ph— [42–44], 2-, 3- and 4-XPh— (X = Me and O_2N [44], FSO_2), 4-XPh— (X = F, Cl, Br, MeO and Me_2N [44]), 2,5-Cl_2Ph—, 2,4-$(O_2N)_2Ph$— [44], F_5Ph—, 2-Th—, Me_2N— [27, 45], $(CD_3)_2N$— [45], O=C=N—, HO— [20, 29, 30], MeO— [20, 46, 47], EtO—, nPrO—, nBuO— and PhO— [47, 48], $ClCH_2O$—, 4-MePhO— and 4-ClPhO— [47], F— [32, 35, 38], Cl— [35, 49].

11.3.4 R′—oxysulfonyl

The spectra of esters of sulfonic acids are dominated by three absorptions with a moderate to strong intensity, due to both SO_2 stretching vibrations and a S—O stretch.

The SO_2 stretching vibrations

Except for $MeSO_2ONa$, which absorbs at 1247 cm^{-1}, the SO_2 antisymmetric stretching vibration lies within the large region 1415 ± 85 cm^{-1}. The highest values of 1500, 1492, 1481, 1480 and 1465 cm^{-1} are attributed in the spectra of derivatives of fluorosulfuric acid, respectively FSO_2OF, FSO_2ONF_2, FSO_2OCl, FSO_2OH and FSO_2OMe (see Section 11.3.2). The lowest wavenumbers come from $EtSSO_2OMe$ (1331), $EtSSO_2OEt$ (1332), $MeSSO_2OEt$ (1337) and $MeSSO_2OMe$ (1338 cm^{-1}). Usually the esters of organic sulfonic acids absorb at 1385 ± 35 cm^{-1}.

The SO_2 symmetric stretching vibration is observed in the region 1195 ± 60 cm^{-1}. Once more the highest values come from FSO_2ONF_2 (1254), FSO_2OF

(1248), FSO_2OCl (1248) and FSO_2OH (1243 cm^{-1}) and the lowest are from $EtSSO_2OMe$ (1135), $EtSSO_2OEt$ (1137), $MeSSO_2OEt$ (1138) and $MeSSO_2OMe$ (1141 cm^{-1}). Most of the esters of sulfonic acids show this ν_sSO_2 at 1200 ± 35 cm^{-1}.

The O—S stretching vibration

With the exception of FSO_2OH (955) and $ClSO_2OH$ (916 cm^{-1}), the O—S stretching vibration is assigned at 795 ± 35 cm^{-1}. As examples, $MeSO_2OMe$ absorbs at 817 cm^{-1} and $MeOSO_2OMe$ gives rise to a ν_aOSO at 828 cm^{-1} and a ν_sOSO at 760 cm^{-1}.

The SO_2 deformations

Usually the SO_2 deformations absorb with a weak to moderate intensity. The SO_2 scissors (565 ± 45 cm^{-1}) absorbs near 610 cm^{-1} in the spectra of F_3CSO_2OR' compounds and near 552 cm^{-1} in those of $MeSO_2OR'$ compounds (R′ = Me, Et). A second SO_2 deformation, often assigned as a wagging vibration, is found in a neighbouring region (535 ± 35 cm^{-1}) but clearly separated from the scissors. The literature does not agree in assigning the absorptions in the regions 425 ± 85 and 345 ± 55 cm^{-1}, which in this work are attributed to the SO_2 twisting and the SO_2 rocking vibration.

Table 11.9 Absorption regions (cm^{-1}) of the normal vibrations of $—SO_2OR'$

Vibration	Region	Vibration	Region
ν_aSO_2	1415 ± 85	τSO_2	425 ± 85
ν_sSO_2	1195 ± 60	ρSO_2	345 ± 55
$\nu O—S$	795 ± 35	$\delta O—S—$	≈250
δSO_2	565 ± 45	torsion	–
ωSO_2	535 ± 35		

$R—S(=O)_2O—R'$ molecules

R	R′
Me—	Me— [20], Et— [20], nPr—, $ClCH_2CH_2$—, HC≡CCH(Me)—, Na— [50],
Et—	Me— [20], Et— [20],
F_3C—	H— [51], Me—, Et—,
4-MePh—	Me— [65], Et—, CF_3CH_2—, $ClCH_2CH_2$—,
4-O_2NPh—	Me—
MeO—	Me— [52, 53],
MeS—	Me— and Et— [54],
EtS—	Me— and Et— [54],

F— H— [29–32], Me— [31], F_2N— [28, 31], F— [31, 37], Cl— [31],
Cl— H— [20, 29, 30], Me— [20, 46, 47], Et—, nPr—, nBu— and Ph— [47, 48], $ClCH_2$—, 4-MePh— and 4-ClPh— [47].

11.3.5 Aminosulfonyl (sulfonamido)

The —S(=O)$_2$NH$_2$ fragment in primary sulfonamides gives 15 normal vibrations. Six vibrations are inherent to the NH_2 group (see Section 9.1): $\nu_a NH_2$, $\nu_s NH_2$, δNH_2, ρNH_2, ωNH_2 and torsion. The remaining nine vibrations belong to the —SO_2—N skeleton: $\nu_a SO_2$, $\nu_s SO_2$, δSO_2, ωSO_2, τSO_2, ρSO_2, νN—S, δN—S— and torsion.

The SO_2 stretching vibrations

The SO_2 antisymmetric stretching vibration in primary sulfonamides is strongly active in the region 1335 ± 25 cm^{-1}, extreme values being 1358 cm^{-1} for $MeHNSO_2NH_2$ and 1310 and 1315 cm^{-1} for $CD_3SO_2NH_2$ and $MeSO_2NH_2$ respectively. Benzenesulfonamides absorb at 1325 ± 15 cm^{-1} [59]. The SO_2 symmetric stretching vibration is observed in the region 1150 ± 15 cm^{-1}.

The N—S stretching vibration

The N—S stretching vibration exhibits a moderate band in the region 905 ± 30 cm^{-1}. Methanesulfonamide absorbs at 881 cm^{-1} and $H_2NSO_2NH_2$ gives rise to a ν_aN—S—N at 931 and a ν_sN—S—N at 904 cm^{-1}.

The SO_2 deformations

Although the region of the SO_2 scissors (570 ± 60 cm^{-1}) and that of the SO_2 wagging vibration (520 ± 40 cm^{-1}) partly overlap, the two vibrations are observed separately.

Table 11.10 Absorption regions (cm^{-1}) of the normal vibrations of —SO_2NH_2

Vibration	Region	Vibration	Region
$\nu_a NH_2$	3355 ± 35	δSO_2	570 ± 60
$\nu_s NH_2$	3250 ± 20	ωSO_2	520 ± 40
δNH_2	1565 ± 15	τSO_2	445 ± 45
$\nu_a SO_2$	1335 ± 25	τNH_2	355 ± 65
ρNH_2	1160 ± 30	ρSO_2	–
$\nu_s SO_2$	1150 ± 15	δ—S—	–
νN—S	905 ± 30	torsion	–
ωNH_2	690 ± 40		

R—S(=O)$_2$NH$_2$ compounds
R = Me— and CD$_3$— [15], Ph— [43, 58, 65], Ph-d$_5$— [43], 2-XPh— (X = Me, Me(O=)C, H$_2$N, O$_2$N), 4-XPh— (X = Me [65], HO(O=)C, H$_2$N [65], O$_2$N, Cl), 2-Th— [60, 62], H$_2$N— [55, 56], MeHN—, Me$_2$N— and (CD$_3$)$_2$N— [57].

11.3.6 R′—aminosulfonyl

The NH vibrations

The NH stretching vibration in secondary sulfonamides is observed in the region 3270 ± 65 cm^{-1}, mostly at 3265 ± 45 cm^{-1}. Laurence *et al.* [61] found 3265 ± 10 cm^{-1} for thirty *N*-(3- and 4-X-phenyl)substituted methanesulfonamides for the dimer and 3390 ± 10 cm^{-1} for the monomer in dilute CCl$_4$ solution. Arcoria *et al.* [60] assigned the νNH at 3240 ± 35 cm^{-1} in the spectra of thirty *N*-substituted 2-thiophenesulfonamides.

The NH in-plane deformation exhibits a moderate absorption band in the region 1395 ± 25 cm^{-1}, usually at 1395 ± 15 cm^{-1}, a region also proposed by Kalová *et al.* [19] for a series of thirty secondary sulfonamides.

The γNH or NH wagging vibration is a broad absorption with low intensity in the region 650 ± 50 cm^{-1}, and is sometimes difficult to detect among the intense aromatic CH out-of-plane deformations.

The SO$_2$ vibrations

The SO$_2$ antisymmetric stretching vibration appears in the region 1330 ± 30 cm^{-1} and the symmetric counterpart in the region 1160 ± 30 cm^{-1}, both with high intensity. In the spectra of a series of thirty 3- and 4-X-PhNHSO$_2$Me compounds the ν_aSO$_2$ is assigned at 1335 ± 10 cm^{-1} and the ν_sSO$_2$ at 1160 ± 10 cm^{-1} [61], and in 2-ThSO$_2$NHR′ compounds at 1340 ± 15 and 1145 ± 15 cm^{-1} [60]. Kalová *et al.* [19] found 1340 ± 20 and 1160 ± 20 cm^{-1} for the series of secondary sulfonamides.

The SO$_2$ scissors is assigned in the region 560 ± 40 cm^{-1} and the SO$_2$ wagging vibration near 500 ± 55 cm^{-1}, both with medium intensity and clearly separated. A compound such as PhSO$_2$NHPh absorbs at the HW side with 580 and 554 cm^{-1} for these deformations and MeSO$_2$NHCD$_3$ at the LW side with 522 and 446 cm^{-1}, followed by MeSO$_2$NHMe with 523 and 457 cm^{-1}.

The SO$_2$ twisting vibration is assigned in the area 440 ± 40 cm^{-1} and the SO$_2$ rock in the neighbourhood of 350 cm^{-1} on the basis of the scarce data available.

The N—S vibrations

The N—S stretching vibration provides a weak to moderate band in the range 905 ± 70 cm^{-1}. In the spectra of the *N*-substituted 2-thiophenesulfonamides, Arcoria *et al.* [60] assigned the νN—S in the region 900 ± 65 cm^{-1}, which narrows to 925 ± 40 cm^{-1} without the low value of 835 cm^{-1} from the spectrum of 2-ThSO_2NHMe, a wavenumber also applicable for the same vibration in MeSO_2NHMe. The majority of the investigated secondary sulfonamides were found to give this stretching vibration at 910 ± 35 cm^{-1}.

The C—N—S deformation absorbs near 280 cm^{-1}.

Table 11.11 Absorption regions (cm^{-1}) of the normal vibrations of R—SO_2NH—R′

Vibration	Region	Vibration	Region
νNH	3270 ± 65	δSO_2	560 ± 40
δNH	1395 ± 25	ωSO_2	500 ± 55
$\nu_a SO_2$	1330 ± 30	τSO_2	440 ± 40
$\nu_s SO_2$	1160 ± 30	ρSO_2	≈350
νN—S	905 ± 70	δC—N—S	≈280
γNH	650 ± 50	torsion	–

R—S(=O)$_2$ NH—R′ molecules

R	R′
Me—	Me— [16, 17], CD_3— [16], iPr— and tBu—, [17], Ph-d_0— and -d_5— [18], H_2N—, 4-XPh— (X = Me, MeC(=O), N≡C, O_2N, MeO, MeS(=O)$_2$, Cl) [19], 3-XPh— (X = Me, O_2N, Cl) [19]
4-O_2NPhCH_2—	Me—
F_3C—	2,6-Et_2Ph—
Ph-d_0— and -d_5—	Me— [17], Ph— [43]
4-MePh—	Me— [17, 19, 65]
2-O_2NPh—	Ph— and 3-XPh— (X = Me, O_2N, MeO, Cl) and 4-XPh— (X = Me, MeC(=O), N≡C, O_2N, MeO, Cl) [19]
4-O_2NPh—	Ph— and 3-XPh— (X = O_2N, Cl) and 4-XPh— (X = O_2N, Meo, Cl) [19]
2-H_2N—4-ClPh—	N≡CCH_2CH_2—

11.3.7 Sulfonyl

The R—SO_2—R′ compounds in which R and R′ represent a saturated or unsaturated carbon atom, generally named sulfones, have been considered. The methylsulfones with the formula MeSO_2—R′ belong also to the methylsulfonyl group (see Section 11.3.1).

The SO_2 stretching vibrations

Except for $(CCl_3)_2S(=O)_2$ (1382) and $(CBr_3)_2S(=O)_2$ (1374 cm^{-1}) the SO_2 antisymmetric stretching vibration appears strongly in the region 1315 ± 45 cm^{-1}. The highest wavenumbers are those due to 4-$ClPhSO_2CF_2CH(F)Cl$ (1360) and 3-O_2N-4-$FPhSO_2Ph$-3-NO_2-4-F (1354 cm^{-1}) and the lowest have been traced in the spectra of $MeSO_2PhNHNH_2$ (1270) and $MeSO_2Et$ (1274 for ν_aSO_2 and 1302 for ωCH_2 [11], but better the reverse ?). The remaining sulfones display the ν_aSO_2 at 1305 ± 25 cm^{-1}, the same region as that found by Flett [59] for 15 sulfones. Schreiber [58] has examined 16 sulfones and located the ν_aSO_2 in the range 1325 ± 25 cm^{-1}.

The SO_2 symmetric stretching vibration is observed at 1150 ± 30 cm^{-1}, a region in good agreement with that proposed by Schreiber (1140 ± 20 cm^{-1}) or by Flett (1155 ± 10 cm^{-1}). For a number of RSO_2R' compounds in CCl_4 solution, Butcher *et al.* [48] found a correlation between the two vibrations: $\nu_s = 0.679\ \nu_a + 244$.

The SO_2 deformations

With the exception of the extreme values 625 and 606 cm^{-1} for respectively 5-O_2N-2-$PySO_2$-2-Py-5-NO_2 and 2-$PySO_2$-2-Py-5-NO_2 [66], the SO_2 scissors gives rise to a moderate band in the region 540 ± 60 cm^{-1}. High values originate also from the spectra of 4-$ClPhSO_2CF_2CH(F)Cl$ (598), 2-$PySO_2$-2-Py (594) and 2-$PySO_2Ph$ (592 cm^{-1}). The lowest wavenumbers are assigned in the spectra of $CD_3SO_2CD_3$ (488) and $MeSO_2Me$ (500 cm^{-1}). Most of the sulfones have their absorptions in the region 550 ± 40 cm^{-1}.

The SO_2 wagging vibration is active in the region 500 ± 70 cm^{-1} with high values for 5-O_2N-2-$PySO_2$-2-Py-5-NO_2 (570), 2-$PySO_2$-2-Py-5-NO_2 (569), 2-$PySO_2$-2-Py (568), 2-$PySO_2Ph$ (566), $PhSO_2Ph$ (563) and 4-$ClPhSO_2CF_2CH(F)Cl$ (561 cm^{-1}) and low wavenumbers for $CD_3SO_2CD_3$ (426) and $nHexSO_2nHex$ (441 cm^{-1}). Disregarding these extreme values, most of the sulfones absorb at 505 ± 45 cm^{-1}.

Ignoring without rejecting the low values of 312 cm^{-1} in the spectrum of $PhSO_2Ph$ and 307 cm^{-1} in that of 2-$PySO_2$-2-Py-5-NO_2, the SO_2 twisting vibration is reported in the range 420 ± 75 cm^{-1}. The twenty available values for the SO_2 rock are located in the region 345 ± 55 cm^{-1}.

Table 11.12 Absorption regions (cm^{-1}) of the normal vibrations of R—SO_2—R′

Vibration	Region	Vibration	Region
ν_aSO_2	1315 ± 45	ωSO_2	500 ± 70
ν_sSO_2	1150 ± 30	τSO_2	420 ± 75
δSO_2	540 ± 60	ρSO_2	345 ± 55

The data are collected from the spectra of the following R—S(=O)$_2$—R′ compounds.

R =	R′ = CD_3— [9], Et—, nPr—, iPr—, nBu—, sBu—, iBu— and nHex— [2], CCl_3— and CBr_3— [68], H_2C=CH— [2, 70, 71], Ph— [43, 58, 64, 66, 70], Ph-d_5— [43], 4-MePh—, 4-MeO(O=)CPh—, 3- and 4-H_2Ph—, 4-HOPh—, 4-FPh—, 4-ClPh—, 3-O_2N—4-FPh—, 2,5-Br_2—4-HOPh—, 2-Py— [66, 67], 5-O_2N—2-Py— [66]
R	R′
Ph—	MeO(O=)CCH_2—, HO(O=)CCH_2CH_2—, $ClCH_2$ and $BrCH_2$— [63], H_2C=CH—, 4-ClPh—, 2-H_2NPh—, 2-Py— [67]
4-MePh—	$ClCH_2$— and $BrCH_2$— [63], N≡CCH_2—
4-ClPh—	$ClCH_2$— and $BrCH_2$— [63], Cl(F)$CHCF_2$—
4-BrPh—	$ClCH_2$— and $BrCH_2$— [63]
4-H_2NPh—	Me$(CH_2)_{15}$—
2-Py—	2-Py—5-NO_2 [66].

References

1. K. Fujimori, *Bull. Chem. Soc. Jpn.*, **32**, 1374 (1959).
2. W.R. Feairheller and J.E. Katon, *Spectrochim. Acta*, **20**, 1099 (1964).
3. R. McLachlan and V. Carter, *Spectrochim. Acta, Part A*, **26A**, 1126 (1970).
4. G. Geiseler and G. Hanschmann, *J. Mol. Struct.*, **8**, 293 (1971).
5. G. Geiseler and G. Hanschmann, *J. Mol. Struct.*, **11**, 283 (1972).
6. T. Uno, K. Machida and K. Hanai, *Spectrochim. Acta, Part A*, **27A**, 107 (1971).
7. K. Machida, Y. Kuroda and K. Hanai, *Spectrochim. Acta, Part A*, **35A**, 835 (1979).
8. G.E. Binder and A. Schmidt, *Spectrochim. Acta, Part A*, **33A**, 816 (1977).
9. G. Chassaing, J. Corset and J. Limouzi, *Spectrochim. Acta, Part A*, **37A**, 721 (1981).
10. M. Spoliti, S.M. Chackalackal and F.E. Stafford, *J. Am. Chem. Soc.*, **89**, 1092 (1967).
11. A.H. Fawcett, S. Fee, M. Stuckey and P. Walkden, *Spectrochim. Acta, Part A*, **43A**, 797 (1987).
12. A.B. Remizov, F.S. Bilalov and I.S. Pominov, *Spectrochim. Acta, Part A*, **43A**, 309 (1987).
13. M. Bouquet, G. Chassaing, J. Corset, J. Favrot and J. Limouzi, *Spectrochim. Acta, Part A*, **37A**, 727 (1981).
14. G. Kresze, E. Popte and B. Schrader, *Spectrochim. Acta*, **21**, 1633 (1965).
15. K. Hanai, T. Okuda, T. Uno and K. Machida, *Spectrochim. Acta, Part A*, **31A**, 1217 (1975).
16. A. Noguchi, K. Hanai and T. Okuda, *Spectrochim. Acta, Part A*, **36A**, 829 (1980).
17. M. Goldstein, M.A. Russel and H.A. Willis, *Spectrochim. Acta, Part A*, **25A**, 1275 (1969).
18. K. Hanai, A. Noguchi and T. Okuda, *Spectrochim. Acta, Part A*, **34A**, 771 (1978).
19. H. Kalová, R. Slechtová, J. Socha and V. Bekárek, *Acta Univ. Palacki. Olomuc. Fac. Rerum Nat.*, **49**, 131 (1976).
20. A. Simon, H. Kriegsmann and H. Dutz, *Chem. Ber.*, **89**, 2378 (1956).
21. G. Geiseler and B.Nagel, *J. Mol. Struct.*, **16**, 79 (1973).

22. S.J. Cyvin, S. Dobos, I. Hargittai, H. Hargittai and E. Augdahl, *J. Mol. Struct.*, **18**, 203 (1973).
23. K. Hanai, T. Okuda and K. Machida, *Spectrochim. Acta, Part A*, **31A**, 1227 (1975).
24. E. Merian, *Helv. Chim. Acta*, **43**, 1122 (1960).
25. T. Momose, Y.Ueda and T. Shoji, *Chem. and Pharm. Bull.*, **7**, 734 (1959).
26. R. Aroca, J. Ali and E.A. Robinson, *J. Mol. Struct.*, **116**, 9 (1984).
27. H. Bürger, K. Burczyk, A. Blaschette and H. Safari, *Spectrochim. Acta, Part A*, **27A**, 1073 (1971).
28. M. Lustig and G.H. Cady, *Inorg. Chem.*, **2**, 388 (1963).
29. R.J. Gillespie and E.A. Robinson, *Can. J. Chem.*, **40**, 644 (1962).
30. R. Savoie and P.A. Giguère, *Can. J. Chem.*, **42**, 277 (1964).
31. K.O. Christe, C.J. Schack and E.C.Curtis, *Spectrochim. Acta, Part A*, **26A**, 2367 (1970).
32. T. Birchall and R.J. Gillespie, *Spectrochim. Acta*, **22**, 681 (1966).
33. A.J. Sumodi and E.L. Pace, *Spectrochim. Acta, Part A*, **28A**, 1129 (1972).
34. R.J. Gillespie and E.A. Robinson, *Can. J. Chem.*, **39**, 2171 (1961).
35. M. Pfeiffer, *Z. Phys. Chem. (Leipzig)*, **240**, 380 (1969).
36. D.R. Lide Jr. and J.J. Comerford, *Spectrochim. Acta*, **21**, 497 (1965).
37. F.B. Dudley, G.H. Cady and D.F. Eggers, *J. Am. Chem. Soc.*, **78**, 290 (1956).
38. N.C. Craig and K. Futamara, *Spectrochim. Acta, Part A*, **45A**, 507 (1989).
39. T.T. Crow and R.T. Lagemann, *Spectrochim. Acta*, **12**, 143 (1958).
40. P.R. Reed and R.W. Lovejoy, *Spectrochim. Acta, Part A*, **24A**, 1795 (1968).
41. J. Ali, R. Aroca and E.A. Robinson, *Spectrochim. Acta, Part A*, **37A**, 819 (1981).
42. H.J. Weigmann and G. Malewski, *Spectrochim. Acta*, **22**, 1045 (1966).
43. T. Uno, K. Machida and K. Hanai, *Spectrochim. Acta, Part A*, **24A**, 1705 (1968).
44. K.Nukada, *Spectrochim. Acta*, **18**, 735 (1962).
45. Y. Tanaka, Y. Tanaka, Y. Saito and K. Machida, *Bull. Chem. Soc. Jpn.*, **51**, 1324 (1978).
46. B. Nagel, J. Stark, J. Fruwert and G. Geiseler, *Spectrochim. Acta, Part A*, **32A**, 1297 (1976).
47. G. Doucet-Baudry, *Bull. Soc. Chim. Fr.*, 218 (1974).
48. F. K. Butcher, J. Charalambous, M.J. Frazer and W. Gerrard, *Spectrochim. Acta, Part A*, **23A**, 2400 (1967).
49. L.W. Herrick and E.L. Wagner, *Spectrochim. Acta*, **21**, 1569 (1965).
50. W. K. Thompson, *Spectrochim. Acta, Part A*, **28A**, 1479 (1972).
51. E.L. Varetti, *Spectrochim. Acta, Part A*, **44A**, 733 (1988).
52. K.O. Christe and E.C.Curtis, *Spectrochim. Acta, Part A*, **28A**, 1889 (1972).
53. A.B. Remizov, A.I. Fishman and I.S. Pominov, *Spectrochim. Acta, Part A*, **35A**, 909 (1979).
54. A. Simon and D. Kunath, *Chem. Ber.*, **94**, 1776 (1961).
55. I.W. Herrick and E.L. Wagner, *Spectrochim. Acta*, **21**, 1569 (1965).
56. T. Uno, K. Machida and K. Hanai, *Spectrochim. Acta*, **22**, 2065 (1966).
57. Y. Tanaka, Y. Tanaka, Y. Saito and K. Machida, *Spectrochim. Acta, Part A*, **39A** 159 (1983).
58. K.C. Schreiber, *Anal. Chem.*, **21**, 1168 (1949).
59. M.St.C. Flett, *Spectrochim. Acta*, **18**, 1537 (1962).
60. A. Arcoria, E. Maccarone, G. Musumarra and G. Tomaselli, *Spectrochim. Acta, Part A*, **30A**, 611 (1974).
61. C. Laurence, M. Berthelot and M.Lucon, *Spectrochim. Acta, Part A*, **38A**, 791 (1982).
62. A. Buzas and J. Teste, *Bull. Soc. Chim. Fr.*, 793 (1960).
63. A.B. Remizov, F.S. Bilalov and I.S. Pominov, *Spectrochim. Acta, Part A*, **43A**, 475 (1987).
64. B. Nagel, T. Steiger, J. Fruwert and G. Geiseler, *Spectrochim. Acta, Part A*, **31A**, 255

(1975).
65. G. Varsányi, *Assignments for Vibrational Spectra of Seven Hundred Benzene Derivatives*, J. Wiley & Sons, New York (1974).
66. A. Bigotto, V. Galasso, G.C. Pappalardo and G. Scarlata, *J. Chem. Soc. Perkin Trans. 2*, 1845 (1976).
67. A. Bigotto, V. Galasso, G.C. Pappalardo and G. Scarlata, *Spectrochim. Acta, Part A*, **34A**, 435 (1978).
68. M. Hargittai, E. Vajda, C.J. Nielsen, P. Klaboe, R. Seip and J. Brunvoll, *Acta Chem. Scand., Ser. A*, **37A**, 341 (1983).
69. G. Chassaing, F. Froment, A. Marquet and J. Corset, *J. Organomet. Chem.*, **232**, 293 (1982).
70. N.L. Allinger and Y. Fan, J. *Comput. Chem.*, **14**, 655 (1993).
71. I. Hargittai, B. Rozsondai, B.Nagel, P. Bulcke, G. Robinet and J.-F. Labarre, *J. Chem. Soc. Dalton Trans.*, 861 (1978).

12

Normal Vibrations and Absorption Regions of Ring Structures

12.1 CYCLOPROPYL

According to the D_{3h} structure, cyclopropane has only 14 normal vibrations [1–5]. Substitution of a hydrogen by X gives rise to a cPr—X molecule with 21 normal vibrations (12a′ + 9 a″), belonging to the C_s group, for which the plane of the molecule is not the symmetry plane:

a′: $\nu_a CH_2$, $\nu_s CH_2$, νCH, δCH_2, δCH, τCH_2, νR, δ_sR, ωCH_2, ρCH_2, νC—X and δ—C—X;
a″: $\nu_a CH_2$, $\nu_s CH_2$, δCH_2, τCH_2, ωCH_2, $\gamma(\omega)$CH, δ_aR, ρCH_2 and γ—C—X.

in which νR represents the cyclopropyl ring breathing, δ_aR the antisymmetric ring deformation and δ_sR the symmetric ring deformation. The methylene vibrations are in-phase (a′) or out-of-phase (a″). Substitution of νC—X (a′) by a torsion (a″) gives the 21 vibrations of the —cPr fragment.

Methylene vibrations

The five CH stretching vibrations absorb at wavenumbers higher than 3000 cm^{-1}, the region in which the absorptions of the sp^2-hybridized CH bonds are active. Because of coincident wavenumbers, usually the spectrum reveals only two bands, a typical pattern for the three-membered ring.

The two methylene deformations, which absorb in a region (1475–1410 cm^{-1}) in which the sp^3-hybridized methylene groups also are active, appear clearly separated.

Most of the investigated molecules show the δCH in the region 1345 $\pm$ 50 cm^{-1}. Although the two methylene twisting vibrations absorb in neighbouring regions (1175 $\pm$ 20 and 1130 $\pm$ 40 cm^{-1}), coinciding wavenumbers are the exception rather than a rule. Both methylene wagging vibrations and the CH wag occurring between 1105 and 975 cm^{-1} are more or less coupled. Because it is not always obvious which vibration makes the greatest contribution to a distinct absorption, these wags are assigned as $\omega CH_2/\omega CH$ or $\omega CH/\omega CH_2$. If it is accepted that the CH_2 rock absorbs at lower wavenumbers than the ring deformations, the regions 830 $\pm$ 40 and 765 $\pm$ 50 cm^{-1} are useful for both rocking vibrations.

Ring vibrations

The ring breathing, absorbing at 1200 $\pm$ 20 cm^{-1}, is a good group vibration for cyclopropyl compounds. The intensity is only moderate in the infrared but strong in the Raman. Although much vibrational analysis reveals the antisymmetric ring deformation at 930 $\pm$ 55 cm^{-1}, sometimes a CH_2 rock is assigned in this region, and the ring vibration is located in the region of the rock (830 $\pm$ 40 cm^{-1}). The symmetric ring deformation absorbs in the region 860 $\pm$ 45 cm^{-1}.

Table 12.1 Absorption regions of the normal vibrations of —cPr

Vibration	Region	Vibration	Region
$\nu_a CH_2$	3090 $\pm$ 25	$\omega CH_2/CH$	1075 $\pm$ 30
$\nu_a CH_2$	3075 $\pm$ 25	$\omega CH_2/CH$	1040 $\pm$ 30
$\nu_s CH_2$	3040 $\pm$ 40	$\omega CH/CH_2$	1010 $\pm$ 35
$\nu_s CH_2$	3020 $\pm$ 20	$\delta_a R$	930 $\pm$ 55
νCH	3015 $\pm$ 45	$\delta_s R$	860 $\pm$ 45
δCH_2	1455 $\pm$ 20	ρCH_2	830 $\pm$ 40
δCH_2	1425 $\pm$ 15	ρCH_2	765 $\pm$ 50
δCH	1330 $\pm$ 90	ext.sk.def.	–
νR	1200 $\pm$ 20	ext.sk.def.	–
τCH_2	1175 $\pm$ 20	torsion	–
τCH_2	1130 $\pm$ 40		

R—cPr molecules

R = D— [6], Me— [7], Et— [8], $cPrCH_2$— [9, 10], N≡CCH_2— [11], $HOCH_2$— [12, 13], $ClCH_2$— [11, 14–17], $BrCH_2$— [11, 17, 18], ICH_2— [11, 17], HO(Me)CH— [12], iPr— [8], cPr— [19], H_2C=C(cPr)— [20], HC≡C— and DC≡C— [21], cPrC≡C— [22, 23], N≡C— [24–30], C=N— [31], H(O=)C— [32], F(O=)C— [33], Cl(O=)C— [5, 25, 34, 35], Br(O=)C— [5, 25], Me(O=)C— [36, 37], HO(O=)C— [38–40], $^{-}O_2C$— [5, 39], MeO(O=)C— [36], cPr(O=)C— [20, 41], Ph(O=)C— [42], 4-XPh(O=)C— (X = F, Cl, Br) [43], H_2N(O=)C— [38],

H_2N— [44, 45], D_2N— [44, 46], O_2N— [78], O=C=N [47], S=C=N— [47–49], Cl— [5, 48, 50], Br— [5, 25, 48, 51, 52], I— [5, 48], F_3Si— [53], $(cPr)_2HSi$— [54].

12.2 OXIRANYL

Oxirane (ethene oxide) belongs to the point group C_{2v} and the 15 normal vibrations are divided among $5a_1 + 3a_2 + 3b_1 + 4b_2$ species of vibration [55–59].The oxiranyl (cC_2H_3O) fragment gives rise to $3N - 6 = 15$ normal vibrations, of which 12 are collected in Table 12.2.

The CH stretching vibrations absorb between 3070 and 2970 cm^{-1}, but the position of the νCH is uncertain. Nyquist *et al.* [63] and Tobin [60] assign the region 1410 ± 35 cm^{-1} rather to the CH_2 twisting vibration and the region 1175 ± 35 cm^{-1} to the CH deformation.

Table 12.2 Absorption regions (cm^{-1}) of the normal vibrations of —Ox

Vibration	Region	Vibration	Region
$\nu_a CH_2$	3055 ± 15	ωCH	1075 ± 35
νCH	3005 ± 30	δ_aR	920 ± 45
$\nu_s CH_2$	2990 ± 20	δ_sR	845 ± 35
δCH_2	1465 ± 35	ρCH_2	775 ± 25
δCH	1410 ± 35	ext.sk.def.	–
νR	1255 ± 10	ext.sk.def.	–
τCH_2	1175 ± 35	torsion	–
ωCH_2	1130 ± 10		

Key: νR: ring breathing; δ_aR: ring antisymmetric in-plane deformation; δ_sR: ring symmetric in-plane deformation.

R—Ox molecules

R = Me— [5, 60–62], Et— [8, 62], $Me(CH_2)_n$— (n = 7, 8, 11, 13, 15) [62], FCH_2— [17, 63, 64], $ClCH_2$— [16, 17, 63], $BrCH_2$— [17, 18, 63], ICH_2— [17, 63], iPr—, iBu—, tBu— [62], Ox— [66], H_2C=CH— [65], Ph—, 2,4-Cl_2Ph—.

12.3 AZIRIDINYL

Aziridine belongs to the point group C_s and the 18 normal vibrations are divided into 10 a′ + 8 a″ species of vibration [67–72]. In *N*-substituted R—Az compounds, the Az fragment (cC_2H_4N—) possesses 18 normal vibrations, of which 15 are collected in Table 12.3.

The in-phase (a′) and out-of-phase (a″) CH_2 antisymmetric and symmetric stretching vibrations coincide but the CH_2 deformations are observed separately. The ring breathing (a′) occurs at 1235 ± 50 cm^{-1} and at 1212 cm^{-1} in aziridine, although Spell [77] assigned this vibration in the region 1285 ± 75 cm^{-1}.

Table 12.3 Absorption regions (cm^{-1}) of the normal vibrations of —Az

Vibration	Region	Vibration	Region
$\nu_a CH_2$	3080 ± 20	ωCH_2	1115 ± 30
$\nu_a CH_2$	3080 ± 20	ωCH_2	1065 ± 40
$\nu_s CH_2$	3000 ± 25	$\delta_a R$	905 ± 20
$\nu_s CH_2$	3000 ± 25	$\delta_s R$	850 ± 30
δCH_2	1470 ± 15	ρCH_2	815 ± 25
δCH_2	1445 ± 20	ρCH_2	765 ± 35
νR	1235 ± 50	ext.sk.def.	–
τCH_2	1210 ± 50	ext.sk.def.	–
τCH_2	1145 ± 40	torsion	

R—Az molecules
R = Me— [5], Az— [75], AzC(=O)— [73], H_2NC(=O)— and D_2NC(=O)— [74], AzS(=O)— [70], Cl— and Br— [69, 76].

References

1. A.W. Baker and R.C. Lord, *J. Chem. Phys.*, **23**, 1636 (1955).
2. C. Brecher, E. Krikorian, J. Blanc and R.S. Halford, *J. Chem. Phys.*, **35**, 1097 (1961).
3. J.W. Russell, C.M. Philips and T. Davidson, *Spectrochim. Acta, Part A*, **37A**, 263 (1981).
4. J.L. Duncan and D.C. McKean, *J. Mol. Spectrosc.*, **27**, 117 (1968).
5. C.J. Wurrey and A.B. Nease, *Vibrational Spectra and Structure*, Vol. **7** (J.R. Durig, Ed.) Elsevier Scientific Publishing Company, Amsterdam, Oxford, New York (1978), pp. 1–235.
6. A.O. Diallo and D.N. Waters, *Spectrochim. Acta, Part A*, **44A**, 1109 (1988).
7. L.M. Sverdlov, N.P. Krainov and N.I. Prokofeva, *Fiz. Probl. Spectrosk. Mater. Soveshch., 13th*, Leningrad, 1960, Vol. **1**, p. 363.
8. A.B. Nease and C.J. Wurrey, *J. Raman Spectrosc.*, **9**, 107 (1080).
9. G. Schrumpf and T. Alshuth, *Spectrochim. Acta, Part A*, **41A**, 1335 (1985).
10. V.F. Kalasinsky, J.L. Pool, N.S. Eymann, J.T. Leahey, M.D. Weakly, Y.Y. Yeh and C.J. Wurrey, *Spectrochim. Acta, Part A*, **42A**, 149 (1986).
11. C.J. Wurrey, Y.Y. Yeh, M.D. Weakly and V.F. Kalasinsky, *J. Raman Spectrosc.*, **15**, 179 (1984).
12. P. Klaboe and D.L. Powell, *J. Mol. Struct.*, **20**, 95 (1974).
13. H.M. Badawi, M.E. Abu-Zeid and Y.A. Yousef, *J. Mol. Struct.*, **240**, 225 (1990).
14. T. Hirokawa and H. Murata, *J. Sci. Hiroshima Univ. Ser. A*, **38A**, 271 (1974).
15. F.G. Fujiwara, J.C. Chang and H. Kim, *J. Mol. Struct.*, **41**, 177 (1977).
16. V.F. Kalasinsky and C.J. Wurrey, *J. Raman Spectrosc.*, **9**, 315 (1980).

17. C.J. Wurrey, Y.Y. Yeh, R. Krishnamoorthi, R.J. Berry, J.E. Dewitt and V.F. Kalasinsky, *J. Phys. Chem.*, **88**, 4059 (1984).
18. C.J. Wurrey, R. Krishnamoorthi, S. Pechsiri and V.F. Kalasinsky, *J. Raman Spectrosc.*, **12**, 95 (1982).
19. M. Spiekermann, B. Schrader, A. De Meijer and W. Luttke, *J. Mol. Struct.*, **77**, 1 (1981).
20. A.B. Nease and C.J. Wurrey, *J. Phys. Chem.*, **83**, 2135 (1979).
21. G. Schrumpf and A.W. Klein, *Spectrochim. Acta, Part A*, **41A**, 1251 (1985).
22. G. Schrumpf and T. Alshuth, *J. Mol. Struct.*, **101**, 47 (1983).
23. V. Mohacek and K. Furić, *J. Mol. Struct.*, **266**, 321 (1992).
24. L.H. Daly and S.E. Wiberley, *J. Mol. Spectrosc.*, **2**, 177 (1958).
25. J.E. Katon, W.R. Feairheller Jr. and J.T. Miller Jr., *J. Chem. Phys.*, **49**, 823 (1968).
26. J. Maillols, V. Tabacik and S. Sportouch, *J. Raman Spectrosc.*, **11**, 312 (1981).
27. G. Schrumpf, *Spectrochim. Acta, Part A*, **39A**, 511 (1983).
28. G. Schrumpf and H. Dunker, *Spectrochim. Acta, Part A*, **42A**, 785 (1986).
29. P.M. Green, C.J. Wurrey, R. Krishnamoorthi and Y.Y. Yeh, *J. Raman Spectrosc.*, **17**, 355 (1986).
30. T.S. Little, W. Zhao and J.R. Durig, *J. Raman Spectrosc.*, **19**, 479 (1988).
31. G. Schrumpf and S. Martin, *Spectrochim. Acta, Part A*, **44A**, 479 (1988).
32. J.R. Durig and T.S. Little, *Croat. Chem. Acta*, **61**, 529 (1988).
33. J.R. Durig, H.D. Bist and T.S. Little, *J. Chem. Phys.*, **77**, 4884 (1982).
34. J.R. Durig, H.D. Bist, S.V. Saari, J.A.S. Smith and T.S. Little, *J. Mol. Struct.*, **99**, 217 (1983).
35. J.R. Durig, A. Wang and T.S. Little, *J. Mol. Struct.*, **269**, 285 (1992).
36. D.L. Powell, P. Klaboe and D.H. Christensen, *J. Mol. Struct.*, **15**, 77 (1973).
37. J.R. Durig, H.D. Bist and T.S. Little, *J. Mol. Struct.*, **116**, 345 (1984).
38. D.L. Powell and P. Klaboe, *J. Mol. Struct.*, **15**, 217 (1973).
39. J. Maillols, *J. Mol. Struct.*, **14**, 171 (1972).
40. V. Tabacik and J. Maillols, *Spectrochim. Acta, Part A*, **34A**, 315 (1978).
41. G. Schrumpf and T. Alshuth, *Spectrochim. Acta, Part A*, **43A**, 939 (1987).
42. W.A. Seth Paul and B.J. Van der Veken, *Can. J. Spectrosc.*, **25**, 38 (1980).
43. W.A. Seth Paul and B.J. Van der Veken, *Can. J. Spectrosc.*, **25**, 57 (1980).
44. V.F. Kalasinsky, D.E. Powers and W.C. Harris, *J. Phys. Chem.*, **83**, 509 (1979).
45. A.O. Diallo, Nguyen-Van-Thanh and I. Rossi, *Spectrochim. Acta, Part A*, **43A**, 415 (1987).
46. D.A.C. Compton, J.J. Rizzolo and J.R. Durig, *J. Phys. Chem.*, **86**, 3746 (1982).
47. J.R. Durig, R.J. Berry and C.J. Wurrey, *J. Am. Chem. Soc.*, **110**, 718 (1988).
48. C.J. Wurrey, R. Krishnamoorthi and A.B. Nease, *J. Raman Spectrosc.*, **9**, 334 (1980).
49. J.R. Durig, J.F.Sullivan, D.T. Durig and S. Cradock, *Can. J. Chem.*, **63**, 2000 (1985).
50. J.R. Durig, P.L. Trowell, Z. Szafran, S.A. Johnston and J.D. Odom, *J. Mol. Struct.*, **74**, 85 (1981).
51. W.G. Rothschild, *J. Chem. Phys.*, **44**, 3875 (1966).
52. J. Maillols and V. Tabacik, *Spectrochim. Acta, Part A*, **35A**, 1125 (1979).
53. T.S. Little, M. Qtaitat, J.R. Durig, M. Dakkouri and A. Dakkouri, *J. Raman Spectrosc.*, **21**, 591 (1990).
54. C.J. Wurrey, A.J. Holder and P.M. Green, *J. Raman Spectrosc.*, **21**, 557 (1990).
55. J.W. Linnett, *J. Chem. Phys.*, **6**, 692 (1938)
56. R.C. Lord and B. Nolin, *J. Chem. Phys.*, **24**, 656 (1956).
57. W.J. Potts, *Spectrochim. Acta*, **21**, 511 (1965).
58. J.E. Bertie and D.A. Othen, *Can. J. Chem.*, **51**, 1155 (1973).
59. N.W. Cant and W.J. Armstead, *Spectrochim. Acta, Part A*, **31A**, 839 (1975).
60. M.C. Tobin, *Spectrochim. Acta*, **16**, 1108 (1960).

61. P.L. Polavarapu, L. Hecht and L.D. Baron, *J. Phys. Chem.*, **97**, 1793 (1993).
62. R.A. Nyquist, *Appl. Spectrosc.*, **40**, 275 (1986).
63. R.A. Nyquist, C.L. Putzig and N.E.Skelly, *Appl. Spectrosc.*, **40**, 821 (1986).
64. S.W. Charles, G.I.L. Jones and N.L. Owen, *J. Mol. Struct.*, **20**, 83 (1974).
65. V.F. Kalasinsky and S. Pechsiri, *J. Raman Spectrosc.*, **9**, 120 (1980).
66. C.F. Su, R.L. Cook, C. Saiwan, J.A.S. Smith and V.F. Kalasinsky, *J. Mol. Spectrosc.*, **127**, 337 (1988).
67. W.J. Potts, *Spectrochim. Acta*, **21**, 511 (1965).
68. R.W. Mitchell, C.J. Burr Jr. and J.A. Merritt, *Spectrochim. Acta, Part A*, **23A**, 195 (1967).
69. J.W. Russell, M. Bishop and J. Limburg, *Spectrochim. Acta, Part A*, **25A**, 1929 (1969).
70. H.L. Spell and J. Laane, *Appl. Spectrocs.*, **26**, 86 (1972).
71. H.W. Thompson and W.T. Cave, *Trans. Faraday Soc.*, **47**, 951 (1951).
72. A.R. Katritzky, *Physical Methods in Heterocyclic Chemistry*, Academic Press, New York (1963).
73. H.L. Spell and J. Laane, *Spectrochim. Acta, Part A*, **28A**, 295 (1972).
74. H.L. Spell and J. Laane, *J. Mol. Struct.*, **14**, 39 (1972).
75. P. Rademacher, *Spectrochim. Acta, Part A*, **28A**, 987 (1972).
76. K. Kalcher, W. Kosmus and K. Faegri Jr., *Spectrochim. Acta, Part A*, **37A**, 889 (1981).
77. H.L. Spell, *Anal. Chem.*, **39**, 185 (1965).
78. J.R. Holtzclaw, W.C. Harris and S.F. Bush, *J. Raman Spectrosc.*, **9**, 257 (1980).

12.4 PHENYL

Benzene, belonging to the point group D_{6h}, has 20 normal vibrations, of which four are infrared active. After substitution of an H atom by an X atom, the symmetry is lowered from D_{6h} to C_{2v} and the number of normal vibrations increases to 30, being CH, CX and ring vibrations. By an uncoupling, the vibrations 6 up to and including 10 and 16 up to and including 20 of benzene are redoubled. This uncoupling does not take place in 1,3,5-trisubstituted benzenes with identical substituents. A further lowering of the symmetry from C_{2v} to C_s leads to a distribution of the normal vibrations in 21a′ and 9a″ types. Table 12.4 shows the normal vibrations of benzene and X—substituted benzene derivatives, corresponding to the Wilson notation [8].

Table 12.4 Normal vibrations of benzene and substituted benzenes

Benzene		X—substituted benzenes				
	D_{6h}	C_{2v} C_s	a_1 a′	a_2 a″	b_1 a″	b_2 a′
νCH	2, 7, 13, 20	νCH, νCX	2, 7a, 13, 20a			7b, 20b
νPh	1, 8, 14, 19		1, 8a, 19a			8b, 14, 19b
δCH	3, 9, 15, 18	δCH, δCX	9a, 18a			3, 9b, 15, 18b
δPh	6, 12		6a, 12			6b
γ CH	5, 10, 11, 17	γ CH, γ CX		10a, 17a	5, 10b, 11, 17b	
γ Ph	4, 16			16a	4, 16b	

Every new substituent leads to three CX vibrations and diminishes the number of CH vibrations by three, but there is some ambiguity in deducing which number of CH vibration, according to Wilson, is replaced by a CX vibration and the literature does not agree concerning this problem. In monosubstituted benzenes six modes of vibration are substituent-sensitive, which means that their wavenumbers shift significantly with mass or inductive or mesomeric effects of the substituent. The remaining 24 can be considered good group vibrations. The stretching and bending of the substituent bond not only give rise to a Ph—X stretching vibration, a Ph—X in-plane and a Ph—X out-of-plane deformation, but also influence the absorption frequency of a ring stretch, an in-plane and an out-of-plane ring deformation. These ring vibrations are coupled with the Ph—X vibrations in such a way as to make it difficult to determine which wavenumber belongs to a Ph—X vibration and which wavenumber to a ring vibration. For this reason, the tables showing absorption regions do not always agree completely with the assignments in the literature: interchange can occur between some substituent-sensitive modes. For disubstituted benzenes 10 or 11, and for trisubstituted benzenes 14 or 15 substituent-sensitive vibrations are taken into account.

The CH and CX stretching vibrations (2, 20a, 20b, 13, 7a, 7b)

The CH stretching vibrations absorb weakly to moderately between 3120 and 3000 cm^{-1}. The highest values near 3120 cm^{-1} are observed in the spectra of benzene substituted with CF_3, F, O_2N and C≡N; examples are 2- and 4-O_2NPhCF_3, 2,4-$(O_2N)_2PhF$, 2-Cl(O=)$CPhCF_3$, 2-N≡$CPhCF_3$, 4-MeOPhC≡N and 2,5-Me_2PhC≡N. In this neighbourhood also overtones or combinations of ring stretching vibrations can occur. Low values (near 3000 cm^{-1}) are assigned in the spectra of (substituted) alkylbenzenes such as Ph—nBu, 4-H_2NPhEt, 4-$MePhCH_2Cl$ and 4-$MePhCH_2Br$. Usually the CH stretching vibrations are active between 3110 and 3010 cm^{-1}.

In monosubstituted benzenes the highest substituent-sensitive mode appears in the region 1175 ± 115 cm^{-1}, in this text called the C—X stretching vibration (in vibrational analysis often 13 or 7a). According to Varsányi and Szöke [1], the absorption in the region 1195 ± 90 cm^{-1} is for the most part due to a C—X stretching vibration in benzenes substituted with a light atom (atomic mass < 25) and that in the region 1090 ± 30 cm^{-1} to a ring stretch in benzenes substituted with a heavy atom (atomic mass > 25). The Ph—C stretching vibration in alkylbenzenes appears weakly in the neighbourhood of 1210 cm^{-1}. This band shifts to higher wavenumbers in benzenes substituted with atoms with an unshared pair of electrons next to the ring (activators) such as nitrogen (≈1280) and oxygen (≈1250 cm^{-1}). Benzenes with multiple bonds next to the ring (deactivators) such as NO_2 and N=C=O (1107), C(=O)OH, C(=O)NH_2 and CO_2^- (≈1150) and C≡CH and C≡N (≈1190 cm^{-1}) absorb at lower wavenumbers. In benzenes substituted with a P atom this substituent-sensitive mode occurs near 1100 cm^{-1}, with an

S atom near 1090 and with an As atom near 1080 cm^{-1}. In halogenated benzenes the wavenumber drops from 1083 cm^{-1} in chlorobenzene, via 1070 in bromobenzene, to 1060 cm^{-1} in iodobenzene. The intensity of this Ph—X stretching vibration increases with the polarity of the X atom. In di- and tri-substituted benzenes the same effects are applicable within the appropriate absorption regions. For substituents in the 1,3-position, one of the Ph—X stretching vibrations absorbs at lower wavenumbers (*meta*-effect) [1].

Ring stretching vibrations (8a, 8b, 19a, 19b, 14, 1)

The benzene ring possesses six ring stretching vibrations, of which the four with the highest wavenumbers (8a, 8b, 19a and 19b, occurring respectively near 1600, 1580, 1490 and 1440 cm^{-1}) are good group vibrations. With heavy substituents the bands tend to shift to somewhat lower wavenumbers. The greater the number of substituents on the ring, the broader the absorption regions will be.

In the absence of ring conjugation, the band at 1580 cm^{-1} is usually weaker than that at 1600 cm^{-1}. In many cases, such as alkyl substitution, the 1580 cm^{-1} absorption is only a shoulder on the 1600 cm^{-1} band. When the ring is substituted at the 1,4-position with identical substituents, both vibrations are infrared inactive. In the case of C=O substitution the band near 1490 cm^{-1} can be very weak. The fifth ring stretching vibration (νPh 14 or Kekulé vibration) is active near 1315 $\pm$ 65 cm^{-1}, a region which overlaps strongly that of the CH in-plane deformation (δCH 3). If further data are not available, the highest {lowest} wavenumber is assigned to the ring stretching vibration (νPh 14) {CH in-plane deformation (δCH 3)}. In phenols this ring stretching vibration is coupled to the OH in-plane deformation by which the region extends to 1390 cm^{-1} [2]. The sixth ring stretching vibration or ring breathing mode (νPh 1) appears as a weak band near 1000 cm^{-1} in mono-,1,3-di- and 1,3,5-tri-substituted benzenes. In the otherwise substituted benzenes, however, this vibration is substituent-sensitive and difficult to distinguish from the ring in-plane deformation (δPh 12). In this text the ring breathing is assigned at a higher wavenumber than the ring in-plane deformation.

The CH and CX in-plane deformation (3, 9a, 9b, 15, 18a and 18b)

Supposing that all the CH in-plane deformations absorb at lower wavenumbers than the ring stretching vibration νPh 14, the CH in-plane deformation with the highest wavenumber (δCH 3) may be assigned at 1265 $\pm$ 65 cm^{-1}. Together with the other CH in-plane deformations this provides a supplementary indicator for the presence of a benzene ring.

Particularly in di- and tri-substituted benzenes the substituent-sensitive Ph—X in-plane deformations are of minor importance as a diagnostic tool. The heaviest substituents give rise to the lowest wavenumbers.

The CH and CX out-of-plane deformations (5, 17a, 17b, 10a, 10b and 11)

The CH out-of-plane deformations are observed between 1000 and 720 cm^{-1}, but 2-H(O=)CPhC(=O)H (1002), 2-O_2NPhC(=O)H (1002) and 2-O_2NPhNO_2 (1005) overstep this upper limit. Generally the CH out-of-plane deformations with the highest wavenumbers have a weaker intensity than those absorbing at lower wavenumbers. The spectral position of the weaker CH out-of-plane deformation near 900 cm^{-1} (in this text named γCH 17b) correlates well with the electron donating or electron attracting properties of the substituent (see Table 12.5). The stronger CH in-phase out-of-plane deformation band occurring in the region 770 ± 50 cm^{-1} (γCH 11 or umbrella mode) also tends to shift to lower {higher} wavenumbers with increasing electron donating {attracting} power of the substituent, but seems to be more sensitive to mechanical interaction effects. The lowest wavenumbers for this umbrella mode are found in the spectra of benzenes substituted with a saturated carbon or an heavy atom such as halogen, sulfur or phosphorus [2, 3, 5, 7].

Table 12.5 Absorption wavenumber (cm^{-1}) of γCH 17b and γCH 11 in monosubstituted benzenes

Substituent	γCH 17b	γCH 11	Substituent	γCH 17b	γCH 11
C(=O)F	938	797	CH=CH_2	909	775
C(=O)Cl	937	777	CH=CHMe	909	735
C(=O)OMe	936	808	NHC(=O)Me	906	755
C(=O)OH	935	812	N=C=S	905	749
NO_2	934	793	tBu	905	765
S(=O)$_2$Me	932	750	nPr	905	743
S(=O)$_2$Cl	930	752	Et	904	746
S(=O)$_2$$NH_2$	928	754	N=C=O	904	751
C(=O)Me	927	760	I	904	730
C≡N	926	758	Br	903	735
C(=O)NH_2	925	807	Cl	902	740
CF_3	924	770	iPr	902	761
S(=O)$_2$Ph	921	758	CH_2Cl	900	768
C(=O)H	920	746	CH_2N=C=S	899	733
CO_2^-	919	819	Me	896	728
S(=O)Me	919	748	F	896	753
C≡CH	918	757	SH	895	727
CH=C(C≡N)$_2$	918	764	SMe	893	738
CCl_3	917	788	OC(=O)Me	892	750
$CHCl_2$	916	780	OEt	884	755
CH_2Br	915	758	OH	883	757
S(=O)Ph	915	754	OMe	882	754
CH_2OH	912	735	NH_2	880	753
C(Me)=CH_2	912	770	NHMe	870	754
$CH_2$$NH_2$	910	736	NMe_2	864	758

The weak substituent-sensitive Ph—X out-of-plane deformations absorb at the LW side of the Ph—X in-plane deformations. They are of little practical use in elucidating the molecular structure.

Ring deformations

Three or more ring deformations are substituent-sensitive and their utility for identification purposes is very limited. They are strongly coupled to the Ph—X stretching vibration and an interchange with the latter is not excluded. The ring out-of-plane deformation occurring near 695 cm^{-1} (γ Ph 4), on the contrary, merits being called a good group vibration. It is, together with the lowest CH out-of-plane deformation, of great importance in distinguishing different types of aromatic ring substitution.

Types of aromatic ring substitution (region 970–650 cm^{-1})

The out-of-plane CH deformation near 770 cm^{-1} (γ CH 11) and the out-of-plane ring deformation absorbing near 690 cm^{-1} (γ Ph 4) form a pair of strong bands characteristic of monosubstituted benzene derivatives [7]. The spectral position of γ CH 17b and γ CH 11 in a number of monosubstituted benzenes is shown in Table 12.5.

In the case of 1,2-disubstitution only one strong absorption in the region 755 $\pm$ 35 cm^{-1} is observed, and is due to the γ CH. The out-of-plane ring deformation γ Ph 4 gives rise to a weak shoulder on the γ CH band or coincides with it, or is even inactive for identical substituents. Colthup [3] calculated the CH out-of-plane deformations near 940 and 755 cm^{-1} with the 900 cm^{-1} phenyl band (γ CH 17b) wavenumbers of benzenes substituted with the same substituents as on the 1- and 2-positions (see Table 12.5).

$$\gamma \text{ CH near } 940 = 0.5\ (\sigma_1 + \sigma_2) + 36 \text{ cm}^{-1}$$
$$\gamma \text{ CH near } 755 = 0.5\ (\sigma_1 + \sigma_2) - 149 \text{ cm}^{-1}$$

in which σ_1 = γ CH 17b wavenumber of benzene substituted with the same group as on the 1-position and σ_2 = γ CH 17b wavenumber of benzene substituted with the same group as on the 2-position.

The 1,3-substitution pattern resembles that of monosubstitution, with the CH out-of-plane deformation near 770 cm^{-1} and the out-of-plane ring deformation (γ Ph 4) in the neighbourhood of 685 cm^{-1} often absorbing with equal intensity. According to the formula of Colthup [3]:

$$\gamma \text{ CH near } 860 = 0.9\ (\sigma_1 + \sigma_3) - 757 \text{ cm}^{-1}$$
$$\gamma \text{ CH near } 770 = 0.5\ (\sigma_1 + \sigma_3) - 120 \text{ cm}^{-1}$$

in which σ_1 $\{\sigma_3\}$ = γCH 17b wavenumber of benzene substituted with the same group as on the 1 {3} position (see Table 12.5).

The strong CH out-of-plane deformation, occurring at 840 $\pm$ 50 cm^{-1}, is typical for a 1,4-disubstitution. This band is situated at a greater distance from the weak ring out-of-plane deformation (γPh 4), which is even infrared inactive with identical substituents. The lower CH out-of-plane deformation absorbs in the same neighbourhood (820 $\pm$ 45 cm^{-1}) but is much weaker or infrared inactive. According to the formula of Colthup *et al.* [4] the γCH near 840 cm^{-1} can be predicted from the γCH 11 wavenumbers of both substituents:

$$\gamma \text{CH near } 840 = 0.6\ (\sigma_1 + \sigma_4) - 75 \text{ cm}^{-1}$$

in which σ_1 $\{\sigma_4\}$ = γCH 11 wavenumber of benzene substituted with the same group as on position 1 {4} (see Table 12.5).

For 1,2,3-trisubstitution the strong CH out-of-plane deformation and the ring out-of-plane deformation (γPh 4) are respectively observed at 790 $\pm$ 40 and 710 $\pm$ 30 cm^{-1}.

$$\gamma \text{CH near } 790 = 0.6\ (\sigma_1 + \sigma_3) - 303 \text{ cm}^{-1}\ [3]$$

in which σ_1 $\{\sigma_3\}$ = γCH 17b wavenumber of benzene substituted with the same group as on position 1 {3}.

In the case of 1,2,4-trisubstitution, two medium to strong γCH bands are observed at 890 $\pm$ 50 and 815 $\pm$ 45 cm^{-1} respectively. The γPh 4 appears as a medium band at 700 $\pm$ 40 cm^{-1}.

$$\gamma \text{CH near } 890 = 0.7\ (\sigma_2 + \sigma_4) - 393 \text{ cm}^{-1}\ [3]$$

in which σ_2 $\{\sigma_4\}$ = γCH 17b wavenumber of benzene substituted with the same group as on position 2 {4}.

$$\gamma \text{CH near } 815 = 0.6\ (\sigma_1 + \sigma_4) - 75 \text{ cm}^{-1}\ [4]$$

in which σ_1 $\{\sigma_4\}$ = γCH 11 wavenumber of benzene substituted with the same group as on position 1 {4}.

1,3,5-trisubstituted benzenes give rise to simple spectra if the three substituents are not too different. Identical substituents reduce the number of normal vibrations to 20. Usually a medium γCH band is observed at 835 $\pm$ 25 cm^{-1} and another at 860 $\pm$ 30 cm^{-1} accompanied by a weaker band at 890 $\pm$ 50 cm^{-1}. The last two coincide if the three substituents are identical or near identical to give one band in the region 885 $\pm$ 55 cm^{-1}. Colthup [3] calculated the position of this γCH with the formula:

$$\gamma \text{CH} = 0.7\ (\sigma_1 + \sigma_3 + \sigma_5) - 1040 \text{ cm}^{-1}$$

in which σ_1, σ_3 and σ_5 respectively are the γCH 17b wavenumbers of benzenes substituted with the same group as on position 1, 3 or 5.

Summation bands (region 2000–1650 cm^{-1})

The infrared spectra of substituted benzenes show a pattern of weak bands that tend to be characteristic for the type of substitution. These bands are produced by overtones and combinations of the CH wagging modes. In C=O substituted benzenes, however, these bands are disturbed. According to Young *et al.* [6] these typical patterns are illustrated in Figure 12.1.

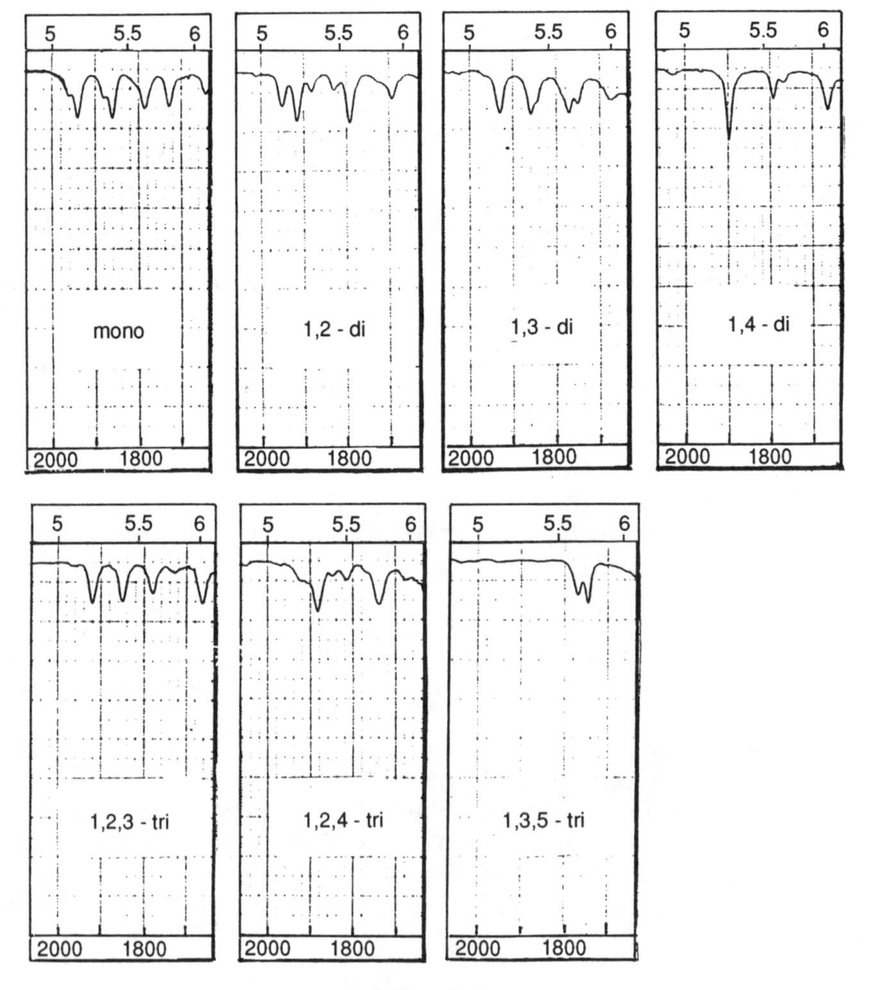

Figure 12.1

12.4.1 Monosubstituted benzene derivatives

Table 12.6 Absorption regions (cm^{-1}) of the normal vibrations of monosubstituted benzenes

Vibration		Wilson	Mono light	Mono heavy
νCH	CH stretching vibration	20a	3085 ± 20	3085 ± 20
νCH	CH stretching vibration	20b	3070 ± 20	3075 ± 15
νCH	CH stretching vibration	2	3060 ± 20	3055 ± 15
νCH	CH stretching vibration	13	3040 ± 20	3040 ± 20
νCH	CH stretching vibration	7b	3020 ± 20	3020 ± 20
νPh	Phenyl ring stretching vibration	8a	1605 ± 15	1590 ± 20
νPh	Phenyl ring stretching vibration	8b	1585 ± 15	1575 ± 15
νPh	Phenyl ring stretching vibration	19a	1485 ± 25	1475 ± 15
νPh	Phenyl ring stretching vibration	19b	1450 ± 20	1435 ± 15
νPh	Phenyl ring stretching vibration	14	1335 ± 35	1310 ± 25
δCH	CH in-plane deformation	3	1295 ± 25	1275 ± 25
νCX	Phenyl—X stretching vibration	7a (X)	1195 ± 90	1090 ± 30
δCH	CH in-plane deformation	9a	1175 ± 20	1180 ± 15
δCH	CH in-plane deformation	9b	1150 ± 20	1165 ± 10
δCH	CH in-plane deformation	15	1070 ± 20	1065 ± 15
δCH	CH in-plane deformation	18a	1020 ± 20	1025 ± 10
νCH	Phenyl ring stretching vibration	1	1000 ± 10	1000 ± 05
γ CH	CH out-of-plane deformation	5	980 ± 20	980 ± 15
γ CH	CH out-of-plane deformation	17a	960 ± 25	965 ± 10
γ CH	CH out-of-plane deformation	17b	900 ± 35	915 ± 20
γ CH	CH out-of-plane deformation	10a	840 ± 25	835 ± 20
γ CH	CH out-of-plane deformation	11	775 ± 45	745 ± 25
δPh	Phenyl ring in-plane deformation	12 (X)	750 ± 80	700 ± 50
γ Ph	Phenyl ring out-of-plane deform.	4	695 ± 15	685 ± 15
δPh	Phenyl ring in-plane deformation	6b	625 ± 15	605 ± 15
γ Ph	Phenyl ring out-of-plane deform.	16b (X)	510 ± 90	475 ± 55
δPh	Phenyl ring in-plane deformation	6a (X)	420 ± 115	355 ± 90
γ Ph	Phenyl ring out-of-plane deform.	16a	405 ± 25	405 ± 15
δCX	Phenyl—X in-plane deformation	18b (X)	300 ± 110	250 ± 60
γ CX	Phenyl—X out-of-plane deform.	10b (X)	205 ± 70	170 ± 30

Mono light = benzenes substituted with a light atom (atomic mass < 25).
Mono heavy = benzenes substituted with a heavy atom (atomic mass > 25).
(X) = substituent-sensitive vibration.

X—Ph compounds

X = Me— [9–11], Et— [2, 9], nPr—, nBu— and iBu— [2], $PhCH_2CH_2$— [12, 85], $ClCH_2CH_2$— [85], $N{\equiv}CCH_2$— [2, 13], H_2NCH_2— [2, 13], $SCNCH_2$— [2], $HOCH_2$— [2], $HSCH_2$— [14], $N{=}C{=}SCH_2$— [15], $PhCH_2SCH_2$— [16], $ClCH_2$— [2, 10], $BrCH_2$— [13, 42], iPr— and sBu— [2], $H_2NCH_2(Ph)HC$— and $HOCH_2(Ph)HC$— [17], Cl_2HC— [10], tBu— [2], Cl_3C— [10, 81, 82], F_3C— [27, 80], $H_2C{=}CH$— [18, 19, 24], PhCH=CH— [29–31], $(N{\equiv}C)_2C{=}CH$— [32], $H_2C{=}C(Me)$— [33], Ph— [29, 34], HC≡C— [2, 48, 49], N≡C— [20, 23, 35, 36], ClC≡C— and BrC≡C— [84], H(O=)C— [22, 37], F(O=)C— [26], Cl(O=)C— [26, 38], Me(O=)C— [26, 39, 40], HO(O=)C— [25, 26, 41, 43], MeO(O=)C— [2, 26, 44], $H_2N(O{=})C$— [26, 45], $^{-}O_2C$— [25,

46], $H_2N(S{=})C$— [86], EtO(O=)C— [26], H_2N— [47], MeHN— [2, 50, 83], Me_2N— [2, 50, 51, 83], Et_2N— [51], Me(O=)CHN— [52], O_2N— [11, 43, 53, 54], O=C=N— [55], S=C=N— [56], CN— [79], HO— [9, 57, 58], MeO— [59, 60], EtO— [9, 21], Me(O=)CO— [2, 61], PhO— [9], F— [34, 36, 62–65], HS— [21], MeS— [66, 67], PhS— [28, 68], Me(O=)S— [69], Ph(O=)S— [28, 68], $Me(O{=})_2S$— [69], $Ph(O{=})_2S$— [68], $Cl(O{=})_2S$— [2, 70], $H_2N(O{=})_2S$— [2, 70], $R_2N(O{=})_2S$— (R = Me, Et, nPr) [71], H_2P— [2, 72], $(Ph)_2P$— [73], H_3Si— [74], Cl— [64, 70, 75, 76], Br— [77, 78], I— [2, 65, 78].

12.4.2 Disubstituted benzene derivatives

Table 12.7 Absorption regions (cm^{-1}) of the normal vibrations of 1,2-disubstituted benzenes

Vibration		1,2-di-light	1,2-light–heavy	1,2-di-heavy
νCH		3085 ± 35	3090 ± 20	3090 ± 30
νCH		3070 ± 30	3070 ± 20	3065 ± 25
νCH		3050 ± 30	3050 ± 20	3050 ± 20
νCH		3030 ± 30	3035 ± 35	3035 ± 35
νPh		1605 ± 20	1600 ± 20	1590 ± 25
νPh		1585 ± 20	1575 ± 15	1570 ± 15
νPh		1495 ± 30	1475 ± 25	1470 ± 25
νPh		1455 ± 25	1435 ± 25	1440 ± 20
νPh		1315 ± 40	1300 ± 40	1280 ± 20
δCH		1275 ± 25	1270 ± 30	1255 ± 25
νCX	(X)	1245 ± 55	1220 ± 80	1105 ± 25
νCX′	(X)	1170 ± 80	1085 ± 55	1050 ± 50
δCH		1150 ± 20	1150 ± 20	1160 ± 10
δCH		1110 ± 35	1125 ± 20	1125 ± 20
δCH		1030 ± 30	1030 ± 20	1030 ± 30
γ CH		975 ± 30	980 ± 15	980 ± 10
γ CH		940 ± 30	935 ± 25	945 ± 15
γ CH		880 ± 40	880 ± 40	860 ± 20
νPh	(X)	810 ± 80	770 ± 70	720 ± 30
γ CH		755 ± 35	755 ± 35	755 ± 15
δPh	(X)	730 ± 80	715 ± 85	685 ± 45
γ Ph		705 ± 35	700 ± 40	695 ± 35
δPh	(X)	560 ± 90	500 ± 150	450 ± 100
γ Ph	(X)	525 ± 65	515 ± 75	500 ± 60
γ Ph		440 ± 30	430 ± 40	420 ± 30
δPh	(X)	380 ± 180	380 ± 180	350 ± 150
δCX	(X)	370 ± 70	360 ± 100	270 ± 70
γ CX	(X)	270 ± 90	270 ± 90	215 ± 35
δCX′	(X)	230 ± 70	210 ± 70	170 ± 50
γ CX′	(X)	180 ± 60	170 ± 70	135 ± 35

di-light = benzenes substituted with two light atoms (atomic mass < 25).
light–heavy = benzenes substituted with a light and a heavy atom.
di-heavy = benzenes substituted with two heavy atoms (atomic mass > 25 *).
*: strong electron withdrawing groups such as NO_2, C(=O)OH, $C(=O)NH_2$, N=C=O give the effect of a heavy substituent.
(X) = X-sensitive vibration.

The compounds include:

2-MePhX′, where X′ = Me [93], Et [2], C(=O)H [22], C(=O)Me [2], C(=O)OH [25, 97], CO_2^- [25], HC=CH_2 [87, 88], C≡N [23], NH_2 [50], NO_2 [94], OH [96], F and I [93], Br and Cl [93, 168];

2-EtPhX′, where X′ = Et [2], NH_2 [2], OH [2];

2-F_3CPhX′, where X′ = C(=O)H [89], C(=O)Cl [100], C≡N [101], F, Cl and Br [27];

2-O_2NPhX′, where X′ = C(=O)OH [2], NH_2 [2], NO_2 [95, 102], OH [103], OEt [104], F [94, 105], Cl [94, 95, 105, 106], Br [94, 106], I [94, 105];

2-HOPhX′, where X′ = CH_2OH [107], C(=O)H [108, 109], C(=O)Me [110], C(=O)Et [110], C(=O)OMe and C(=O)OEt [108], OH [111, 112], F, Cl, Br and I [96];

2-MeOPhX′, where X′ = C(=O)H [113, 114], C(=O)Me [2], NH_2 [115], OMe [2], F [60], Cl [60, 116, 117], Br [60, 116, 117];

2-FPhX′, where X′ = CH_2Br, C(=O)H [22], C(=O)OH [25], CO_2^- [25], HC=CH_2 [87], C≡N [23], NH_2 [118], OEt [119], F [93, 120–123], Cl and Br [93, 124, 125], I [93];

2-ClPhX′, where X′ = CH_2Cl [170], C(=O)H [22], C(=O)OH [25, 97], CO_2^- [25], HC=CH_2 [87], C≡N [23], NH_2 [126], OEt [116], Cl [93, 95], Br and I [93];

2-BrPhX′, where X′ = C(=O)H [22], C(=O)OH [25], CO_2^- [25], HC=CH_2 [87], C≡N [23], OEt [116], Br [93, 127, 128], I [93];

2-IPhX′, where X′ = I [93];

2-HO(O=)CPhX′, where X′ = C(=O)OH [97, 129], NH_2 [97], NHMe and NMe_2 [130];

2-H(O=)CPhX′, where X′ = OEt [131], NH_2 [132];

2-HC≡CPhX′, where X′ = C≡CH [133];

2-N≡CPhX′, where X′ = CH_2Br [169], C≡N [134].

Table 12.8 Absorption regions (cm^{-1}) of the normal vibrations of 1,3-disubstituted benzenes

Vibration		1,3-di-light	1,3-light–heavy	1,3-di-heavy
νCH		3090 ± 30	3090 ± 20	3095 ± 25
νCH		3075 ± 25	3075 ± 25	3065 ± 15
νCH		3050 ± 30	3050 ± 20	3055 ± 15
νCH		3030 ± 30	3030 ± 30	3035 ± 25
νPh		1605 ± 20	1595 ± 20	1590 ± 25
νPh		1585 ± 20	1575 ± 20	1570 ± 20
νPh		1485 ± 25	1480 ± 20	1475 ± 25
νPh		1440 ± 30	1440 ± 30	1420 ± 20
νPh		1305 ± 35	1295 ± 25	1285 ± 15
δCH		1280 ± 35	1270 ± 30	1255 ± 15
νCX	(X)	1230 ± 80	1185 ± 80	1105 ± 30
δCH		1170 ± 30	1160 ± 20	1170 ± 10

(continued)

Table 12.8 (*continued*)

Vibration		1,3-di-light	1,3-light–heavy	1,3-di-heavy
δCH		1115 ± 45	1115 ± 45	1100 ± 30
δCH		1070 ± 30	1070 ± 30	1065 ± 20
νPh		1000 ± 10	1000 ± 10	1000 ± 10
γ CH		960 ± 30	960 ± 30	965 ± 15
γ CH		900 ± 50	900 ± 50	910 ± 40
γ CH		860 ± 50	860 ± 50	855 ± 40
νCX′	(X)	840 ± 120	805 ± 95	760 ± 50
γ CH		770 ± 45	770 ± 40	770 ± 30
δPh	(X)	720 ± 80	715 ± 85	685 ± 55
γ Ph		685 ± 25	685 ± 25	680 ± 20
γ Ph	(X)	530 ± 90	510 ± 110	480 ± 80
δPh	(X)	480 ± 90	440 ± 130	330 ± 100
γ Ph		440 ± 35	440 ± 35	430 ± 20
δPh	(X)	430 ± 90	390 ± 130	300 ± 100
δCX	(X)	390 ± 100	370 ± 120	265 ± 105
δCX′	(X)	290 ± 100	230 ± 90	170 ± 60
γ CX	(X)	235 ± 55	235 ± 55	200 ± 20
γ CX′	(X)	180 ± 60	170 ± 50	145 ± 45

Notes as in Table 12.7.

Compounds include:

3-MePhX, where (X = Me [92], Et [2], C(=O)H [22], C(=O)OH [25, 98], CO_2^- [25], CH=CH_2 [87, 88], C≡N [23], NH_2 [50], NO_2 [94], OH [96], OMe [2, 135], F and I [92], Cl and Br [92, 168];

3-EtPhX′, where X′ = Et [2], OH [2];

3-F_3CPhX′, where X′ = CF_3 [136], C(=O)H [89], C(=O)Cl [100], C≡N [101], NH_2 [137], F, Cl and Br [27];

3-O_2NPhX′, where X′ = C(=O)H [2], C(=O)OH [43], CH=CH_2 [87], NO_2 [95, 102], OH [103], F [94, 105], Cl [94, 95, 105, 106], Br [105, 106], I [94, 105];

3-HOPhX′, where X′ = CH_2OH [107], C(=O)H [2], C(=O)OH [2, 98], OH [111], F, Cl, Br and I [96];

3-MeOPhX′, where (X′ = C(=O)H [113, 114], C(=O)Me [2], C(=O)OH [98], C≡N [138], NH_2 [115], MeO [2], F [60, 117], Cl and Br [60];

3-FPhX′, where (X′ = C(=O)H [22], C(=O)OH [25], CO_2^- [25], CH=CH_2 [87], C≡N [23], NH_2 [118, 139], F [92, 120], Cl and Br [92, 124, 125], I [92];

3-ClPhX′, where X′ = C(=O)H [22], C(=O)OH [25, 98], CO_2^- [25], CH=CH_2 [87], C≡N [23], NH_2 [126], Cl [92, 95], Br and I [92];

3-BrPhX′, where X′ = nPr [140], C(=O)H [22], C(=O)Me [2], C(=O)OH [25], CO_2^- [25], CH=CH_2 [87], C≡N [23], Br [92, 127], I [92];

3-IPhX′, where X′ = I [92];

3-HO(O=)CPhX′, where (X′ = C(=O)OH [98, 129], NH_2 [98];

3-HC≡CPhX′, where X′ = C≡CH [133];

3-N≡CPhX′, where X′ = C≡N [134], NH_2 [141];
3-H_2NPhX′, where X′ = C(=O)H [132], C(=O)Me [2], S(=O)$_2$OH and S(=O)$_2$ONa [142].

Table 12.9 Absorption regions (cm^{-1}) of the normal vibrations of 1,4-disubstituted benzenes

Vibration		1,4-di-light	1,4-light–heavy	1,4-di-heavy
νCH		3090 ± 30	3090 ± 25	3095 ± 25
νCH		3075 ± 30	3070 ± 15	3080 ± 20
νCH		3055 ± 35	3050 ± 20	3070 ± 20
νCH		3045 ± 40	3025 ± 20	3050 ± 30
νPh		1610 ± 20	1600 ± 20	1590 ± 40
νPh		1585 ± 30	1580 ± 25	1585 ± 40
νPh		1500 ± 30	1495 ± 25	1495 ± 35
νPh		1430 ± 45	1415 ± 35	1405 ± 30
νPh		1325 ± 45	1320 ± 40	1305 ± 25
δCH		1280 ± 45	1270 ± 45	1265 ± 40
νCX	(X)	1245 ± 50	1210 ± 90	1085 ± 25
νCX′	(X)	1160 ± 90	1075 ± 45	1055 ± 35
δCH		1165 ± 25	1170 ± 20	1175 ± 15
δCH		1105 ± 25	1110 ± 30	1120 ± 20
δCH		1015 ± 20	1010 ± 15	1010 ± 15
γ CH		970 ± 30	965 ± 25	960 ± 10
γ CH		940 ± 35	940 ± 30	940 ± 20
γ CH		840 ± 50	830 ± 40	825 ± 35
γ CH		825 ± 45	825 ± 35	825 ± 25
νPh	(X)	790 ± 70	780 ± 80	745 ± 65
γ Ph		705 ± 30	700 ± 35	700 ± 25
δPh	(X)	660 ± 120	640 ± 140	510 ± 110
δPh		635 ± 25	625 ± 15	620 ± 15
γ Ph	(X)	540 ± 80	505 ± 65	480 ± 30
δPh	(X)	455 ± 80	365 ± 150	270 ± 115
γ Ph		400 ± 20	400 ± 20	395 ± 15
δCX	(X)	370 ± 100	365 ± 85	310 ± 60
δCX′	(X)	290 ± 140	245 ± 105	220 ± 80
γ CX	(X)	260 ± 120	250 ± 110	230 ± 90
γ CX′	(X)	195 ± 90	155 ± 70	115 ± 30

Notes as in Table 12.7.

Compounds include:
4-MePhX′, where X′ = Me [90], Et [2], C(=O)H [22], C(=O)Me [2], C(=O)OH [25, 99], CO_2^- [25], CH=CH_2 [88], C≡N [23, 143], NO_2 [94], OH [96, 144], OMe [135], SH [91], F and I [90], Cl and Br [90, 168];
4-EtPhX′, where X′ = Et [2], OH [2];
4-tBuPhX′, where X′ = tBu [145];
4-F_3CPhX′, where X′ = C(=O)H [89], C(=O)Cl [100], C≡N [101], NH_2 [146], F, Cl and Br [27];

4-O_2NPhX′, where X′ = C(=O)OH [2, 43], NH_2 [147], NO_2 [95, 102], OH [103], OEt [104], F [94, 105], Cl [95, 105], Br [105, 106], I [94, 105];

4-HOPhX′, where X′ = CH_2OH [107], C(=O)H [2], C(=O)Me [2], C(=O)OH [2, 99], C≡N [148], NH_2 [149], OH [58, 111], F, Cl, Br and I [96];

4-MeOPhX′, where X′ = CH_2OH [150], C(=O)H [113, 114], C(=O)Me [2], C(=O)OH [99], C≡N [138], NH_2 [115], F [60, 151, 152, 153], Cl and Br [60, 116, 117, 152, 153], I [153];

4-FPhX′, where X′ = CH_2OH [154], CH_2Cl [155], C(=O)H [22], C(=O)Et [156], C(=O)CH_2CH_2Cl [157], C(=O)CH_2Cl [158], C(=O)OH [25], CO_2^- [25], C≡N [23], NH_2 [118], SH [91], F [90, 120], Cl and Br [90, 124, 125], I [90];

4-ClPhX′, where X′ = CH_2C≡N [150], CH_2Cl [170], C(=O)H [22, 159], C(=O)Me [2, 160], C(=O)Et [156], C(=O)CH_2CH_2Cl [157], C(=O)OH [25, 99], CO_2^- [25], C≡N [23, 161], NH_2 [126, 162], NHC(=O)Me [163], OEt [2, 116], SH and SMe [91], Cl [90, 95], Br and I [90];

4-BrPhX′, where X′ = C(=O)H [22], C(=O)Me [2], C(=O)Et [156], C(=O)CH_2CH_2Cl [157], C(=O)OH [25], CO_2^- [25], C≡N [23], NH_2 [164], NHC(=O)Me [52], OEt [2, 116], SH and SMe [91], Br [90, 127], I [90];

4-IPhX′, where X′ = C(=O)OH [25], CO_2^- [25], SH [91], I [90];

4-HO(O=)CPhX′, where X′ = C(=O)H [131], C(=O)OH [99, 129], NH_2 [2, 99], NMe_2 [165];

4-H(O=)CPhX′, where X′ = NH_2 [132], NMe_2 [165], OEt [131];

4-HC≡CPhX′, where X′ = C≡CH [133, 167];

4-N≡CPhX′, where X′ = CH_2Br [169], C≡N [134, 166], C(=O)H [131], NH_2 [141, 148], NMe_2 [161, 165];

4-H_2NPhX′, where X′ = C(=O)Me [2], NH_2 [165], NHC(=O)Me [52], S$(=O)_2$OH and S$(=O)_2$ONa [142].

12.4.3 Trisubstituted benzene derivatives

Table 12.10 Absorption regions (cm^{-1}) of the normal vibrations of 1,2,3-trisubstituted benzenes

Vibration		tri-light	di-light, heavy	light, di-heavy	tri-heavy
νCH		3085 ± 25	3085 ± 15	3085 ± 15	3075 ± 25
νCH		3055 ± 25	3055 ± 25	3060 ± 20	3055 ± 15
νCH		3025 ± 25	3020 ± 20	3035 ± 25	3035 ± 25
νPh		1610 ± 25	1600 ± 20	1600 ± 30	1585 ± 20
νPh		1575 ± 45	1575 ± 15	1570 ± 20	1555 ± 15
νPh		1505 ± 45	1480 ± 30	1475 ± 35	1440 ± 20
νPh		1455 ± 35	1440 ± 30	1430 ± 30	1415 ± 25
νPh		1290 ± 40	1285 ± 35	1280 ± 30	1280 ± 20
δCH		1265 ± 40	1260 ± 40	1260 ± 40	1265 ± 25
νCX	(X)	1255 ± 55	1245 ± 55	1235 ± 55	1180 ± 20

Table 12.10 (*continued*)

Vibration		tri-light	di-light, heavy	light, di-heavy	tri-heavy
$\nu CX'$	(X)	1160 ± 60	1160 ± 60	1080 ± 50	1030 ± 30
δCH		1155 ± 35	1165 ± 25	1155 ± 25	1155 ± 25
δCH		1085 ± 35	1085 ± 35	1085 ± 35	1075 ± 25
$\nu CX''$	(X)	955 ± 65	930 ± 90	840 ± 90	770 ± 40
γ CH		970 ± 25	970 ± 25	970 ± 25	970 ± 20
γ CH		890 ± 30	890 ± 30	890 ± 30	890 ± 20
γ CH		790 ± 40	790 ± 40	775 ± 25	770 ± 20
νPh	(X)	780 ± 50	770 ± 60	755 ± 55	715 ± 35
γ Ph		715 ± 25	710 ± 20	700 ± 20	700 ± 20
δPh	(X)	650 ± 70	630 ± 50	630 ± 50	470 ± 50
γ Ph	(X)	520 ± 60	520 ± 60	545 ± 45	520 ± 20
δPh	(X)	505 ± 95	495 ± 95	485 ± 105	445 ± 65
γ Ph	(X)	480 ± 70	470 ± 50	500 ± 50	480 ± 40
δPh	(X)	435 ± 65	430 ± 60	390 ± 50	330 ± 50
δCX	(X)	405 ± 75	405 ± 75	405 ± 75	370 ± 50
$\delta CX'$	(X)	330 ± 60	310 ± 60	260 ± 70	210 ± 50
γ CX	(X)	285 ± 55	285 ± 55	245 ± 35	235 ± 25
$\delta CX''$	(X)	270 ± 50	250 ± 50	220 ± 60	180 ± 50
γ CX′	(X)	210 ± 50	210 ± 50	180 ± 40	180 ± 40
γ CX″	(X)	140 ± 40	140 ± 40	120 ± 40	90 ± 30

light = benzenes substituted with light atom (atomic mass < 25).
heavy = benzenes substituted with heavy atom (atomic mass > 25 *).
* Strong electron withdrawing groups such as NO_2, C(=O)OH, C(=O)NH_2, N=C=O have the effect of a heavy substituent.

Compounds include:

Me—Ph-2X′-3-X″ (X′, X″ = Me, Me [2, 171, 209], Me, C≡N [232], Me, NH_2 [200, 221], Me, OH [173], Me, F [2, 171], Me, Br [207, 254], C≡N, Cl [184], NH_2, Me [2, 244], NH_2, Cl [199, 247], OH, Me [2, 173], OMe, OMe [182, 255], F, Me [2, 171], F, Cl [2], Cl, Me [2, 171], Br, Me [207, 254]);

H(O=)C—Ph-2X′-3-X″ (X′, X″ = OH, OH [248], OH, OMe [2, 188], OMe, OMe [223], F, F [253]);

N≡C—Ph-2-X′-3-X″ (X′, X″ = F, F [202]);

H_2N—Ph-2-X′-3-X″ (X′, X″ = Me, NH_2 [181], Me, Cl [199, 236], Cl, Cl [177, 230]);

O_2N—Ph-2-X′-3-X″ (X′, X″ = Me, NH_2 [192], Me, NO_2 [205], Me, Cl [210, 239], Cl, Cl [102, 233]);

HO—Ph-2-X′-3-X″ (X′, X″ = OH, OH [2, 174], OMe, OMe [196]);

MeO—Ph-2-X′-3-X″ (X′, X″ = OH, OMe [2], OMe, OMe [2], Cl, Cl [179]);

F—Ph-2-X′-3-X″ (X′, X″ = Me, Cl [171, 212], C≡N, F [202], C≡N, Cl [141, 229], NH_2, F);

Cl—Ph-2-X′-3-X″ (X′, X″ = Me, Cl [2, 171], CH_2Cl, Cl [242, 256], C(=O)H, Cl [219], C(=O)Cl, Cl, C≡N, Cl, NH_2, Cl [177, 187, 210, 230], OH, Cl [2, 173], OMe, Cl [2], F, Cl [2, 171], Cl, Cl [171, 191, 203], Cl, Br [203], Br, Cl [203, 226]).

Table 12.11 Absorption regions (cm^{-1}) of the normal vibrations of 1,2,4-trisubstituted benzenes

Vibration		tri-light	di-light, heavy	light, di-heavy	tri-heavy
νCH		3085 ± 40	3085 ± 35	3085 ± 25	3085 ± 25
νCH		3065 ± 35	3065 ± 25	3065 ± 25	3065 ± 25
νCH		3030 ± 30	3030 ± 30	3030 ± 30	3030 ± 30
νPh		1615 ± 30	1615 ± 25	1585 ± 25	1570 ± 15
νPh		1575 ± 30	1575 ± 25	1575 ± 25	1555 ± 15
νPh		1480 ± 50	1490 ± 40	1475 ± 35	1450 ± 20
νPh		1440 ± 50	1440 ± 50	1410 ± 50	1370 ± 20
νPh		1295 ± 35	1295 ± 35	1280 ± 30	1275 ± 25
δCH		1255 ± 40	1255 ± 35	1250 ± 30	1240 ± 25
νCX	(X)	1250 ± 60	1240 ± 60	1220 ± 80	1095 ± 25
νCX′	(X)	1160 ± 80	1120 ± 95	1060 ± 50	1025 ± 25
δCH		1150 ± 35	1140 ± 25	1140 ± 25	1140 ± 20
δCH		1075 ± 55	1095 ± 45	1095 ± 45	1110 ± 20
νCX″	(X)	940 ± 60	900 ± 100	830 ± 90	780 ± 40
γ CH		950 ± 40	950 ± 40	935 ± 35	945 ± 15
γ CH		890 ± 50	880 ± 40	880 ± 40	860 ± 20
γ CH		820 ± 40	810 ± 40	820 ± 40	810 ± 15
νPh	(X)	755 ± 65	760 ± 40	720 ± 50	665 ± 25
δPh	(X)	700 ± 80	660 ± 60	630 ± 60	520 ± 70
γ Ph		700 ± 40	700 ± 40	695 ± 35	680 ± 20
γ Ph	(X)	580 ± 50	565 ± 45	565 ± 45	530 ± 30
δPh	(X)	505 ± 85	495 ± 75	495 ± 75	430 ± 50
γ Ph	(X)	465 ± 65	460 ± 70	450 ± 50	430 ± 30
δPh	(X)	430 ± 70	410 ± 70	410 ± 70	350 ± 50
δCX	(X)	370 ± 80	365 ± 75	355 ± 65	280 ± 60
δCX′	(X)	320 ± 70	285 ± 65	285 ± 65	190 ± 30
γ CX	(X)	290 ± 90	290 ± 90	275 ± 75	260 ± 60
δCX″	(X)	235 ± 65	225 ± 55	225 ± 55	160 ± 40
γ CX′	(X)	220 ± 70	210 ± 60	200 ± 50	165 ± 35
γ CX″	(X)	160 ± 60	140 ± 60	140 ± 60	110 ± 30

Notes as for Table 12.10.

Compounds include:

Me—Ph-2-X′-4-X″ (X′, X″ = Me, Me [2, 172, 209], Me, NH_2 [221], Me, OH [173], Me, F [2, 172], C(=O)H, Me [249], C≡N, Me [232], NH_2, Me [2, 244], NH_2, NH_2 [181], NH_2, NO_2 [243], NH_2, Cl [236], NO_2, NH_2 [243], NO_2, NO_2

[197, 205], NO_2, Cl [214], OH, Me [173], F, Me [2], F, NH_2 [220], F, F [172, 216, 250], F, Cl [2, 172], Cl, NH_2 [236], Cl, F [2, 172], Cl, Cl [2, 172, 216], Br, Me [207], I, F [180], Br, Br [254]);

$ClCH_2$—Ph-2-X′-4-X″ (X′, X″ = Cl, Cl [256]);

HO(O=)CCH_2—Ph-2-X′-4-X″ (X′, X″ = NO_2, NO_2 [205]);

MeO(O=)CCH_2—Ph-2-X′-4-X″ (X′, X″ = NO_2, NO_2 [205]);

F_3C—Ph-2-X′-4-X″ (X′, X″ = Cl, Cl [185]);

H(O=)C—Ph-2-X′-4-X″ (X′, X″ = Me, Me [249], OMe, OMe [223], Cl, Cl [219], F, F [253]);

HO(O=)C—Ph-2-X′-4-X″ (X′, X″ = Cl, Cl [237]);

Cl(O=)C—Ph-2-X′-4-X″ (X′, X″ = Cl, Cl [183, 186]);

N≡C—Ph-2-X′-4-X″ (X′, X″ = Me, Me [232], F, F [202]);

H_2N—Ph-2-X′-4-X″ (X′, X″ = Me, Me [198, 221], Me, NO_2 [243], Me, OH [193, 196], Me, Cl [236], Me, Br [211], NH_2, Me [181], NO_2, Me [2, 243], NO_2, NO_2 [2], NO_2, Cl [208, 218], OH, Me [176], F, F [225], Cl, NO_2 [208, 218], Cl, Cl [2, 245], Br, Me [195]);

H_2NHN—Ph-2-X′-4-X″ (X′, X″ = NO_2, NO_2 [2, 197]);

Me(O=)CHN—Ph-2-X′-4-X″ (X′, X″ = Cl, Cl [241]:

O_2N—Ph-2-X′-4-X″ (X′, X″ = Me, OH [193, 196], HO, C(=O)H [2], Cl, Cl [102]:

HO—Ph-2-X′-4-X″ (X′, X″ = Me, Me [173], Me, Cl [231], $BrCH_2$, NO_2 [238], C(=O)Me, NO_2 [2], NH_2, Me [178], NH_2, NO_2 [193], NH_2, Cl [178], NO_2, C(=O)Me [2], NO_2, NO_2 [2, 197, 234], OH, C(=O)H [2, 248], OMe, C(=O)H [188], Cl, Cl [2, 215]);

MeO—Ph-2-X′-4-X″ (X′, X″ = $BrCH_2$, NO_2 [238], NH_2, Cl [227], OH, C(=O)H [188], OMe, C(=O)H [2, 223], OMe, OMe [235], Cl, NH_2 [227], Br, NO_2 [224]);

F—Ph-2-X′-4-X″ (X′, X″ = Me, F [2, 172, 250], Me, Cl [2, 172, 251], Me, I [180], C≡N, F [202], NH_2, Me [220], NH_2, F [2, 217], NO_2, NO_2 [2, 175, 206], F, C≡N [202], F, NH_2 [202], F, OMe [202], F, C(=O)H [253], C(=O)H, F [253], F, F [2, 172], F, Br [252], Cl, NH_2, [ ] Cl, Cl [184], Br, F [252]);

Cl—Ph-2-X′-4-X″ (X′, X″ = Me, OH [231], Me, F [2, 172, 251], Me, Cl [2, 172], F_3C, Cl, [ ] C(=O)OH, NO_2 [237], NH_2, Cl [177, 217, 245], NO_2, NO_2 [2, 106, 206], NO_2, Cl [102, 228], NHC(=O)Me, Cl [246], OH, Me [231], OH, Cl [2, 173], F, NH_2 [222], Cl, Me [2, 172], Cl, CF_3, [ ] Cl, C(=O)H [219], Cl, C(=O)Cl [183, 186], Cl, NH_2 [201], Cl, NHC(=O)Me [246], Cl, NO_2 [102, 213], Cl, OH [173], Cl, F [2, 172], Cl, Cl [2, 172, 191], Cl, Br [203, 240], Br, Cl [203, 240]);

Br—Ph-2-X′-4-X″ (X′, X″ = Me, Me [207], Me, NH_2 [195, 211], Me, Br [2, 212, 254], NO_2, Br [213], OMe, NO_2 [224], F, F [252], Cl, Cl [203, 240], Br, Br [203]).

Table 12.12 Absorption regions (cm^{-1}) of the normal vibrations of 1,3,5-trisubstituted benzenes

Vibration		tri-light	di-light, heavy	light, di-heavy	tri-heavy
νCH		3080 ± 30	3080 ± 30	3075 ± 25	3075 ± 25
νCH		3045 ± 45	3045 ± 45	3045 ± 45	3045 ± 45
νCH		3045 ± 45	3045 ± 45	3045 ± 45	3045 ± 45
νPh		1610 ± 20	1600 ± 20	1590 ± 20	1570 ± 20
νPh		1610 ± 20	1580 ± 20	1570 ± 20	1570 ± 20
νPh		1480 ± 40	1470 ± 30	1460 ± 40	1430 ± 30
νPh		1450 ± 40	1440 ± 30	1410 ± 40	1410 ± 30
νPh		1300 ± 30	1300 ± 30	1295 ± 25	1295 ± 25
δCH		1260 ± 60	1260 ± 40	1240 ± 40	1230 ± 30
νCX	(X)	1245 ± 65	1240 ± 50	1230 ± 60	1130 ± 30
δCH		1140 ± 40	1150 ± 30	1130 ± 30	1090 ± 30
δCH		1115 ± 55	1100 ± 30	1090 ± 30	1090 ± 30
νPh		995 ± 15	995 ± 15	1000 ± 10	990 ± 10
νCX′	(X)	940 ± 60	880 ± 100	880 ± 100	770 ± 50
νCX″	(X)	940 ± 60	850 ± 100	800 ± 80	770 ± 50
γ CH		865 ± 25	870 ± 30	910 ± 30	865 ± 25
γ CH		865 ± 25	860 ± 30	860 ± 30	865 ± 25
γ CH		835 ± 25	835 ± 25	835 ± 25	835 ± 25
γ Ph		685 ± 25	680 ± 20	685 ± 25	665 ± 15
δPh	(X)	560 ± 40	560 ± 40	490 ± 110	319 ± 90
γ Ph	(X)	550 ± 50	540 ± 60	520 ± 60	520 ± 40
γ Ph	(X)	550 ± 50	540 ± 60	520 ± 60	520 ± 40
δPh	(X)	480 ± 50	470 ± 50	450 ± 50	390 ± 50
δPh	(X)	480 ± 50	470 ± 50	450 ± 50	390 ± 50
δCX	(X)	440 ± 60	380 ± 80	380 ± 80	440 ± 60
δCX′	(X)	290 ± 50	250 ± 90	250 ± 90	160 ± 50
δCX″	(X)	290 ± 50	250 ± 90	250 ± 90	160 ± 50
γ CX	(X)	220 ± 40	220 ± 40	190 ± 40	200 ± 30
γ CX′	(X)	220 ± 40	220 ± 40	190 ± 40	200 ± 30
γ CX″	(X)	170 ± 50	160 ± 40	140 ± 40	140 ± 40

Notes as Table 12.10.

Compounds include: Me—Ph-3-X′-5-X″ (X′, X″ = Me, Me [2, 171, 209], Me, NH_2 [2, 171], Me, OH [173], OH, OH [2], Me, F [2, 171, 173], Me, Cl [2, 171], Me, Br [2, 171], Cl, Cl [2]);

Et—Ph-3-X′-5-X″ (X′, X″ = Et, Et [2]);

H(O=)C—Ph-3-X′-5-X″ (X′, X″ = Cl, Cl);

HO(O=)C—Ph-3-X′-5-X″ (X′, X″ = NO[2], NO_2 [2, 237]);

Cl(O=)C—Ph-3-X′-5-X″ (X′, X″ = OMe, OMe, [ ] Cl, Cl [183]);

N≡C—Ph-3-X′-5-X″ (X′, X″ = Cl, Cl);

H_2N—Ph-3-X′-5-X″ (X′, X″ = Cl, Cl [177]);

O_2N—Ph-3-X′-5-X″ (X′, X″ = Cl, Cl [205]);
HO—Ph-3-X′-5-X″ (X′, X″ = OMe, OMe [194]);
MeO—Ph-3-X′-5-X″ (X′, X″ = OMe, OMe [2, 235], OMe, Cl, [ ] Cl, Cl [151]);
F—Ph-3-X′-5-X″ (X′, X″ = F, F [2, 171, 189, 204], F, Br [252]);
Cl—Ph-3-X′-5-X″ (X′, X″ = Cl, Cl [2, 171, 190], Cl, Br [203]);
Br—Ph-3-X′-5-X″ (X′, X″ = Br, Br [2, 171, 203]).

References

1. G. Varsányi and S. Szöke, *Vibrational Spectra of Benzene Derivatives*, Academic Press, New York, London (1969).
2. G. Varsányi, *Assignments for Vibrational Spectra of Seven Hundred Benzene Derivatives*, John Wiley & Sons, New York (1974).
3. N.B. Colthup, *Appl. Spectrosc.*, **30**, 589 (1976).
4. N.B. Colthup, L.H. Daly and S.E. Wiberley, *Introduction to Infrared and Raman Spectroscopy*, Academic Press, Boston, 3rd edn. (1990).
5. A. Kuwae and K. Machida, *Spectrochim. Acta, Part A*, **34A**, 785 (1978).
6. C.W. Young, R.B. Duvall and N. Wright, *Anal. Chem.*, **23**, 709 (1951).
7. S. Higuchi, H. Tsuyama, S. Tanaka and H. Kamada, *Spectrochim. Acta, Part A*, **30A**, 463 (1974).
8. E.B. Wilson Jr. *Phys. Rev.*, **45**, 427 (1934).
9. J.E. Katon, W.R. Feairheller Jr. and E.R. Lippincot, *J. Mol. Spectrosc.*, **13**, 72 (1964).
10. R.J.A. Ribeiro-Claro, A.M.D'A Rocha Gonsalves and J.J.C. Teixeira-Dias, *Spectrochim. Acta, Part A*, **41A**, 1055 (1985).
11. J.J.P. Stewart, S.R. Bosco and W.R. Carper, *Spectrochim. Acta, Part A*, **42A**, 13 (1986).
12. M.S. Mathur and N.A. Weir, *J. Mol. Struct.*, **14**, 303 (1972).
13. S. Chattopadhyay, *Indian J. Phys.*, **41**, 759 (1967).
14. P.K. Mallick, S. Chattopadhyay and S.B. Banerjee, *Indian J. Pure Appl. Phys.*, **11**, 609 (1973).
15. C.E. Sjøgren, *Acta Chem. Scand., Ser. A*, **38A**, 657 (1984).
16. U.C. Joshi, M. Joshi and R.N. Singh, *Spectrochim. Acta, Part A*, **37A**, 592 (1981).
17. S. Chakravorti, R. De, P.K. Mallick and S.B. Banerjee, *Spectrochim. Acta, Part A*, **49A**, 543 (1993).
18. A. Marchand and J.P. Quintard, *Spectrochim. Acta, Part A*, **36A**, 941 (1980).
19. T.R. Gilson, J.M. Hollas, E. Khalilipour and J.V. Warrington, *J. Mol. Spectrosc.*, **73**, 234 (1978).
20. J.H.S. Green, *Spectrochim. Acta*, **17**, 607 (1961).
21. J.H.S. Green, *Spectrochim. Acta*, **18**, 39 (1962).
22. J.H.S. Green and D.J. Harrison, *Spectrochim. Acta, Part A*, **32A**, 1265 (1976).
23. J.H.S. Green and D.J. Harrison, *Spectrochim. Acta, Part A*, **32A**, 1279 (1976).
24. J.H.S. Green and D.J. Harrison, *Spectrochim. Acta, Part A*, **33A**, 249 (1977).
25. J.H.S. Green, *Spectrochim. Acta, Part A*, **33A**, 575 (1977).
26. J.H.S. Green and D.J. Harrison, *Spectrochim. Acta, Part A*, **33A**, 583 (1977).
27. J.H.S. Green and D.J. Harrison, *Spectrochim. Acta, Part A*, **33A**, 837 (1977).
28. J.H.S. Green, *Spectrochim. Acta, Part A*, **24A**, 1627 (1968).
29. A. Bree and R. Zwarich, *J. Mol. Struct.*, **75**, 213 (1981).
30. Z. Meić and H. Güsten, *Spectrochim. Acta, Part A*, **34A**, 101 (1978).
31. K. Palmo, *Spectrochim. Acta, Part A*, **44A**, 341 (1988).
32. R. Buguene-Hoffmann, *Spectrochim. Acta, Part A*, **45A**, 1227 (1989).

33. R.M.P. Jaiswal and P.P. Garg, *Indian J. Phys.*, **58B**, 307 (1984).
34. D. Steele and E.R. Lippincott, *J. Mol. Spectrosc.*, **6**, 238 (1961).
35. A.G. Császár and G. Fogarasi, *Spectrochim. Acta, Part A*, **45A**, 845 (1989).
36. R.J. Jakobsen, *Spectrochim. Acta*, **21**, 128 (1965).
37. R. Zwarich, J. Smolarek and L. Goodman, *J. Mol. Spectrosc.*, **38**, 336 (1971).
38. D. Condit, S.M. Craven and J.E. Katon, *Appl. Spectrosc.*, **28**, 420 (1974).
39. A. Gambi, S. Giorgianni, A. Passerini, R. Visinoni and S. Ghersetti, *Spectrochim. Acta, Part A*, **36A**, 871 (1980).
40. W.D. Mross and G. Zundel, *Spectrochim. Acta, Part A*, **26A**, 1097 (1970).
41. Y. Kim and K. Machida, *Spectrochim. Acta, Part A*, **42A**, 881 (1986).
42. C.W. Bird, *Spectrochim. Acta, Part A*, **24A**, 1666 (1968).
43. H. Ratinen and M. Kiviharju, *Spectrochim. Acta, Part A*, **45A**, 732 (1989).
44. F.J. Boerio and S.K. Bahl, *Spectrochim. Acta, Part A*, **32A**, 987 (1976).
45. S. Weckherlin and W. Lüttke, *Z. Elektrochem.*, **64**, 1228 (1960).
46. K. Machida, A. Kuwae, Y. Saito and T. Uno, *Spectrochim. Acta, Part A*, **34A**, 793 (1978).
47. J.C. Evans, *Spectrochim. Acta*, **16**, 428 (1960).
48. J.C. Evans and R.A. Nyquist, *Spectrochim. Acta*, **16**, 918 (1960).
49. G.W. King and S.P. So, *J. Mol. Spectrosc.*, **36**, 468 (1970).
50. A.K. Ansari and P.K. Verma, *India J. Pure Appl. Phys.*, **16**, 454 (1978).
51. V. Guichard, A. Bourkba, M.-F. Lautie and O. Poizat, *Spectrochim. Acta, Part A*, **45A**, 187 (1989).
52. S. Tariq, N. Ali and P.K. Verma, *Indian J. Pure Appl. Phys.*, **22**, 265 (1984).
53. J.D. Laposa, *Spectrochim. Acta, Part A*, **35A**, 65 (1979).
54. R.E. Clavijo-Campos and B. Weiss-Löpez, *Spectrosc. Lett.*, **23**, 137 (1990).
55. G.W. Chantry, E.A. Nicol, D.J. Harrison, A. Bouchy and G. Roussy, *Spectrochim. Acta, Part A*, **30A**, 1717 (1974).
56. C.V. Stephenson, W.C. Coburn and W.S. Wilcox, *Spectrochim. Acta*, **17**, 933 (1961).
57. H.D. Bist, J.C.D. Brand and D.R. Williams, *J. Mol. Spectrosc.*, **21**, 76 (1966).
58. M. Kubinyi, F. Billes, A. Grofcsik and G. Keresztury, *J. Mol. Struct.*, **266**, 339 (1992).
59. W.J. Balfour, *Spectrochim. Acta, Part A*, **39A**, 795 (1983).
60. N.L. Owen and R.E. Hester, *Spectrochim. Acta, Part A*, **25A**, 343 (1969).
61. Y. Kim, H. Noma and K. Machida, *Spectrochim. Acta, Part A*, **42A**, 891 (1986).
62. G. Fogarasi and A.G. Császár, *Spectrochim. Acta, Part A*, **44A**, 1067 (1988).
63. E.D. Lipp and C.J. Seliskar, *J. Mol. Spectrosc.*, **73**, 290 (1978).
64. T.R. Nanney, R.T. Baily and E.R. Lippincott, *Spectrochim. Acta*, **21**, 1500 (1965).
65. H.J.K. Koser, *Spectrochim. Acta, Part A*, **40A**, 125 (1984).
66. G. Paliani and S. Santini, *J. Raman Spectrosc.*, **19**, 161 (1988).
67. W.J. Balfour, K.S. Chandrasekhar and S.P. Kyca, *Spectrochim. Acta, Part A*, **42A**, 39 (1986).
68. B. Nagel, T. Steiger, J. Fruwert and G. Geiseler, *Spectrochim. Acta, Part A*, **31A**, 255 (1975).
69. M. Bouquet, G. Chassaing, J. Corset, J. Favrot and J. Limouzi, *Spectrochim. Acta, Part A*, **37A**, 727 (1981).
70. T. Uno, K. Machida and K. Hanai, *Spectrochim. Acta, Part A*, **24A**, 1705 (1968).
71. M. Goldstein, M.A. Russell and H.A. Willis, *Spectrochim. Acta, Part A*, **25A**, 1275 (1969).
72. H. Stenzenberger and H. Schindlbauer, *Spectrochim. Acta, Part A*, **26A**, 1713 (1970).
73. H.G.M. Edwards, A.F. Johnson and I.R. Lewis, *Spectrochim. Acta, Part A*, **49A**, 707 (1933).

74. J.R. Durig, K.L. Hellams and J.H. Mulligan, *Spectrochim. Acta, Part A*, **28A**, 1039 (1972).
75. T. Uno, K. Machida and K. Hanai, *Spectrochim. Acta, Part A*, **24A**, 1705 (1968).
76. H.D. Bist, V.N. Sarin, A. Ojha and Y.S. Jain, *Spectrochim. Acta, Part A*, **26A**, 841 (1970).
77. T. Uno, A. Kuwae and K. Machida, *Spectrochim. Acta, Part A*, **33A**, 607 (1977).
78. D.H. Whiffen, *J. Chem. Soc.*, 1350 (1956).
79. R.A. Nalepa and J.D. Laposa, *J. Mol. Spectrosc.*, **50**, 106 (1974).
80. R.D. Cunha and V.B. Kartha, *Can. J. Spectrosc.*, **20**, 18 (1975).
81. A.L. Smith, *Spectrochim. Acta, Part A*, **24A**, 695 (1968).
82. C.V. Stephenson and W.C. Coburn, *J. Chem. Phys.*, **42**, 35 (1965).
83. A. Perrier-Datin and J.M. Lebas, *J. Chim. Phys.*, **69**, 591 (1972).
84. R.D. McLachlan, *Spectrochim. Acta, Part A*, **26A**, 919 (1970).
85. A.M. North, R.A. Pethrick and A.D. Wilson, *Spectrochim. Acta, Part A*, **30A**, 1317 (1974).
86. A.J. Aarts, H.O. Desseyn, B.J. Van der Veken and M.H. Herman, *Can. J. Spectrosc.*, **24**, 29 (1979).
87. W.G. Fateley, G.L. Carlson and F.E. Dickson, *Appl. Spectrosc.*, **22**, 651 (1968).
88. P.P. Garg and R.M.P. Jaiswal, *Indian J. Pure Appl. Phys.*, **27**, 75 (1989).
89. R.A. Yadav and I.S. Singh, *Indian J. Phys.*, **58B**, 556 (1984).
90. J.H.S. Green, *Spectrochim. Acta, Part A*, **26A**, 1503 (1970).
91. J.H.S. Green, D.J. Harrison, W. Kynaston and D.W. Scott, *Spectrochim. Acta, Part A*, **26A**, 1515 (1970).
92. J.H.S. Green, *Spectrochim. Acta, Part A*, **26A**, 1523 (1970).
93. J.H.S. Green, *Spectrochim. Acta, Part A*, **26A**, 1913 (1970).
94. J.H.S. Green and D.J. Harrison, *Spectrochim. Acta, Part A*, **26A**, 1925 (1970).
95. J.H.S. Green and H.A. Lauwers, *Spectrochim. Acta, Part A*, **27A**, 817 (1971).
96. J.H.S. Green, D.J. Harrison and W. Kynaston, *Spectrochim. Acta, Part A*, **27A**, 2199 (1971).
97. E. Sánchez de la Blanca, J.L. Nunez and P. Martinez, *J. Mol. Struct.*, **142**, 45 (1986).
98. E. Sánchez de la Blanca, J.L. Nunez and P. Martinez, *An. Quim.*, **82**, 490 (1986).
99. E. Sánchez de la Blanca, J.L. Nunez and P. Martinez, *An. Quim.*, **82**, 480 (1986).
100. R. Shanker, R.A. Yadav, I.S. Singh and O.N. Singh, *J. Raman Spectrosc.*, **23**, 141 (1992).
101. R.A. Yadav and I.S. Singh, *Proc. Indian Acad. Sci.* (*Chem. Sci.*), **95**, 471 (1985).
102. V. Suryanarayana, A.P. Kumer, G.R. Rao and G.C. Pandey, *Spectrochim. Acta, Part A*, **48A**, 1481 (1992).
103. Y. Kishore, S.N. Sharma and C.P.D. Dwivedi, *Indian J. Phys.*, **48**, 412 (1974).
104. K.M. Mathur, D.P. Juyal and R.N. Singh, *Indian J. Pure Appl. Phys.*, **9**, 756 (1971).
105. P.M. Rao and G.R. Rao, *J. Raman Spectrosc.*, **20**, 529 (1989).
106. E.F. Mooney, *Spectrochim. Acta*, **20**, 1021 (1964).
107. N.M.D. Brown, B.J. Meenan and G.M. Taggart, *Spectrochim. Acta, Part A*, **48A**, 939 (1992).
108. M.M. Radhi and M.F. El-Bermani, *Spectrochim. Acta, Part A*, **46A**, 33 (1990).
109. A.P. Upadhyay and K.N. Upadhyay, *Indian J. Phys.*, **55B**, 232 (1981).
110. W.A.L.K. Al-Rashid and M.F. El-Bermani, *Spectrochim. Acta, Part A*, **47A**, 35 (1991).
111. H.W. Wilson, *Spectrochim. Acta, Part A*, **30A**, 2141 (1974).
112. S.J. Greaves and W.P. Griffith, *Spectrochim. Acta, Part A*, **47A**, 133 (1991).
113. M.P. Srivastava, O.N. Singh and I.S. Singh, *Curr. Sci.*, **37**, 100 (1968).
114. C.P.D. Dwivedi, *Indian J. Pure Appl. Phys.*, **6**, 440 (1968).
115. V.B. Singh and A.K. Sinha, *Indian J. Phys.*, **61B**, 344 (1987).

116. E.F. Mooney, *Spectrochim. Acta*, **19**, 877 (1963).
117. B. Lakshmaiah and G.R. Rao, *J. Raman Spectrosc.*, **20**, 449 (1989).
118. M.A. Shashidhar, K.S. Rao and E.S. Jayadevappa, *Spectrochim. Acta, Part A*, **26A**, 2373 (1970).
119. R.C. Maheshwari and M.M. Shukla, *Indian J. Pure Appl. Phys.*, **13**, 135 (1975).
120. O.P. Singh, J.S. Yadav and R.A. Yadav, *Proc. Indian Acad. Sci.* (*Chem. Sci.*), **99**, 159 (1987).
121. B. Lunelli and M.G. Giorgini, *J. Mol. Spectrosc.*, **64**, 1 (1977).
122. B. Lunelli and M.G. Giorgini, *J. Mol. Spectrosc.*, **104**, 203 (1984).
123. D.A. Thornton and G.M. Watkins, *Bull. Soc. Chim. Belg.*, **100**, 221 (1991).
124. N.A. Narasimham and C.V.S. Ramachandra Rao, *J. Mol. Spectrosc.*, **28**, 44 (1968).
125. N.A. Narasimham and C.V.S. Ramachandra Rao, *J. Mol. Spectrosc.*, **30**, 192 (1969).
126. V.B. Singh, R.N. Singh and I.S. Singh, *Spectrochim. Acta*, **22**, 927 (1966).
127. H.F. Shurvell, B. Dulaurens and P. Pesteil, *Spectrochim. Acta*, **22**, 334 (1966).
128. G. Joshi and N.L. Singh, *Spectrochim. Acta*, **22**, 1502 (1966).
129. J.F. Arenas and J.I. Marcos, *Spectrochim. Acta, Part A*, **36A**, 1075 (1980).
130. A. Tramer, *J. Mol. Struct.*, **4**, 313 (1969).
131. P. Venkoji, *Spectrochim. Acta, Part A*, **42A**, 1301 (1986).
132. M.P. Srivastava, B.B. Bal and I.S. Singh, *Indian J. Pure Appl. Phys.*, **10**, 50 (1972).
133. G.W. King and A.A.G. Van Putten, *J. Mol. Spectrosc.*, **70**, 53 (1978).
134. M.C. Castro-Pedrozo and G.W. King, *J. Mol. Spectrosc.*, **73**, 386 (1978).
135. C.P.D. Dwivedi and S.N. Sharma, *Indian J. Pure Appl. Phys.*, **11**, 787 (1973).
136. R.A. Yadav, S. Hyampati, N.P. Singh and I.S. Singh, *Indian J. Pure Appl. Phys.*, **20**, 674 (1982).
137. R.A. Amma, K.P.R. Nair and M.P. Srivastava, *Indian J. Pure Appl. Phys.*, **10**, 58 (1972).
138. R.K. Goel and M.L. Agarwal, *Spectrochim. Acta, Part A*, **38A**, 583 (1982).
139. P.K. Verma, *Indian J. Phys.*, **51B**, 58 (1977).
140. S. Chattopadhyay, L. Chakravorti and G.S. Kastha, *Indian J. Pure Appl. Phys.*, **25**, 456 (1987).
141. E.V. Huded, N.H. Ayachit, M.A. Shashidhar and K.S. Rao, *Indian J. Pure Appl. Phys.*, **23**, 470 (1985).
142. W.H. Evans, *Spectrochim. Acta, Part A*, **30A**, 543 (1974).
143. C.L. Chatterjee, P.P. Garg and R.M.P. Jaiswal, *Spectrochim. Acta, Part A*, **34A**, 943 (1978).
144. H. Takeuchi, N. Watanabe and I. Harada, *Spectrochim. Acta, Part A*, **44A**, 749 (1988).
145. S. Dobos, A. Szabo and B. Zelei, *Spectrochim. Acta, Part A*, **32A**, 1401 (1976).
146. R.A. Yadav and I.S. Singh, *Spectrochim. Acta, Part A*, **41A**, 191 (1985).
147. M. Harrand, *J. Raman Spectrosc.*, **4**, 53 (1975).
148. H.W. Wilson and J.F. Bloor, *Spectrochim. Acta*, **21**, 45 (1965).
149. V.N. Verma and D.K. Rai, *Appl. Spectrosc.*, **24**, 445 (1970).
150. S. Chakravorti, A.K. Sarkar, P.K. Mallick and S.B. Banerjee, *Indian J. Phys.*, **56B**, 96 (1982).
151. B. Lakshmaiah and G.R. Rao, *J. Raman Spectrosc.*, **20**, 439 (1989).
152. J.N. Rai and K.N. Upadhya, *Spectrochim. Acta*, **22**, 1428 (1966).
153. M. Horak, E.R. Lippincott and R.K. Khanna, *Spectrochim. Acta*, **23**, 1111 (1967).
154. S. Tariq, N. Ali and P.K. Verma, *Indian J. Pure Appl. Phys.*, **21**, 220 (1983).
155. W.A. Seth Paul and H. Shino, *Spectrochim. Acta, Part A*, **31A**, 1605 (1975).
156. W.A. Seth Paul and J. Meeuwesen, *Bull. Soc. Chim. Belg.*, **90**, 127 (1981).
157. W.A. Seth Paul, B.J. Van der Veken and M.A. Herman, *Can. J. Spectrosc.*, **27**, 21 (1982).

158. W.A. Seth Paul, *Bull. Soc. Chim. Belg.*, **85**, 187 (1976).
159. S.H.W. Hankin, O.S. Khalil and L. Goodman, *J. Mol. Spectrosc.*, **72**, 383 (1978).
160. A. Gambi, S. Giorgiani, A. Passerini and R. Visinoni, *Spectrochim. Acta, Part A*, **38A**, 871 (1982).
161. N.P. Gates, D. Steele and R.A.R. Pearce, *J. Chem. Soc.*, Perkin Trans II, 1607 (1972).
162. M.M. Szostak, *Croat. Chem. Acta*, **61**, 633 (1988).
163. P. Venkatacharyulu, I.V.S.R.M. Sarma and D. Premaswarup, *Indian J. Pure Appl. Phys.*, **20**, 670 (1982).
164. R.M.P. Jaiswal and J.E. Katon, *Spectrochim. Acta, Part A*, **39A**, 275 (1983).
165. J.G. Rosencrance and P.W. Jagodzinski, *Spectrochim. Acta, Part A*, **42A**, 869 (1986).
166. J.F. Arenas, J.I. Marcos and F.J. Ramirez, *Spectrochim. Acta, Part A*, **44A**, 1045 (1988).
167. J.F. Arenas, J.I. Marcos and F.J. Ramirez, *Spectrochim. Acta, Part A*, **45A**, 781 (1989).
168. E.F. Mooney, *Spectrochim. Acta*, **20**, 1343 (1964).
169. T.V.K. Sarma, *Acta Phys. Pol. A*, **66A**, 185 (1984).
170. C.J. Cattanach and E.F. Mooney, *Spectrochim. Acta, Part A*, **24A**, 407 (1968).
171. J.H.S. Green, D.J. Harrison and W. Kynaston, *Spectrochim. Acta, Part A*, **27A**, 793 (1971).
172. J.H.S. Green, D.J. Harrison and W. Kynaston, *Spectrochim. Acta, Part A*, **27A**, 807 (1971).
173. J.H.S. Green, D.J. Harrison and W. Kynaston, *Spectrochim. Acta, Part A*, **28A**, 33 (1972).
174. A.K. Ansari and P.K. Verma, *Indian J. Phys.*, **53B**, 136 (1979).
175. A.K. Ansari and P.K. Verma, *Spectrochim. Acta, Part A*, **35A**, 35 (1979).
176. R.K. Goel, K.P. Kansal and S.N. Sharma, *Indian J. Pure Appl. Phys.*, **17**, 778 (1979).
177. R.K. Goel, S.K. Gupta, R.M.P. Jaiswal and P.P. Garg, *Indian J. Pure Appl. Phys.*, **18**, 223 (1980).
178. R.K. Goel, S. Sharma, K.P. Kansal and S.N. Sharma, *Indian J. Pure Appl. Phys.*, **18**, 281 (1980).
179. R.K. Goel and S.K. Mathur, *Proc. Natl. Acad. Sci. India, Sect. A*, **51A**, 190 (1981).
180. R.K. Goel, *Spectrochim. Acta, Part A*, **40A**, 723 (1984).
181. R.K. Goel and M.L. Agarwal, *Indian J. Pure Appl. Phys.*, **21**, 752 (1983).
182. R.K. Goel, S. Sharma and A. Gupta, *Indian J. Phys.*, **60B**, 375 (1986).
183. U.C. Joshi, R.N. Singh and S.N. Sharma, *Spectrochim. Acta, Part A*, **38A**, 205 (1982).
184. R.K. Goel and S.D. Sharma, *Indian J. Pure Appl. Phys.*, **17**, 55 (1979).
185. U.C. Joshi, R.N. Singh and S.N. Sharma, *Indian J. Pure Appl. Phys.*, **19**, 1123 (1981).
186. U.C. Joshi, M. Joshi, R.N. Singh and S.N. Sharma, *Indian J. Phys.*, **55B**, 220 (1981).
187. P.K. Bishui, *Indian J. Pure Appl. Phys.*, **10**, 637 (1972).
188. S.P. Gupta, C. Gupta, S. Sharma and R.K. Goel, *Indian J. Pure Appl. Phys.*, **24**, 111 (1986).
189. J.R. Scherer, J.C. Evans and W.W. Muelder, *Spectrochim. Acta*, **18**, 1579 (1962).
190. J.R. Scherer, J.C. Evans, W.W. Muelder and J. Overend, *Spectrochim. Acta*, **18**, 57 (1962).
191. J.R. Scherer and J.C. Evans, *Spectrochim. Acta*, **19**, 1739 (1963).
192. N.S. Sundar, *Spectrochim. Acta, Part A*, **41A**, 905 (1985).

193. N.S. Sundar, *Spectrochim. Acta, Part A*, **41A**, 1449 (1985).
194. N.S. Sundar, *Indian J. Phys.*, **60B**, 490 (1986).
195. N.S. Sundar, *Can. J. Chem.*, **62**, 2238 (1984).
196. N.S. Sundar, *Indian J. Phys.*, **61B**, 464 (1987).
197. J. Shukla and K.N. Upadhya, *Indian J. Pure Appl. Phys.*, **11**, 787 (1973).
198. A.R. Shukla, C.M. Pathak, N.G. Dongre, B.P. Asthana and J. Shamir, *Proc. Indian Acad. Sci.* (*Chem. Sc.*), **97**, 97 (1986).
199. A.R. Shukla, C.M. Pathak, N.G. Dongre, B.P. Asthana and J. Shamir, *Proc. Indian Acad. Sci.* (*Che. Sc.*), **97**, 593 (1986).
200. A.R. Shukla, C.M. Pathak, N.G. Dongre, B.P. Asthana and J. Shamir, *J. Raman Spectrosc.*, **17**, 299 (1986).
201. M.V. Fraile Dotes, C. Siguenza and P.F. Gonzalez-Diaz, *Spectrochim. Acta, Part A*, **42A**, 1029 (1986).
202. R. Rao, M.K. Aralakkanavar, K.S. Rao and M.A. Shashidhar, *Spectrochim. Acta, Part A*, **45A**, 103 (1989).
203. R.A. Nyquist, B.R. Lov and R.W. Chrisman, *Spectrochim. Acta, Part A*, **37A**, 319 (1981).
204. H.F. Shurvell, T.E. Cameron and D.B. Baker, *Spectrochim. Acta, Part A*, **35A**, 757 (1979).
205. G. Varsányi, E. Molnár-Páal, K. Kósa and G. Keresztury, *Acta Chim. Acad. Sci. Hung.*, **100**, 481 (1979).
206. K.C. Medhi, *Indian J. Phys.*, **35**, 583 (1961).
207. R.N. Singh, S.C. Prasad and R.K. Prasad, *Spectrochim. Acta, Part A*, **34A**, 39 (1978).
208. V.N. Verma, *Spectrosc. Lett.*, **6**, 23 (1973).
209. V.N. Verma, *Spectrosc. Lett.*, **8**, 349 (1975).
210. R.B. Singh and D.K. Rai, *Indian J. Phys.*, **53B**, 144 (1979).
211. M. Rangacharyuhu and D. Premaswarup, *Indian J. Phys.*, **54B**, 567 (1980).
212. N.K. Sanyal, R.K. Goel, S.D. Sharma, S.M. Sharma and K.P. Kansal, *Indian J. Phys.*, **55B**, 67 (1981).
213. A.P. Upadhya and K.N. Upadhyay, *Indian J. Phys.*, **55B**, 213 (1981).
214. R.B. Singh, N.P. Singh and D.K. Rai, *Indian J. Phys.*, **56B**, 62 (1982).
215. M.P. Srivastava, *Indian J. Pure Appl. Phys.*, **5**, 189 (1967).
216. G. Thakur, V.B. Singh and N.L. Singh, *Indian J. Pure Appl. Phys.*, **7**, 107 (1969).
217. S.N. Singh and N.L. Singh, *Indian J. Pure Appl. Phys.*, **7**, 250 (1969).
218. V.N. Verma and K.P.R. Nair, *Indian J. Pure Appl. Phys.*, **8**, 682 (1970).
219. H.S. Singh and N.K. Sanyal, *Indian J. Pure Appl. Phys.*, **10**, 545 (1972).
220. S.N. Sharma and C.P.D. Dwivedi, *Indian J. Pure Appl. Phys.*, **13**, 570 (1975).
221. M. Prasad, *Indian J. Pure Appl. Phys.*, **13**, 718 (1975).
222. N.K. Sanyal, S.L. Srivastava and R.K. Goel, *Indian J. Pure Appl. Phys.*, **16**, 719 (1978).
223. S.J. Singh and R. Singh, *Indian J. Pure Appl. Phys.*, **16**, 939 (1978).
224. M. Rangacharyulu and D. Premaswarup, *Indian J. Pure Appl. Phys.*, **19**, 166 (1981).
225. R.B. Singh, N.P. Singh and D.K. Rai, *Indian J. Pure Appl. Phys.*, **19**, 740 (1981).
226. R.N. Singh and S.K. Singh, *Indian J. Pure Appl. Phys.*, **19**, 599 (1981).
227. P. Venkatacharyulu, V.L.N. Prasad, Nallgonda and D. Premaswarup, *Indian J. Pure Appl. Phys.*, **19**, 1178 (1981).
228. K. Singh and R.N. Singh, *Indian J. Pure Appl. Phys.*, **22**, 112 (1984).
229. E.V. Huded, N.H. Ayachit, M.A. Shashidhar and K.S. Rao, *Indian J. Pure Appl. Phys.*, **25**, 289 (1987).
230. R.P. Singh and R.N. Singh, *Pramana*, **30**, 217 (1988).

231. R.B. Singh and D.K. Rai, *Proc. Indian Acad. Sci.* (*Chem. Sci.*), **89**, 163 (1980).
232. T.V.K. Sarma, *Appl. Spectrosc.*, **40**, 933 (1986).
233. R.P. Singh and R.N. Singh, *Indian J. Phys.*, **62B**, 502 (1988).
234. R.B. Singh, M. Mukul and S.K. Sharma, *Indian J. Phys.*, **62B**, 556 (1988).
235. A.K. Sarkar, S. Chakravorti and S.B. Banerjee, *Indian J. Phys.*, **51B**, 71 (1977).
236. S.N. Sharma and C.P.D. Dwivedi, *Indian J. Phys.*, **50**, 25 (1976).
237. U.K. Rastogi, M.P. Rajpoot and S.N. Sharma, *Indian J. Phys.*, **58B**, 311 (1984).
238. T.V.K. Sarma, *Indian J. Phys.*, **59B**, 478 (1985).
239. R.B. Singh and D.K. Rai, *Indian J. Phys.*, **60B**, 404 (1986).
240. S.R. Tripathi and G.N.R. Tripathi, *Indian J. Pure Appl. Phys.*, **18**, 143 (1980).
241. P. Venkatacharyulu, D.V. Ramanamurti and D. Premaswarup, *Indian J. Pure Appl. Phys.*, **20**, 328 (1982).
242. S.N. Sharma, *Indian J. Pure Appl. Phys.*, **20**, 562 (1982).
243. R.B. Singh and D.K. Rai, *Indian J. Pure Appl. Phys.*, **20**, 812 (1982).
244. S.K. Singh and R.N. Singh, *Indian J. Pure Appl. Phys.*, **21**, 163 (1983).
245. S.K. Singh and R.N. Singh, *Indian J. Pure Appl. Phys.*, **21**, 744 (1983).
246. P. Venkatacharyulu, N.V.L.N. Prasad, I.V.S.R.M. Sarma, J.V. Rao and D. Premaswarup, *Indian J. Pure Appl. Phys.*, **23**, 383 (1985).
247. R. Shanker, R.A. Yadav, I.S. Singh and O.N. Singh, *Indian J. Pure Appl. Phys.*, **23**, 339 (1985).
248. P. Venkoji, *Indian J. Pure Appl. Phys.*, **24**, 166 (1986).
249. P. Venkoji, *Proc. Indian Acad. Sci.* (*Chem. Sci.*), **93**, 105 (1984).
250. R.P. Singh and R.N. Singh, *Indian J. Pure Appl. Phys.*, **26**, 644 (1988).
251. S. Mohan and F. Payami, *Indian J. Pure Appl. Phys.*, **24**, 570 (1986).
252. M.K. Aralakkanavar, R. Rao, N.R. Katti and M.A. Shashidhar, *Spectrochim. Acta, Part A*, **47A**, 149 (1991).
253. M.K. Aralakkanavar, N.R. Katti, P.R. Jeergal, G.B. Kalkoti, R. Rao and M.A. Shashidhar, *Spectrochim. Acta, Part A*, **48A**, 983 (1992).
254. R.N. Singh and K.N. Upadhya, *Indian J. Phys.*, **51A**, 88 (1977).
255. O.P. Singh, R.P. Singh and R.N. Singh, *Spectrochim. Acta, Part A*, **49A**, 517 (1993).
256. U.C. Joshi, G.G. Manchanda, N.K. Naithani and S.N. Sharma, *Can. J. Spectrosc.*, **32**, 148 (1987).

12.5 PYRIDYL

The 27 normal vibrations of pyridine are divided into five CH stretching vibrations (νCH), five CH in-plane deformations (δCH), five CH out-of-plane deformations (γ CH), six ring stretching vibrations (νPy), three ring in-plane deformations (δPy) and three ring out-of-plane deformations (γ Py) [1–5, 10]. The infrared spectrum of pyridine looks like that of a monosubstituted benzene and the spectra of substituted pyridines resemble those of substituted benzenes, counting the ring nitrogen as a substituted carbon. Table 12.13 gives the absorption regions of the vibrations of monosubstituted pyridines. In this text we accept that the ring stretching vibration νPy 14 always has a higher wavenumber than the CH in-plane deformation δCH 3. The ring out-of-plane deformation γ Py 16b and the C—X out-of-plane deformation 11 are strongly coupled, so that the assignments of the two substituent-sensitive modes are often interchanged.

Table 12.13 Absorption regions (cm^{-1}) of the normal vibrations of Monosubstituted pyridines

Vibration	Wilson	2-X-substituted pyridines		3-X-substituted pyridines		4-X-substituted pyridines	
		X = C, N, O, F	X=S, Cl, Br, I	X = C, F	X = Cl, Br, I	X = C, N	X = Cl, Br
νCH	20a	3085 ± 15	3080 ± 10	3085 ± 15	3075 ± 10	3075 ± 15	3075 ± 10
νCH	20b	3075 ± 15	3070 ± 15	3070 ± 15	3070 ± 10	3055 ± 15	3070 ± 10
νCH	2	3055 ± 15	3055 ± 15	3045 ± 15	3045 ± 10	3040 ± 15	3040 ± 10
νCH	7b	3020 ± 20	3035 ± 15	3020 ± 20	3035 ± 10	3025 ± 15	3035 ± 15
νPy	8a	1595 ± 15	1575 ± 10	1590 ± 10	1570 ± 10	1600 ± 10	1570 ± 05
νPy	8b	1570 ± 10	1560 ± 10	1570 ± 10	1560 ± 10	1565 ± 20	1565 ± 02
νPy	19a	1480 ± 20	1455 ± 10	1480 ± 20	1462 ± 02	1490 ± 20	1483 ± 02
νPy	19b	1435 ± 15	1425 ± 10	1420 ± 10	1415 ± 05	1410 ± 10	1405 ± 05
νPy	14	1305 ± 40	1325 ± 40	1305 ± 35	1320 ± 02	1320 ± 40	1350 ± 10
δCH	3	1270 ± 30	1260 ± 30	1260 ± 40	1215 ± 15	1265 ± 30	1316 ± 02
νCX	13 (X)	1215 ± 35	1110 ± 25	1215 ± 35	1090 ± 15	1225 ± 25	1095 ± 10
δCH	9a	1155 ± 10	1150 ± 05	1180 ± 15	1190 ± 02	1210 ± 20	1218 ± 02
δCH	15	1085 ± 15	1080 ± 10	1110 ± 15	1095 ± 15	1095 ± 20	1077 ± 03
δCH	18a	1040 ± 15	1040 ± 10	1040 ± 10	1030 ± 10	1060 ± 20	1063 ± 02
νPy	1	995 ± 10	990 ± 10	1025 ± 05	1010 ± 05	1000 ± 10	994 ± 02
γ CH	17a	980 ± 15	965 ± 15	970 ± 20	975 ± 05	965 ± 25	960 ± 05
γ CH	5	930 ± 35	925 ± 25	935 ± 10	945 ± 05	905 ± 35	914 ± 02
γ CH	10a	855 ± 50	865 ± 15	905 ± 20	915 ± 02	850 ± 30	848 ± 12
γ CH	10b	770 ± 25	760 ± 10	795 ± 15	790 ± 05	805 ± 25	808 ± 04
δPy	12 (X)	740 ± 85	735 ± 55	785 ± 45	705 ± 25	785 ± 50	700 ± 20
γ Py	4	725 ± 20	730 ± 10	710 ± 10	700 ± 10	730 ± 20	722 ± 02
δPy	6b	620 ± 10	620 ± 10	620 ± 10	613 ± 02	660 ± 10	662 ± 02
δPy	6a (X)	460 ± 100	390 ± 130	495 ± 50	345 ± 85	520 ± 60	365 ± 55
γ Py	16b (X)	470 ± 50	470 ± 20	450 ± 20	445 ± 15	490 ± 60	485 ± 10
γ Py	16a	410 ± 10	400 ± 10	400 ± 10	400 ± 10	395 ± 20	390 ± 02
δCX	18b (X)	320 ± 70	270 ± 40	300 ± 65	260 ± 40	290 ± 70	280 ± 25
γ CX	11 (X)	175 ± 65	185 ± 25	205 ± 45	190 ± 10	180 ± 40	180 ± 05

(X) = substituent-sensitive vibration.
Wilson: Notation according to Wilson (see Section 12.4 for equivalent numbers).

X—Py compounds

X = 2-Me— [8–10], 3-Me— [6–10], 4-Me— [9–13], 4-Et— [11], 2- and 4-$H_2C{=}CH$— [15], 2-, 3- and 4-N≡C— [16], 2-(2′Py)— [17, 18], 2-, 3- and 4-H(O=)C— [16], 2-, 3- and 4-Me(O=)C— [14], 2-, 3- and 4-HO(O=)C— [19], 2-H(S=)CNH— and 2-Me(S=)CNH— [21], 2- and 4-Me_2N— [22], 2-MeO— [23], 2-PhS—, 2-PyS—, 2-PhS$(=O)_2$— and 2-PyS$(=O)_2$— [24], 2- and 3-F— [9, 16, 20], 2-, 3- and 4-Cl— [9, 20], 2-, 3- and 4-Br— [9, 20], 2- and 4-I— [20].

References

1. L. Corrsin, B.J. Fax and R.C. Lord, *J. Chem. Phys.*, **21**, 1170 (1953).
2. L. Harsányi and F. Kilár, *J. Mol. Struct.*, **65**, 141 (1980).

3. A.J. Ashe, G.L. Jones and F.A. Miller, *J. Mol. Struct.*, **78**, 169 (1982).
4. H.K.J. Koser, *Spectrochim. Acta, Part A*, **40A**, 125 (1984).
5. G. Pongor, P. Pulay, G. Fogarasi and J.E. Boggs, *J. Am. Chem. Soc.*, **106**, 2765 (1984).
6. D. Gandolfo and J. Zarembowitch, *Spectrochim. Acta, Part A*, 33A, 615 (1977).
7. D.W. Scott, W.D. Good, G.B. Guthrie, S.S. Todd, I.A. Hossenlopp, A.G. Osborn and J.P. McCullough, *J. Phys. Chem.*, **67**, 685 (1963).
8. D.W. Scott, W.N. Hubbard, J.F. Messerly, S.S. Todd, I.A. Hossenlopp, W.D. Good, D.R.Douslin and J.P. McCullough, *J. Phys. Chem.*, **67**, 680 (1963).
9. J.H.S. Green, W. Kynaston and H.M. Paisly, *Spectrochim. Acta*, **19**, 549 (1963).
10. J.A. Draeger, *Spectrochim. Acta, Part A*, **39A**, 809 (1983).
11. D.L. Cummings and J.L. Wood, *J. Mol. Struct.*, **20**, 1 (1974).
12. E. Allenstein, W. Podszun, P. Kiemle, H.J. Mauk, E. Schlipf and J. Weidlein, *Spectrochim. Acta, Part A*, **32A**, 777 (1976).
13. D.A.Thornton, P.F.M. Verhoeven, G.M. Watkins, H.O. Desseyn and B.J. Van der Veken, *Bull. Soc. Chim. Belg.*, **100**, 211 (1991).
14. K.C. Medhi, *Indian J. Phys.*, **51A**, 399 (1977).
15. J.H.S. Green and D.J. Harrison, *Spectrochim. Acta, Part A*, **33A**, 249 (1977).
16. J.H.S. Green and D.J. Harrison, *Spectrochim. Acta, Part A*, **33A**, 75 (1977).
17. D.A. Thornton and G.M. Watkins, *Bull. Soc. Chim. Belg.*, **100**, 221 (1991).
18. M. Muniz-Miranda, E. Castelluci, N. Neto and G. Sbrana, *Spectrochim. Acta, Part A*, **39A**, 107 (1983).
19. S. Chattopadhyay and S.K. Brahma, *Spectrochim. Acta, Part A*, **49A**, 589 (1993).
20. H. Abdel-Shafy, H. Perlmutter and H. Kimmel, *J. Mol. Struct.*, **42**, 37 (1977).
21. D.N. Sathyanarayana and S.V.K. Raja, *J. Mol. Struct.*, **157**, 399 (1987).
22. S.P. Gupta, S. Ahmad and R.K. Goel, *Indian J. Phys.*, **61B**, 427 (1987).
23. K.C. Medhi, *Bull. Chem. Soc. Jpn.*, **57**, 261 (1984).
24. A. Bigotto, V. Galasso, G.C. Pappalardo and G. Scarlata, *Spectrochim. Acta, Part A*, **34A**, 435 (1978).

12.6 PYRIMIDINYL

Pyrimidine (1,3-diazine) has 24 normal vibrations: four CH stretching modes (νCH), four CH in-plane deformations (δCH), four CH out-of-plane deformations (γ CH), six ring stretching vibrations (νPym), three ring in-plane deformations (δPym) and three ring out-of-plane deformations (γ Pym) [1–5]. Kartha interpreted the spectra of 4-methyl- and 5-methyl-pyrimidine [12] and the 2-X-pyrimidines have been studied by several authors [6–11]. Table 12.14 shows the absorption regions of 2-X-pyrimidines, based on a few interpretations in the literature. Just as for substituted benzenes (Section 12.4), the HW side of the extensive regions of the substituent-sensitive modes (X) is occupies by pyrimidines substituted with an electron donating atom (N, O) and the LW side by pyrimidines substituted with a heavy atom (S, Cl, Br, I). The normal vibrations are also defined by a Wilson number. Table 12.4 gives the possible equivalent numbers.

Table 12.14 Absorption regions (cm^{-1}) of the normal vibrations of 2-substituted pyrimidines

Vibration	Wilson	Region	Vibration	Wilson	Region
νCH	20a	3100 ± 20	νPym	1	985 ± 10
νCH	2	3070 ± 15	γCH	5	920 ± 70
νCH	7b	3040 ± 20	γCH	10a	820 ± 20
νPym	8a	1570 ± 20	δPym	12 (X)	795 ± 75
νPym	8b	1555 ± 15	γCH	17b	780 ± 20
νPym	19a	1415 ± 40	γPym	4	770 ± 20
νPym	19b	1400 ± 30	δPym	6b	630 ± 10
νPym	14	1300 ± 50	δPym	6a (X)	380 ± 110
νCX	13 (X)	1180 ± 120	γPym	16b (X)	485 ± 40
δCH	3	1185 ± 35	γPym	16a	400 ± 15
δCH	15	1165 ± 30	δCX	18b (X)	350 ± 120
δCH	9a	1080 ± 20	γCX	11 (X)	195 ± 40

2-X—Pym compounds

X = Me— [7], H_2N— [8, 11], D_2N— [11], MeO— [11], 2-RS— (R = Me, Et, iPr, nBu, $CH_2{=}CHCH_2$, $PhCH_2$, $PhCH_2CH_2$, $PhO(CH_2)_3$) [9], F—, Br— and I— [6], Cl— [5, 6, 10, 11].

References

1. G. Sbrana, G. Adembri and S. Califano, *Spectrochim. Acta*, **22**, 1839 (1966).
2. L. Bokobza-Sebach and J. Zarembowitch, *Spectrochim. Acta, Part A*, **32A**, 797 (1976).
3. S.G. Stepanian, G.G. Sheina, E.D. Radchenko and Yu.P.. Blagoi, *J. Mol. Struct.*, **131**, 333 (1985).
4. Y.A. Sarma, *Spectrochim. Acta, Part A*, **34A**, 825 (1978).
5. Y.A. Sarma, *Spectrochim. Acta, Part A*, **30A**, 1801 (1974).
6. E. Allenstein, P. Kiemle, J. Weidlein and W. Podszun, *Spectrochim. Acta, Part A*, **33A**, 189 (1977).
7. E. Allenstein, W. Podszun, P. Kiemle, H.-J. Mauk, E. Schlipf and J. Weidlein, *Spectrochim. Acta, part A*, **32A**, 777 (1976).
8. M. Maehara, S. Nakama, Y. Nibu, H. Shimada and R. Shimada, *Bull. Chem. Soc. Jpn.*, **60**, 2769 (1987)
9. G. Mille, M. Guiliano, J. Kister, J. Chouteau and J. Metzger, *Spectrochim. Acta, Part A*, **36A**, 713 (1980).
10. H. Gauthier and J.M. Lebas, *Spectrochim. Acta, Part A*, **35A**, 787 (1979).
11. A.J. Lafaix and J.M. Lebas, *Spectrochim. Acta, Part A*, **26A**, 1243 (1970).
12. S.B. Kartha, *Spectrochim. Acta, Part A*, **38A**, 859 (1982).

12.7 THIENYL

The 21 normal vibrations of thiophene are divided into four ring stretching vibrations (νTh), three ring in-plane deformations (δTh), two ring out-of-plane deformations (γTh), four CH stretching vibrations (νCH), four CH in-plane deformations (δCH) and four CH out-of-plane deformations (γCH) [1–6]. Table 12.15 shows the 21 normal vibrations of 2-monosubstituted thiophenes. The absorption regions of the ring stretching vibrations are in good agreement with those found by Angelelli *et al.* [12] from the spectra of twenty-four 2-substituted thiophenes. The high value (1554 cm^{-1}) for the ring stretch of 2-fluorothiophene however, falls outside the region. Just like in benzenes, the substituent-sensitive mode νC—X (1110 ± 170 cm^{-1}) is partly a ring vibration. The spectral position of this vibration depends upon the resonance, inductive and mass effects of the substituent (see Section 12.4). The compound 2-MeTh absorbs at 1236 cm^{-1}, 2-O_2NTh at 1126, 2-HSTh at 1022, 2-ClTh at 1000 and 2-ITh at 945 cm^{-1}. The absorption regions of 3-X-thiophenes resemble those of the 2-X-thiophenes. Scott interpreted the spectrum of 3-methylthiophene [10], Paliani and Cataliotti the spectra of 3-halogenothiophenes [13] and Alberghina *et al.* those of 3-thiophenecarboxamides [11].

Table 12.15 Absorption regions (cm^{-1}) of the normal vibrations of 2-substituted thiophenes

Vibration	2-X-substituted thiophenes		Vibration	2-X-substituted thiophenes	
	X = C, N, O, F	X = S, Cl, Br, I		X = C, N, O, F	X = S, Cl, Br, I
νCH	3120 ± 10	3115 ± 05	νTh	855 ± 15	850 ± 10
νCH	3095 ± 15	3095 ± 10	γCH	825 ± 30	830 ± 15
νCH	3080 ± 10	3080 ± 10	δTh	755 ± 15	745 ± 10
νTh	1520 ± 20	1510 ± 10	γCH	700 ± 25	695 ± 25
νTh	1425 ± 25	1405 ± 10	δTh	665 ± 25	645 ± 20
νTh	1355 ± 10	1342 ± 07	γTh	560 ± 15	560 ± 10
δCH	1230 ± 10	1222 ± 07	γTh (X)	495 ± 55	460 ± 20
δCH	1080 ± 05	1077 ± 07	δTh (X)	395 ± 80	305 ± 25
δCH	1040 ± 15	1047 ± 07	δCX (X)	295 ± 45	220 ± 40
νCX (X)	1200 ± 80	985 ± 45	γCX (X)	220 ± 30	200 ± 30
γCH	910 ± 20	900 ± 10			

2-X—Th compounds

X = Me— [7, 10], Et—, N≡CH_2—, HON=HC—, HO(O=)CC≡C—, N≡C—, H(O=)C—, Me(O=)C—, Et(O=)C—, ClCH_2(O=)C—, Cl(O=)C—, HO(O=)C—, EtO(O=)C—, O_2N—, MeO—, HS— and Cl$(O=)_2$S— [7], Cl and Br— [7–9], I— [7, 8].

References

1. J.M. Orza, M. Rico and J.F. Biarge, *J. Mol. Spectrosc.*, **19**, 188 (1966).
2. J. Loisel and V. Lorenzelli, *Spectrochim. Acta, Part A*, **23A**, 2903 (1967).
3. J. Loisel, J.-R. Pinan-Lucarre and V. Lorenzelli, *J. Mol. Struct.*, **17**, 341 (1973).
4. G. Paliani, R. Cataliotti and A. Poletti, *Spectrochim. Acta, Part A*, **32A**, 1089 (1976).

5. G. Paliani, A. Poletti and R. Cataliotti, *Chem. Phys. Lett.*, **18**, 525 (1973).
6. M. Kofranek, T. Kovar, H. Lischka and A. Karpfen, *J. Mol. Struct.*, **259**, 181 (1992).
7. J.J. Peron, P. Saumagne and J.M. Lebas, *Spectrochim. Acta, part A*, **26A**, 1651 (1970).
8. M. Horak, I.J. Hyams and E.R. Lippincott, *Spectrochim. Acta*, **22**, 1355 (1966).
9. G. Paliani and R. Cataliotti, *Spectrochim. Acta, Part A*, **37A**, 707 (1981).
10. D.W. Scott, *J. Mol. Spectrosc.*, **31**, 451 (1969).
11. G. Alberghina, S. Fisichella and S. Occhipinti, *Spectrochim. Acta, Part A*, **36A**, 349 (1980).
12. J.M. Angelelli, A.R. Katritzky, R.F. Pinzelli and R.D. Topsom, *Tetrahedron*, **28**, 2037 (1972).
13. G. Paliani and R. Cataliotti, *Spectrochim. Acta, Part A*, **38A**, 751 (1982).

12.8 FURYL

The 21 normal vibrations of furan can be divided into four ring stretching modes (νFu), three ring in-plane deformations (δFu), two ring out-of-plane deformations (γ Fu), four CH stretching vibrations (νCH), four CH in-plane deformations (δCH) and four CH out-of-plane deformations [γ CH) [1–6]. The 21 normal vibrations of 2-monosubstituted furans are listed in Table 12.16. The absorption regions of the three ring stretching vibrations with the highest wavenumbers agree well with those found by Angelelli *et al.* [18] in the spectra of sixteen 2-substituted furans. The extreme values assigned in the spectra of 2′-Fu—2-Fu (1645R and *1542*), 2-MeOFu (1520) and 2-IFu (1457 cm^{-1}) are outside these regions. The fourth ring stretch is the breathing and occurs at 1015 ± 15 cm^{-1}. The extensive regions 1200 ± 105 and 485 ± 180 cm^{-1} for the substituent-sensitive modes: respectively νC—X, partly a ring vibration (see Section 12.4) and δFu, narrow to 1215 ± 90 and 550 ± 115 cm^{-1} for C-substituted furans. Furans substituted with Cl, Br or I absorb at lower wavenumbers, 1110 ± 15 and 400 ± 95 cm^{-1} respectively. The spectra of 3-substituted furans have been studied by Alberghina *et al.* [16] and by Volka *et al.* [17] and those of 2,5-disubstituted furans by Green *et al.* [8].

Table 12.16 Absorption regions (cm^{-1}) of the normal vibrations of 2-substituted furans

Vibration	C_s	Region	Vibration	C_s	Region
νCH	a′	3155 ± 15	δFu (X)	a′	920 ± 35
νCH	a′	3135 ± 15	δFu	a′	875 ± 15
νCH	a′	3115 ± 15	γ CH	a″	865 ± 20
νFu	a′	1575 ± 30	γ CH	a″	790 ± 40
νFu	a′	1485 ± 25	γ CH	a″	740 ± 30
νFu	a′	1380 ± 20	γ Fu	a″	625 ± 30
δCH	a′	1235 ± 20	δFu (X)	a′	485 ± 180
νCX (X)	a′	1200 ± 105	γ Fu	a″	590 ± 10
δCH	a′	1170 ± 25	δCX (X)	a′	280 ± 65
δCH	a′	1075 ± 20	γ CX (X)	a″	200 ± 55
νFu	a′	1010 ± 15			

X—Fu compounds

X = Me— [5–8], Et—, H_2NCH_2— [7], $HOCH_2$— [7, 9], $HSCH_2$— [7], 2′-Fu—, N≡C— [7, 17], H(O═)C— [7, 8, 10–13], Me(O═)C [7], Cl(O═)C— [7, 14], HO(O═)C— [7, 8], CO_2^- [8, 15], MeO(O═)C— and EtO(O═)C— [7], N≡C(O═)C—, H_2N(O═)C—, MeHN(O═)C— and Me_2N(O═)C— [16], R—PhHN(O═)C— (R = H, 3- and 4-Me, 3- and 4-MeO, 3- and 4-NO_2, 3- and 4-Cl), MeO— [7], Cl— [8], Br— [5, 7, 8], I— [8].

References

1. M. Rico, M. Barrachina and J.M. Orza, *J. Mol. Spectrosc.*, **24**, 133 (1967).
2. J. Loisel and V. Lorenzelli, *Spectrochim. Acta, Part A*, **23A**, 2903 (1967).
3. J. Loisel, J.P. Pinan-Lucarre and V. Lorenzelli, *J. Mol. Struct.*, **17**, 341 (1973).
4. G. Paliani, R. Cataliotti and A. Poletti, *Spectrochim. Acta, Part A*, **32A**, 1089 (1976).
5. C. Pouchan, J. Raymond, H. Sauvaitre and M. Chaillet, *J. Mol. Struct.*, **21**, 253 (1974).
6. D.W. Scott, *J. Mol. Spectrosc.*, **37**, 77 (1971).
7. M. Sénéchal and P. Saumagne, *J. Chim. Phys.*, **69**, 1246 (1972).
8. J.H.S. Green and D.J. Harrison, *Spectrochim. Acta, Part A*, **33A**, 843 (1977).
9. L. Strandman-Long and J. Murto, *Spectrochim. Acta, Part A*, **37A**, 643 (1981).
10. J. Bánki, F. Billes, M. Gál, A. Grofcsik, G. Jalsovszky and L. Sztraka, *Acta Chim. Hung.*, **123**, 115 (1986).
11. J. Bánki, F. Billes, M. Gál, A. Grofcsik, G. Jalsovszky and L. Sztraka, *J. Mol. Struct.*, **142**, 351 (1986).
12. G. Allen and H.J. Bernstein, *Can. J. Chem.*, **33**, 1055 (1955).
13. P. Adámek, K. Volka, Z. Ksandr and I. Stibor, *J. Mol. Spectrosc.*, **47**, 252 (1973).
14. G. Cassanas-Fabre and L. Bardet, *J. Mol. Struct.*, **25**, 281 (1975).
15. L. Bardet, J. Maillols and G. Fabre, *J. Chim. Phys.*, **68**, 984 (1971).
16. G. Alberghina, S. Fisichella and S. Occhipinti, *Spectrochim. Acta, Part A*, **36A**, 349 (1980).
17. K. Volka, P. Adámek, I. Stibor and Z. Ksadr, *Spectrochim. Acta, Part A*, **32A**, 397 (1976).
18. J.M. Angelelli, A.R. Katritzky, R.F. Pinzelli and R.D. Topsom, *Tetrahedron*, **28**, 2037 (1972).

Index

Acetates, $—OC(=O)Me$, 22, 157, 274
Acetamides, $—NHC(=O)Me$, 22, 157, 242
Acetanilides, $—PhNHC(=O)Me$, 245
Acetoxy, $—OC(=O)Me$, 22, 157, 274
Acetyl, $—C(=O)Me$, 20, 155
Acetylamino, $—NHC(=O)Me$, 242
Acetyloxy, $—OC(=O)Me$, 22, 157, 274
Acids, $—C(=O)OH$, 163, 261
Acid bromides, $—C(=O)Br$, 152
Acid chlorides, $—C(=O)Cl$, 148
Acid fluorides, $—C(=O)F$, 145
Alcohols
  primary, $—CH_2OH$, 64, 98, 261
  secondary, $—CH(OH)—$, 128, 261
  tertiary, $>C(OH)—$, 261
Aldehydes, $—C(=O)H$, 137
Alkane fragments, 13, 89, 100
Alkenes, 15, 204
Alkynes, 15, 104, 225
  halogen substituted, $—C≡CX$, 227
Amides
  primary, $—C(=O)NH_2$, 179, 236
  secondary, $—C(=O)NHMe$, 17, 190, 241
  tertiary, $—C(=O)NMe_2$, 246
Amines
  primary, $—NH_2$, 80, 233
  secondary, $—NHMe$, 17, 239
  tertiary, $—NMe_2$, 246
Amine salts, $—CH_2NH_3^+$, 83
Amino acids, $^+H_3N—R—CO_2^-$, 83, 198
Aminocarbonyl, $—C(=O)NH_2$, 179
Aminomethyl, $—CH_2NH_2$, 80
Aminosulfonyl, $—SO_2NH_2$, 294
Aminothiocarbonyl, $—C(=S)NH_2$, 184
Ammoniomethyl, $—CH_2NH_3^+$, 83
Aromatic hydrocarbons, 306, 329
Aromatic substitution, 310
Aziridinyl, $—cNC_2H_4$, 303

Benzene derivatives, 306
  monosubstituted, 313
  disubstituted, 314
  trisubstituted, 318
Benzylbromides, $—PhCH_2Br$, 59
Benzylchlorides, $—PhCH_2Cl$, 53
Bromocarbonyl, $—C(=O)Br$, 152
Bromoethynyl, $—C≡CBr$, 228
Bromoformyl, $—C(=O)Br$, 152
Bromomethyl, $—CH_2Br$, 57
Bromomethylene, $—CHBr—$, 126
Butyl, tertiary, $—C(Me)_3$, 40

Carbamoyl, $—C(=O)NH_2$, 179
Carbamoylchlorides, $>NC(=O)Cl$, 150
Carbonyl compounds, 137
Carboxyl, $—C(=O)OH$, 163, 261
Carboxylates, $—CO_2^-$, 198
Carboxylic acids, $—C(=O)OH$, 163, 261
Carboxylic acids salts, $—CO_2^-$, 198
Chlorocarbonyl, $—C(=O)Cl$, 148
Chlorocarbonyloxy, $—OC(=O)Cl$, 150, 273
Chloroethyl, $—CH_2CH_2Cl$, 95

Chloroethynyl, $—C≡CCl$, 227
Chloroformates, $—OC(=O)Cl$, 150, 273
Chloroformyl, $—C(=O)Cl$, 148
Chloroformyloxy, $—OC(=O)Cl$, 150, 273
Chloromethyl, $—CH_2Cl$, 51
Chloromethylene, $—CHCl—$, 126
Chlorosulfonyl, $—SO_2Cl$, 291
Cyanomethyl, $—CH_2C≡N$, 108
Cyanomethylene, $—CH(C≡N)—$, 127
Cyclopropyl, $—cC_3H_5$, 301

Dibromomethyl, $—CHBr_2$, 117
Dibromomethylene, $—CBr_2—$, 134
Dichloromethyl, $—CHCl_2$, 115
Dichloromethylene, $—CCl_2—$, 133
Difluoromethyl, $—CHF_2—$, 114
Difluoromethylene, $—CF_2—$, 132
Dihalogenomethyl, $—CHX_2—$ 113
Dihalogenomethylene, $—CX_2—$ 132
Dimethylamino, $—NMe_2$, 246
Dipolar ions, $^+H_3N—R—CO_2^-$, 83, 198
Dissimilar, CX_3 bonds, 11, 28
Disulfides, —SSMe, 19, 279
Dithioesters, —C(=S)SMe, 279

Esters
ethyl, —C(=O)OEt, 174, 271
methyl, —C(=O)OMe, 23, 169
of sulfonic acids, $—SO_2OR'$, 292
Ethenyl, $—CH=CH_2$, 204
Ethenylene, —CH=CH—, 218
Ethenylidene, $>C=CH_2$, 212
Ethers
ethyl, —OEt, 91, 269
methyl, —OMe, 24, 68, 263
Ethoxy, —OEt, 91, 269
Ethoxycarbonyl, —C(=O)OEt, 174, 271
Ethyl, $—CH_2CH_3$, 13, 89
Ethyl esters, —C(=O)OEt, 174, 271
Ethyl ethers, —OEt, 269
Ethylthio, —SEt, 91, 281
Ethynyl, $—C≡CH$, 225

Fluorocarbonyl, —C(=O)F, 145
Fluoroformates, —OC(=O)F, 146
Fluoroformyl, —C(=O)F, 145
Fluoromethyl, $—CH_2F$, 48
Fluoromethylene, —CHF— , 125
Fluorosulfonyl, $—SO_2F$, 289
Formamides, 140
Formates, —OC(=O)H, 140, 272
Formyl, —C(=O)H, 137
Formyloxy, —OC(=O)H, 140, 272
Fundamental vibrations
of a molecular fragment, 1
of a molecule, 2
Furan derivatives, 334
Furyl, Fu, 334

Halogen compounds, 28, 48, 94, 113, 125, 132, 148, 152, 227, 273, 291
Halogenomethyl, $—CH_2X$, 48
Halogenomethylene, —CHX—, 125
Hydroxy compounds, —OH, 64, 98, 128, 258
Hydroxyethyl, $—CH_2CH_2OH$, 98
Hydroxylamines, >N—OH, 261
Hydroxymethyl, $—CH_2OH$, 64
Hydroxymethylene, —CH(OH)—, 128

Interpretation of spectra, 3
Iodoethynyl, $—C≡CI$, 229
Iodomethyl, $—CH_2I$, 62
Isocyanates, $—CH_2N=C=O$, 87
Isocyanatomethyl, $—CH_2N=C=O$, 87
Isopropyl, $—CH(Me)_2$, 119
Isothiocyanates, $—CH_2N=C=S$, 88
Isothiocyanatomethyl, $—CH_2N=C=S$, 88

Ketones
methyl, —C(=O)Me, 20, 155
ethyl, —C(=O)Et, 91, 160

Mercaptomethyl, $—CH_2SH$, 72
Methanesulfonamides, $>NSO_2Me$, 289
N-phenyl $—PhHNSO_2Me$, 289
Methanesulfonates, $—OSO_2Me$, 289
Methanesulfonylhalides, XSO_2Me, 289
Methoxy, —OMe, 24, 68, 263
Methoxycarbonyl, —C(=O)OMe, 23, 169
Methoxymethyl, $—CH_2OMe$, 68
Methoxythiocarbonyl, 24
Methyl, $—CH_3$, 10
Methylamino, —NHMe, 17, 239
Methylaminocarbonyl, —C(=O)NHMe, 17, 190, 241

Methylaminosulfonyl, $-SO_2NHMe$, 17, 241, 296
Methylaminothiocarbonyl, $-C(=S)NHMe$, 17, 195, 241
Methylcarbamoyl, $-C(=O)NHMe$, 17, 190, 241
Methyl esters, $-C(=O)OMe$, 23, 169
Methyl ethers, $-OMe$, 24, 68, 263
Methylphosphinates, $>P(=O)OMe$, 24, 266
Methylphosphonates, $-P(=O)(OMe)_2$, 266
Methylsilanes, $>Si(Me)-$, 20
Methylsulfinyl, $-S(=O)Me$, 20, 284
Methylsulfonyl, $-SO_2Me$, 20, 286
Methylsulfoxide, $-S(=O)Me$, 20, 284
Methylthio, $-SMe$, 17, 75, 277
Methylthiocarbamoyl, $-C(=S)NHMe$, 17, 195, 241
Methylthiomethyl, $-CH_2SMe$, 75
Methylthiophosphinates, $>P(=S)OMe$, 24, 266
Methylthiophosphonates, $-P(=S)(OMe)_2$, 266

Nitrates, $-ONO_2$, 254
Nitriles
  cyanomethyl, $-CH_2C{\equiv}N$, 108
  branched, $-CH(C{\equiv}N)-$, 127
Nitro, $-NO_2$, 250
Nitroamines, $>N-NO_2$, 254
Normal vibrations
  of a molecular fragment, 1
  of a molecule, 2

Oximes, $>N-OH$, 261
Oxiranyl, $-cC_2H_3O$, 303
Oxy compounds, 258
Oxymethyl compounds, 64
Oxysulfonyl, $-SO_2OR'$, 292

Phenols, $-PhOH$, 259
Phenyl, Ph, 306
  methyl substituted, 13
  monosubstituted, 313
  disubstituted, 314
  trisubstituted, 318
Phosphinates, $>P(=O)OMe$, 24, 266
Phosphonates, $-P(=O)(OMe)_2$, 266
Propanoyl, $-C(=O)Et$, 91, 160
Propionyl, $-C(=O)Et$, 91, 160
n-Propyl, $-CH_2CH_2CH_3$, 13, 100
Propynyl, $-CH_2C{\equiv}CH$, 104
Pyridine derivatives, 329
Pyridyl, $-Py$, 329
Pyrimidine derivatives, 331
Pyrimidinyl, $-Pym$, 331

Silanols, $>Si(OH)-$, 261
Substituent sensitive mode, 307
Sulfides, 17, 75, 277
Sulfinates, 24, 266
Sulfonamides
  primary, $-SO_2NH_2$, 236, 294
  N-methyl, $-SO_2NHMe$, 17, 241, 296
  N-phenyl $MeSO_2NHPh-$, 289
  secondary, $-SO_2NH-$, 295
Sulf(on)ates $-SO_2O-$, 292
Sulfones
  $-SO_2-$, 296
  $-SO_2Me$, 286
Sulfonyl, $-SO_2-$, 296
Sulfonylchlorides, $-SO_2Cl$, 291
Sulfonylfluorides, $-SO_2F$, 289
Sulfoxides, $-S(=O)Me$, 20, 284
Sulfur compounds, 277

Tertiary butyl, $-C(Me)_3$, 40
Thienyl, $-Th$, 332
Thioacetates, $-SC(=O)Me$, 22, 157
Thioamides
  primary, $-C(=S)NH_2$, 184, 236
  secondary, *N*-methyl, $-C(=S)NHMe$, 17, 195, 241
Thiocarbamoyl, $-C(=S)NH_2$, 184
Thiocyanates, $-CH_2SCN$, 78
Thiocyanatomethyl, $-CH_2SCN$, 78
Thioesters, $-C(=O)SMe$, 279
Thioethers, $-SMe$, 17, 75, 277
Thioethyl compounds, $-SEt$, 91, 281
Thioformates, $-SC(=O)H$, 140
Thiols, $-CH_2SH$, 72
Thiomethyl compounds, $-SMe$, 17, 75, 277
Thiophene derivatives, 332
Tribromomethyl, $-CBr_3$, 36
Trichloroethyl, $-CH_2CCl_3$, 94
Trichloromethyl, $-CCl_3$, 34
Trifluoroacetyl, $-C(=O)CF_3$, 30
Trifluoromethyl, $-CF_3$, 29

Trifluoromethylbenzenes, $—Ph—CF_3$, 33
Trihalogenomethyl, $—CX_3$, 28

Ureas, $>N—C(=O)NH_2$, 182

Vinyl, $—CH=CH_2$, 204
Vinylene, $—CH=CH—$, 218
Vinylidene, $>C=CH_2$, 212

Zwitter-ions, $^+H_3N—R—CO_2^-$, 83, 198